MATHEMATICS FOR PHYSICISTS

Philippe Dennery
André Krzywicki

Université Paris XI
Campus d'Orsay

DOVER PUBLICATIONS, INC.
Mineola, New York

Copyright

Bibliographical Note

This Dover edition, first published in 1996, is an unabridged, corrected republication of the work first published by Harper & Row, New York, 1967, in the "Harper's Physics Series."

Library of Congress Cataloging-in-Publication Data

Dennery, Philippe.
Mathematics for physicists / Philippe Dennery, André Krzywicki. — Dover ed.
p. cm.
Originally published: New York : Harper & Row, 1967.
Includes bibliographical references and index.
ISBN-13: 978-0-486-69193-0 (pbk.)
ISBN-10: 0-486-69193-4 (pbk.)
1. Mathematical physics. I. Krzywicki, André. II. Title.
QC20.D39 1996
515—dc20 96-10774
CIP

Manufactured in the United States by Courier Corporation
69193410
www.doverpublications.com

CONTENTS

PREFACE

So how do you go about teaching them something new? By mixing what they know with what they don't know. Then, when they see in their fog something they recognize they think, "Ah I know that!" And then it's just one more step to "Ah, I know the whole thing." And their mind thrusts forward into the unknown and they begin to recognize what they didn't know before and they increase their powers of understanding."*

PICASSO

The content of this book includes the material traditionally covered in a senior or first year graduate course, usually called "Mathematical Physics," but which should more appropriately be entitled (using A. Sommerfeld's description) "Physical Mathematics." Our book, however, differs somewhat from those that a student is likely to encounter in his parallel readings. It differs in its presentation from a book of pure mathematics and it differs in spirit from those books on mathematical physics where the methods for solving specific problems are given priority over the exposition of the underlying mathematical concepts.

We assume that the reader is familiar only with elementary differential and integral calculus, with vector analysis, and with the theory of systems of algebraic equations. Naturally, with these relatively limited means, we are unable to develop the various mathematical topics in their most general and abstract formulation. However, we do not wish that these limitations should serve as a pretext to do away with rigor. We believe that semirigorous arguments can only demoralize a reader because an understanding of what constitutes a mathematical demonstration is in itself very important.

A warning is given to the reader when rigorous demonstrations cannot be given, except of course in those cases where a lack of rigor is due to inadvertence on the part of the authors.

In other words, what we hope the reader will acquire is not only a certain knowledge of the basic results he will need in applications, but also a minimum of what one could call "mathematical culture," which, among other things, will make the mathematical literature more accessible to him.

What distinguishes "physical mathematics" from pure mathematics is essen-

* From F. Gilot and C. Lake, *Life with Picasso*, New York: McGraw-Hill Book Co., Inc., 1964.

tially a choice of language and a choice of topics. The difference in language between a physicist and a mathematician, which often makes communication between the two very difficult, is based on the quite different roles that intuition plays in the two fields.

A mathematician, i.e., one who creates mathematics, must get away from purely intuitive concepts if he wants to go beyond what he already knows. Only what exists can be given an intuitive interpretation; what is unknown is not intuitively obvious and cannot be compared to what is known. On the other hand, a physicist does not create mathematics, but uses it to describe phenomena, and therefore he is bound to gain if he uses a more intuitive language.

The choice of subjects reflects the difference in needs between a physicist and a mathematician. This choice, which is dictated by reasons that are foreign to the mathematician and which forces us to develop within the same book very different fields of mathematics, makes a certain amount of fluctuation in the level of presentation inevitable. Thus, we are obliged to discuss elementary properties of complex numbers and also to introduce the sophisticated notion of infinite-dimensional vector spaces. At times the book will probably sound to the reader something like this: "We will discuss the properties of Hilbert space but first we must recall a few notions. The letter 'H' in the name Hilbert is the eighth letter of the alphabet and should be familiar to the reader from his first grade studies."

We have tried to emphasize the intuitive aspect of mathematics by establishing a link, wherever possible, between the abstract notions introduced and the notions of elementary mathematics that are obvious to the reader. We do not think, however, that these notions gain in clarity by consistently giving examples borrowed from physics, since mathematical ideas should be understood in themselves. Thus, we have concentrated here almost exclusively on the mathematics that a physicist will need and we have tried to avoid as much as possible a discussion of problems that properly belong to physics and which the reader will encounter in his physics courses.

The book is intended as a guide, not as an encyclopedia. This is particularly true in the sections devoted to special functions, but the same spirit prevails throughout the book. Since the publication by Erdelyi *et al.* of the excellent "Bateman Manuscript Project," where the properties of the transcendental functions are exhaustively listed (as are the references to the original works), it seems no longer necessary to drown the reader in details which he need not memorize and which he can easily find. Therefore, we have emphasized mainly the interrelations between the special functions and have illustrated their general properties so that a reader can refer without any difficulty to the Erdelyi *et al.* work.

Many of the ideas introduced are illustrated by means of simple examples that have been set in smaller type. The reader may skip these examples if his understanding of the subject is sufficient. We have also used smaller type to discuss more specialized topics, which may be read by the interested reader only, for these topics are completely separate from the main body of the text. We have omitted certain topics in mathematics that are of importance in applications because we wanted the book to be "digestible," and therefore of reasonable length. However, many topics are presented in such a manner that it should be easy for the reader to fill in some of the gaps by himself. For example, many of the results of the theory of Fredholm integral equations will be easily understood by anyone who is familiar with the properties of completely continuous linear operators discussed in Chapter III.

The book is presented in order of decreasing completeness. Chapter IV is very short compared to the many volumes that have been written on its subject matter and in fact on elliptic equations alone. The first chapter deals mainly with analysis and the second chapter with algebra. The order of these two chapters is practically interchangeable except for the first few sections of Chapter I, which should be read first. Chapters III and IV, however, should be read in this order and since they discuss freely the analytical and algebraic aspects of a problem, they presuppose a knowledge of the first two chapters.

The proofs of the theorems presented are, of course, not original. The choice of the proofs was made according to our taste. We did not consistently quote the names of the authors responsible for the theorems and their proofs. References to these authors can be found in the bibliography given.

We wish to thank Dr. R. Stora for a critical reading of the manuscript. Thanks are also due to Miss H. Noir and Miss M. M. Rançon for their patient and energetic assistance in putting together the manuscript.

Paris, 1966

P. DENNERY
A. KRZYWICKI

MATHEMATICS FOR PHYSICISTS

CHAPTER I

THE THEORY OF ANALYTIC FUNCTIONS

1 · ELEMENTARY NOTIONS OF SET THEORY AND ANALYSIS

1.1 Sets

The notion of a **set** is basic to all of modern mathematics. We shall mean by set a collection of objects, hereafter called **elements** of the set. For example, the integers 1, 2, 3, ⋯, 98, 99, 100 form a set of 100 elements. Another example of a set is given by the collection of all points on a line segment; here the number of elements is clearly infinite.

As in the case of other fundamental notions of mathematics (for instance, that of a geometrical point), it is impossible to give a truly rigorous definition of a set. We simply do not have more basic notions at our disposal. Thus, we stated that a set is a "collection of objects," but of course we would be very embarrassed if we were asked to clarify the meaning of the word "collection."

The standard way to circumvent the difficulty of defining fundamental mathematical objects is to formulate a certain number of axioms, which are the "rules of the game," and which form the basis of a deductive theory. The axioms are fashioned upon the intuitive properties of very familiar objects, such as the integers or the real numbers, but once these axioms have been adopted, we need no longer appeal to our intuition. In other words, when the "rules" have been specified, the question of knowing exactly what these objects represent is no longer relevant to the construction of a rigorous theory.

It is possible to develop a rigorous theory of sets based on an axiomatic formulation, but this is completely outside the scope of this book. However, since the theory of sets is now involved in almost all branches of mathematics the use, albeit very limited, of certain concepts and notations of this theory will be very useful to us. It will be quite sufficient for the reader to understand the notion of a set in its most intuitive sense.

1.2 Some Notations of Set Theory

We shall usually denote sets by capital letters; e.g., A, B. Sometimes, however, other symbols will also be used. For instance, (a,b) will denote the set of real numbers satisfying the inequality

$$a < x < b \tag{1.1}$$

The symbol $\in$ will frequently be used. It is an abbreviation for "belongs to."

For example, the real number x satisfying the inequality (1.1) belongs to the set (a,b); therefore, we shall write

$$x \in (a,b)$$

In general,

$$a \in S$$

should be read "a belongs to the set S."

Consider now two sets, A and B. When A and B are identical, we write

$$A = B$$

When each element of A is necessarily also an element of B, we say that A is included in B, or that A is a **subset** of B, and we write*

$$A \subset B \qquad \text{or} \qquad B \supset A$$

EXAMPLE 1

Let A be the sequence of numbers 80, 81, $\cdots$, 99, 100, and B the sequence 1, 2, 3, $\cdots$, 99, 100. Then

$$A \subset B$$

The set that contains all elements of A and all elements of B, but counted only once, is called the **sum** or the **union** of A and B and is denoted by

$$A + B$$

EXAMPLE 2

Let A be the sequence 1, 2, $\cdots$, 29, 30, and B the sequence 10, 11, $\cdots$, 49, 50. Then A + B is the sequence 1, 2, $\cdots$, 49, 50 in which the numbers 10, 11, $\cdots$, 29, 30, which are common to A and B, appear only once.

From the definition of the sum of sets it follows that

$$A + B = B \quad \text{if} \quad A \subset B$$

The set of all elements common to both A and B is called the **intersection** or sometimes the **product**, of A and B and is denoted by

$$A \cap B$$

EXAMPLE 3

The intersection $A \cap B$ of the two sets A and B of the preceding example is the sequence 10, 11, $\cdots$, 29, 30.

The set of all elements of A that are not included in B is called the **difference** between A and B and is denoted by

$$A - B$$

---

* Notice the formal similarity between the relation

$$A \subset B$$

and the inequality

$$a < b$$

The set "which is included in" occupies a position with respect to the symbol $\subset$, similar to the position the smaller member in an inequality occupies with respect to the symbol $<$.

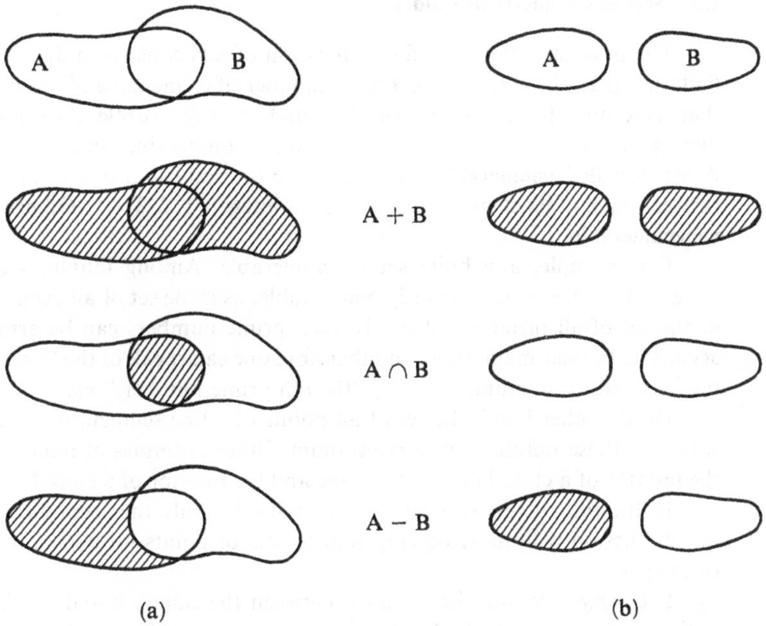

Fig. 1. The shaded area represents the set resulting from the indicated operation. In the case (b), where A and B do not overlap, the set A ∩ B is empty.

It is convenient to introduce the notion of an empty set, i.e., a "set" that has no elements at all. It plays the role of the number 0 in algebra and it is also denoted by 0 in this text.

EXAMPLE 4

The difference A − B between the sets A and B of Example 2 is the sequence 1, 2, ⋯, 8, 9, and the difference B − A is the sequence 31, 32, ⋯, 49, 50.

The operations with sets can be visualized with the aid of the diagrams of Fig. 1, where the sets resulting from the addition, intersection, and subtraction of A and B correspond to the shaded areas.

To end this subsection, we summarize in Table 1 the meaning of the symbols that have been introduced.

TABLE 1

Notation	Significance
$a \in A$	a belongs to A
$A \subset B$	A is included in B
$A = B$	A is identical to B
$A + B$	The union of A and B
$A \cap B$	The set of common elements of A and B
$A - B$	The set of elements of A not included in B

1.3 Sets of Geometrical Points

The most straightforward classification of sets consists in distinguishing between finite sets (i.e., those that have a finite number of elements) and infinite sets (i.e., those that have an infinite number of elements). A more subtle classification consists in distinguishing among infinite sets between enumerable and nonenumerable ones. A set is called **enumerable** if it is possible to establish a one-to-one correspondence between each of its elements and the set of integers 1, 2, 3, $\cdots$. Otherwise a set is called **nonenumerable**.

For example, any finite set is enumerable. Among infinite sets, the set of all integers 1, 2, 3, $\cdots$, is obviously enumerable, as is the set of all even (or odd) integers, or the set of all prime numbers. In fact, prime numbers can be arranged in a series according to their magnitude, and therefore one can speak of the "first prime number," the "second prime number," $\cdots$, "the nth prime number," etc.

On the other hand, the set of all points of a line segment is nonenumerable; one says that these points form a continuum. Other examples of nonenumerable sets are the interior of a closed curve in a plane and the interior of a closed surface in space.

In the rest of this section we shall consider only the sets of geometrical points; our discussion will apply equally well to sets of points located on a line, in a plane, or in space.

Let $\rho(p,p')$ denote the distance between the points p and p'. A very important notion is that of a **neighborhood** of a point. By a neighborhood of a given point p, we shall mean a set of points p' satisfying.

$$\rho(p,p') < R \tag{1.2}$$

where R is an arbitrary positive number.

If we consider exclusively the sets of points in a plane or on a line, then we restrict p' in (1.2) to lie on the plane or on the line. For example, the set of all points in a plane lying in the interior of an arbitrary circle centered at the point p is a neighborhood of p, whereas the interior of an arbitrary sphere centered at p stands for the neighborhood of p in space.

Using the concept of a neighborhood of a point, we can give a classification of the point sets.

Consider a set S of geometrical points. A point $p \in S$ is called an **isolated** point of the set if there exists a neighborhood of p which does not contain any other

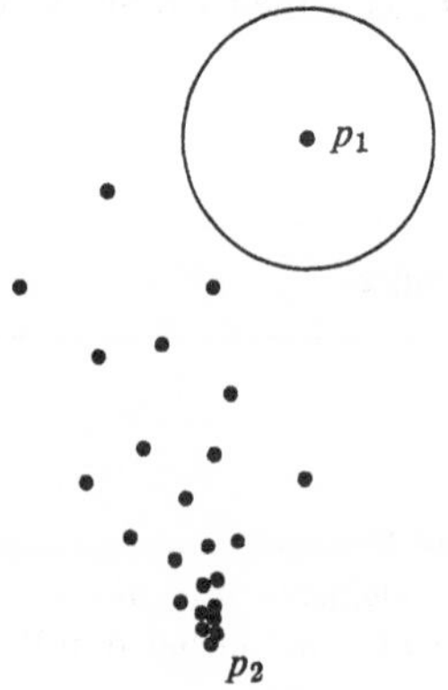

Fig. 2. A set of points in a plane: p_1 is an isolated point, since there exists a circle centered at p_1 and which does not contain any of the points of the set. On the contrary, p_2 is an accumulation point.

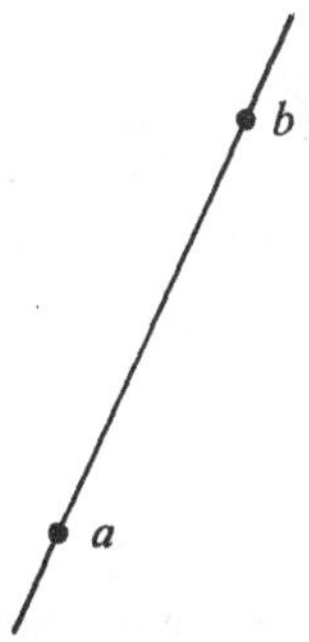

Fig. 3. The points of the segment ab form a closed or an open set, depending upon whether the end points belong or do not belong to the set.

point belonging to S. This is in agreement with the intuitive meaning of the word "isolated"; effectively, if p is an isolated point, then every element of S is located at a finite distance from p (see Fig. 2). A point, **every** neighborhood of which contains at least one element of S, which is not identical with the point itself, is called an **accumulation point** of the set. If not only a given point but also **all** points of some neighborhood of p belong to S, then p is called an **interior point** of S. Every interior point of a set is an accumulation point. The converse is not true. Moreover, an accumulation point of a set need not necessarily belong to the set.

For example, consider a set of points on a line located between two points a and b, and suppose that a and b do not belong to the set. It is obvious that any point of the set is an interior point, and consequently an accumulation point. However, the points a and b are also accumulation points of the set, since points of the set come arbitrarily close to a and b. We can now distinguish between two important classes of point sets: A set is called an **open set** if all its points are interior points; a set is called a **closed set** if it contains all its accumulation points.

Arbitrary point sets are neither open nor closed. For example, all the points lying within, but not on, a closed curve in a plane form an open set in the plane. If the points lying on the boundary curve are added to the set, it becomes closed. However, the interior points together with several isolated points in the plane form a set which is neither open nor closed.

The set of interior points of a segment ab of a line is an open one-dimensional set. But if the points a and b are added to the set, we get a closed set (see Fig. 3). There is a one-to-one correspondence between points on a line and the real numbers, and similarly one can distinguish between an open interval (a,b) that does not contain the numbers a and b and a closed interval $[a,b]$ that does:

$$x \in (a,b) \qquad \text{if} \quad a < x < b$$

$$x \in [a,b] \qquad \text{if} \quad a \leq x \leq b$$

To end this section, let us define what we shall later mean by a **region**: A region is an open set, any two points of which can be connected by a continuous line that is contained entirely within the set (see Fig. 4).

1.4 The Complex Plane

It is assumed that the reader is familiar with complex numbers; nevertheless we shall start with a short summary of their properties and of the notation that will be used throughout this chapter.

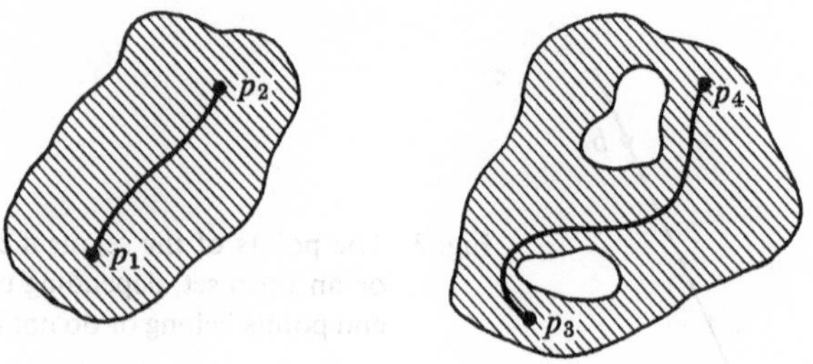

Fig. 4. The points belonging to either one of the shaded areas, but not lying on the boundary curves, form a region. The points p_1 and p_2 as well as the points p_3 and p_4 can be connected by a continuous curve lying within the shaded areas. When the points p_1 and p_3 are connected by a curve, a part of this curve necessarily lies outside the shaded areas.

A complex number z is completely specified by a pair of real numbers x and y.

$$z = x + iy$$

Manipulations with complex numbers are carried out using the usual rules of arithmetic, remembering, however, that by convention

$$i^2 = -1$$

The real numbers x and y are called, respectively, the **real** and **imaginary** parts of z and are denoted by Re z and Im z:

$$\text{Re } z \equiv x$$

$$\text{Im } z \equiv y$$

The numbers x and y may be considered as the Cartesian coordinates of a point in a plane. Thus, any complex number can be represented by a point in a plane, hereafter called the **complex plane**. One can also represent a complex number by a pair of polar

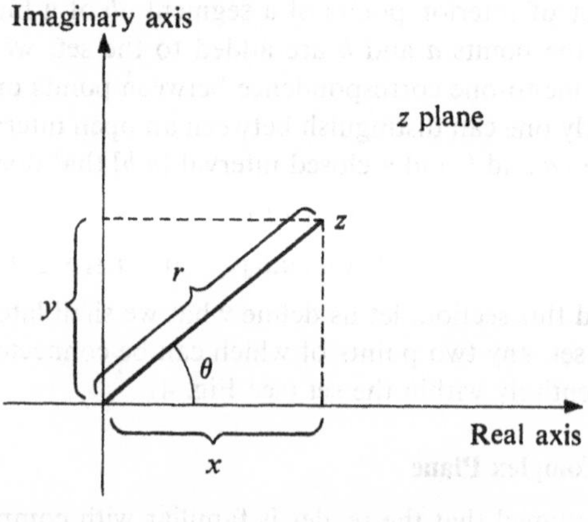

Fig. 5. The complex number $z = x + iy$, represented as a point, with coordinates x and y or r and θ in the complex plane.

coordinates r and θ defined in the complex plane. They are related to the Cartesian coordinates (see Fig. 5) by the equations

$$x = r\cos\theta$$

$$y = r\sin\theta$$

Hence, z can also be written as

$$z = x + iy = r(\cos\theta + i\sin\theta) \tag{1.3}$$

where

$$r = \sqrt{x^2 + y^2}$$

$$\theta = \tan^{-1}\frac{y}{x}$$

r is called the **modulus** of z and is denoted by $|z|$. The concept of the modulus of a complex number is simply a straightforward generalization of the concept of the absolute value of a real number. θ is called the **argument** of z and is denoted by $\arg z$. In fact the polar coordinate θ is determined only up to an integer multiple of 2π. Hence

$$\arg z = \theta + 2\pi k \qquad k = 0, \pm 1, \pm 2, \cdots \tag{1.4}$$

Given two complex numbers z_1 and z_2, we have

$$\begin{aligned} z_1 \cdot z_2 &= (x_1 + iy_1)(x_2 + iy_2) \\ &= r_1 r_2(\cos\theta_1 + i\sin\theta_1)(\cos\theta_2 + i\sin\theta_2) \\ &= r_1 r_2[\cos(\theta_1 + \theta_2) + i\sin(\theta_1 + \theta_2)] \end{aligned} \tag{1.5}$$

Comparing with Eq. 1.3 we obtain

$$\begin{aligned} |z_1 \cdot z_2| &= |z_1| \cdot |z_2| \\ \arg(z_1 \cdot z_2) &= \arg z_1 + \arg z_2 \end{aligned} \tag{1.6}$$

Each complex number z can also be represented by a vector of length $|z|$, with components x and y, in the complex plane. The reader may easily verify that the ruie of addition of complex numbers

$$\begin{aligned} (z_1 + z_2) &= (x_1 + iy_1) + (x_2 + iy_2) \\ &= (x_1 + x_2) + i(y_1 + y_2) \end{aligned} \tag{1.7}$$

can be represented graphically as the familiar geometrical rule of vector addition (see Fig. 6). Since the sum of the lengths of two sides of a triangle is larger than the length of the third side, one immediately gets the "triangle inequalities"

$$\begin{aligned} |z_1 + z_2| &\le |z_1| + |z_2| \\ |z_1 + z_2| &\ge |z_1| - |z_2| \\ |z_1 + z_2| &\ge |z_2| - |z_1| \end{aligned} \tag{1.8}$$

The numbers $a = \text{Re}\, a + i\,\text{Im}\, a$ and $\bar{a} = \text{Re}\, a - i\,\text{Im}\, a$ are called **complex conjugates** of each other and are represented by points whose positions are symmetrical with respect to the real axis. In this book a bar over any quantity will always mean that we take the complex conjugate of that quantity. The following relation is evident

$$a \cdot \bar{a} = (\text{Re}\, a)^2 + (\text{Im}\, a)^2 = |a|^2 \tag{1.9}$$

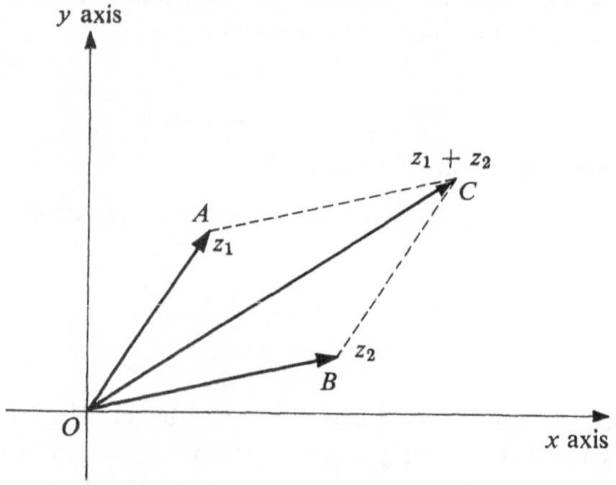

Fig. 6. OA is parallel to BC and OB is parallel to AC. To derive inequalities 1.8, we consider the triangle OBC (or OAC).

1.5 Functions

One says that a function has been defined on a set A if one has associated a number (in general, complex) with every element $p \in$ A. A is then called the **set of arguments** of the function.

As a familiar example, one can take A to be the set of real numbers $x \in [a,b]$, and then associate with every x a real number $f(x)$. Such a function can be represented graphically as a curve in a plane (see Fig. 7); the point with Cartesian coordinates x and $f(x)$ describes this curve as x goes from a to b.

A more general definition of a function would be to say that we associate not simply one number but some set F(p) with every $p \in$ A.

As an example, consider a vector field (for example, an electric field) in space. With every point of some region in space, one associates a set of three real numbers, the components of the vector, which change from point to point.

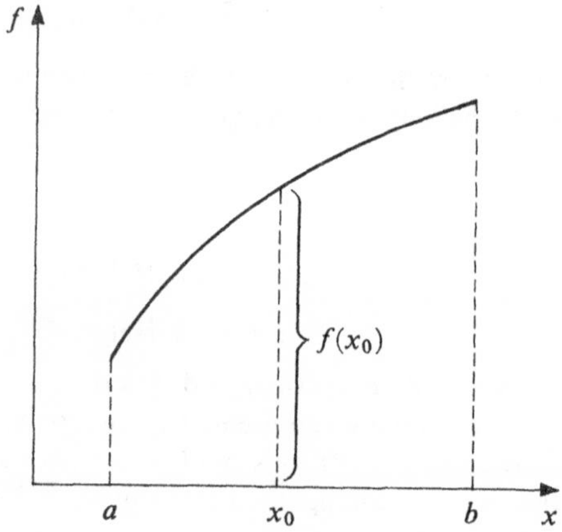

Fig. 7.

Let a function $f(p)$ be defined in a region R; it is irrelevant to our discussion whether R is a set of points located on a line, in a plane, or in space.

The function $f(p)$ is said to be **continuous** at a point $p \in \mathrm{R}$ if for an arbitrary number $\varepsilon > 0$, one can find a number $\delta > 0$ such that, provided the distance $\rho(p,p')$ between the points p and p' is smaller than δ

$$\rho(p,p') < \delta \tag{1.10}$$

one has

$$|f(p) - f(p')| < \varepsilon \tag{1.11}$$

Stated less precisely, one can say that a function $f(p)$ is continuous at a point p if, when p' is sufficiently close to p, $f(p')$ is arbitrarily "close" to $f(p)$.

Consider now an infinite sequence of functions defined in R

$$f_1(p), f_2(p), \cdots, f_n(p), \cdots \tag{1.12}$$

One says that this sequence converges in R to the function $f(p)$, if for any point $p \in \mathrm{R}$ and for an arbitrary $\varepsilon > 0$ one can find a number $N = N(p,\varepsilon)$ such that, provided $n > N$, one has

$$|f(p) - f_n(p)| < \varepsilon \tag{1.13}$$

As we have indicated, N will in general depend on both ε and p. If, however, N is independent of p **throughout the set**, then we say that the sequence 1.12 converges **uniformly** to $f(p)$ in R.

The meaning of the uniform convergence of a sequence of functions can be easily illustrated if one considers the simple case of a real function defined on a line or (what is equivalent because of the one-to-one correspondence between points on a line and real numbers) on an interval $[a,b]$. If the sequence $f_n(x)$ $(n = 1, 2 \cdots)$ converges uniformly to $f(x)$, then for n large enough, all curves representing the functions $f_n(x)$ may be enclosed within an arbitrarily thin strip containing the curve representing $f(x)$ (see Fig. 8).

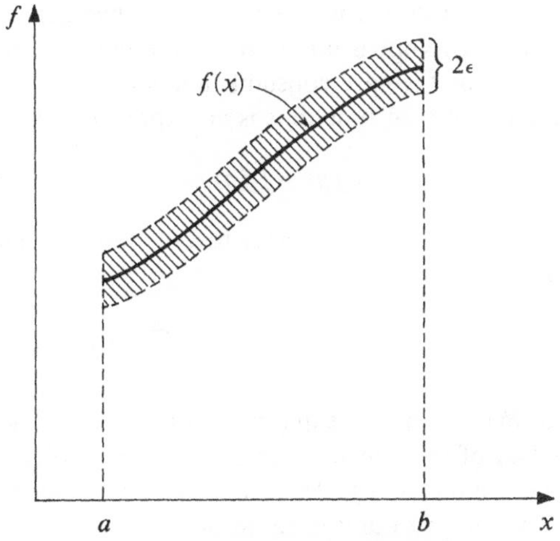

Fig. 8. For an $\varepsilon > 0$, we can enclose all the curves $f_n(x)$ within the shaded strip, provided n is large enough and the sequence of functions $f_n(x)$ converges uniformly.

EXAMPLE

It is worthwhile to give an example of a sequence which does not converge uniformly. Let

$$f_n(x) = \frac{nx}{1 + n^2x^2}$$

For any fixed $x \in (-\infty, +\infty), f_n(x) \to 0$ as $n \to \infty$. However,

$$f_n\left(\frac{1}{n}\right) = \frac{1}{2}$$

Thus, one cannot make $f_n(x)$ arbitrarily small in the whole interval $(-\infty, +\infty)$ simply by choosing n large enough. The convergence to zero of the functions $f_n(x)$ is therefore not uniform in $(-\infty, +\infty)$.

There exists an important criterion to decide whether or not a sequence of functions is uniformly convergent. This criterion is certainly known to the reader from elementary calculus. It is, however, more generally valid and holds also when the function is defined on an arbitrary set. The proof is analogous to the one given in the elementary case.

The Cauchy Criterion

A sequence of functions $f_n(p)$ $(n = 1, 2, \cdots)$ converges uniformly on a set R if and only if for any $\varepsilon > 0$, one can find a real number N independent of $p \in$ R and such that, provided

$$m > n > N$$

one has

$$|f_n(p) - f_m(p)| < \varepsilon \tag{1.14}$$

The preceding criterion for the uniform convergence of a sequence of functions reduces to the well-known criterion for the convergence of a sequence of numbers when the functions $f_n(p)$ are constant functions.

An infinite series of functions is an expression of the type

$$f_1(p) + f_2(p) + \cdots + f_n(p) + \cdots \tag{1.15}$$

One says that such a series converges to a function $S(p)$ if the sequence of partial sums $S_1(p), S_2(p), \cdots$, where

$$S_n(p) = \sum_{k=1}^{n} f_k(p)$$

converges to $S(p)$. Hence, the question of the convergence of an infinite series reduces to the problem of the convergence of a sequence of partial sums. In particular, one says that the infinite sum (1.15) converges uniformly to $S(p)$ if the sequence $S_n(p)$ $(n = 1, 2, \cdots)$ converges uniformly to $S(p)$.

One also has the Weierstrass criterion for the uniform convergence of the series (1.15).

The Weierstrass Criterion

The infinite series

$$\sum_{k=1}^{\infty} f_k(p)$$

converges uniformly if $|f_n(p)| \leq a_n$ and if the series

$$\sum_{k=1}^{\infty} a_n$$

is convergent. The proof is again analogous to the one known from elementary calculus.

2 · FUNCTIONS OF A COMPLEX ARGUMENT

We shall consider in this chapter those functions whose arguments are complex numbers. Defining a function $f(z)$ over a set of complex numbers amounts to defining a function over a set of points in a plane, the complex plane, that is a function of two real variables. Since $f(z)$ is assumed to take on complex values in general, it can always be written as

$$f(z) = u(x,y) + iv(x,y) \tag{2.1}$$

where $u(x,y)$ and $v(x,y)$ are real functions of the real arguments x and y. It is therefore evident that the theory of functions of a complex variable would reduce trivially to the theory of functions of two real variables if the theory of functions of a complex argument was considered in its whole generality.

The theory of analytic functions deals, however, only with a restricted class of functions, namely, those functions that satisfy certain smoothness requirements or, to be specific, that are "differentiable." We shall explain presently what is meant by the differentiability of a function of a complex variable, but we may state now that although the condition of differentiability places a severe limitation on the functions that one is allowed to consider, it leads nevertheless to a theory of these functions that is both elegant and extremely powerful.

Before defining what we mean by the differentiability of a function of a complex variable, we shall make a few brief comments about the corresponding problem in the theory of functions of a real variable.

Let $g(x)$ be a function of a real variable and suppose that it is continuous in the neighborhood of the point $x = x_0$. Consider the two limits

$$D^{\pm} g(x_0) = \lim_{|\Delta x| \to 0} \frac{g(x_0 \pm |\Delta x|) - g(x_0)}{\pm |\Delta x|} \tag{2.2}$$

Either of or both limits may not exist at all, even though $g(x)$ may be continuous.* It may also happen that the two limits do exist, but that they are different. The

* One can construct continuous functions that are nondifferentiable everywhere. It is difficult to visualize such a function, but roughly speaking, one may describe it by saying that it makes infinitely rapid oscillations. This illustrates the fact that what one intuitively understands by smoothness is not at all guaranteed by the continuity of a function.

simplest example is provided by the function $g(x) = |x|$. At $x = 0$, we have $D^+g(0) = 1$, whereas $D^-g(0) = -1$. More generally, $D^+g(x_0) \neq D^-g(x_0)$ when $g(x)$ has a "cusp" (i.e., a sharp turning point) at $x = x_0$. In that case, the tangent to the curve, $g = g(x)$, tends to different limits, depending upon whether the point $x = x_0$ is approached from the left or from the right. When the two limits are finite and equal, then

$$D^+g(x_0) = D^-g(x_0) \tag{2.3}$$

In that case, the function $g(x)$ is said to be **differentiable** and the (unique) limit

$$\lim_{\Delta x \to 0} \frac{g(x_0 + \Delta x) - g(x_0)}{\Delta x} \tag{2.4}$$

is called the derivative of $g(x)$ at $x = x_0$ and denoted by

$$\left.\frac{dg(x)}{dx}\right|_{x=x_0}$$

The derivative is a **local** characteristic of the function in the sense that it determines its behavior only in the infinitesimal neighborhood of a single point.

We now turn to a consideration of the differentiability of a function of a complex variable.

3 · THE DIFFERENTIAL CALCULUS OF FUNCTIONS OF A COMPLEX VARIABLE

The derivative of a function of a complex variable with respect to its argument z is formally defined in the same way as it is for functions of a real variable.

$$\frac{df}{dz} = \lim_{\Delta z \to 0} \frac{f(z + \Delta z) - f(z)}{\Delta z} \tag{3.1}$$

Here the limit $\Delta z \to 0$ is actually a double limit inasmuch as both the real part Δx and the imaginary part Δy of Δz must each tend separately to zero. Since there are an infinite number of ways by which Δz can tend to zero (even if for each of these ways the corresponding limit exists), there are in general an infinite number of possible values that the limits can assume. For example, to achieve the limit in Eq. 3.1, we could let $|\Delta z| \to 0$ for any value of arg Δz (Fig. 9); but what is more likely is that the resulting derivative will depend on the particular value of arg Δz. Similarly, the limit in Eq. 3.1 will, in general, depend on the order in which $\Delta x, \Delta y$ tend to zero.

Consider as a simple example the function $f(z) = x + 2iy$. We shall show that this function does not have a well-defined derivative at the origin. Its derivative at $z = 0$ is given by

$$\left.\frac{df}{dz}\right|_{z=0} = \lim_{z \to 0} \frac{f(z) - f(0)}{z} = \lim_{\substack{x \to 0 \\ y \to 0}} \frac{x + 2iy}{x + iy} = \lim_{\substack{x \to 0 \\ y \to 0}} \frac{x^2 + 2y^2 + ixy}{x^2 + y^2}$$

The value of

$$\left.\frac{df}{dz}\right|_{z=0}$$

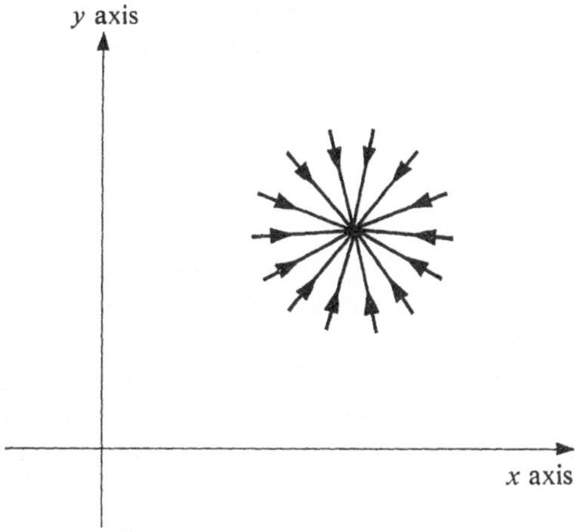

Fig. 9. Ways of achieving the limit (Eq. 3.1) by letting $|\Delta z| \to 0$ while letting arg Δz take on arbitrary values.

will obviously depend on the order in which the two limits are taken. If x is first held fixed while $y \to 0$, we obtain

$$\left.\frac{df}{dz}\right|_{z=0} = 1$$

whereas if y is first held fixed while $x \to 0$, we obtain

$$\left.\frac{df}{dz}\right|_{z=0} = 2$$

Again, suppose that x and y tend to zero along some arbitrary line, $y = \alpha x$. Then

$$\left.\frac{df}{dz}\right|_{z=0} = \frac{1 + 2\alpha^2 + i\alpha}{1 + \alpha^2}$$

and the value of the derivative depends in this case on

$$\arg z = \frac{y}{x} = \alpha$$

In analogy to the case of functions of a real variable, we shall say that a function of a complex argument z is **differentiable** at a given point $z = z_0$ if the limit

$$\lim_{\Delta z \to 0} \frac{f(z_0 + \Delta z) - f(z_0)}{\Delta z} \tag{3.2}$$

exists, is finite, and does not depend on the manner in which one takes the limit or, in other words, does not depend on the way one approaches the point $z = z_0$. Whereas in the case of a real variable, one can approach a given point only in two ways (either from the left or from the right), a point in the complex plane can be reached from an infinite number of directions. Thus, instead of one requirement (Eq. 2.3), an infinite number of such requirements has to be satisfied in order to ensure the differentiability of a function of a complex argument. One can expect, therefore, that the property of being differentiable is, in the case of functions of a complex variable, very much more restrictive than it is for functions of a real variable. This is indeed the case. Without entering into the details, which will be discussed later on, we can state now that although the derivative

$$\left.\frac{df(z)}{dz}\right|_{z=z_0}$$

can still be considered as a local characteristic of the function $f(z)$ at $z = z_0$, the condition that a function be differentiable within some region of the complex plane implies that the local behavior of the function in that region governs its behavior at different and distant points of the region.

The formal rules of differentiation, which follow from the basic definition (Eq. 3.1), are the same as in the case of a function of a real variable. Thus, provided the derivatives on the right-hand side (RHS) exist, we have

$$\begin{aligned} \frac{d}{dz}(f \pm g) &= \frac{df}{dz} \pm \frac{dg}{dz} \\ \frac{d}{dz}(f \cdot g) &= \frac{df}{dz} g + f\frac{dg}{dz} \\ \frac{d}{dz}\left(\frac{f}{g}\right) &= \frac{\frac{df}{dz} g - f\frac{dg}{dz}}{g^2} \\ \frac{d}{dz} f[g(z)] &= \frac{df}{dg}\frac{dg}{dz} \end{aligned} \tag{3.3}$$

The proofs are the same as in the real variable case.

It is, of course, important to have a criterion that will allow us to decide whether a function is differentiable. The necessary condition for a function to be differentiable at a given point is that, at that point, it obeys the Cauchy-Riemann conditions, which we shall now proceed to derive.

4 · THE CAUCHY-RIEMANN CONDITIONS

Let $u(v,y)$ and $v(x,y)$ denote as before the real and imaginary parts of a function $f(z)$ of the complex variable z

$$f(z) = u(x,y) + iv(x,y) \tag{4.1}$$

We shall suppose that at a point z of the complex plane, $u(x,y)$ and $v(x,y)$ possess first-order partial derivatives with respect to x and y. According to Eq. 3.1, the derivative $df(z)/dz$ is

$$\begin{aligned} \frac{df}{dz} &= \lim_{\Delta z \to 0} \frac{f(z + \Delta z) - f(z)}{\Delta z} \\ &= \lim_{\substack{\Delta x \to 0 \\ \Delta y \to 0}} \frac{u(x + \Delta x, y + \Delta y) - u(x,y) + i[v(x + \Delta x, y + \Delta y) - v(x,y)]}{\Delta x + i\,\Delta y} \end{aligned} \tag{4.2}$$

We now impose the condition that the right-hand side of (4.2) should yield the same result, whatever the order in which the limit $\Delta x, \Delta y \to 0$ is taken. By first setting $\Delta y = 0$ and then taking the limit $\Delta x \to 0$, we find

$$\frac{df(z)}{dz} = \frac{\partial u(x,y)}{\partial x} + i\frac{\partial v(x,y)}{\partial x} \tag{4.3}$$

In the other case, in which we first set $\Delta x = 0$, and then take the limit $\Delta y \to 0$, we find

$$\frac{df(z)}{dz} = -i\frac{\partial u(x,y)}{\partial y} + \frac{\partial v(x,y)}{\partial y} \tag{4.4}$$

By equating the real and imaginary parts of (4.3) and (4.4), one obtains the **Cauchy-Riemann conditions**

$$\begin{aligned} \frac{\partial u(x,y)}{\partial x} &= \frac{\partial v(x,y)}{\partial y} \\ \frac{\partial u(x,y)}{\partial y} &= -\frac{\partial v(x,y)}{\partial x} \end{aligned} \tag{4.5}$$

Differentiating Eq. 4.5, first with respect to x and then with respect to y, one easily obtains

$$\begin{aligned} \frac{\partial^2 u}{\partial x^2} + \frac{\partial^2 u}{\partial y^2} &= 0 \\ \frac{\partial^2 v}{\partial x^2} + \frac{\partial^2 v}{\partial y^2} &= 0 \end{aligned} \tag{4.6}$$

A function $h(x_1, x_2, \cdots, x_N)$ of N variables satisfying the equation

$$\sum_{j=1}^{N} \frac{\partial^2 h}{\partial x_j^2} = 0 \tag{4.7}$$

is called an **harmonic function**, and the differential equation, Eq. 4.7, is known as Laplace's equation. Thus, the real and imaginary parts of a differentiable function separately satisfy the Laplace equation (with $N = 2$) and are therefore harmonic functions of two variables. The converse is not true, however; a pair of harmonic functions does not, in general, define a differentiable function. For example, it is easy to verify that the function $f(z) = x + 2iy$, which does not have a well-defined derivative at the origin, also does not satisfy the Cauchy-Riemann conditions; the real and imaginary parts of this function, however, trivially satisfy Laplace's equation.

The Cauchy-Riemann conditions have been derived under rather restrictive assumptions, for of the many possible limiting processes that could have been used to deduce Eq. 4.5 from Eq. 3.1, only two specific ones were considered in which the increment of the variable z approached zero either along lines parallel to the x axis or parallel to the y axis. The result, however, is much more general than the derivation would indicate, as we shall now demonstrate by considering the sufficiency condition for $f(z)$ to be differentiable.

Theorem. Let the real and imaginary parts $u(x,y)$ and $v(x,y)$ of a function of a complex variable $f(z)$ obey the Cauchy-Riemann equations and also possess continuous first partial derivatives with respect to the two variables x and y at all points of some region of the complex plane. Then $f(z)$ is differentiable throughout this region.

Proof. Since $u(x,y)$ and $v(x,y)$ have continuous first partial derivatives, there exist four positive numbers $\varepsilon_1,\varepsilon_2,\delta_1,\delta_2$, which can be made arbitrarily small as Δx and Δy tend to zero, and such that

$$\begin{aligned} u(x+\Delta x,y+\Delta y)-u(x,y)&=\frac{\partial u}{\partial x}\Delta x+\frac{\partial u}{\partial y}\Delta y+\varepsilon_1\,\Delta x+\delta_1\,\Delta y\\ v(x+\Delta x,y+\Delta y)-v(x,y)&=\frac{\partial v}{\partial x}\Delta x+\frac{\partial v}{\partial y}\Delta y+\varepsilon_2\,\Delta x+\delta_2\,\Delta y \end{aligned}\tag{4.8}$$

Using the relations (4.8), we easily deduce

$$\left|\frac{f(z+\Delta z)-f(z)}{\Delta z}-\left(\frac{\partial u}{\partial x}+i\frac{\partial v}{\partial x}\right)\right|\le\left|\frac{\Delta x}{\Delta z}(\varepsilon_1+i\varepsilon_2)\right|+\left|\frac{\Delta y}{\Delta z}(\delta_1+i\delta_2)\right|\tag{4.9}$$

But since

$$\left|\frac{\Delta x}{\Delta z}\right|=\frac{\Delta x}{[(\Delta x)^2+(\Delta y)^2]^{1/2}}\le 1$$

$$\left|\frac{\Delta y}{\Delta z}\right|=\frac{\Delta y}{[(\Delta x)^2+(\Delta y)^2]^{1/2}}\le 1$$

we obtain from (4.9), on taking the limit $\Delta z\to 0$,

$$\frac{df}{dz}=\frac{\partial u(x,y)}{\partial x}+i\frac{\partial v(x,y)}{\theta x}\tag{4.10}$$

which shows that $f(z)$ is differentiable.

To give a more intuitive meaning to the Cauchy-Riemann conditions, suppose that the two real functions $\operatorname{Re} f(z)=u(x,y)$ and $\operatorname{Im} f(z)=v(x,y)$ can be expanded in a double Taylor series about a point with coordinates x_0 and y_0:

$$u(x,y)+iv(x,y)$$

$$=\sum_{n=0}^{\infty}\frac{1}{n!}\sum_{k=0}^{n}\frac{n!}{k!(n-k)!}(x-x_0)^{n-k}(y-y_0)^k\times\frac{\partial^n}{\partial^{n-k}x_0\,\partial^k y_0}[u(x_0,y_0)+iv(x_0,y_0)]\tag{4.11}$$

According to Eq. 4.5

$$\frac{\partial}{\partial y_0}[u(x_0,y_0)+iv(x_0,y_0)]=i\frac{\partial}{\partial x_0}[u(x_0,y_0)+iv(x_0,y_0)]$$

Hence

$$\frac{\partial^k}{\partial y_0^k}[u(x_0,y_0)+iv(x_0,y_0)]=i^k\frac{\partial^k}{\partial x_0^k}[u(x_0,y_0)+iv(x_0,y_0)]$$

Inserting this in Eq. 4.11 we obtain

$$u(x,y)+iv(x,y)=\sum_{n=0}^{\infty}\frac{1}{n!}\sum_{k=0}^{\infty}\frac{n!}{k!(n-k)!}(x-x_0)^{n-k}(y-y_0)^k\; i^k\frac{\partial^n}{\partial x_0^n}[u(x_0,y_0)+iv(x_0,y_0)]$$

$$=\sum_{n=0}^{\infty}\frac{1}{n!}[(x+iy)-(x_0+iy_0)]^n\frac{\partial^n}{\partial x_0^n}[u(x_0,y_0)+iv(x_0,y_0)]\tag{4.12}$$

We see that because of the Cauchy-Riemann conditions, the two real variables x and y enter into the function in the unique combination $x+iy$. Thus, these conditions have as a consequence that a mathematical expression defining a differentiable function can depend **explicitly** only on $z = x + iy$ but not on $\bar{z} = x - iy$. The preceding result was based on the assumption that a function can be expanded in a Taylor series. In fact, it should be noted that merely because an expression depends on z only and not on $\bar{z}$ does not ensure the differentiability of the function. However, as we shall see later, any function that is differentiable within a neighborhood of a point can be expanded in a Taylor series about this point. Therefore the expression that defines such a function can explicitly depend only on z. For example, one immediately sees that the function $f(z) = x + 2iy$, considered in Sec. 3, cannot be differentiable because $x + 2iy$ cannot be explicitly expressed in terms of z alone

$$x + 2iy = \tfrac{3}{2}z - \tfrac{1}{2}\bar{z}$$

The use of both z and $\bar{z}$ is unavoidable here.

Let $w(x,y)$ be a real function of two real arguments x and y. One has

$$dw = \frac{\partial w}{\partial x}\,dx + \frac{\partial w}{\partial y}\,dy \tag{4.13}$$

A useful notation is obtained by introducing a symbolic "vector" $\vec{\nabla}$ with components along the x and y axis given by

$$(\vec{\nabla})_x = \frac{\partial}{\partial x} \quad \text{and} \quad (\vec{\nabla})_y = \frac{\partial}{\partial y} \tag{4.14}$$

respectively.

The vector $\vec{\nabla} w$ with components

$$(\vec{\nabla} w)_x = \frac{\partial w}{\partial x} \quad \text{and} \quad (\vec{\nabla} w)_y = \frac{\partial w}{\partial y} \tag{4.15}$$

is called the **gradient** of w at a given point. Since dx and dy can also be considered as components of a vector $d\vec{r}$

$$(d\vec{r})_x = dx \quad \text{and} \quad (d\vec{r})_y = dy \tag{4.16}$$

Equation 4.13 can be rewritten as

$$dw = (\vec{\nabla} w)_x(d\vec{r})_x + (\vec{\nabla} w)_y(d\vec{r})_y = \vec{\nabla} w \cdot d\vec{r} \tag{4.17}$$

The expression on the RHS is the usual scalar product of vectors $\vec{\nabla} w$ and $d\vec{r}$.

In the case when the points x and y lie on the same curve

$$w(x,y) = \text{constant} \tag{4.18}$$

one has

$$dw = (\vec{\nabla} w) \cdot d\vec{r} = 0 \tag{4.19}$$

Since $d\vec{r}$ is now tangent to the curve (4.18) equation 4.19 shows that the gradient $\vec{\nabla} w$ at a point x_0,y_0 is perpendicular to the curve $w(x,y) = w(x_0,y_0)$.*

Consider now a differentiable function

$$f(z) = u(x,y) + iv(x,y)$$

* The foregoing results also hold in the case of a function defined in space. Hence, the three-dimensional gradient $\vec{\nabla} w$ at a point p_0 is perpendicular to the surface $w(p) = w(p_0)$.

The gradients $\vec{\nabla} u$ and $\vec{\nabla} v$ are perpendicular at an arbitrary point x_0,y_0 to the curves $u(x,y) = u(x_0,y_0)$ and $v(x,y) = v(x_0,y_0)$, respectively. Furthermore, using the Cauchy-Riemann conditions, it is easy to see that $\vec{\nabla} u$ and $\vec{\nabla} v$ are perpendicular to each other

$$(\vec{\nabla} u) \cdot (\vec{\nabla} v) = \frac{\partial u}{\partial x}\frac{\partial v}{\partial x} + \frac{\partial u}{\partial y}\frac{\partial v}{\partial y} = 0 \tag{4.20}$$

Hence the curves

$$u(x,y) = u(x_0,y_0) \tag{4.21}$$

and

$$v(x,y) = v(x_0,y_0) \tag{4.22}$$

make a right angle with each other at the point x_0,y_0.

In other words, the tangents to the curves

$$\operatorname{Re} f(z) = \operatorname{Re} f(z_0) \tag{4.23}$$

and

$$\operatorname{Im} f(z) = \operatorname{Im} f(z_0) \tag{4.24}$$

are perpendicular at the point z_0 if $f(z)$ is differentiable in a neighborhood of this point.

5 · THE INTEGRAL CALCULUS OF FUNCTIONS OF A COMPLEX VARIABLE

The definition of the integral of a function of a complex variable is a straightforward generalization of the definition of the Riemann integral of a function of a real variable. Let $f(z)$ be a function of the complex variable $z = x + iy$, and let us consider a curve C in the complex plane, with end points a and b. We shall suppose that the curve C is a regular one, by which will be meant that it may be described by a parametric equation

$$z = z(t) \equiv x(t) + iy(t) \qquad t_a \leq t \leq t_b \tag{5.1}$$

where t is a real parameter and where $x(t)$ and $y(t)$ are real, single-valued functions that have continuous first-order derivatives. Our discussion will also be valid for a piecewise, regular curve C, i.e., for a continuous curve consisting of a finite number of regular arcs.

As shown in Fig. 10, we first subdivide the arc ab into n intervals by introducing the $n + 1$ points

$$z_0,z_1,z_2 \cdots z_{n-1},z_n \tag{5.2}$$

On this arc z_0 and z_n will be taken to coincide with the end points a and b, respectively. Next, we refine further the subdivision of C by introducing an additional series of points $\zeta_1,\zeta_2 \cdots \zeta_n$, taken along C and such that ζ_k lies between z_{k-1} and z_k.

We now form the sum

$$I_n = \sum_{k=1}^{n} f(\zeta_k)(z_k - z_{k-1}) \tag{5.3}$$

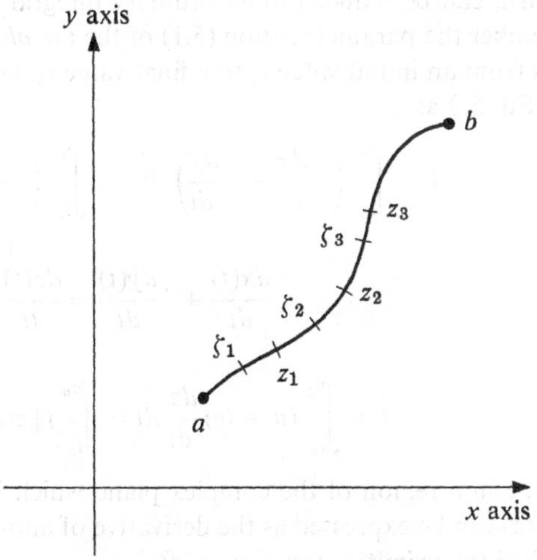

Fig. 10.

and take the limit $n \to \infty$ in such a manner that

$$|z_k - z_{k-1}| \to 0 \qquad \text{for all } k \tag{5.4}$$

This limit, provided it exists and provided it is independent of the manner in which we have chosen the points z_j and ζ_j, is called the **contour integral** of $f(z)$ along C and is written as

$$I = \int_C f(z)\, dz \tag{5.5}$$

One also may write

$$I = \int_a^b f(z)\, dz \tag{5.6}$$

but it must be remembered that the value of the contour integral depends in general on the path connecting the points a and b.

Separating $f(z)$ and z into their real and imaginary parts, we can also write the sum I_n as

$$I_n = \sum_{k=1}^{n} \{[u(\zeta_k)(x_k - x_{k-1}) - v(\zeta_k)(y_k - y_{k-1})] + i[v(\zeta_k)(x_k - x_{k-1}) + u(\zeta_k)(y_k - y_{k-1})]\} \tag{5.7}$$

The limiting transition (5.4) implies

$$\begin{aligned} |x_k - x_{k-1}| &\to 0 \\ |y_k - y_{k-1}| &\to 0 \end{aligned} \tag{5.8}$$

for all k, and the integral I can be expressed in terms of real curvilinear integrals as

$$I = \int_C (u\, dx - v\, dy) + i \int_C (v\, dx + u\, dy) \tag{5.9}$$

This, in turn, can be reduced to an ordinary integral with respect to the parameter t if we remember the parametrization (5.1) of the arc ab. Assuming for definiteness that t increases from an initial value t_a to a final value t_b, as we go along C from a to b, we can write Eq. 5.9 as

$$I = \int_{t_a}^{t_b} \left(u \frac{dx}{dt} - v \frac{dy}{dt}\right) dt + i \int_{t_a}^{t_b} \left(v \frac{dx}{dt} + u \frac{dy}{dt}\right) dt \tag{5.10}$$

Since

$$\frac{dx(t)}{dt} + i \frac{dy(t)}{dt} = \frac{dz(t)}{dt}$$

we have

$$I = \int_{t_a}^{t_b} (u + iv) \frac{dz}{dt} dt = \int_{t_a}^{t_b} f[z(t)] \frac{dz(t)}{dt} dt \tag{5.11}$$

Assume in a region of the complex plane which includes the contour C that the function $f(z)$ can be expressed as the derivative of another function $F(z)$. The function $F(z)$ is called the **primitive** function of $f(z)$

$$f(z) = \frac{dF(z)}{dz} \tag{5.12}$$

Then, for a z situated on the contour

$$f(z) \frac{dz}{dt} = \frac{dF(z)}{dz} \frac{dz}{dt} = \frac{dF[z(t)]}{dt} \tag{5.13}$$

Inserting Eq. 5.13 into Eq. 5.11, we get

$$\int_C f(z)\, dz = \int_{t_a}^{t_b} \frac{dF[z(t)]}{dt} dt = F(b) - F(a) \tag{5.14}$$

Equation 5.14 forms the content of the so-called fundamental theorem of integral calculus.

The formal properties of the contour integral are analogous to those of the familiar integral with respect to a real variable. This is because those properties follow directly from the corresponding properties of a sum. Thus,

$$\int_C af(z)\, dz = a \int_C f(z)\, dz \tag{5.15}$$

where a is a constant; one also has

$$\int_C \{f(z) + g(z)\}\, dz = \int_C f(z)\, dz + \int_C g(z)\, dz \tag{5.16}$$

If c is a point that divides the arc ab into two arcs ac and cb, then

$$\int_a^b f(z)\, dz = \int_a^c f(z)\, dz + \int_c^b f(z)\, dz \tag{5.17}$$

It is clear that the sign of a contour integral is determined by the choice of a direction along the contour. Thus, we have (compare Eq. 5.14)

$$\int_a^b f(z)\, dz = -\int_b^a f(z)\, dz \tag{5.18}$$

Finally, integrating the relation (all the derivatives are supposed to exist)

$$\frac{d(f \cdot g)}{dz} = f\frac{dg}{dz} + \frac{df}{dz}g \tag{5.19}$$

and using Eq. 5.16, one gets (as in the real variable case) the formula for integration by parts

$$\int_a^b \frac{df}{dz} g \, dz = (f \cdot g)\Big|_a^b - \int_a^b f\frac{dg}{dz} \, dz \tag{5.20}$$

6 · THE DARBOUX INEQUALITY

It is very often useful to consider the upper bounds of certain contour integrals. To this end, consider the integral (which is supposed to exist)

$$I = \int_C f(z) \, dz \tag{6.1}$$

where C is a piecewise regular path in the complex plane. We shall furthermore assume that $|f(z)|$ is bounded on C. As discussed in the preceding section, the integral I is the limit, as $n \to \infty$, of the sum

$$I_n = \sum_{k=1}^{n} (z_k - z_{k-1})f(\zeta_k) \tag{6.2}$$

Denoting by max $|f|$ the maximum modulus of $f(z)$ on C, we find

$$|I_n| \leq \sum_{k=1}^{n} |z_k - z_{k-1}| \, |f(\zeta_k)| \leq \max|f| \sum_{k=1}^{n} |z_k - z_{k-1}| \tag{6.3}$$

The sum on the RHS of inequality 6.3 is the length of a polygon inscribed in the curve C and is therefore smaller than the arc length L of the curve itself. Hence, for all n

$$|I_n| \leq \max|f| \cdot L \tag{6.4}$$

In particular, as $n \to \infty$

$$\left| \int_C f(z) \, dz \right| \leq \max|f| \cdot L \tag{6.5}$$

Equation 6.5 is called **Darboux's inequality**. It will be frequently used in subsequent sections.

7 · SOME DEFINITIONS

Until now we have assumed that a function $f(z)$ of a complex variable z is defined in such a way that to every element z of a set of complex numbers there is associated a single number $f(z)$. A straightforward generalization consists in allowing more than one number to be associated with each element of a set of complex numbers. Hence, one can distinguish between **single-valued** and **multivalued** functions. We shall discuss multivalued functions in Sec. 24. Unless stated to the contrary, we shall suppose throughout this chapter that the functions with which we are dealing are single-valued.

We shall say that a function is **analytic** at a given point of the complex plane if there exists a neighborhood of this point such that the function is single-valued and differentiable at all points of this neighborhood. If a function is analytic at all points of some region of the complex plane, it is called analytic throughout this region, and the set of all points where a function is analytic is called the **domain of analyticity** of the function. If the domain of analyticity is the entire complex plane, the function is called an **entire** function.

A point where a function is analytic is a **regular** point of the function. If a function is not analytic at some point, the point is called a **singular point** of the function. It may happen that a function is analytic in a complex environment of a point without being single-valued or differentiable (or both) at the point itself. For example, the single-valued function $f(z) = 1/z$ is differentiable everywhere except at $z = 0$, since its derivative at that point is infinite. Such an exceptional point is called an **isolated singular point** of the function, since it is an isolated point among the set of singular points of the function.

8 · EXAMPLES OF ANALYTIC FUNCTIONS

8.1 Polynomials

The simplest, although trivial, example of an analytic function is

$$p_0(z) = \text{constant}$$

Consider now the function

$$p_1(z) = z$$

It is analytic in the entire complex plane, since its derivative

$$\frac{dp_1(z)}{dz} = \lim_{\Delta z \to 0} \frac{(z + \Delta z) - z}{\Delta z} = 1$$

exists for any z.

From the rules of differentiation (Eqs. 3.3), we immediately see that by adding and multiplying analytic functions, one again gets analytic functions. Hence, an arbitrary polynomial

$$p_n(z) = \sum_{k=0}^{n} a_k z^k$$

is an entire function. This can also be explicitly verified. We have

$$\frac{dp_n(z)}{dz} = \lim_{\Delta z \to 0} \frac{1}{\Delta z} \sum_{k=0}^{n} a_k[(z + \Delta z)^k - z^k]$$

but

$$(z + \Delta z)^k = z^k + kz^{k-1}\,\Delta z + \cdots + (\Delta z)^k$$

and therefore

$$\frac{dp_n(z)}{dz} = \sum_{k=1}^{n} a_k k z^{k-1}$$

for any z.

8.2 Power Series

One can ask what happens if one takes (instead of a polynomial that has a finite number of terms) a power series with an infinite number of terms

$$f(z) = \sum_{k=0}^{\infty} a_k(z - z_0)^k \tag{8.1}$$

Suppose that the series converges for

$$|z - z_0| < R \tag{8.2}$$

R is called the **radius of convergence** of the corresponding power series because Eq. 8.2 defines the interior of a circle of radius R centered at the point z_0.

It will be shown later that any power series is an analytic function within its radius of convergence.

8.3 Exponential and Related Functions

The exponential function is defined by its power series expansion

$$e^z = \sum_{k=0}^{\infty} \frac{z^k}{k!} \tag{8.3}$$

The series reduces to the well-known power series when z is a real variable. Equation 8.3 has an infinite radius of convergence, since for any z one has ($|z|$ being a real number)

$$\sum_{k=0}^{\infty} \frac{|z|^k}{k!} = e^{|z|}$$

which, by virtue of Weierstrass' criterion, implies the convergence of the sum in Eq. 8.3. Therefore, e^z is an entire function.

As in the real variable case, one can prove that

$$e^{z+w} = e^z e^w \tag{8.4}$$

From Eq. 8.3 we have, if y is a real variable

$$e^{iy} = \sum_{k=0}^{\infty} \frac{(iy)^k}{k!} = \left(1 - \frac{y^2}{2!} + \frac{y^4}{4!} + \cdots\right) + i\left(y - \frac{y^3}{3!} + \frac{y^5}{5!} - \cdots\right)$$

The first series is simply the series for $\cos y$ and the second series is the series for $\sin y$. Hence, we have

$$e^{iy} = \cos y + i \sin y \tag{8.5}$$

from which it follows that

$$\cos y = \frac{1}{2}(e^{iy} + e^{-iy}) \tag{8.6}$$

$$\sin y = \frac{1}{2i}(e^{iy} - e^{-iy}) \tag{8.7}$$

Equation 8.5 allows one to write z (Eq. 1.3) in the useful form

$$z = re^{i\theta} \tag{8.8}$$

Using Eqs. 8.4 and 8.5, we can also re-express the exponential function

$$e^z = e^{x+iy} = e^x(\cos y + i \sin y) \tag{8.9}$$

The relations 8.6 and 8.7 suggest the following definition of the trigonometric functions for complex arguments

$$\cos z = \frac{1}{2}(e^{iz} + e^{-iz}) \tag{8.10}$$

$$\sin z = \frac{1}{2i}(e^{iz} - e^{-iz}) \tag{8.11}$$

Using Eq. 8.3 we deduce from Eqs. 8.10 and 8.11 the expressions

$$\cos z = \sum_{k=0}^{\infty} (-1)^k \frac{z^{2k}}{(2k)!} \tag{8.12}$$

$$\sin z = \sum_{k=0}^{\infty} (-1)^k \frac{z^{2k+1}}{(2k+1)!} \tag{8.13}$$

The trigonometric relations that hold with real variables also hold when the variable is complex. For example, using Eqs. 8.10 and 8.11, we readily obtain the relation

$$\sin^2 z + \cos^2 z = 1 \tag{8.14}$$

The hyperbolic functions are defined as follows

$$\cosh z = \cos iz \tag{8.15}$$

$$\sinh z = -i \sin iz \tag{8.16}$$

Using Eqs. 8.15 and 8.16, we deduce from Eqs. 8.10, 8.11, and 8.14 the relations

$$\cosh^2 z - \sinh^2 z = 1 \tag{8.17}$$

$$\cosh z = \tfrac{1}{2}(e^z + e^{-z}) \tag{8.18}$$

$$\sinh z = \tfrac{1}{2}(e^z - e^{-z}) \tag{8.19}$$

The functions tan z, tanh z, sec z, sech z, etc., are defined as in the case of real variables, and expressions for them can be readily obtained from the previous formulae.

The logarithm $\ln z$ of a complex variable is, by definition, that function $f(z)$ which satisfies the equation

$$e^{f(z)} = z \tag{8.20}$$

Setting $f(z) = u + iv$ and using Eq. 8.5, we obtain from Eq. 8.20

$$e^u(\cos v + i \sin v) = r(\cos \theta + i \sin \theta) \tag{8.21}$$

Equating real and imaginary parts in Eq. 8.21, we obtain

$$e^u = r \qquad \text{or} \qquad u = \ln r$$

(since u and v are real) and

$$v = \theta + 2\pi n \qquad (n = 0, \pm 1, \pm 2, \cdots)$$

Thus, we have

$$\ln z = \ln r + i(\theta + 2\pi n) \qquad (n = 0, \pm 1, \pm 2, \cdots) \tag{8.22}$$

The imaginary part of $\ln z$ is equal to the argument of z; we have already mentioned that the argument of a complex number is determined up to a constant multiple of 2π. From Eq. 8.22 we see that an infinite number of different values of $\ln z$ correspond to the same point in the z plane. We have in the logarithmic function an example of a multivalued function.

It is not difficult to define (at least in a limited part of the complex plane) a function that satisfies the preceding relation (Eq. 8.20) and is single-valued. It is called the **principal logarithm** and is defined as

$$\operatorname{Ln} z = \ln r + i\theta \qquad -\pi < \theta < \pi \tag{8.23}$$

$\operatorname{Ln} z$ has a discontinuity across the negative real axis

$$\lim_{\varepsilon \to 0} \operatorname{Ln}[re^{i(\pi \pm \varepsilon)}] = \ln r \pm i\pi \tag{8.24}$$

In its region of definition (i.e., throughout the entire complex plane with the exception of the negative real axis), the principal logarithm is an analytic function, as we shall prove in Sec. 24.

9 · CONFORMAL TRANSFORMATIONS

9.1 Conformal Mapping

Consider a function

$$z'(z) = u(x,y) + iv(x,y) \tag{9.1}$$

of the complex variable $z = x + iy$, which is analytic in a region D of the complex z plane. With every point $z = x + iy$ of D, one can associate via Eq. 9.1 another point $z' = x' + iy'$, where

$$x' = u(x,y) \qquad \text{and} \qquad y' = v(x,y) \tag{9.2}$$

Geometrically, one can draw in addition to the z plane another plane, the z' plane with x' and y' along the abscissa and ordinate, respectively; then Eq. 9.1 can be regarded as defining a continuous mapping of the points of D on the z plane onto a domain consisting of the "image" points x',y' of the z' plane. One says simply that $z'_0 = x'_0 + iy'_0 = u(x_0,y_0) + iv(x_0,y_0)$ is the image in the $x'y'$ plane of the point $z_0 = x_0 + iy_0$ in the xy plane.

Since the mapping is continuous, two continuous curves C_1 and C_2 in the z plane will be mapped into two other continuous curves, C'_1 and C'_2 (Figs. 11(a) and (b)) in the z' plane and the point of intersection $z_0 = x_0 + iy_0$ of C_1 and C_2 will go over into the point of intersection $z'_0 = x'_0 + iy'_0 = u(x_0,y_0) + iv(x_0,y_0)$ of C'_1 and C'_2. Let ν be the angle between the tangents to C_1 and C_2 at z_0, and let ν' be the angle between the tangents to C'_1 and C'_2 at z'_0. We shall show that for all points $z \in D$ such that

$$\frac{dz'(z)}{dz} \neq 0$$

the mapping (9.1) is angle-preserving; i.e., $\nu = \nu'$.

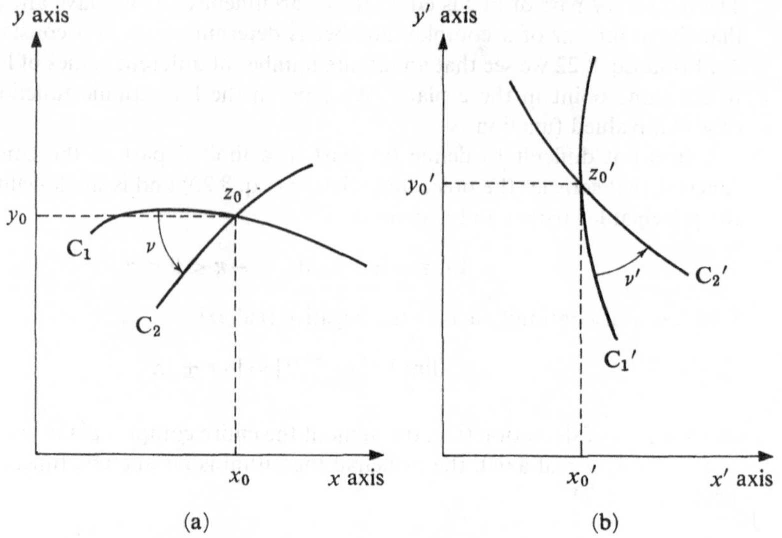

Fig. 11. The point $z_0 = x_0 + iy_0$ is transferred onto the point z'_0 with coordinates $x'_0 = u(x_0,y_0)$ and $y'_0 = v(x_0,y_0)$.

Let z be an arbitrary point on C_1 and z' the image of z on C'_1. We put

$$z - z_0 = re^{i\theta} \tag{9.3}$$

$$z' - z'_0 = r'e^{i\theta'} \tag{9.4}$$

$$\frac{dz'(z_0)}{dz_0} = \left|\frac{dz'(z_0)}{dz_0}\right| e^{i\psi_0} \tag{9.5}$$

Consider now the quantity

$$\frac{z' - z'_0}{z - z_0} = \frac{r'}{r} e^{i(\theta' - \theta)} \tag{9.6}$$

From the foregoing relations (Eqs. 9.3 and 9.4), we see that as $z \to z_0$ along C_1, θ tends to the angle α_1, which is the angle to the real axis made by the tangent to the curve C_1 at the point z_0. Similarly, θ' tends to the angle β_1, which is the angle to the real axis made by the tangent to the curve C'_1 at the point z_0. Therefore, as $z \to z_0$, Eqs. 9.5 and 9.6 yield

$$\psi_0 = \alpha_1 - \beta_1 \tag{9.7}$$

Since $z'(z)$ is, by hypothesis, analytic at z_0 and

$$\left.\frac{dz'}{dz}\right|_{z=z_0} \neq 0$$

the argument ψ_0 of

$$\left.\frac{dz'}{dz}\right|_{z=z_0}$$

has a definite value. The same reasoning (but assuming now that $z \to z_0$ along C_2) leads to

$$\psi_0 = \alpha_2 - \beta_2 \tag{9.8}$$

where α_2 and β_2 are defined analogously to α_1 and β_1.

Hence

$$\alpha_1 - \beta_1 = \alpha_2 - \beta_2 \tag{9.9}$$

and so

$$\nu = \alpha_2 - \alpha_1 = \beta_2 - \beta_1 = \nu' \tag{9.10}$$

We have proved therefore, that the angle between the tangents at a point of intersection of any two regular curves $C_1 \in D$ and $C_2 \in D$ is preserved under the mapping,

$$\nu = \nu' \tag{9.11}$$

when

$$\frac{dz'}{dz} \neq 0 \qquad \text{for } z \in D \tag{9.12}$$

Such a mapping is called a **conformal mapping.**

Without entering into the details, let us merely mention that where $dz'/dz \neq 0$, the points of the z plane and their images in the z' plane are in one-to-one correspondence, and the inverse mapping $z = z(z')$ is well defined. This is no longer the case when at a given point z_0, one has

$$\left.\frac{dz'}{dz}\right|_{z=z_0} = 0$$

Then the transformation $z' = z'(z)$ no longer establishes a one-to-one correspondence between the points of the neighborhood of z_0 and the points of the environment of its image $z'(z_0)$, and the mapping is not conformal.

9.2 Homographic Transformations

The transformation

$$z' = \frac{1}{z} \tag{9.13}$$

is called an **inversion**. It effects a conformal mapping of any region in the z plane that does not include the point $z = 0$.

We shall prove that the transformation 9.13 maps circles in the z plane into circles or straight lines* in the z' plane. The points in the z plane, lying on the circumference of a circle with radius r and centered at the point a, satisfy the relation

$$|z - a| = r \tag{9.14}$$

Therefore, their images in the z' plane obey the condition

$$\left|\frac{1}{z'} - a\right| = r \tag{9.15}$$

* However, a straight line may also be considered as a "circle," but with an infinite radius.

We assume first that

$$a \neq r \tag{9.16}$$

In other words, the circle (9.14) does not pass through the origin. Equation 9.15 is equivalent to

$$\frac{|1 - az'|}{|z'|} = r$$

Since $|z'| > 0$, one has

$$|1 - az'| = r|z'| \tag{9.17}$$

Taking the square of both sides above, one easily obtains

$$|z'|^2 - \frac{2}{|a|^2 - r^2} \operatorname{Re}(az') + \frac{1}{|a|^2 - r^2} = 0 \tag{9.18}$$

We leave to the reader the verification that Eq. 9.18 is equivalent to the equation

$$|z' - A| = R \tag{9.19}$$

where

$$A = \frac{a}{|a|^2 - r^2}$$
$$R = \left|\frac{r}{|a|^2 - r^2}\right| \tag{9.20}$$

The points in the z' plane satisfying Eq. 9.19 again lie on the circumference of a circle, with radius R and centered at the point A.

In the case when

$$|a| = r \tag{9.21}$$

one obtains from Eq. 9.13 (instead of Eq. 9.18), the equation

$$2 \operatorname{Re}(az') - 1 = 0 \tag{9.22}$$

Putting $z' = x' + iy'$, one gets from Eq. 9.22

$$2x' \operatorname{Re} a - 2y' \operatorname{Im} a - 1 = 0 \tag{9.23}$$

which is the equation of a line in the z' plane.

The transformation $z' = 1/z$ establishes a one-to-one correspondence between the points in the z plane and those in the z' plane. The only question is, "what happens to the singular point $z = 0$?"

The image of the origin is defined to be "the point at infinity." The image of every line that reaches the point $z = 0$ is a line that extends to infinity. Therefore, every straight line in the z' plane reaches the "point at infinity" independently of its orientation. Since the transformation inverse to $z' = 1/z$ is the transformation $z = 1/z'$, the point $z = 0$ in the z plane can itself be considered as the image of the point $z' = \infty$. Similarly, one can speak of the point $z = \infty$ as the image of the point $z' = 0$.

Consider now the series of transformations

$$
\begin{aligned}
&\text{(i)} && z_1 = z + \frac{d}{c} \\
&\text{(ii)} && z_2 = c^2 z_1 \\
&\text{(iii)} && z_3 = \frac{1}{z_2} \\
&\text{(iv)} && z_4 = (bc - ad)z_3 \\
&\text{(v)} && z' = \frac{a}{c} + z_4
\end{aligned}
\tag{9.24}
$$

where

$$c \neq 0 \qquad \text{and} \qquad (bc - ad) \neq 0 \tag{9.25}$$

In Eq. 9.24, i and v represent translations; ii and iv are similarity transformations (i.e., they reproduce the same figure in the z' plane as was originally in the z plane but on a different scale) and iii is an inversion.

All these transformations map circles in the z plane into circles (or straight lines) in the z' plane. Therefore, the successive application of these transformations

$$z' = \frac{a}{c} + z_4 = \frac{a}{c} + (bc - ad)z_3 = \cdots = \frac{az + b}{cz + d} \tag{9.26}$$

also maps circles into circles (or straight lines). Eq. 9.26 is called a **homographic** transformation.*

Using the well-known rules of differentiation, one obtains

$$\frac{dz'}{dz} = \frac{(ad - bc)}{(cz + d)^2} \tag{9.27}$$

Thus, the condition (9.25) ensures that $dz'/dz \neq 0$, and therefore a homographic transformation (9.26) maps conformally every region in the z plane that does not include the point $z = -(d/c)$. In the case where $c = 0$, (9.26) reduces to a linear transformation, which may be regarded as a particular case of a homographic transformation.

9.3 Change of Integration Variable

Among the many applications of conformal transformations, the simplest is the change of the integration variable in a contour integral. Consider the integral $\int_C f(z)\, dz$ taken along a regular curve C, with a parametric equation

$$z = z(t) \qquad t_a \leq t \leq t_b$$

* The reader can consult many textbooks where a number of useful mappings have been tabulated; for example, R. Churchill, *Introduction to Complex Variables and Applications*, (McGraw-Hill, New York, 1960.) Among the more important transformations which we have not discussed here is the **Schwarz-Christoffel** transformation, which maps a polygon into a half-plane.

Suppose now that the transformation

$$z' = z'(z) \tag{9.28}$$

is conformal in the region of the z plane, which contains the curve C. Let Eq. 9.28 map C into a curve C′.

A conformal mapping establishes a one-to-one correspondence between points in the z plane and their images in the z' plane, and consequently there must exist a function $z = z(z')$ inverse to $z' = z'(z)$. As in elementary differential calculus, one can show that

$$\left(\frac{dz(z')}{dz'}\right)_{z'=z'(z)} = \left(\frac{dz'(z)}{dz}\right)^{-1} \tag{9.29}$$

Therefore, $z(z')$ is analytic, since $dz'(z)/dz \neq 0$. The parametric equation of C′ is

$$z' = z'[z(t)] \qquad a < t < b \tag{9.30}$$

Reducing contour integrals to Riemann integrals, we easily obtain

$$\int_C f(z)\,dz = \int_{C'} f[z(z')]\left(\frac{dz(z')}{dz'}\right)dz' \tag{9.31}$$

The fact that the transformation (9.28) is conformal ensures the existence of the integral on the RHS.

10 · A SIMPLE APPLICATION OF CONFORMAL MAPPING

Suppose that we are given a problem with a certain geometry. This problem may be difficult to solve, but it may turn out that by a conformal transformation, the problem can be reduced to one with a much simpler geometry. One can then solve this simpler problem and, by transforming back to the original geometry, obtain the solution to the more difficult problem. We shall present in this section an example to illustrate the method.

Consider the following electrostatic problem: Given two parallel conducting cylinders of infinite length, what is the electric field at any point of space due to these cylinders if their surfaces are at given potentials?

For simplicity, we shall solve the problem in the case when the diameters of the cylinders are equal. The problem is in fact a two-dimensional one, since (because of symmetry reasons) the electric field cannot depend upon the coordinate directed along the axes of the cylinders; the field vector $\vec{E}$ depends on two Cartesian coordinates, x and y, in any plane perpendicular to the axes of the cylinders

$$\vec{E} = \vec{E}(x,y) \tag{10.1}$$

Let

$$U = U(x,y) \tag{10.2}$$

be the electrostatic potential. In empty space $U(x,y)$ satisfies Laplace's equation

$$\frac{\partial^2 U(x,y)}{\partial x^2} + \frac{\partial^2 U(x,y)}{\partial y^2} = 0 \tag{10.3}$$

$U(x,y)$ determines $\vec{E}(x,y)$ by the well-known relation

$$\vec{E}(x,y) = -\vec{\nabla} U(x,y) \tag{10.4}$$

or in terms of coordinates

$$E_x(x,y) = -\frac{\partial U(x,y)}{\partial x}$$
$$E_y(x,y) = -\frac{\partial U(x,y)}{\partial y} \tag{10.5}$$

Since $U(x,y)$ satisfies Laplace's equation, this suggests the introduction of another function $V(x,y)$ such that the pair of functions $U(x,y)$ and $V(x,y)$ satisfy the Cauchy-Riemann conditions

$$\frac{\partial V(x,y)}{\partial x} = -\frac{\partial U(x,y)}{\partial y}$$
$$\frac{\partial V(x,y)}{\partial y} = \frac{\partial U(x,y)}{\partial x} \tag{10.6}$$

It is easy to verify that in terms of $V(x,y)$, $\vec{E}(x,y)$ is given by

$$\vec{E}(x,y) = -\vec{\nabla} \times \vec{m}V(x,y) \tag{10.7}$$

where $\vec{m}$ is a unit vector normal to the x,y plane. $U(x,y)$ and $V(x,y)$ together define a function

$$F(z) = U(x,y) + iV(x,y) \tag{10.8}$$

called the **complex potential**, which is analytic at all points of empty space where $U(x,y)$ and $V(x,y)$ satisfy Laplace's equation.

Let the choice of the coordinate system be such that in the xy plane, the cylinders are represented by two circles of radius r, the first centered at $x = c$, $y = 0$, and the second at $x = -(r^2/c)$, $y = 0$. Here, $c \neq r$ is a positive parameter which determines the distance between the circles. As

$$(c - r)^2 > 0$$

one has

$$c + \frac{r^2}{c} > 2r$$

Therefore, with our parametrization, the requirement that the distance between the centers of the two circles is larger than the diameter of a circle is automatically satisfied.

It is not difficult to see that one of the circles contains the origin in its interior. Without any loss of generality, we may suppose that

$$c > r$$

(Compare Fig. 12.) One can now formulate the electrostatic problem in the language of the theory of analytic functions.

We seek a function $F(z)$ which is analytic outside the two circles and whose real part [i.e., the electrostatic potential $U(x,y)$] is constant on the circumference of each of the two circles. Performing the inversion

$$z' = \frac{1}{z}$$

and using the formulae 9.20 of the preceding section, one obtains in the z' plane two concentric circles with radii

$$R_1 = \frac{r}{c^2 - r^2}$$
$$R_2 = \frac{c^2}{r(c^2 - r^2)} \tag{10.9}$$

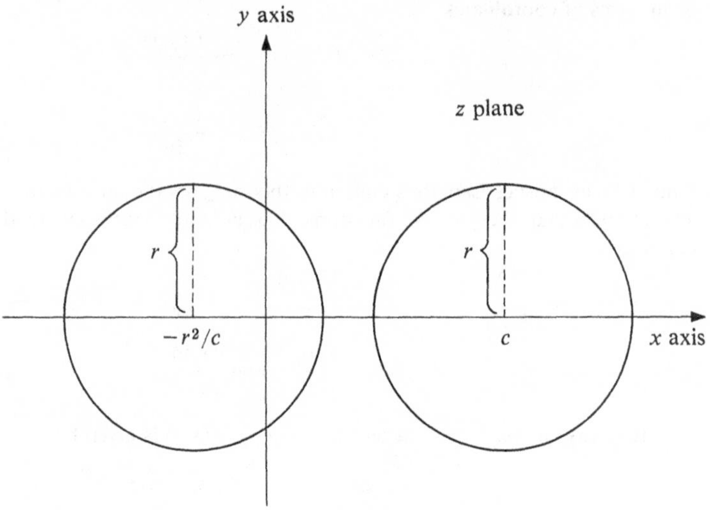

Fig. 12. The complex potential $F(z)$ is analytic outside the two circles, and its real part is constant on the circumference of each of the circles.

Now that one has $R_2 > R_1$, it is easy to verify that the outside of the circles in the z plane is mapped onto the annular region

$$R_1 < \left| z' - \frac{c}{c^2 - r^2} \right| < R_2 \tag{10.10}$$

in the z' plane (Fig. 13). Set, for simplicity of notation, $w = z' - \dfrac{c}{c^2 - r^2}$.

The function satisfying our requirements is

$$F(w) = A \operatorname{Ln} w + B \tag{10.11}$$

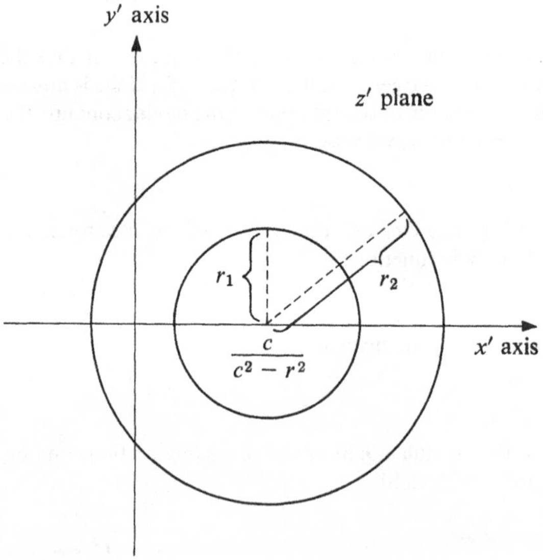

Fig. 13. The two circles of Fig. 12 are transformed by Eq. 9.20 into two concentric circles in the z' plane.

where A and B are real constants. Indeed

$$\operatorname{Re} \ln w = \ln |w|$$

is constant for

$$|w| = R_1, R_2$$

and Ln w is analytic in the annular region (10.10) except for arg $w = \pm \pi$, where it is double-valued. However, it is essentially the derivative of the potential that has a physical meaning, and it will be shown in Sec. 24 that

$$\frac{d \operatorname{Ln} z}{dz} = \frac{1}{z}$$

Thus, the derivative of $F(z)$ exists and is single-valued, so that our solution is perfectly acceptable.

From Eq. 10.11 we have

$$\operatorname{Re} F(w) = A \ln |w| + B \tag{10.12}$$

The constants A and B are determined from the boundary conditions

$$\operatorname{Re} F(w) = \begin{cases} U_1 & \text{for } |w| = R_1 \\ U_2 & \text{for } |w| = R_2 \end{cases} \tag{10.13}$$

U_1 and U_2 denote the potentials of the first and of the second cylinders, respectively. Hence

$$A = \frac{U_1 - U_2}{\ln(R_1/R_2)}$$
$$B = \frac{U_2 \ln R_1 - U_1 \ln R_2}{\ln(R_1/R_2)} \tag{10.14}$$

After performing simple algebra, we obtain the solution of the problem

$$U(x,y) = A \ln \left| \frac{1}{x + iy} - \frac{c}{c^2 - r^2} \right| + B$$

11 · THE CAUCHY THEOREM

Before developing further the theory of analytic functions, it should be mentioned that there exists a very simple and important classification of regions in the complex plane. For example, a region could consist of the entire domain contained within a closed curve, or it could consist of the entire domain with the exclusion of a number of "holes" punched out of it. A region is said to be **simply connected** if it is such that all closed paths within it contain only points that belong to the region (no holes!). Otherwise, the region is said to be **multiply connected**.

EXAMPLE

The region between two concentric circles is not simply-connected, since there exist closed paths within it which contain points that are outside the ring-shaped region (for instance, the center of the circles).

We have seen in Sec. 5 that when the contour of integration lies in a region where $f(z)$, the function to be integrated, possesses a primitive function $F(z)$ (Eq. 5.12), the value of the contour integral is determined by the values of that primitive function

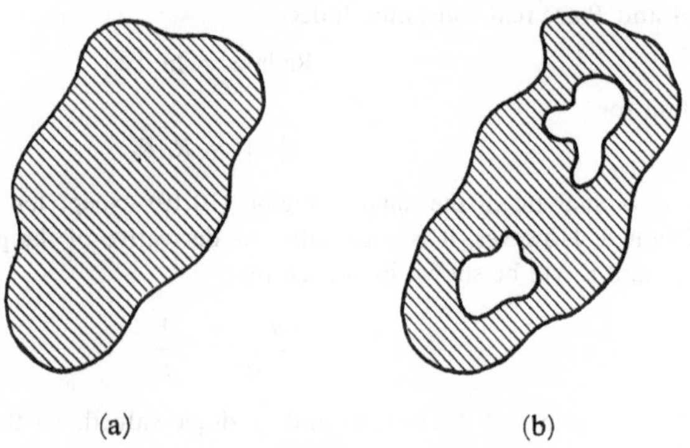

Fig. 14. (a) The interior points of the shaded area form a simply-connected region. (b) The interior points of the shaded area form a multiply-connected region.

at the end points of the path of integration. Therefore, it is independent of the choice of the contour. In general, however, an arbitrary function of a complex variable will not possess a single-valued primitive function, and the result of the contour integration will almost invariably depend upon the path linking the two end points. This is similar to the case of a general function of two real variables. On the other hand, analytic functions have a remarkable property. We shall see that within simply-connected regions (Fig. 14), they always possess single-valued primitive functions, so that the result of their integration is independent of the choice of the path (this statement is, however, no longer correct in the case when the region is multiply-connected). Stated differently, the contour integrations of an analytic function along two piecewise, regular curves, both lying in the domain of analyticity of the function and having the same end points, yield the same result, provided one can bring one contour into the other by deforming it continuously without crossing any singular point of the function in question. This follows immediately from the following famous Cauchy integral theorem, which plays a central role in the theory of analytic functions.

Theorem. Let C denote a piecewise, regular closed curve in the complex plane and let $f(z)$ be analytic on C and within the whole region enclosed by C. Then

$$\int_C f(z)\,dz = 0 \tag{11.1}$$

The theorem in its original form required not only that the derivative of $f(z)$ exist, but also that it be continuous. Goursat has shown that this latter requirement is unnecessary; we shall follow his proof of the theorem.

The Proof of Cauchy's Theorem. Let us begin by showing that the theorem is valid for $f(z) = z^n$, with $n \geq 0$. For $n \geq 0$, z^n is an entire function, and therefore any finite curve C lies in its domain of analyticity; furthermore

$$z^n = \frac{d}{dz}\left[\frac{1}{n+1}\,z^{n+1}\right]$$

so that according to (5.14)

$$\int_C z^n\,dz = 0 \tag{11.2}$$

where C is a closed contour.

We now consider the general case. Let ε be an arbitrary positive constant and let A denote the set of all points, either lying on C or enclosed by C. Consider the function

$$g(z,z_0) = f(z) - f(z_0) - (z - z_0)\left(\frac{df}{dz}\right)_{z=z_0} \tag{11.3}$$

We shall first demonstrate that A can be subdivided into a finite number (n, say) of subsets A_j

$$A = A_1 + A_2 + \cdots + A_n \tag{11.4}$$

such that whenever

$$|z - z_0| < \delta_j(\varepsilon) \qquad z,z_0 \in A_j \tag{11.5}$$

one has

$$|g(z,z_0)| < \varepsilon|z - z_0| \tag{11.6}$$

In 11.5, $\delta_j(\varepsilon)$ denotes a positive number which depends on ε and on the choice of A_j but not on the points z and z_0.

To prove that the decomposition (11.4) is always possible, we cover A by a network of small squares B_j (Fig. 15), some of which may overlap with C. Then one has an alternative: either B_j is entirely contained within A or it is crossed by C. In the latter case, C divides B_j into two parts, each of which has a segment of C as part of its boundary, and only one of which belongs to A. Then $B_j \cap A$ denotes the region common to both B_j and A (Sec. 1).

For each B_j we can check to find whether 11.5 ȧnd 11.6 are satisfied when

$$z,z_0 \in B_j \cap A \tag{11.7}$$

Suppose that 11.5 and 11.6 are **not** satisfied for some set $B_j \cap A$. Then we subdivide the **square** B_j into four equal squares B'_j, reject those squares that lie entirely outside A, and check once more to find whether 11.5 and 11.6 are satisfied when

$$z,z_0 \in B'_j \cap A \tag{11.8}$$

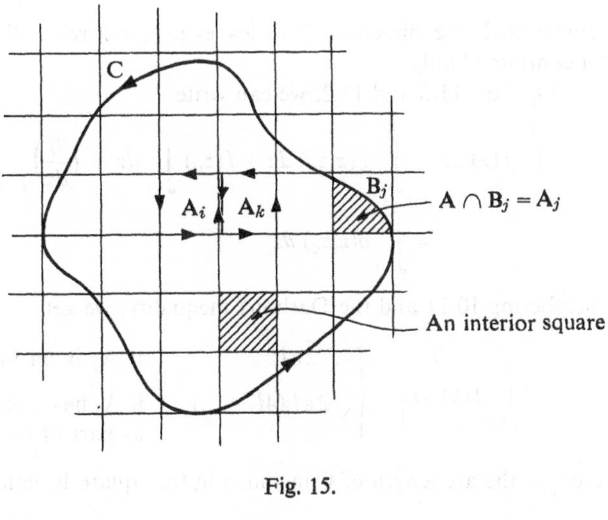

Fig. 15.

If 11.5 and 11.6 are violated in the domain defined by 11.8, we repeat the subdivision process for this rebellious domain, obtaining smaller and smaller squares.

However, this subdivision process cannot continue indefinitely, since it would imply that there exists at least one "rebellious" point z_0 for which no finite neighborhood would exist where 11.5 and 11.6 could be satisfied. This would contradict the fact that $f(z)$ is differentiable, since 11.6 is equivalent to

$$\left|\left(\frac{df}{dz}\right)_{z=z_0} - \frac{f(z)-f(z_0)}{z-z_0}\right| < \varepsilon \tag{11.9}$$

and 11.9 is surely satisfied when $|z - z_0|$ is small enough. We see that A can be covered by a fine network of squares, B_j, such that for every B_j, 11.5 and 11.6 are satisfied within that part of B_j which is included in A. The decomposition of A (Eq. 11.4) into a sum of regions A_i is then obtained simply by putting

$$A_j = B_j \cap A \tag{11.10}$$

Now let l_j denote the length of the edge of the square B_j. If we choose $\delta_j(\varepsilon) = \sqrt{2}l_j$, then obviously 11.5 and 11.6 will be satisfied whenever $z,z_0 \in A_j$. ($\sqrt{2}l_j$ is simply the length of the diagonal of B_j; it is therefore the longest distance in B_j and thus also in A_j).

By combining 11.5 and 11.6, we get

$$|g(z,z_0)| < \sqrt{2}\,\varepsilon\, l_j \qquad \text{for } z,z_0 \in A_j \tag{11.11}$$

We now notice that the integral $\int_C f(z)\,dz$ can be replaced by a sum of mesh integrals, where the meshes correspond to the boundaries C_j of the areas A_j

$$\int_C f(z)\,dz = \sum_{j=1}^{n} \int_{C_j} f(z)\,dz \tag{11.12}$$

This procedure is legitimate because the common boundary of two adjacent subregions gives equal and opposite contributions to the mesh integrals in each of the adjacent subregions, provided all the integrals, including $\int_C f(z)\,dz$ are done in, say, a counterclockwise direction. This leaves as a net result the contribution from the outer contour C only.

Using Eqs. 11.3 and 11.2, we can write

$$\begin{aligned}\int_{C_j} f(z)\,dz &= \int_{C_j} g(z,z_0)\,dz + f(z_0)\int_{C_j} dz + \left(\frac{df}{dz}\right)_{z=z_0}\int_{C_j}(z-z_0)\,dz \\ &= \int_{C_j} g(z,z_0)\,dz\end{aligned}$$

Remembering 10.11 and the Darboux inequality, we get

$$\left|\int_{C_j} f(z)\,dz\right| < \begin{cases} 4\sqrt{2}\,\varepsilon\, l_j^2 & \text{if } A_j \text{ is an interior square} \\ \sqrt{2}\,\varepsilon\, l_j(4l_j + s_j) & \text{if } A_j \text{ has a segment of C as part of its boundary} \end{cases}$$

where s_j is the arc length of C included in the square B_j contiguous to C.

Let l be the length of the edge of a square that contains the entire region A, and let s be the length of the contour C. Then

$$l_j < l$$

$$\sum_{j=1}^{n} s_j = s$$

$$\sum_{j=1}^{n} l_j^2 < l^2$$

The last inequality simply means that the sum of the areas of all the B_j (and consequently of all the A_j) is not larger than l^2. Finally, we have

$$\left|\int_C f(z)\,dz\right| < \sum_{j=1}^{n}\left|\int_{C_j} f(z)\,dz\right| < \varepsilon\{4\sqrt{2}\,l^2 + \sqrt{2}\,ls\} \underset{\varepsilon\to 0}{\longrightarrow} 0$$

This completes the proof of Cauchy's theorem. It should be noted carefully that this proof relies heavily on the fact that the function $f(z)$ has a derivative in the region enclosed by the contour C.

A consequence of the preceding theorem, which has already been mentioned at the beginning of this section, is the following: Suppose that C_1 and C_2 are two curves that have the same end points, and suppose further that $f(z)$ has no singularities in the region enclosed by C_1 and C_2. Then $C_1 + (-C_2)$ is a closed contour,* and therefore, using Cauchy's theorem together with the relations 5.17 and 5.18, we find

$$0 = \int_{C_1+(-C_2)} f(z)\,dz = \int_{C_1} f(z)\,dz + \int_{-C_2} f(z)\,dz$$

$$= \int_{C_1} f(z)\,dz - \int_{C_2} f(z)\,dz$$

Hence

$$\int_{C_1} f(z)\,dz = \int_{C_2} f(z)\,dz \tag{11.13}$$

12 · CAUCHY'S INTEGRAL REPRESENTATION

From Cauchy's theorem, it is possible to derive an integral formula that is very important for the further development of the theory of analytic functions as well as for a wide variety of applications to physical problems. This formula is contained in the following

Theorem. Let $f(z)$ be analytic throughout a simply-connected region R. If C is a closed, piecewise, regular curve within R, and z a point not on C, then

$$\frac{1}{2\pi i}\int_C \frac{f(z')}{z'-z}\,dz' = \begin{cases} f(z) & \text{if } z \text{ is interior to C} \\ 0 & \text{if } z \text{ is exterior to C} \end{cases} \tag{12.1}$$

where the integration along C is taken in the counterclockwise direction.

* $(-C_2)$ denotes the contour, which differs from C_2 by the sense of integration.

Proof. Consider the function

$$\frac{f(z') - f(z)}{z' - z} \tag{12.2}$$

Since $f(z)$ is continuous,* we have

$$|f(z') - f(z)| < \varepsilon \tag{12.3}$$

whenever

$$|z' - z| < \delta(\varepsilon) \tag{12.4}$$

Let Γ be a circle in R centered at z and of radius $r < \delta(\varepsilon)$. Suppose, for definiteness, that we integrate along Γ in the counterclockwise direction. Then the parametric equation for Γ is

$$z'(\theta) = z + re^{i\theta} \qquad 0 \le \theta \le 2\pi \tag{12.5}$$

and for z' on Γ, we have

$$\left|\frac{f(z') - f(z)}{z' - z}\right| < \frac{\varepsilon}{r} \tag{12.6}$$

Hence, using Eq. 12.6 and Darboux's inequality

$$\left|\int_\Gamma \frac{f(z') - f(z)}{z' - z}\, dz'\right| \le \frac{\varepsilon}{r} 2\pi r = 2\pi\varepsilon \tag{12.7}$$

Therefore, in the limit $\varepsilon \to 0$, the RHS of Eq. 12.7 tends to zero, and hence

$$\int_\Gamma \frac{f(z') - f(z)}{z' - z}\, dz' = 0 \tag{12.8}$$

It should be noted that even though 12.2 tends to df/dz as $z' \to z$, and is therefore finite, Eq. 12.8 could not have been written down directly, since we have not yet proved that df/dz is itself analytic.

Using Eq. 12.5, Eq. 12.8 yields

$$\int_\Gamma \frac{f(z')}{z' - z}\, dz' = f(z) \int_\Gamma \frac{dz'}{z' - z} = f(z) \int_0^{2\pi} \frac{ire^{i\theta}\, d\theta}{re^{i\theta}} = 2\pi i f(z) \tag{12.9}$$

The function $f(z')/(z' - z)$ is an analytic function of z', except in general at the point $z' = z$; therefore, according to the results of the preceding section, the boundary circle Γ can be deformed into an arbitrary closed path C within R, encircling the point z and not passing through it. Hence, we finally obtain the general form of Cauchy's integral formula

$$f(z) = \frac{1}{2\pi i} \int_C \frac{f(z')}{z' - z}\, dz' \tag{12.10}$$

where z is an interior point of an otherwise arbitrary curve C.

Naturally, if z is exterior to C, the integrand in 12.10 is analytic and the RHS of 12.10 is no longer equal to $f(z)$, but vanishes.

---

* Differentiability implies continuity, of course, since $\lim\limits_{z \to z_0} \dfrac{f(z) - f(z'_0)}{z - z_0}$ may exist only if $|f(z) - f(z_0)| \to 0$ whenever $|z - z_0| \to 0$.

Note. Most of the results that we shall derive will be based upon the Cauchy formula. To avoid ambiguity, we shall herewith adopt the convention that every integration along a closed contour will be taken in the counterclockwise direction. More generally, given a contour enclosing some region in the complex plane, we shall always choose the direction of the integration path so that the interior of C is on the left-hand side. Any integration performed in a clockwise direction will bear an opposite sign.

The Cauchy's integral formula expresses completely the value of an analytic function at any point z within a contour C, once its values on the boundary curve have been specified. This is a very powerful result, rich in implications, as we shall see, and a consequence of the differentiability of $f(z)$. It illustrates the statement we made in Sec. 3 about the nonlocal implications of the differentiability of a function with respect to a complex argument. A function of a real variable, i.e., a function of a point in a one-dimensional continuum, may be differentiable everywhere; nevertheless, the specification of its values at the end points of an interval does not at all fix its behavior in the interior of this interval unless the function satisfies a differential equation. But a function of a complex variable, i.e., a function of a point in a plane, must satisfy a differential equation, namely, the Laplace equation, in order for it to have a derivative!

Equation 12.10 is the first encounter with so-called integral representation of a function. One says that one has established an integral representation for a function $f(z)$ if one has succeeded in writing down an integral of the form

$$f(z) = \int_C K(z,z')g(z')\,dz' \tag{12.11}$$

where the argument z of the function plays the role of a parameter in the integrand, and which is (at least in some range of values of this parameter z) equal to the function in question. The function $K(z,z')$, which depends on both the parameter z and the integration variable z', is called the **kernel** of the integral representation. The integral representations of functions are very useful. In particular, when the integrand is an analytic function and the integral is or may be reduced to a contour integral, the Cauchy theorem offers the possibility, by a judicious deformation of the integration path, to derive approximate expressions for the function.

13 · THE DERIVATIVES OF AN ANALYTIC FUNCTION

We shall demonstrate that any function $f(z)$ which can be represented as

$$f(z) = \frac{1}{2\pi i}\int_C \frac{g(z')}{z'-z}\,dz' \tag{13.1}$$

where C is a piecewise, regular curve of finite length (not necessarily closed) and $g(z)$ a function that is continuous on C, is analytic at any point z which does not lie on C. To this end, let us consider the expression

$$\Delta \equiv \left|\frac{f(z+\Delta z)-f(z)}{\Delta z} - \frac{1}{2\pi i}\int_C \frac{g(z')}{(z'-z)^2}\,dz'\right| \tag{13.2}$$

Using the integral formula (Eq. 13.1) for $f(z)$ and $f(z+\Delta z)$, one finds from Eq. 13.2

$$\Delta = \left| \frac{\Delta z}{2\pi} \int_C \frac{g(z')}{(z'-z-\Delta z)(z'-z)^2}\, dz' \right| \tag{13.3}$$

Since z is not on C, the above integrand is bounded, and therefore by what should now be a familiar argument

$$\Delta \to 0 \quad \text{as} \quad \Delta z \to 0 \tag{13.4}$$

This proves the differentiability of $f(z)$ and also demonstrates (see Eq. 13.2) that

$$\frac{df(z)}{dz} = \frac{1}{2\pi i} \int_C \frac{g(z')}{(z'-z)^2}\, dz' \tag{13.5}$$

An analogous result holds for the nth derivative of $f(z)$

$$\frac{d^n f(z)}{dz^n} = \frac{n!}{2\pi i} \int_C \frac{g(z')}{(z'-z)^{n+1}}\, dz' \tag{13.6}$$

We leave the proof to the reader.

In particular, a function $f(z)$ that is analytic in some region can be expressed in that region by Cauchy's integral formula, which has exactly the form of Eq. 13.1; C is now an arbitrary closed contour encircling the point z and $g(z) \equiv f(z)$. We have, therefore, the following

Theorem 1. The derivatives of all order of an analytic function are themselves analytic.

This is a remarkable result, and one that has no counterpart in the theory of functions of a real variable. We write Eq. 13.6 as

$$\frac{d^n f(z)}{dz^n} = \frac{n!}{2\pi i} \int_C \frac{f(z')}{(z'-z)^{n+1}}\, dz' \tag{13.7}$$

which holds when C lies in a simply-connected region of analyticity of $f(z)$ and encircles the point z.

Equation 13.1 is a particular example of an integral representation of a function, where the kernel of the representation is

$$K(z,z') = \frac{1}{2\pi i} \frac{1}{z'-z} \tag{13.8}$$

The differentiation formula (Eq. 13.5) shows that, for the integral representation 13.1, one can differentiate with respect to z under the integral sign. This result can be generalized.

Theorem 2. Given an integral representation

$$f(z) = \int_C K(z,z') g(z')\, dz', \quad z \in \text{R} \tag{13.9}$$

the derivative of $f(z)$ is given by

$$\frac{df(z)}{dz} = \int_C \frac{\partial K(z,z')}{\partial z} g(z')\, dz' \tag{13.10}$$

provided the following conditions are fulfilled:

(i) For $z \in$ R, $K(z,z')$ is an analytic function of z for any z' on the contour C.
(ii) For any $z \in$ R, $K(z,z')g(z')$ is a continuous function of z'.

Proof. Because of (i) we can write

$$K(z,z') = \frac{1}{2\pi i}\int_{C'} \frac{K(t,z')}{t-z}\,dt \qquad C' \subset R \tag{13.11}$$

Inserting this expression in Eq. 13.9 and interchanging the order of integrations,* we obtain

$$f(z) = \frac{1}{2\pi i}\int_{C'} \frac{\displaystyle\int_C K(t,z')g(z')\,dz'}{t-z}\,dt \tag{13.12}$$

This representation has the form of Eq. 13.1, since the integral in the numerator is a continuous function of t. Thus, by Eq. 13.5, we have

$$\frac{df(z)}{dz} = \frac{1}{2\pi i}\int_{C'} \frac{\displaystyle\int_C K(t,z')g(z')\,dz'}{(t-z)^2}\,dt \tag{13.13}$$

Again, changing the order of integrations and comparing with

$$\frac{\partial K(z,z')}{\partial z}$$

as obtained from Eq. 13.11, we arrive at Eq. 13.10.

14 · LOCAL BEHAVIOR OF AN ANALYTIC FUNCTION

An important property of analytic functions is given in the following theorem.

Theorem. The modulus of an analytic function $f(z)$ cannot have a local maximum within the region of analyticity of the function.

Proof. Consider an arbitrary regular point z_0 of $f(z)$. Then, provided r is small enough, $f(z)$ is analytic within and on a circle Γ of radius r centered at $z = z_0$. According to the Cauchy integral formula, one has

$$f(z_0) = \frac{1}{2\pi i}\int_\Gamma \frac{f(z)}{z-z_0}\,dz$$

Using the Darboux inequality, we obtain

$$|f(z_0)| \leq \frac{1}{2\pi}\max\left|\frac{f(z)}{z-z_0}\right|_{z\in\Gamma} \cdot 2\pi r$$

That is

$$|f(z_0)| \leq \max|f(z)|_{z\in\Gamma}$$

since for $z \in \Gamma$

$$|z - z_0| = r$$

* The order of integrations can be interchanged because of the continuity of the integrand, as for ordinary integrals. Remember that a contour integral can always be converted into an ordinary integral.

Thus, there exists on Γ at least one point where

$$|f(z)| \geq |f(z_0)|$$

But since one may choose r arbitrarily small, this means that in an arbitrarily small neighborhood of z_0, there always exists at least one point where

$$|f(z)| \geq |f(z_0)|$$

Hence $|f(z)|$ cannot have a local maximum at z_0.

An analogous result holds for the real and imaginary parts of an analytic function $f(z)$. In fact, applying the preceding theorem to the functions $e^{f(z)}$ and $e^{-if(z)}$, we find that neither

$$|e^{f(z)}| \equiv e^{\text{Re} f(z)}$$

nor

$$|e^{-if(z)}| \equiv e^{\text{Im} f(z)}$$

can have a local maximum at any regular point of $f(z)$. This implies, since the exponential function is monotonic, that $\text{Re} f(z)$ and $\text{Im} f(z)$ also cannot have a local maximum.

One can also show that $|f(z)|$ cannot have a local minimum at a regular point $z = z_0$ except where $f(z_0) = 0$, for at a regular point $z = z_0$ such that $f(z_0) \neq 0$, $1/f(z)$ is an analytic function of z. Therefore, $1/|f(z)|$ cannot have a maximum at $z = z_0$ and so $|f(z)|$ cannot have a minimum at this point. The same is evidently true for $\text{Re} f(z)$ and $\text{Im} f(z)$.

15 · THE CAUCHY-LIOUVILLE THEOREM

As one of the consequences of the Cauchy integral formula one also has the following:

Theorem. A bounded, entire function must be a constant.

Proof. We start with the derivative formula (see Sec. 13)

$$\frac{df}{dz} = \frac{1}{2\pi i} \int_C \frac{f(z')}{(z'-z)^2}\, dz' \tag{15.1}$$

Since $f(z)$ is an entire function, the closed contour C may be chosen to be a very large circle of radius R centered at z. Darboux's inequality applied to Eq. 15.1 yields

$$\left|\frac{df}{dz}\right| \leq \frac{\{\max|f|\}}{R}$$

Now letting $R \to \infty$, we have, since $\max|f|$ is finite

$$\frac{df}{dz} = 0$$

and therefore $f(z)$ is a constant.*

---

* $df/dz = 0$ implies f = constant also for functions of a complex variable. In fact, using Eq. 5.14, one has

$$0 = \int_C \frac{df}{dz}\, dz = f(b) - f(a)$$

a and b being the end points of the (arbitrary) contour C. Thus, $f(b) = f(a)$.

This theorem has as an implication that all "genuine" functions of a complex variable that are bounded at infinity must have at least one singularity in the complex plane.

16 · THE THEOREM OF MORERA

The following theorem due to Morera is a converse of Cauchy's theorem. It follows from the fact that the derivative of an analytic function is itself analytic.

Theorem. If the integral

$$\int_C f(z)\,dz$$

of a function, which is continuous in some region, vanishes for **any** closed contour C lying within this region, then $f(z)$ is analytic in that region.

Proof. As we explained at the end of Sec. 11, the vanishing of an integral along any closed path lying within some region means that the value of an integral along a path connecting any two points in the region does not depend on the path. Thus, if a is a fixed point and z an arbitrary point of the region in question, the integral

$$F(z) = \int_a^z f(z')\,dz' \tag{16.1}$$

depends only on the choice of the points a and z. We have

$$\begin{aligned}\frac{F(z+\Delta z)-F(z)}{\Delta z} &= \frac{1}{\Delta z}\int_a^{z+\Delta z} f(z')\,dz' - \frac{1}{\Delta z}\int_a^z f(z')\,dz' \\ &= \frac{1}{\Delta z}\int_z^{z+\Delta z} f(z')\,dz' \\ &\equiv \frac{f(z)}{\Delta z}\int_z^{z+\Delta z} dz + \frac{1}{\Delta z}\int_z^{z+\Delta z} [f(z')-f(z)]\,dz' \end{aligned} \tag{16.2}$$

However, using Eq. 5.14

$$\int_z^{z+\Delta z} dz = (z+\Delta z) - z = \Delta z$$

Taking the second integral along a straight line connecting the points z and $z + \Delta z$, and using Darboux's inequality, we obtain

$$\left|\frac{1}{\Delta z}\int_z^{z+\Delta z} [f(z')-f(z)]\,dz'\right| \le |\max[f(z')-f(z)]| \tag{16.3}$$

The RHS of Eq. 16.3 tends to zero as $\Delta z \to 0$ because $f(z)$ is continuous and therefore

$$\frac{dF(z)}{dz} = f(z) \tag{16.4}$$

Equation 16.4 shows that the derivative of $F(z)$ exists, and therefore (since z was an arbitrary point) that $F(z)$ is analytic throughout the region. Thus, $F(z)$ also has a second derivative (theorem of Sec. 13) and this together with Eq. 16.4 implies that $f(z)$ is differentiable throughout the region.

Every function that is analytic in a simply-connected region satisfies there (by virtue of Cauchy's theorem) the conditions of Morera's theorem. Therefore, every such function possesses a primitive function, given by the integral formula (16.1). We stated this result without proof at the beginning of Sec. 11.

17 · MANIPULATIONS WITH SERIES OF ANALYTIC FUNCTIONS

Consider a sequence of functions of the complex variable z

$$f_1(z), f_2(z), \cdots, f_n(z), \cdots \tag{17.1}$$

defined in some region R and such that

$$\sum_{n=1}^{\infty} f_n(z) = f(z) \tag{17.2}$$

We assume that along a smooth curve C of length L, the convergence of the sum in Eq. 17.2 is uniform. Then, provided all integrals exist, we can show that

$$\int_C f(z)\,dz \equiv \int_C \sum_{n=1}^{\infty} f_n(z)\,dz = \sum_{n=1}^{\infty} \int_C f_n(z)\,dz \tag{17.3}$$

That is, the operations of the addition of an infinite number of terms and of integration commute. Putting

$$s_n(z) = \sum_{j=1}^{n} f_j(z) \tag{17.4}$$

we can rewrite Eq. 17.3 as

$$\int_C f(z)\,dz \equiv \int_C \lim_{n\to\infty} s_n(z)\,dz = \lim_{n\to\infty} \int_C s_n(z)\,dz \tag{17.5}$$

The proof is almost immediate. Using again the inestimable Darboux inequality, we have

$$\left| \int_C [f(z) - s_n(z)]\,dz \right| \leq \max|f(z) - s_n(z)| \cdot L \tag{17.6}$$

and because the sum in Eq. 17.2 is assumed to converge uniformly, $\max|f(z) - s_n(z)|$ can be made arbitrarily small by taking n large enough.

We now assume that the functions $f_n(z)$ are analytic throughout R. Analogously to the case of functions of a real variable, one can prove that $f(z)$ is also analytic and that

$$\frac{df(z)}{dz} = \sum_{n=1}^{\infty} \frac{df_n(z)}{dz} \qquad z \in \text{R}' \subset \text{R} \tag{17.7}$$

provided the sum on the RHS converges uniformly in R′.

Actually, a stronger result, due to Weierstrass, follows from the analyticity of the functions $f_n(z)$.

Theorem. If $f_n(z)$ $(n = 1, 2, \cdots)$ is a sequence of analytic functions and if the infinite sum

$$\sum_{n=1}^{\infty} f_n(z) = f(z) \tag{17.8}$$

converges uniformly to $f(z)$ in any region $R' \subset R$, then $f(z)$ is analytic in R and

$$\frac{df(z)}{dz} = \sum_{n=1}^{\infty} \frac{df_n(z)}{dz} \tag{17.9}$$

Notice now that we no longer assume the uniform convergence of the sum of the derivatives $df_n(z)/dz$, for this can be shown to follow from the conditions of the theorem for any region $R' \subset R$.

Proof. $f(z)$ is analytic by virtue of Morera's theorem. In fact

$$\int_C f(z)\, dz = \sum_{n=1}^{\infty} \int_C f_n(z)\, dz = 0 \tag{17.10}$$

for any closed contour $C \subset R$, since $f_n(z)$ are analytic functions.

From Eq. 17.8 we have

$$\frac{1}{2\pi i} \frac{f(z)}{(z - z_0)^2} = \sum_{n=1}^{\infty} \frac{1}{2\pi i} \frac{f_n(z)}{(z - z_0)^2} \qquad z_0 \in R \tag{17.11}$$

Integrating along a closed contour C encircling the point z_0, we get

$$\frac{1}{2\pi i} \int_C \frac{f(z)}{(z - z_0)^2}\, dz = \sum_{n=1}^{\infty} \frac{1}{2\pi i} \int_C \frac{f_n(z)}{(z - z_0)^2}\, dz \tag{17.12}$$

or, using the results of Sec. 13 (differentiation formula 13.7)

$$\left.\frac{df(z)}{dz}\right|_{z=z_0} = \sum_{n=1}^{\infty} \left.\frac{df_n(z)}{dz}\right|_{z=z_0} \tag{17.13}$$

Since z_0 is arbitrary, we have proved the theorem. In order to save space we shall skip the proof of the uniform convergence of the sum in Eq. 17.9.

Of course Eq. 17.9 can also be written as

$$\frac{d}{dz}\left[\lim_{n\to\infty} s_n(z)\right] = \lim_{n\to\infty} \frac{ds_n(z)}{dz} \tag{17.14}$$

with $s_n(z)$ defined in Eq. 17.4.

18 · THE TAYLOR SERIES

A power series expansion, which is a direct generalization of the well-known Taylor expansion, holds for analytic functions.

Theorem. Let $f(z)$ be a function, analytic within and on a circle Γ centered at $z = z$. The value of this function at any point z within Γ is given by the uniformly convergent power series

$$f(z) = \sum_{n=0}^{\infty} a_n(z - z_0)^n \tag{18.1}$$

where

$$a_n = \frac{1}{n!} \frac{d^n f(z)}{dz^n}\bigg|_{z=z_0} = \frac{1}{2\pi i} \int_\Gamma \frac{f(z')}{(z' - z_0)^{n+1}} dz' \tag{18.2}$$

Proof. Since $f(z)$ is analytic, by Eq. 12.1 we have

$$f(z) = \frac{1}{2\pi i} \int_\Gamma \frac{f(z')}{z' - z} dz' \tag{18.3}$$

We expand the denominator in Eq. 18.3 as

$$\frac{1}{z' - z} = \frac{1}{(z' - z_0) - (z - z_0)} = \frac{1}{(z' - z_0)} \cdot \frac{1}{\left(1 - \dfrac{z - z_0}{z' - z_0}\right)} = \frac{1}{z' - z_0} \sum_{n=0}^{\infty} \left(\frac{z - z_0}{z' - z_0}\right)^n \tag{18.4}$$

The expansion is justified, since

$$\left|\frac{z - z_0}{z' - z_0}\right| < 1 \tag{18.5}$$

z' being on the circumference of Γ and z within Γ. As we shall see later in this section, every power series $\sum_{n=0}^{\infty} a_n(z - z_0)^n$ which is convergent for $|z - z_0| < R$ is uniformly convergent for $|z - z_0| < R$. The geometric series (18.4) is therefore also uniformly convergent, provided the condition (18.5) is satisfied. Putting (18.4) into Eq. 18.3, we find

$$f(z) = \frac{1}{2\pi i} \sum_{n=0}^{\infty} (z - z_0)^n \int_\Gamma \frac{f(z')}{(z' - z_0)^{n+1}} dz' \tag{18.6}$$

and the theorem is established.

One might ask: What is the radius of convergence of the Taylor series? The answer is given immediately by inspection of Cauchy's formula (Eq. 18.3), on which the proof of the Taylor expansion rests. Indeed, this formula breaks down when Γ goes through or encircles a singularity of $f(z)$. Therefore, we are led to the conclusion that the radius of convergence of the power series cannot be greater than the distance from the point $z = z_0$ to the nearest singularity of $f(z)$. As an example, the series $1 + z + z^2 + \cdots = 1/(1 - z)$ converges in the interior of the circle $|z| = 1$.

We have demonstrated that an analytic function can be expanded in a Taylor series at any of its regular points. The converse statement is also true; i.e., a function that can be expanded in a power series

$$f(z) = \sum_{n=0}^{\infty} a_n(z - z_0)^n \tag{18.7}$$

which is convergent in some neighborhood of the point $z = z_0$ (for example, for $|z - z_0| < R$) is necessarily analytic. Indeed, the convergence of the infinite sum in Eq. 18.7 for $|z - z_0| = r < R$ implies the existence of a constant A such that

$$|a_n| r^n < A \tag{18.8}$$

for any n. Therefore, the sum of terms with $M \leq n \leq N$ satisfies

$$\left|\sum_{n=M}^{N} a_n(z-z_0)^n\right| \leq \sum_{n=M}^{N} |a_n| \cdot |z-z_0|^n$$
$$< A \sum_{n=M}^{N} \left(\frac{|z-z_0|}{r}\right)^n$$
$$= A\left(\frac{|z-z_0|}{r}\right)^M \frac{1-\left(\dfrac{|z-z_0|}{r}\right)^{N-M}}{1-\left(\dfrac{|z-z_0|}{r}\right)} \tag{18.9}$$

For $|z - z_0| < r$, the expression on the RHS of the preceding inequality can be made smaller than any arbitrary constant, independent of z, by choosing M large enough. Hence, the convergence of the power series in Eq. 18.7 is uniform for $|z - z_0| < r$, and so it can be integrated term by term to give

$$\int_C f(z)\, dz = \sum_{n=0}^{\infty} a_n \int_C (z-z_0)^n\, dz = 0 \tag{18.10}$$

Here C is an arbitrary closed path lying within the circle of radius r and centered at $z = z_0$. Therefore, by virtue of Morera's theorem, $f(z)$ is analytic for $|z - z_0| < r$. But r can be arbitrarily close to R, and therefore the power series in Eq. 18.7 is an analytic function of z for $|z - z_0| < R$. This result was anticipated in Sec. 8.2.

19 · POISSON'S INTEGRAL REPRESENTATION

Consider the Taylor series

$$f(z) = \sum_{n=0}^{\infty} a_n z^n \tag{19.1}$$

where the complex constants a_n are given by Eq. 18.2, with $z_0 = 0$. Equation 18.2 is also meaningful for $n < 0$, but naturally $a_{-|n|}$ (and hence $\bar{a}_{-|n|}$) vanishes, since the integrand in $a_{-|n|}$ is analytic. Letting $z' = Re^{i\theta'}$ in Eq. 18.2, and using the notation $f(R,\theta) \equiv f(Re^{i\theta})$, one has

$$a_n = \frac{1}{2\pi R^n}\int_0^{2\pi} [\mathrm{Re} f(R,\theta') + i\, \mathrm{Im}\, f(R,\theta')] e^{-in\theta'}\, d\theta' \tag{19.2}$$

The equation $\bar{a}_{-|n|} = 0$ can thus be written as

$$\int_0^{2\pi} \mathrm{Re}\, f(R,\theta') e^{-in\theta'} d\theta' = i\int_0^{2\pi} \mathrm{Im}\, f(R,\theta') e^{-in\theta'}\, d\theta' \qquad (n > 0) \tag{19.3}$$

Hence

$$a_n = \frac{1}{\pi R^n}\int_0^{2\pi} \mathrm{Re}\, f(R,\theta') e^{-in\theta'}\, d\theta' \qquad (n > 0) \tag{19.4}$$

For $n = 0$, Eq. 19.2 yields

$$\mathrm{Re}\, a_0 + i\, \mathrm{Im}\, a_0 = \frac{1}{2\pi}\int_0^{2\pi} [\mathrm{Re}\, f(R,\theta') + i\, \mathrm{Im}\, f(R,\theta')]\, d\theta' \tag{19.5}$$

Combining Eqs. 19.1, 19.4, and 19.5, and setting $z = re^{i\theta}$ with $r < R$, we find

$$f(r,\theta) = i\,\mathrm{Im}\,a_0 + \frac{1}{2\pi}\int_0^{2\pi} \mathrm{Re}\,f(R,\theta')\,d\theta' + \frac{1}{\pi}\sum_{n=1}^{\infty}\left(\frac{r}{R}\right)^n \int_0^{2\pi} \mathrm{Re}\,f(R,\theta')e^{in\theta-\theta'}d\theta' \tag{19.6}$$

But, for $r < R$

$$\sum_{n=1}^{\infty}\left(\frac{r}{R}\right)^n e^{in(\theta-\theta')} = \frac{(r/R)e^{i(\theta-\theta')}}{1-(r/R)e^{i(\theta-\theta')}} = \frac{re^{i\theta}}{Re^{i\theta'}-re^{i\theta}} \tag{19.7}$$

Collecting Eqs. 19.6 and 19.7, we have

$$f(r,\theta) = i\,\mathrm{Im}\,a_0 + \frac{1}{2\pi}\int_0^{2\pi} \mathrm{Re}\,f(R,\theta')\,\frac{Re^{i\theta'}+re^{i\theta}}{Re^{i\theta'}-re^{i\theta}}\,d\theta' \qquad r < R \tag{19.8}$$

Taking the real parts of both sides of Eq. 19.8 leads to

$$\mathrm{Re}\,f(r,\theta) = \frac{1}{2\pi}\int_0^{2\pi} \mathrm{Re}\,f(R,\theta')\,\frac{R^2-r^2}{R^2-2Rr\cos(\theta-\theta')+r^2}\,d\theta' \qquad r < R \tag{19.9}$$

The preceding equation is known as Poisson's formula. Analogously, one can derive

$$\mathrm{Im}\,f(r,\theta) = \frac{1}{2\pi}\int_0^{2\pi} \mathrm{Im}\,f(R,\theta')\,\frac{R^2-r^2}{R^2-2Rr\cos(\theta-\theta')+r^2}\,d\theta' \qquad r<R \tag{19.10}$$

Equations 19.9 and 19.10 are examples of the so-called Dirichlet principle applied to a circle, whereby the value of an harmonic function at an interior point of a closed curve can be determined, once the values of this function along the boundary are given.

20 · THE LAURENT SERIES

If a function $f(z)$ is not analytic throughout the whole interior of a circle (as it was assumed to be in the derivation of the Taylor series), but only throughout the annular region between two concentric circles Γ_1 and Γ_2, it is possible to generalize the Taylor expansion of $f(z)$, which now becomes an expansion in both positive and negative powers of $(z - z_0)$. Such an expansion is called a **Laurent series.*** We state this as a theorem.

Theorem. Let $f(z)$ be analytic in the annular region between and on two concentric circles Γ_1 and Γ_2 centered at $z = z_0$. The value of $f(z)$ at any point z within the annular region is given by the uniformly convergent power series

$$f(z) = \sum_{n=-\infty}^{\infty} d_n(z-z_0)^n \tag{20.1}$$

where

$$d_n = \frac{1}{2\pi i}\int_C \frac{f(z')}{(z'-z_0)^{n+1}}\,dz' \tag{20.2}$$

and C is any contour within the annular region encircling the point z_0. Note that the sum in Eq. 20.1, in contradistinction to the Taylor expansion (18.1), extends from $n = -\infty$ to $n = +\infty$.

* In this section we are clearly dealing with an extension of Taylor's theorem to the case of a multiply-connected region.

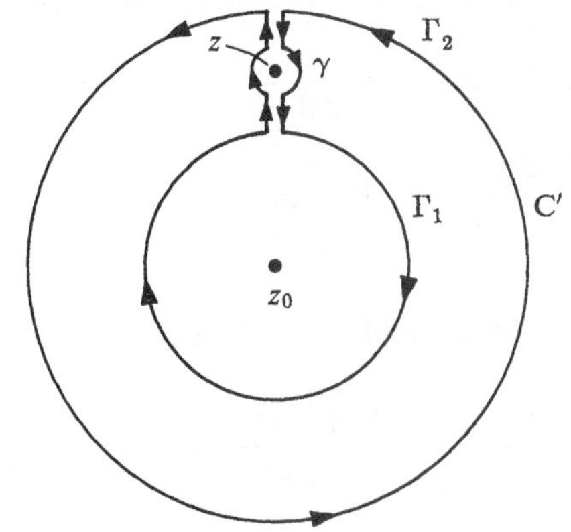

Fig. 16.

Proof. We draw a small circle γ centered at z and contained entirely within the annular region of analyticity of $f(z)$. Consider the closed contour C′, shown in Fig. 16. According to Cauchy's theorem

$$\int_{C'} \frac{f(z')}{z' - z}\, dz' = 0 \tag{20.3}$$

since the integrand is analytic in the region enclosed by C′. However, the parallel straight segments of the path can lie arbitrarily close to each other and thus give contributions equal in magnitude but of opposite sign; therefore Eq. 20.3 can be rewritten as*

$$\int_{\gamma} \frac{f(z')}{z' - z}\, dz' = \int_{\Gamma_2} \frac{f(z')}{z' - z}\, dz' - \int_{\Gamma_1} \frac{f(z')}{z' - z}\, dz' \tag{20.4}$$

By Cauchy's integral formula, the integral on the left-hand side (LHS) of Eq. 20.4 is simply $2\pi i f(z)$. Therefore

$$f(z) = \frac{1}{2\pi i} \int_{\Gamma_2} \frac{f(z')}{z' - z}\, dz' - \frac{1}{2\pi i} \int_{\Gamma_1} \frac{f(z')}{z' - z}\, dz \tag{20.5}$$

The first integral in Eq. 20.5 can be expended in positive powers of $(z - z_0)$, exactly as in the case of the Taylor series, with the result

$$\frac{1}{2\pi i} \int_{\Gamma_2} \frac{f(z')}{z' - z}\, dz' = \sum_{n=0}^{\infty} a_n (z - z_0)^n \tag{20.6}$$

where

$$a_n = \frac{1}{2\pi i} \int_{\Gamma_2} \frac{f(z')}{(z' - z_0)^{n+1}}\, dz' \tag{20.7}$$

* Remember the convention of Sec. 12.

The expansion of the second integral is done in a similar way. We have

$$\frac{1}{z'-z} = -\frac{1}{(z-z_0)-(z'-z_0)} = -\frac{1}{(z-z_0)} \cdot \frac{1}{\left(1-\dfrac{z'-z_0}{z-z_0}\right)}$$
$$= -\frac{1}{z-z_0}\sum_{n=1}^{\infty}\left(\frac{z'-z_0}{z-z_0}\right)^{n-1} \tag{20.8}$$

The sum in Eq. 20.8 is uniformly convergent for $\left|\dfrac{z'-z_0}{z-z_0}\right| < 1$, and therefore

$$\frac{1}{2\pi i}\int_{\Gamma_1}\frac{f(z')}{z'-z}\,dz' = -\sum_{n=1}^{\infty}\frac{b_n}{(z-z_0)^n} \tag{20.9}$$

where

$$b_n = \frac{1}{2\pi i}\int_{\Gamma_1}(z'-z_0)^{n-1}f(z')\,dz' \tag{20.10}$$

Using Eqs. 20.6, 20.7, 20.9, and 20.10, Eq. 20.5 can be rewritten as

$$f(z) = \sum_{n=-\infty}^{\infty} d_n(z-z_0)^n \tag{20.11}$$

where, for all n

$$d_n = \frac{1}{2\pi i}\int_C \frac{f(z')}{(z'-z_0)^{n+1}}\,dz' \tag{20.12}$$

C being a contour within the annular region of analyticity of $f(z)$, encircling the point z_0.

The sum in Eq. 20.11 converges uniformly to $f(z)$ for

$$R_1 < |z-z_0| < R_2 \tag{20.13}$$

R_1 and R_2 denoting respectively the radii of the circles Γ_1 and Γ_2, since this is the condition for the simultaneous convergence of the expansions, in Eqs. 20.6 and 20.9.

Again, the annular region throughout which the series converges can be enlarged until the first singular point of the function $f(z)$ is reached.

21 · ZEROS AND ISOLATED SINGULAR POINTS OF ANALYTIC FUNCTIONS

21.1 Zeros

If a function $f(z)$ vanishes at a point $z = z_0$, this point is called a **zero** of $f(z)$. A function is said to have a zero of order n at $z = z_0$ if

$$f(z_0) = \left.\frac{df(z)}{dz}\right|_{z=z_0} = \cdots = \left.\frac{d^{n-1}f(z)}{dz^{n-1}}\right|_{z=z_0} = 0 \tag{21.1}$$

but

$$\left.\frac{d^n f(z)}{dz^n}\right|_{z=z_0} \neq 0 \tag{21.2}$$

Then the first n coefficients in the Taylor expansion of $f(z)$ about $z = z_0$ vanish so that

$$\begin{aligned} f(z) &= a_n(z - z_0)^n + a_{n+1}(z - z_0)^{n+1} + \cdots \\ &= (z - z_0)^n \sum_{k=0}^{\infty} a_{n+k}(z - z_0)^k \\ &= (z - z_0)^n h(z) \end{aligned} \tag{21.3}$$

Here $h(z)$ is analytic and nonvanishing at $z = z_0$. It is clear that since $h(z)$ is continuous, it must also differ from zero in some finite neighborhood of $z = z_0$, and the same is true for $f(z)$. This in turn implies that the zeros of an analytic function are isolated. Thus, the set of zeros of an analytic function cannot have an accumulation point except when $f(z) \equiv 0$; this is an important fact, the consequences of which will be seen in Sec. 26.

21.2 Isolated Singular Points

In order to give a classification of isolated singular points of analytic functions, it is very convenient to consider the Laurent expansion of the function $f(z)$ about a point z_0

$$f(z) = \sum_{n=0}^{\infty} a_n(z - z_0)^n + \frac{b_1}{z - z_0} + \frac{b_2}{(z - z_0)^2} + \cdots \tag{21.4}$$

Suppose that $f(z)$ has an isolated singularity at the point $z = z_0$ and is analytic within a circle centered at this point. Then the annular region where the preceding expansion converges will reduce to the whole interior of the circle with the point $z = z_0$ taken out.

It is clear that $f(z)$ may have a singularity at $z = z_0$ if at least one of the b_j is not equal to zero. If the coefficient b_n does not vanish, but all higher coefficients b_j do vanish

$$b_{n+1} = b_{n+2} = \cdots = 0 \tag{21.5}$$

then the function $f(z)$ is said to have a **pole of order** n at $z = z_0$. The sum

$$\frac{b_1}{z - z_0} + \frac{b_2}{(z - z_0)^2} + \cdots + \frac{b_n}{(z - z_0)^n}$$

is then referred to as the **principle part** of $f(z)$ at $z = z_0$. If $b_1 \neq 0$, but $b_2 = b_3 = \cdots = 0$, the function is said to have a **simple pole.**

It is easy to verify whether or not a function has a pole of order n: $f(z)$ has a pole of order n at $z = z_0$ if $1/f(z)$ has a zero of order n at that point and the condition for that to occur is as given by Eqs. 21.1 and 21.2.

A function that is analytic in a region of the complex plane, except at a set of points of the region where the function has poles, is called a **meromorphic** function in this region.

When there is an infinite number of coefficients that do not vanish in the Laurent expansion of $f(z)$ about z_0, the function is said to have an **isolated essential singularity** at $z = z_0$. The very peculiar nature of an isolated essential singularity is made manifest by the following theorem due to Weierstrass.

Theorem. If a function $f(z)$ has an isolated essential singularity at a point $z = z_0$, then for arbitrary positive numbers ε and δ and for **any** complex number a, one has

$$|f(z) - a| < \varepsilon \tag{21.6}$$

for some point z satisfying $|z - z_0| < \delta$.

Expressed differently: The Weierstrass theorem states that in an arbitrary neighborhood of an essential singularity, a function oscillates so rapidly that it comes arbitrarily close to any possible complex number.

Proof. We first convince ourselves that the sum of an infinite number of singular terms in the Laurent expansion cannot be bounded in the neighborhood of $z = z_0$. Applying the Darboux inequality to the integral

$$b_n = \frac{1}{2\pi i} \int_\Gamma (z' - z)^{n-1} f(z')\, dz'$$

where Γ is a circle of radius r centered at $z = z_0$, we get for $n > 1$,* (assuming that $f(z)$ is bounded)

$$\begin{aligned} |b_n| &\le \frac{r^{n-1}}{2\pi} \max|f(z)| \cdot 2\pi r \\ &= r^n \max|f(z)| \Big|_{r \to 0} \to 0 \end{aligned} \tag{21.7}$$

Thus, $b_n = 0$ for $n > 1$, in contradiction with the assumption that there exists an essential singularity at $z = z_0$. Hence, $f(z)$ cannot be bounded.

Take an arbitrary complex number a. The point $z = z_0$ is or is not an accumulation point of the zeros of the function $f(z) - a$. If it is, then (since in an arbitrary neighborhood of z_0 one has at least one point where $f(z) - a = 0$), the theorem follows immediately. If z_0 is not an accumulation point of the zeros of $f(z)$, then for some η, $f(z) - a \neq 0$, provided

$$|z - z_0| < \eta \tag{21.8}$$

Consider the function

$$g(z) = \frac{1}{f(z) - a} \tag{21.9}$$

which is well defined in the region (21.8) enclosing the point z_0. Since

$$f(z) = \frac{1}{g(z)} + a$$

the function $g(z)$ must also have an essential singularity at $z = z_0$, since otherwise $f(z)$ would either be analytic [if $g(z)$ were analytic and nonvanishing or had a pole at $z = z_0$] or it would have a pole [if $g(z)$ had a zero at $z = z_0$] at this point. Therefore, $g(z)$ cannot be bounded within any circle $|z - z_0| < \delta$, and in particular there must be a point where

$$|g(z)| > \frac{1}{\varepsilon}$$

This inequality is, by virtue of Eq. 21.9, equivalent to (21.6.)

---

* To estimate the integral here we use polar coordinates as in Sec. 14.

22 · THE CALCULUS OF RESIDUES

22.1 Theorem of Residues

Let $f(z)$ be a function of a complex variable which is analytic everywhere within and on a closed curve C, with the exception of a point z_0 in the interior of C where $f(z)$ may have an isolated singularity. Then the integral

$$\frac{1}{2\pi i}\int_C f(z')\,dz' \tag{22.1}$$

which vanished by virtue of Cauchy's theorem when z_0 was a regular point of $f(z)$, need no longer vanish if z_0 is a singular point of that function. We can therefore generalize Cauchy's theorem by setting (22.1) equal to a quantity that may or may not vanish, depending upon the nature of the point z_0. This quantity is called the **residue** of $f(z)$ at the point z_0 and is denoted by $\operatorname{Res} f(z_0)$:

$$\operatorname{Res} f(z_0) \underset{\text{def}}{=} \frac{1}{2\pi i}\int_C f(z')\,dz' \tag{22.2}$$

If z_0 is a regular point of $f(z)$, then evidently $\operatorname{Res} f(z_0) = 0$; in all other cases we need to evaluate $\operatorname{Res} f(z_0)$.

When the closed curve C encloses (instead of one isolated singularity), say, m isolated singularities of $f(z)$, we can proceed in the way we did when deriving the Laurent expansion. That is, we can enclose each singularity z_j $(j = 1, 2, \cdots, m)$ within a small circle γ_j contained within C, and join each γ_j to C by a pair of infinitesimally separated parallel paths. Considering now the contours C, γ_j, and the pairs of corresponding parallel paths as parts of a single contour, with the direction of integration fixed by the convention of Sec. 12, we easily arrive at the result

$$\int_C f(z')\,dz' = \sum_{j=1}^{m}\int_{\gamma_j} f(z')\,dz' \tag{22.3}$$

since $f(z)$ is analytic within the complete contour and the contributions from each pair of parallel paths cancel each other (see Fig. 17).

By the definition of the residue of $f(z)$

$$\int_C f(z')\,dz' = 2\pi i \sum_{j=1}^{m} \operatorname{Res} f(z_j) \tag{22.4}$$

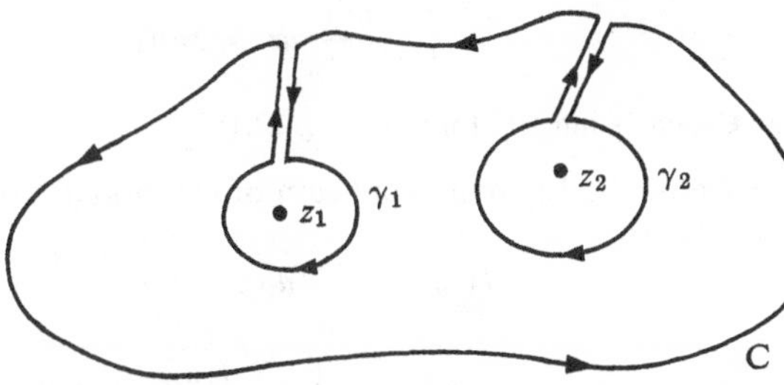

Fig. 17. The derivation of Eq. 22.3 in the case of $m = 2$.

Equation 22.4, which expresses the so-called **theorem of residues**, will be used frequently. It states that in order to find the value of the contour integral of a function along a certain closed path, it is only necessary to find the residues of the function at its singularities within the contour, to add them, and to multiply the result by $2\pi i$, provided all singularities are isolated ones. The problem of evaluating a contour integral of a function that has only isolated singularities is therefore reduced to a problem of calculating the residues of that function. The following considerations will simplify this task.

Let us calculate the residue of a function $f(z)$ at a pole of order n. Now, if $f(z)$ has a pole of order n at $z = z_0$, there must exist a function $g(z)$ that is analytic and nonzero at $z = z_0$, and such that

$$f(z) = \frac{g(z)}{(z - z_0)^n} \tag{22.5}$$

Putting Eq. 22.5 into Eq. 22.2, we find

$$\operatorname{Res} f(z_0) = \frac{1}{2\pi i} \int_C \frac{g(z')}{(z' - z_0)^n}\, dz' = \frac{1}{(n-1)!} \left. \frac{d^{n-1} g(z)}{dz^{n-1}} \right|_{z=z_0} \tag{22.6}$$

The last step is a consequence of the derivative formula for analytic functions (Eq. 13.7). Equation 22.6 can be rewritten, using Eq. 22.5, as

$$\operatorname{Res} f(z_0) = \lim_{z \to z_0} \frac{1}{(n-1)!} \left\{ \frac{d^{n-1}}{dz^{n-1}} [(z - z_0)^n f(z)] \right\} \tag{22.7}$$

In the special but very important case where $f(z)$ has a simple pole at z_0, we find from Eq. 22.7

$$\operatorname{Res} f(z_0) = \lim_{z \to z_0} (z - z_0) f(z) \tag{22.8}$$

Consider the integral

$$\int_\Gamma \frac{f(z')}{z' - z_0}\, dz'$$

where Γ is a closed contour encircling the point z_0 and $f(z)$ is a function that is analytic within and on Γ. Then $f(z)/(z - z_0)$ has a simple pole at $z = z_0$, and according to Eq. 22.8, its residue there is

$$\lim_{z \to z_0} (z - z_0) \left[\frac{f(z)}{z - z_0} \right] = f(z_0)$$

Hence, from Eq. 22.4 we have

$$\int_\Gamma \frac{f(z')}{z' - z_0}\, dz' = 2\pi i f(z_0)$$

which is just Cauchy's integral formula (Eq. 12.1).

Consider now the Laurent expression of $f(z)$ about a point $z = z_0$:

$$f(z) = \sum_{k=-\infty}^{\infty} d_k (z - z_0)^k$$

$$d_k = \frac{1}{2\pi i} \int_C \frac{f(z')}{(z' - z_0)^{k+1}}\, dz' \tag{22.9}$$

Equation 22.9 immediately shows that

$$\operatorname{Res} f(z_0) = d_{-1} \tag{22.10}$$

This follows from the definition of the residue (Eq. 22.2).

We have therefore arrived at the result that the residue of a function $f(z)$ at a pole of order n located at $z = z_0$ can be calculated in either of two ways:

(i) By using formula 22.7
(ii) By finding the coefficient of the inverse first power of $(z - z_0)$ in the Laurent expansion of $f(z)$ about the pole whose residue we are seeking.

The second method applies equally well when, instead of a pole, one has an essential singularity at $z = z_0$.

EXAMPLE 1

Let

$$f(z) = \frac{z^2 + 5z + 3}{(z-1)(z+2)^2}$$

By writing

$$f(z) = \frac{1}{(z+2)^2} + \frac{1}{z-1}$$

we can immediately read off the values of the residues of $f(z)$ at $z = 1$ and $z = -2$, since the first term is analytic at $z = 1$ and the second at $z = -2$; they are

$$\operatorname{Res} f(1) = 1 \qquad \text{and} \qquad \operatorname{Res} f(-2) = 0$$

These results can also be verified by using Eq. 22.7; for example,

$$\operatorname{Res} f(-2) = \lim_{z \to -2} \left\{ \frac{d}{dz} \left[\frac{z^2 + 5z + 3}{z - 1} \right] \right\} = 0$$

EXAMPLE 2

Suppose that a function $h(z)$ has a simple zero at $z = z_0$ and that the function $g(z)$ is analytic there.

$$h(z_0) = 0 \qquad \left.\frac{dh(z)}{dz}\right|_{z=z_0} \neq 0$$

The residue of the function $f(z) = g(z)/h(z)$ at $z = z_0$ is $\operatorname{Res} f(z_0) = \lim_{z \to z_0} \frac{(z - z_0)g(z)}{h(z)}$.

Expanding $h(z)$ about $z = z_0$ and using Eq. 18.2, we find

$$\operatorname{Res} f(z_0) = \lim_{z \to z_0} \frac{(z - z_0)g(z)}{a_1(z - z_0) + a_2(z - z_0)^2 + \cdots}$$

$$= \frac{g(z_0)}{\left.\dfrac{dh(z)}{dz}\right|_{z=z_0}}$$

As an example, $\cos z$ has zeros at $z = (2n + 1)(\pi/2)$ for all integers n. These are simple zeros since

$$\cos\left[(2n+1)\frac{\pi}{2}\right] = 0$$

and

$$\left[\frac{d}{dz}\cos z\right]_{z=(2n+1)(\pi/2)} = -\sin\left[(2n+1)\frac{\pi}{2}\right] \neq 0$$

Therefore, the residues of $1/(\cos z)$ at the points $z = (2n + 1)(\pi/2)$ are

$$\operatorname{Res}\left\{\frac{1}{\cos z}\right\}_{z=(2n+1)(\pi/2)} = -\frac{1}{\sin(2n+1)(\pi/2)} = \begin{cases} -1 & \text{for } n \text{ even} \\ +1 & \text{for } n \text{ odd} \end{cases}$$

22.2 Evaluation of Integrals

Equation 22.4 and the prescriptions given in the preceding section for calculating the residues of a function provide an elegant and powerful method for the evaluation of many definite integrals. Another possibility for the evaluation of integrals is afforded by making an appropriate deformation of the contour of integration. We shall give examples of how these evaluations are actually carried out in specific instances. Before doing this, however, we shall prove a lemma, due to Jordan, that will be useful in this section.

Jordan's Lemma. Let Γ_R denote a semicircle in the upper half of the complex plane, of radius R, and centered at the origin. Let $f(z)$ be a function that tends uniformly to zero with respect to arg z as $|z| \to \infty$ when arg z lies in the interval

$$0 \leq \arg z \leq \pi$$

Then, if α is a real, non-negative number

$$\lim_{R\to\infty} I_R \equiv \lim_{R\to\infty} \int_{\Gamma_R} e^{i\alpha z'} f(z')\, dz' = 0 \tag{22.11}$$

Proof. In polar coordinates, the integral I_R is

$$I_R = i \int_0^\pi f(Re^{i\theta})\, e^{i\alpha R\cos\theta - \alpha R\sin\theta + i\theta} R\, d\theta \tag{22.12}$$

Since, by hypothesis, $f(z)$ tends uniformly to zero as $R \to \infty$ we must have

$$|f(Re^{i\theta})| < \varepsilon(R) \tag{22.13}$$

where $\varepsilon(R)$ is some positive number, which depends on R only and which tends to zero as $R \to \infty$. Therefore

$$|I_R| < \varepsilon(R) \cdot R \cdot \int_0^\pi e^{-\alpha R\sin\theta}\, d\theta = 2\varepsilon(R) R \int_0^{\pi/2} e^{-\alpha R\sin\theta}\, d\theta \tag{22.14}$$

where the last step follows from the symmetry of $\sin\theta$ about $\theta = \pi/2$. But it is a property of $\sin\theta$ that

$$\sin\theta \geq \frac{2\theta}{\pi} \qquad \text{when} \quad 0 \leq \theta \leq \frac{\pi}{2} \tag{22.15}$$

Hence

$$|I_R| < 2\varepsilon(R) \cdot R \int_0^{\pi/2} e^{-(2\alpha R\theta/\pi)}\, d\theta = \frac{\pi\varepsilon(R)}{\alpha}(1 - e^{-\alpha R}) \tag{22.16}$$

and therefore

$$\lim_{R\to\infty} I_R = 0$$

thus establishing the lemma.

Note. If $\alpha < 0$, the lemma remains valid, provided the semicircle Γ_R is taken in the lower half of the complex plane, and provided $f(z)$ tends uniformly to zero for $\pi \leq \arg z \leq 2\pi$.

We are now prepared to show how the calculus of residues can be applied to the evaluation of a number of types of definite integrals. We illustrate the procedure with certain characteristic examples.

EXAMPLE 1

$$I_1 = \int_0^\infty \frac{x^2\,dx}{(x^2+1)(x^2+4)}$$

It is convenient to write this integral as

$$I_1 = \frac{1}{2}\int_{-\infty}^{\infty} \frac{x^2\,dx}{(x^2+1)(x^2+4)}$$

so that the range of integration is the entire real axis. Consider also the integral

$$I'_1 = \frac{1}{2}\int_C \frac{z^2\,dz}{(z^2+1)(z^2+4)}$$

where z is complex and C is a contour that consists of a semicircle of radius R centered at the origin, extending in the upper half of the complex plane, and of the interval $(-R,R)$ of the real axis (see Fig. 18). The integral I'_1 can be written as the sum of two integrals, one extending over the real axis from $x = -R$ to $x = R$ and the other extending over the circumference Γ_R of the semicircle

$$\frac{1}{2}\int_C \frac{z^2\,dz}{(z^2+1)(z^2+4)} = \frac{1}{2}\int_{-R}^{R} \frac{x^2\,dx}{(x^2+1)(x^2+4)} + \frac{1}{2}\int_{\Gamma R} \frac{z^2\,dz}{(z^2+1)(z^2+4)} \tag{22.17}$$

By going over to polar coordinates, it is very easy to see that the last integral on the RHS of Eq. 22.17 vanishes as $R \to \infty$. Therefore

$$I_1 = \frac{1}{2}\int_{-\infty}^{\infty} \frac{x^2\,dx}{(x^2+1)(x^2+4)} = \frac{1}{2}\int_C \frac{z^2\,dz}{(z^2+1)(z^2+4)}$$

$$= \frac{1}{2}\cdot 2\pi i \sum_j \text{Res}\, f(z_j)$$

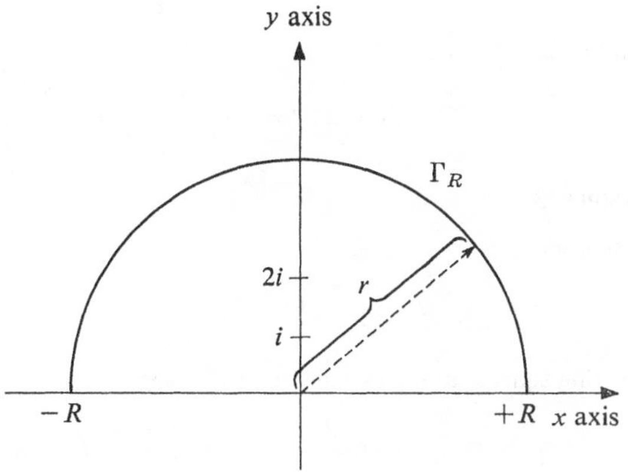

Fig. 18.

where the last step follows from 22.4 and the sum is over the residues of those poles of

$$f(z) = \frac{z^2}{(z^2+1)(z^2+4)}$$

that lie in the upper half of the complex plane. Since

$$\frac{1}{(z^2+1)(z^2+4)} = \frac{1}{(z+i)(z-i)(z+2i)(z-2i)}$$

$f(z)$ has two simple poles in the upper half of the complex plane, one at $z = i$ and the other at $z = 2i$. According to Eq. 22.8, the residues at these poles are

$$\operatorname{Res} f(i) = \frac{i^2}{(2i)(3i)(-i)} = \frac{i}{6}$$

$$\operatorname{Res} f(2i) = \frac{(2i)^2}{(3i)(4i)(i)} = -\frac{i}{3}$$

Hence

$$I_1 = \frac{1}{2} \cdot 2\pi i \left(\frac{i}{6} - \frac{i}{3}\right) = \frac{\pi}{6}$$

EXAMPLE 2

$$I_2 = \int_0^{2\pi} \frac{d\theta}{1 + a \sin \theta} \qquad (0 \leq a^2 < 1)$$

It is always convenient in the integrals involving trigonometric functions to express these functions in terms of exponentials. We evaluate I_2 on the unit circle Γ.

$$z = e^{i\theta}, \quad \sin \theta = \frac{z - z^{-1}}{2i} \qquad \text{for } z \in \Gamma$$

Then

$$I_2 = 2 \int_\Gamma \frac{dz}{az^2 + 2iz - a}$$

Since $0 \leq a^2 < 1$, the integrand has only one simple pole within the unit circle at

$$z = \frac{i(\sqrt{1-a^2} - 1)}{a}$$

Hence, using Eq. 22.4, we find

$$I_2 = 2 \,.\, 2\pi i \,.\, \frac{1}{2i\sqrt{1-a^2}} = \frac{2\pi}{\sqrt{1-a^2}}$$

EXAMPLE 3

Gauss' Integral.

$$I_3 = \int_{-\infty}^{+\infty} e^{iax - bx^2}\, dx, \qquad a,b \text{ real}, \quad b > 0$$

By completing squares in the exponential we have

$$I_3 = e^{-(a^2/4b)} \int_{-\infty - i(a/2b)}^{+\infty - i(a/2b)} e^{-bz^2}\, dz \tag{22.18}$$

where we have set $z = x - ia/2b$.

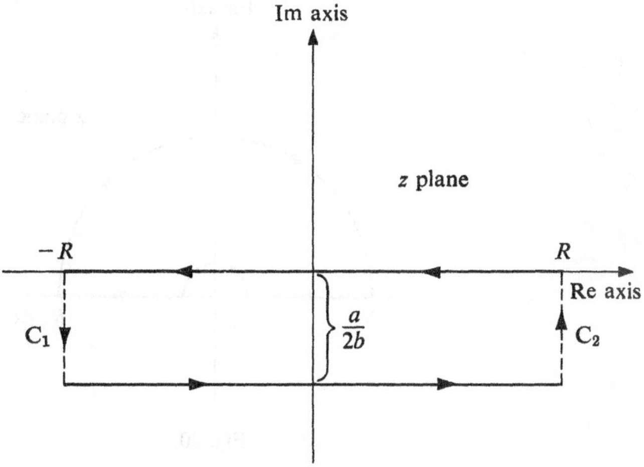

Fig. 19. As $R \to \infty$, the integration of e^{-bz^2} from $-R-(ia/2b)$ to $R-(ia/2b)$ yields the same result as the integration from $-R$ to R.

The function e^{-bz^2} can be integrated along the contour of Fig. 19, inside of which there are no poles of the integrand. Since the integrals along C_1 and C_2 make no contribution, we may as well write I_3 as

$$I_3 = e^{-(a^2/4b)} \int_{-\infty}^{+\infty} e^{-bz^2}\, dz \tag{22.19}$$

The integral in Eq. 22.19 can be evaluated by the following successive transformations

$$\begin{aligned}\int_{-\infty}^{+\infty} e^{-bz^2}\, dz &= \left[\int_{-\infty}^{+\infty} e^{-bx^2}\, dx\right]^{1/2} \left[\int_{-\infty}^{+\infty} e^{-by^2}\, dy\right]^{1/2} \\ &= \left[\int_{-\infty}^{+\infty}\int_{-\infty}^{+\infty} e^{-b(x^2+y^2)}\, dx\, dy\right]^{1/2} \\ &= \left[\frac{1}{2}\int_0^{2\pi}\int_0^{\infty} e^{-br^2}\, d(r^2)\, d\theta\right]^{1/2} \\ &= \sqrt{\frac{\pi}{b}}\end{aligned}$$

The next to the last step follows upon transforming the preceding integral to polar coordinates; whence

$$\int_{-\infty}^{+\infty} e^{iax-bx^2}\, dx = \sqrt{\frac{\pi}{b}}\, e^{-(a^2/4b)} \tag{22.20}$$

for a,b real and $b > 0$.

The integral in Eq. 22.18 has been evaluated without using the theorem of residues. The analyticity of the integrand allowed us to deform conveniently the contour of integration, as explained in Fig. 19.

EXAMPLE 4

$$I_4 = \int_0^{\infty} \frac{\sin x}{x}\, dx$$

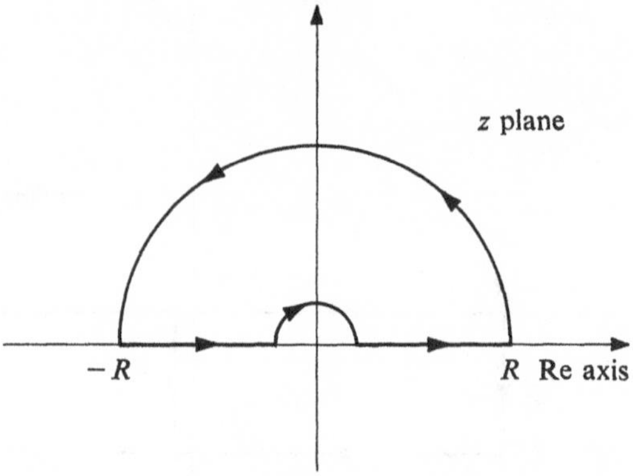

Fig. 20.

Consider the integral

$$I'_4 = \int_C \frac{e^{iz}}{z}\,dz$$

taken along the contour shown in Fig. 20. Since e^{iz}/z is analytic within this contour, we have

$$\int_r^R \frac{e^{ix}}{x}\,dx + \int_{\Gamma_R} \frac{e^{iz}}{z}\,dz + \int_{-R}^{-r} \frac{e^{ix}}{x}\,dx + \int_{\Gamma_0} \frac{e^{iz}}{z}\,dz = 0 \tag{22.21}$$

By Jordan's lemma, the integral over the semicircle Γ_R vanishes in the limit as $R \to \infty$.

The integral along Γ_0 can be evaluated in the limit as $r \to 0$ (notice that the path is followed in the clockwise direction).

$$\begin{aligned}\lim_{r\to 0} \int_{\Gamma_0} \frac{e^{iz}}{z}\,dz &= \lim_{r\to 0} \int_{C_0} \frac{1}{z} \sum_{n=0}^{\infty} \frac{(iz)^n}{n!}\,dz \\ &= -i\int_0^{\pi} d\theta - \lim_{r\to 0} \sum_{n=1}^{\infty} \frac{i^{n+1}r^n}{n!} \int_0^{\pi} e^{in\theta}\,d\theta \\ &= -i\pi \end{aligned} \tag{22.22}$$

The sum in Eq. 22.22 tends to zero as $r \to 0$, since this sum converges uniformly as $n \to \infty$ and therefore the operations $\lim\limits_{r\to 0}$ and $\sum\limits_{n=1}^{\infty}$ commute.

Hence, from Eq. 22.21, making an appropriate change of variables in the third integral

$$\lim_{\substack{r\to 0\\ R\to\infty}} \int_r^R \frac{e^{ix} - e^{-ix}}{x}\,dx = i\pi$$

Therefore

$$\int_0^{\infty} \frac{\sin x}{x}\,dx = \frac{\pi}{2}$$

23 · THE PRINCIPAL VALUE OF AN INTEGRAL

Until now we have considered contour integrals of functions in two different cases. In the first case, the integral was taken along a curve inside, or on which there were no singularities of the function; this led us to Cauchy's theorem (Eq. 11.1). In the second case, the boundary curve enclosed, but did not go through, singularities of the function, and this led us to the theorem of residues (Eq. 22.4). There remains to consider the case when the path of integration actually passes **through** a singularity of the integrand. In this case, strictly speaking, the integral does not exist. To give it meaning, one must choose a path that circumvents the singularity. We shall show that the result of the integration will then depend on how the path is chosen to avoid the singularity. We shall restrict ourselves to the case where the singularity is a simple pole on the real axis.

To be specific, consider the integral

$$\int_{-\infty}^{+\infty} \frac{f(x)}{x - x_0}\, dx \tag{23.1}$$

where x_0 in **on** the path of integration, i.e., on the real axis, and $f(x)$ is analytic at $x = x_0$. We shall suppose that

$$|x^\alpha f(x)| \to \text{constant} \qquad \text{as} \quad |x| \to \infty$$

where $\alpha > 0$. We may indent the path, as in Fig. 21, and follow the contour which encircles, rather than goes through, the singularity x_0. The integral (23.1) along this indented contour can be written as the sum of three integrals

$$\int_{\rightarrow} \frac{f(z)}{z - x_0}\, dz = \int_{-\infty}^{x_0-\varepsilon} \frac{f(x)}{x - x_0}\, dx + \int_{x_0+\varepsilon}^{+\infty} \frac{f(x)}{x - x_0}\, dx + \int_{\Gamma_0} \frac{f(z)}{z - x_0}\, dz \tag{23.2}$$

In the limit as $\varepsilon \to 0$, the first two terms define what is called the **principal value** of the integral 23.1; this is written as

$$\lim_{\varepsilon \to +0} \left[\int_{-\infty}^{x_0-\varepsilon} \frac{f(x)}{x - x_0}\, dx + \int_{x_0+\varepsilon}^{+\infty} \frac{f(x)}{x - x_0}\, dx\right] \equiv P \int_{-\infty}^{+\infty} \frac{f(x)}{x - x_0}\, dx \tag{23.3}$$

The integral over Γ_0 in Eq. 23.2 is similar to an integral that has already been evaluated (see Example 4, Sec. 22) and can be evaluated analogously

$$\lim_{\varepsilon \to 0} \int_{\Gamma_0} \frac{f(z)}{z - x_0}\, dz = -i\pi f(x_0) \tag{23.4}$$

Hence

$$\int_{\rightarrow} \frac{f(z)}{z - x_0}\, dz = P \int_{-\infty}^{+\infty} \frac{f(x)}{x - x_0}\, dx - i\pi f(x_0) \tag{23.5}$$

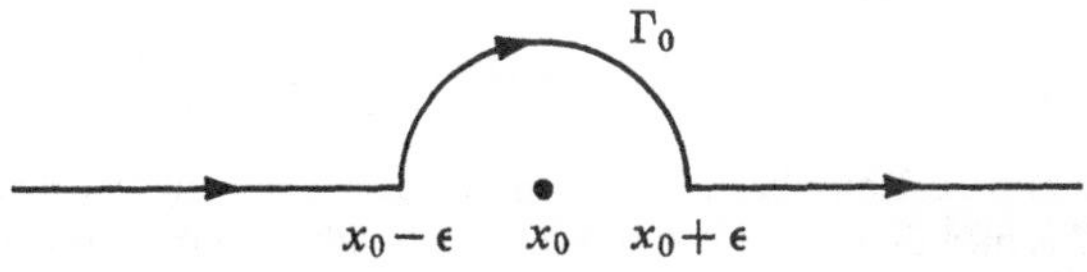

Fig. 21.

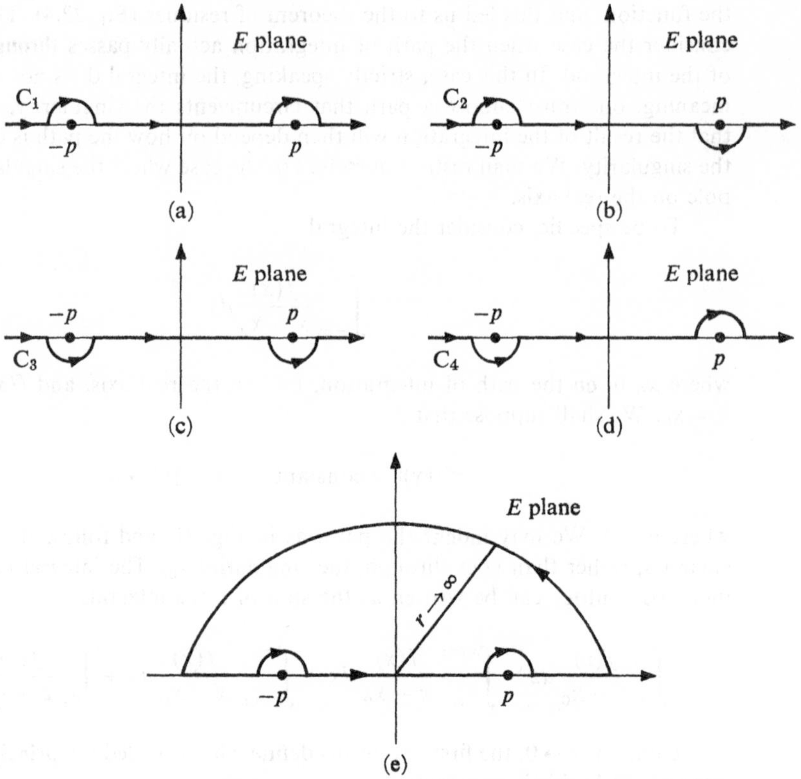

Fig. 22. Parts (a), (b), (c), and (d) show the contours defining the functions $\Delta_1(t)$, $\Delta_2(t)$, $\Delta_3(t)$, and $\Delta_4(t)$. Part (e) shows the contour used to calculate $\Delta_1(t)$ for $t < 0$.

On the other hand, if Γ_0 had been taken **below** the singularity instead of **above** it, Eq. 23.5 would have been replaced by

$$\int_{\rightarrow} \frac{f(z)}{z - x_0}\, dz = P \int_{-\infty}^{+\infty} \frac{f(x)}{x - x_0}\, dx + i\pi f(x_0) \tag{23.6}$$

Equations 23.5 and 23.6 display the very important point that the value of integrals such as those considered in this section depends upon the path chosen to circumvent the singularity. In applications, the physical situation will always determine the path that must be chosen. The choice will usually depend on what "boundary conditions" are required to solve the problem under consideration.

EXAMPLE.

Take the integral

$$\Delta(t) = \int_{-\infty}^{+\infty} \frac{e^{-iEt}}{p^2 - E^2}\, dE \tag{23.7}$$

where p is a real number. The integrand has poles on the real axis at $E = \pm p$, and hence Eq. 23.7 is meaningless as it stands and only takes on meaning once the way of circumventing the singularities has been specified. Corresponding to the four paths C_1, C_2, C_3, C_4 of Figs. 22(a), (b), (c), and (d), one obtains four **different** functions $\Delta_1(t)$, $\Delta_2(t)$, $\Delta_3(t)$, and $\Delta_4(t)$, which are, however, not completely independent. By "substracting" the contours C_3 and C_1 and also C_4 and C_2, it is easy to see that

$$\frac{1}{2\pi i}[\Delta_3(t) - \Delta_1(t)] = \text{Res}\left\{\frac{e^{-iEt}}{p^2 - E^2}\right\}_{E=-p} + \text{Res}\left\{\frac{e^{-iEt}}{p^2 - E^2}\right\}_{E=p}$$

$$= \frac{i}{p}\sin pt$$

$$\frac{1}{2\pi i}[\Delta_4(t) - \Delta_2(t)] = \text{Res}\left\{\frac{e^{-iEt}}{p^2 - E^2}\right\}_{E=-p} - \text{Res}\left\{\frac{e^{-iEt}}{p^2 - E^2}\right\}_{E=p}$$

$$= \frac{1}{p}\cos pt \tag{23.8}$$

The explicit forms of the functions $\Delta_j(t)$ can be found, using the methods of the calculus of residues and Jordan's lemma.

Consider the function $\Delta_1(t)$. For $t < 0$, one can close the contour of integration in the upper half-plane by a very large circle as shown in Fig. 22(e); the integration along this circle gives a vanishing contribution in the limit of an infinite radius. Since the singularities of the integrand are not enclosed by this contour, we have

$$\Delta_1(t) = 0 \qquad t < 0 \tag{23.9}$$

For $t > 0$, we similarly close the contour but now in the lower half-plane. Therefore

$$\frac{1}{2\pi i}\Delta_1(t) = -\text{Res}\left\{\frac{e^{-iEt}}{p^2 - E^2}\right\}_{E=-p} - \text{Res}\left\{\frac{e^{-iEt}}{p^2 - E^2}\right\}_{E=p}$$

$$= \frac{-i}{p}\sin pt \qquad t > 0 \tag{23.10}$$

We leave to the reader the verification that

$$\frac{1}{2\pi i}\Delta_2(t) = \begin{cases} -\dfrac{e^{-ipt}}{2p} & t < 0 \\[2ex] -\dfrac{e^{ipt}}{2p} & t > 0 \end{cases}$$

$$\frac{1}{2\pi i}\Delta_3(t) = \begin{cases} \dfrac{i}{p}\sin pt & t < 0 \\[2ex] 0 & t > 0 \end{cases} \tag{23.11}$$

$$\frac{1}{2\pi i}\Delta_4(t) = \begin{cases} +\dfrac{e^{ipt}}{2p} & t < 0 \\[2ex] +\dfrac{e^{-ipt}}{2p} & t > 0 \end{cases}$$

There is another equivalent way of getting around the difficulty of having a pole on the path of integration. Instead of lowering the contour, say, we can also give the singularity an infinitesimally small, positive, imaginary part; similarly, instead of raising the contour, we can give the singularity an infinitesimally small, negative, imaginary part; i.e., we "raise" or "lower" the singularity.

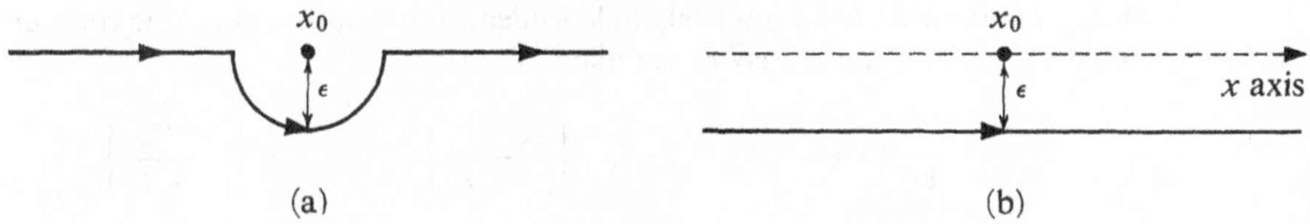

Fig. 23.

To see this, suppose for simplicity that $f(z)$ is analytic in an environment of the real axis. Let us integrate the function

$$\frac{f(z)}{z - x_0} \tag{23.12}$$

where x_0 is real, along a path that goes along the real axis, and circumvents x_0 from below, as in Fig. 23(a).

According to the preceding results, we have

$$\int_{\rightsquigarrow} \frac{f(z)}{z - x_0}\, dz = P \int_{-\infty}^{+\infty} \frac{f(x)}{x - x_0}\, dx + i\pi f(x_0) \tag{23.13}$$

On the other hand, the contour of Fig. 23(a) is completely equivalent to the contour of Fig. 23(b), which is obtained by "stretching out" the former contour below the singularity. Hence, one may write

$$\int_{\rightsquigarrow} \frac{f(z)}{z - x_0}\, dz = \lim_{\varepsilon \to +0} \int_{-\infty - i\varepsilon}^{+\infty - i\varepsilon} \frac{f(z)}{z - x_0}\, dz \tag{23.14}$$

We now perform the change of integration variable

$$z \to z - i\varepsilon$$

in the integral on the RHS of Eq. 23.14, and obtain

$$\begin{aligned} \int_{\rightsquigarrow} \frac{f(z)}{z - x_0}\, dz &= \lim_{\varepsilon \to +0} \int_{-\infty}^{+\infty} \frac{f(z - i\varepsilon)}{z - x_0 - i\varepsilon}\, dz \\ &= \lim_{\varepsilon \to +0} \int_{-\infty}^{+\infty} \frac{f(x)}{x - x_0 - i\varepsilon}\, dx \end{aligned} \tag{23.15}$$

The foregoing relation expresses the equivalence between the prescription of lowering the contour around a singularity and that of raising the singularity, which now appears at $z = x_0 + i\varepsilon$. Comparing Eqs. 23.15 and 23.13, we find

$$P \int_{-\infty}^{+\infty} \frac{f(x)}{x - x_0}\, dx + i\pi f(x_0) = \lim_{\varepsilon \to +0} \int_{-\infty}^{+\infty} \frac{f(x)}{x - x_0 - i\varepsilon}\, dx \tag{23.16}$$

Similarly, one can show that raising the contour around a singularity is equivalent to giving the singularity an infinitesimally small, negative imaginary part.*

$$P \int_{-\infty}^{+\infty} \frac{f(x)}{x - x_0}\, dx - i\pi f(x_0) = \lim_{\varepsilon \to +0} \int_{-\infty}^{+\infty} \frac{f(x)}{x - x_0 + i\varepsilon}\, dx \tag{23.17}$$

---

* The relations 23.16 and 23.17, due to Plemelj, actually hold under much less restrictive conditions on $f(x)$ than those assumed here, and also for contour integrals (see N. I. Muskhelishvili, *Singular Integral Equations*, P. Noordhoff N. V., 1953).

24 · MULTIVALUED FUNCTIONS; RIEMANN SURFACES

24.1 Preliminaries

The entire theory of analytic functions developed so far hinged upon the assumption that the functions considered, and their derivatives, were single-valued. At first, it would seem that multivalued functions should occupy a separate chapter in the theory of functions of a complex variable and lead to entirely different theorems. Fortunately, as will be shown in this section, one can extend the theory of analytic functions to include a wide class of multivalued functions, by making use of an ingenious geometrical construction due to Riemann and known as a Riemann surface.

We show by means of a simple example the kind of contradiction one can get into if one applies some of the preceding results without care. At the same time, we show that multivalued functions may have a physical significance.

Let us return to the case of the two coaxial cylinders, which was discussed in Sec. 10. There we showed that one could define a complex potential

$$F(z) = U + iV$$

which was an analytic function of z.* Since $U(x,y)$ and $V(x,y)$ satisfy the Cauchy-Riemann equations, $\vec{\nabla} U$ and $\vec{\nabla} V$ are orthogonal and equal in magnitude. (See the end of Sec. 4.) It follows that

$$\vec{\nabla} U \cdot \vec{n} = \vec{\nabla} V \cdot \vec{t} \tag{24.1}$$

where n and t are mutually orthogonal unit vectors (Fig. 24).

Consider now Gauss' theorem in two dimensions

$$\varphi = \oint \vec{E} \cdot \vec{n}\, dl = 4\pi\sigma \tag{24.2}$$

where $\vec{n}$ is a unit vector along the radius of the concentric circles, σ is the charge per unit length on the cylinder, and $dl = r\, d\theta$ is an element of length of the circumference of the circle with radius r

$$R_1 < r < R_2$$

We have shown in Sec. 10 that

$$U(r,\theta) \equiv \operatorname{Re} F(z) = A \ln r + B \tag{24.3}$$

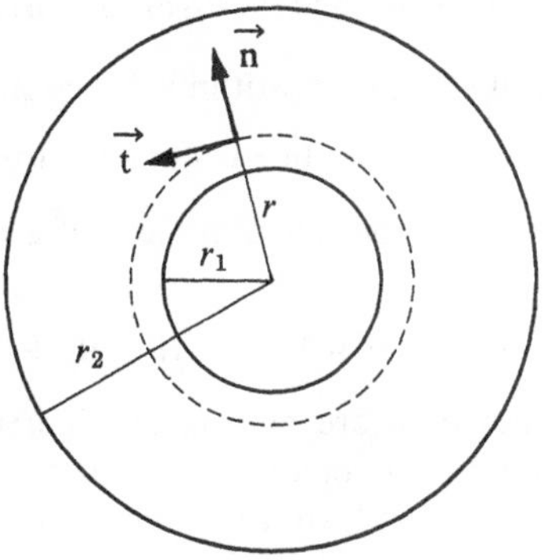

Fig. 24

* Here we use the variables z instead of z'.

With the help of this expression, we can relate the constants A and σ. From (24.2) we obtain

$$\varphi = -\int_0^{2\pi} \frac{A}{r} \cdot r\, d\theta = 4\pi\sigma \tag{24.4}$$

whence

$$A = -2\sigma \tag{24.5}$$

Now, using (24.1) and (24.2), we have

$$\begin{aligned} \varphi &= \oint \vec{E} \cdot \vec{n}\, dl = -\oint \vec{\nabla} U \cdot \vec{n}\, dl \\ &= -\oint \vec{\nabla} V \cdot \vec{t}\, dl \end{aligned} \tag{24.6}$$

In polar coordinates

$$\vec{\nabla} V = \frac{\partial V}{\partial r} \vec{n} + \frac{1}{r} \frac{\partial V}{\partial \theta} \vec{t} \tag{24.7}$$

Hence

$$\varphi = -\int_0^{2\pi} \frac{1}{r} \frac{\partial V}{\partial \theta} r\, d\theta = -V(r,\theta)\Big|_0^{2\pi} \tag{24.8}$$

Now, if $V(r,\theta)$ were a single-valued function, the RHS of Eq. 24.8 would be zero, since $\theta = 2\pi$ and $\theta = 0$ correspond to the same point in the complex plane; this would contradict Gauss' theorem (Eq. 24.2). The contradiction is removed if $V(r,\theta)$ is a multivalued function, i.e., if

$$V(r,2\pi) \neq V(r,0)$$

In fact, we showed in Sec. 10 that $V(r,\theta)$ is indeed a multivalued function, since we had

$$V(r,\theta) \equiv \operatorname{Im} F(z) = A\theta \tag{24.9}$$

Thus, from Eq. 24.5 we have

$$V(r,\theta) = -2\sigma\theta$$

so that, using Eq. 24.8

$$\varphi = 2\sigma\theta\Big|_0^{2\pi} = 4\pi\sigma$$

in agreement with Gauss' theorem.

24.2 The Logarithmic Function and Its Riemann Surface

In Sec. 8 we introduced the logarithmic function

$$\ln z = \ln|z| + i \arg z \tag{24.10}$$

We have already mentioned that the argument of a complex number is defined only up to a multiple of 2π

$$\arg z = \theta + 2\pi n \qquad (n = 0, \pm 1, \pm 2, \cdots)$$

To different values of n are associated different values of the function $\ln z$, although different values of n correspond to supposedly equivalent angles $\theta, \theta \pm 2\pi$, $\theta \pm 4\pi, \cdots$. Expressed differently, to the same point in the complex plane correspond different values of the function $\ln z$. These peculiar properties of the logarithmic function can also be visualized by considering a closed path C encircling the point $z = 0$ (Fig. 25). Suppose that we start at a point $z = z_0$ lying on C and that we follow

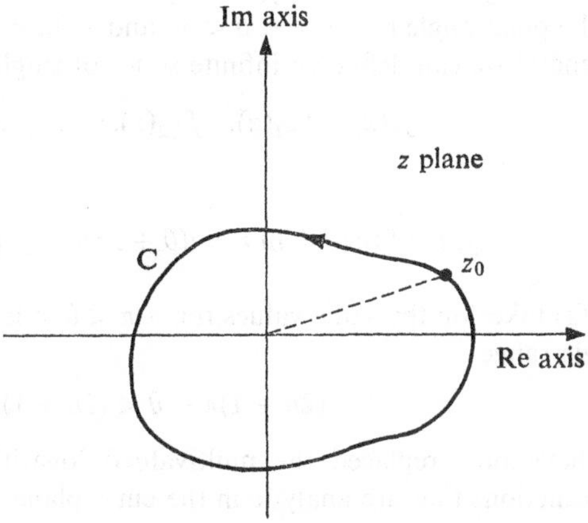

Fig. 25. After encircling the origin by following the closed path C, the argument of z_0 increases by 2π.

the path in the counterclockwise direction, say, until we come back to the starting point $z = z_0$. The function $\ln z$ changes continuously as we follow C, but after the completion of a full cycle, $\ln z_0$ will have increased by $2\pi i$, since $\arg z$ will have increased by 2π

$$(\ln z_0)_{\text{final}} = (\ln z_0)_{\text{initial}} + 2\pi i$$

A point of the complex plane having the property that, after the completion of any cycle around it, a given function is not restored to its initial value (or, more precisely, to the value we have assigned to it initially) is called a **branch point** of the function.

Thus, the point $z = 0$ is a branch point of $f(z) = \ln z$. By considering the behavior of the function $f(1/z') = \ln(1/z')$ at $z' = 0$, we find similarly that "the point at infinity" is also a branch point of $\ln z$. We leave to the reader the verification that $\ln z$ has no other branch points.

Let us draw a curve joining the two branch points of $\ln z$; it can be taken as, for instance, a line starting from the origin and extending out to infinity in an arbitrary direction. Suppose now that we remove from the domain of definition of the function all the points that lie on this curve. Using a more descriptive language, we say that we **cut** the complex plane along the curve, which hereafter will be called a **branch cut** or, more briefly, a **cut**.

For definiteness, we shall assume that the z plane is cut along the negative half of the real axis. One can then define the single-valued function

$$f_0(z) = f_0(r,\theta) = \ln r + i\theta \qquad \begin{pmatrix} -\pi < \theta < \pi \\ 0 < r \end{pmatrix} \tag{24.11}$$

One can also define the single-valued functions

$$f_1(z) = f_1(r,\theta) = \ln r + i(\theta + 2\pi) \qquad \begin{pmatrix} -\pi < \theta < \pi \\ 0 < r \end{pmatrix} \tag{24.12}$$

and

$$f_{-1}(z) = f_{-1}(r,\theta) = \ln r + i(\theta - 2\pi) \qquad \begin{pmatrix} -\pi < \theta < \pi \\ 0 < r \end{pmatrix} \tag{24.13}$$

In their range of definition, $f_1(z)$ and $f_{-1}(z)$ take on the same values the logarithm takes in the polar angle range: $\pi < \theta < 3\pi$ and $-3\pi < \theta < -\pi$, respectively.

In general we can define an infinite series of single-valued functions

$$f_0(z), \quad f_{\pm 1}(z), \quad f_{\pm 2}(z), \cdots, \quad f_{\pm |n|}(z), \cdots$$

where

$$f_n(z) = f_n(r,\theta) = \ln r + i(\theta + 2\pi n) \qquad \begin{pmatrix} -\pi < \theta < \pi \\ 0 < r \end{pmatrix} \tag{24.14}$$

so that $f_n(z)$ takes on the same values for $-\pi < \theta < \pi$ that the logarithm takes in the polar angle range

$$(2n - 1)\pi < \theta < (2n + 1)\pi$$

We have now replaced the multivalued logarithmic function by a series of **different** functions that are analytic in the cut z plane.

To show that $f_n(z)$ is analytic in the cut plane, consider the function defined by the integral

$$g(z) = \int_a^z \frac{1}{z'}\, dz' \qquad a = r_a e^{i\theta_a} \tag{24.15}$$

This integral can be taken as defining the logarithmic function. We choose the cut along the negative real axis and assume that the path of integration joining the points a and z is arbitrary except that it does not intersect the cut. In that case, the integral is independent of the path because two paths that do not cross the cut cannot enclose the only singularity of the integrand located at the origin. But we have already shown (see the proof of Morera's theorem) that the integral $\int_a^z f(z')\, dz'$ is an analytic function of z in regions where it does not depend on the path connecting a and z. Hence, $g(z)$ is analytic in the cut z plane. Choosing the path of integration as shown in Fig. 26, which runs first along the segment ab and then along the arc bz, we easily find that

$$g(z) = \ln r + i\theta - (\ln r_a + i\theta_a)$$

Thus, $g(z)$ differs from $f_n(z)$ by a constant only, which proves the analyticity of $f_n(z)$ in the cut complex plane.

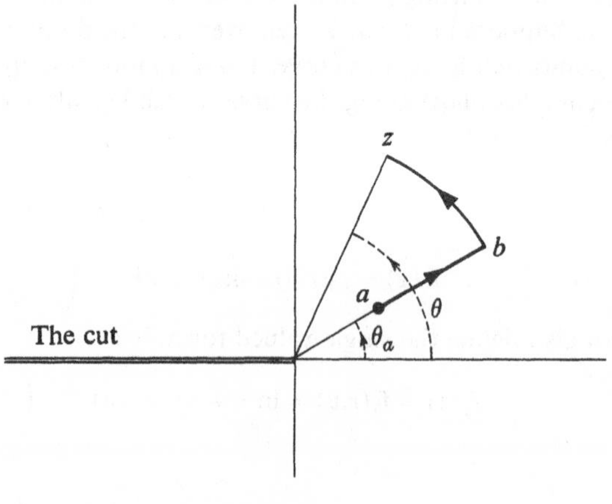

Fig. 26.

The contradiction brought about by the existence of branch points has been removed, although somewhat artificially, for any of the functions $f_n(z)$, since it is now no longer possible to encircle a branch point without crossing the cut; i.e., without leaving the domain of analyticity of that function.

Let us now observe that each function $f_n(z)$ suffers a discontinuity across the cut; for example, the value of the function $f_n(z)$ just above the cut is very different from its value just below it. Above the cut we have

$$f_n(r,\pi - \varepsilon) = \ln r + i[(2n + 1)\pi - \varepsilon] \qquad \varepsilon > 0$$

while below the cut

$$f_n(r, -\pi + \varepsilon) = \ln r + i[(2n - 1)\pi + \varepsilon] \qquad \varepsilon > 0$$

The discontinuity of $f_n(z)$ across the cut is therefore given by

$$\lim_{\varepsilon \to +0} \left[f_n(r,\pi - \varepsilon) - f_n(r, -\pi + \varepsilon) \right] = 2\pi i \tag{24.16}$$

On the other hand, the value of the function $f_n(z)$ just above the cut is the same as the value of the function $f_{n+1}(z)$ just below it.

$$\lim_{\varepsilon \to +0} f_n(\varepsilon,\pi - \varepsilon) = \lim_{\varepsilon \to +0} f_{n+1}(r, -\pi + \varepsilon) \tag{24.17}$$

This suggests the following geometrical construction: We superpose an infinite series of cut complex planes one on top of the other, each plane corresponding to a different value of n $(=0, \pm 1, \pm 2, \cdots)$. The adjacent planes are connected along the cut; the upper lip of the cut in the nth plane is connected to the lower lip of the cut in the $(n + 1)$st plane; the branch points are common to all the planes. Hence, a crossing of the cut is equivalent to going to one of the two adjacent complex planes (Fig. 27).

The geometrical surface obtained from this helix-like superposition of planes is called a **Riemann surface**, and each plane is called a **Riemann sheet** of the function.

A single-valued function defined on a Riemann sheet is called a **branch** of the complete multivalued function.

What we have achieved by this construction is the following: From a sequence of single-valued functions defined in a **single** complex plane, we have obtained one continuous (see Eq. 24.17) single-valued function defined on a Riemann surface. In fact, throughout the Riemann surface we have just constructed, the logarithmic function is analytic except at the branch points, which play the role of singular points.

As for the notion of a branch point, it now gets a simple geometrical interpretation. Performing a complete cycle around a branch point, we move to another Riemann sheet where the function takes on different values. On the other hand, a complete cycle that does not include a branch point brings us to the starting point on the Riemann surface and hence restores the function to its initial value.

One can give a classification of branch points of a function by introducing the notion of the order of a branch point. We say that a branch point is of the nth order if after making $(n + 1)$ complete cycles around it (but not less), we restore the function to its initial value. Otherwise, a branch point is said to be of an infinite order, as is the case for the two branch points of the logarithmic function; by performing successive rotations around the origin, we move farther and farther away from the initial Riemann sheet.

One considers branch points as singular points of an analytic function. The nature of the singularity is different from that of a pole or of an essential singularity.

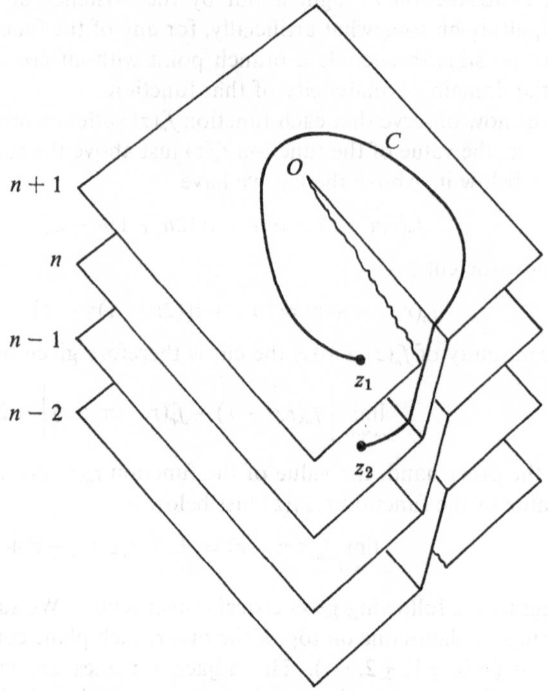

Fig. 27. A part of the Riemann surface of the logarithmic function. z_1 and z_2 are points located on two adjacent Riemann sheets. The path C, which joins z_1 and z_2, encircles the branch point at the origin.

Here, it is not only the differentiability of the function at the branch point which is relevant, but also (and especially) the fact that the function that has branch points, is multivalued. Without the concept of a Riemann surface, the proofs of the fundamental theorems on analytic functions fail because of the ambiguity inherent in multivalued functions. Extending the theory of functions that are analytic on a plane to the case of functions analytic on a Riemann surface, one finds that it is necessary to include in the definition of a singular point not only the points where the function is not differentiable, but also the branch points. On the other hand, the location and type of the branch points of a function completely determine the geometry of its Riemann surface.

24.3 The Functions $f(z) = z^{1/n}$ and Their Riemann Surfaces

Setting $z = re^{i\theta}$ in $f(z) = z^{1/2}$, we have

$$f(z) = f(r,\theta) = r^{1/2}e^{i\theta/2} \tag{24.18}$$

One easily finds that the points $z = 0$ and $z = \infty$ are branch points of $f(z) = z^{1/2}$. For example, a cycle around $z = 0$ changes θ by 2π, and from Eq. 24.18 we see that this results in changing the sign of the function at a given point

$$f(r,\theta) = -f(r,\theta + 2\pi)$$

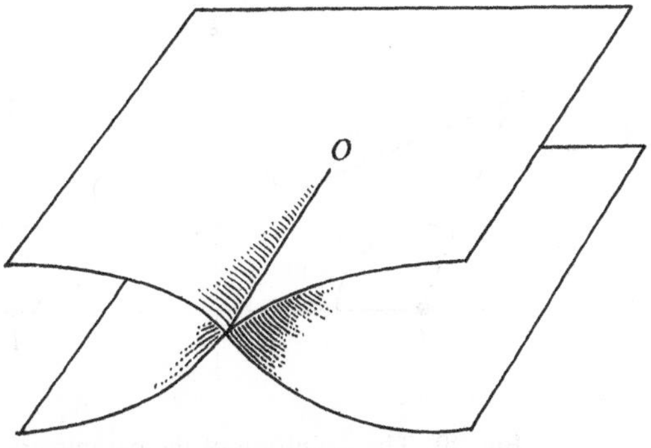

Fig. 28. The Riemann surface of the function $f(z) = z^{1/2}$.

The branch cut of $f(z) = z^{1/2}$ can be chosen to connect the branch points $z = 0$ and $z = \infty$ along the positive real axis.

We define the first Riemann sheet* by fixing the values of $f(z)$ on the upper lip of the cut

$$f(z) = f(r,0) = r^{1/2} \qquad \text{for Re } z > 0, \quad \text{Im } z = +0$$

Then on the lower lip of the cut we have

$$f(z) = f(r,2\pi) = -r^{1/2} \qquad \text{for Re } z > 0, \quad \text{Im } z = -0$$

Crossing the cut, we move to the next sheet, where the values of $f(z)$ on the upper lip of the cut are the same as the values of $f(z)$ on the lower lip on the first sheet, while the values of $f(z)$ on the lower lip are obtained by adding 2π to θ

$$f(z) = f(r,4\pi) = r^{1/2} \qquad \text{for Re } z > 0, \quad \text{Im } z = -0$$

on the second sheet. Hence, a second crossing of the cut does not yield a new value for $f(z)$, since

$$f(r,0) = f(r,4\pi) \tag{24.19}$$

Therefore, the Riemann surface of $f(z) = z^{1/2}$, constructed so as to make this function continuous everywhere, has two sheets that are connected along the cut; because of Eq. 24.19, the lower lip of the second sheet must be reconnected to the upper lip of the first sheet; i.e., if the cut is crossed from the second sheet, the function returns to the first sheet. In other words, the Riemann surface is closed. (Fig. 28.)

In general, the Riemann surface of the function $f(z) = z^{1/n}$ ($n = 2, 3, 4, \cdots$) is a closed, n-sheeted surface, the nth sheet being reconnected to the first sheet.

24.4 The Function $f(z) = (z^2 - 1)^{1/2}$ and Its Riemann Surface

Writing

$$z + 1 = r_- e^{i\theta_-}$$

$$z - 1 = r_+ e^{i\theta_+}$$

* Obviously, when there are no external conditions imposed, (for example, physical conditions), we are free to define the first sheet at will. The values of the function on the other sheets will then follow.

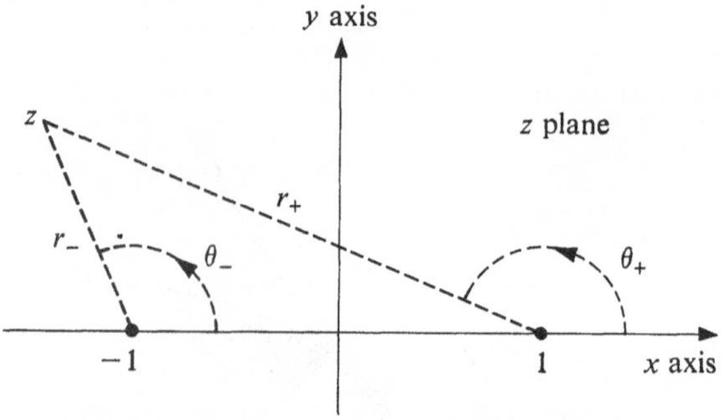

Fig. 29. The definition of the parameters $r_\pm$ and $\theta_\pm$.

where the parameters $r_\pm$, $\theta_\pm$ are defined in Fig. 29, we have

$$f(z) = (r_+ r_-)^{1/2} \exp\left[i\frac{\theta_+ + \theta_-}{2}\right] \tag{24.20}$$

It is easy to see that $f(z)$ has two branch points at $z = \pm 1$. For example, after performing a cycle around the point $z = 1$, which does not include the point $z = -1$, θ_+ changes by 2π while θ_- alternately increases and decreases (passing through negative values) and finally returns to its initial value. Therefore, after a full cycle, $f(z)$ changes its sign, and thus $z = 1$ is a branch point of $f(z)$. A similar argument holds for $z = -1$.

It is worthwhile to mention that a cycle which surrounds both branch points at $z = \pm 1$, will change both angles θ_+ and θ_- by 2π, and therefore, after such a cycle, $f(z)$ will return to its initial value. In a sense, the effects due to two branch points cancel each other.

On the other hand, the point at infinity is not a branch point of $f(z)$, for letting $z = 1/z'$, one has

$$f\left(\frac{1}{z'}\right) = \left(\frac{1}{z'^2} - 1\right)^{1/2} \sim \frac{1}{z'} \qquad \text{as } z' \to 0$$

Thus, $z' = 0$ is a simple pole of $f(1/z')$ and therefore $z = \infty$ is not a branch point of $f(z)$.

We take the cut along the segment of the real axis $-1 \le x \le 1$, as in Fig. 30(a). Thus, in order to remain on the same Riemann sheet, θ_- should vary, for example, from $-\pi$ to $+\pi$ and θ_+ from $-\pi$ to $+\pi$. This choice for the range of variation of the angles $\theta_\pm$ defines the first Riemann sheet:

(i) $f(z) = i\sqrt{1 - x^2}$ for $-1 \le x \le 1$ $y = +0$

since these values of x and y correspond to $\theta_- = 0$ and $\theta_+ = \pi$.

(ii) $f(z) = -i\sqrt{1 - x^2}$ for $-1 \le x \le 1$ $y = -0$

since $\theta_- = 0$ and $\theta_+ = -\pi$.

(iii) $f(z) = -\sqrt{x^2 - 1}$ for $x < -1$ $y = \pm 0$

since $\theta_- = \theta_+ = \pm\pi$.

(iv) $f(z) = +\sqrt{x^2 - 1}$ for $x > 1$ $y = \pm 0$

since $\theta_- = \theta_+ = 0$.

Comparing i and ii, it is seen that $f(z)$ is discontinuous across the cut, as it should be. On the other hand, it is not difficult to see that a second crossing of the cut restores $f(z)$ to its initial value. Hence, the Riemann surface is a closed, two-sheeted surface constructed in such a manner that $f(z)$ changes sheets after any cycle that surrounds **one** of its branch points only, whereas it restores $f(z)$ to its initial value after the completion of any cycle that surrounds **both** branch points.

The cut in the complex plane could equally well be chosen as in Fig. 30(b). In a sense, this corresponds to joining the points $z = \pm 1$ by a path going through the point at infinity. In that case, a sheet is defined by letting θ_+ vary from 0 to 2π and θ_- from $-\pi$ to $+\pi$. We leave it to the reader to find the values that $f(z)$ takes near the cuts of this Riemann sheet, as we have done for the other choice of the cut.

24.5 Concluding Remarks

With the help of a Riemann surface we obtain a unique description of multivalued functions. The trick consists in replacing a multivalued function defined on a simple set of arguments (the usual complex plane) by a function, that is single-valued but defined over a set of arguments that has a complicated geometrical structure (the Riemann surface).

In practice, it is frequently sufficient to focus one's attention on a particular sheet of the Riemann surface, i.e., on a particular branch of the function. This amounts to treating that part of the Riemann surface as if it were a cut complex plane. The values that the function assumes on this cut plane are then those values that correspond to the particular sheet considered (i.e., to the particular branch of the function).

From a general point of view there is no reason to prefer one sheet to another, and in a sense we are allowed to cut the complex plane in any number of ways; this corresponds only to choosing different Riemann sheets. In physical applications,

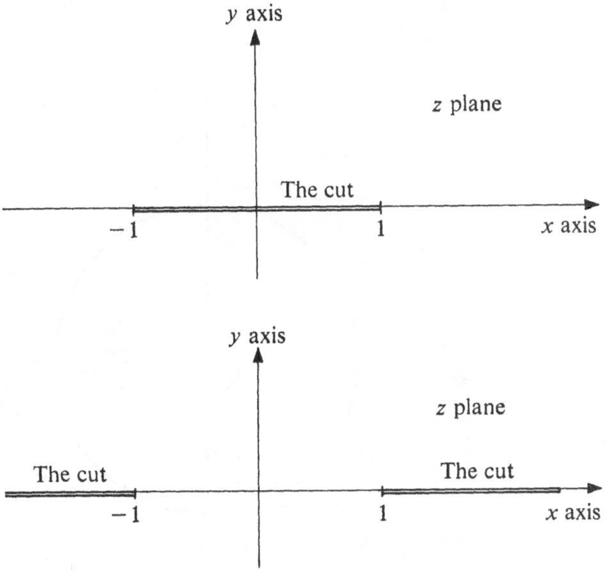

Fig. 30. Two possible ways of choosing cuts in the z plane for the function $f(z) = (z^2 - 1)^{1/2}$.

however, a preference is often given to certain sheets, but for reasons that belong solely to physics and not to mathematics. In purely mathematical applications (for example, in evaluating contour integrals of multivalued functions), we make a particular (arbitrary) choice of cuts in the z plane and we determine in a self-consistent manner the behavior of the function on the lips of the cuts. We must always bear in mind, however, that a simple contour surrounding a branch point is in general **not** a closed contour. The following example will illustrate these points.

25 · EXAMPLE OF THE EVALUATION OF AN INTEGRAL INVOLVING A MULTIVALUED FUNCTION

We shall evaluate the integral

$$I = \int_0^\infty \frac{x^{p-1}}{x^2+1}\,dx \qquad 0 < p < 2 \tag{25.1}$$

When p is not an integer ($p \neq 1$), the integrand has a branch point at $x = 0$. We take the cut along the positive real axis. Let us consider the contour integral

$$I' = \int_C \frac{z^{p-1}}{z^2+1}\,dz \tag{25.2}$$

where C is the path shown in Fig. 31. It consists of a small circle γ of radius ε surrounding the branch point, of a large circle Γ of radius r, and of two straight lines L_1 and L_2 lying respectively just above and just below the branch cut. Thus, C does not cross the cut, and the integrand in Eq. 25.2 is single-valued within this contour. There remains to choose a particular branch of the function. We choose the branch of the function z^{p-1} as

$$z^{p-1} = |z|^{p-1}e^{i(p-1)\theta} \qquad 0 \le \theta < 2\pi \tag{25.3}$$

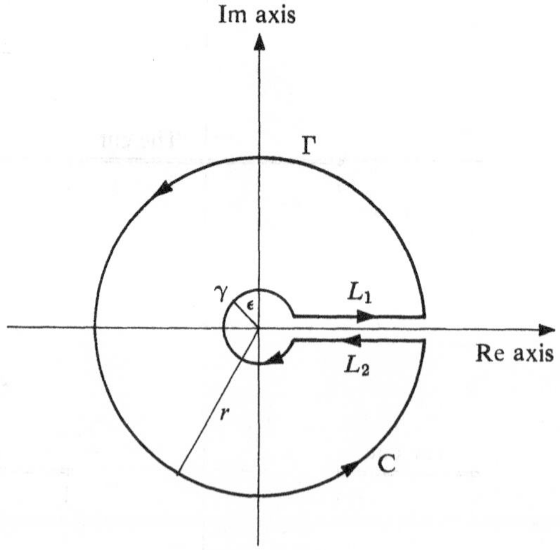

Fig. 31.

Everything is now properly specified, and we can proceed with the evaluation of I'. We first show that the circles γ and Γ do not contribute to I' as $\varepsilon \to 0$ and $R \to \infty$. Let the parametric equation of either one of these circles be

$$z = \rho e^{i\theta}$$

Then for either circle

$$\int_{\gamma \text{ or } \Gamma} \frac{z^{p-1}}{z^2+1}\,dz = \int_0^{2\pi} \frac{\rho^{p-1} e^{i(p-1)\theta}(i\rho e^{i\theta})}{\rho^2 e^{i2\theta}+1}\,d\theta \tag{25.4}$$

and thus, when either $\rho \to 0$ (circle γ) or when $\rho \to \infty$ (circle Γ) (25.4) tends to zero. There remains the evaluation of the contribution to I' from the integrations along L_1 and L_2. Because a branch cut separates L_1 from L_2, the values of the integrand along these lines will not be the same. Along L_1, $z = x$, and therefore

$$z^{p-1} = x^{p-1} \qquad \text{along } \text{L}_1 \tag{25.5}$$

Along L_2, $z = xe^{i2\pi}$, and therefore

$$z^{p-1} = x^{p-1} e^{i2\pi(p-1)} \qquad \text{along } \text{L}_2 \tag{25.6}$$

Thus, we have (as $r \to 0$, $R \to \infty$)

$$\begin{aligned}
\int_{\text{C}} \frac{z^{\,p-1}}{z^2+1}\,dz &= \int_0^\infty [x^{p-1} - x^{p-1}e^{i2\pi(p-1)}]\frac{1}{x^2+1}\,dx \\
&= e^{i\pi(p-1)}[e^{-i\pi(p-1)} - e^{i\pi(p-1)}]\int_0^\infty \frac{x^{p-1}}{x^2+1}\,dx \\
&= e^{i\pi(p-1)}[-2i\sin\pi(p-1)]\int_0^\infty \frac{x^{p-1}}{x^2+1}\,dx \\
&= 2ie^{i\pi(p-1)}\sin\pi p \int_0^\infty \frac{x^{p-1}}{x^2+1}\,dx
\end{aligned} \tag{25.7}$$

On the other hand, the contour integral in Eq. 25.7 is equal to $2\pi i$ times the sum of the residues of the integrand at the simple poles $z = \pm i$. Since

$$i^{p-1} = e^{i\pi(p-1)/2}$$

$$(-i)^{p-1} = e^{i3\pi(p-1)/2}$$

we have

$$\int_{\text{C}} \frac{z^{p-1}}{z^2+1}\,dz = 2\pi i\left[\frac{e^{i\pi(p-1)/2}}{2i} - \frac{e^{i3\pi(p-1)/2}}{2i}\right] \tag{25.8}$$

By combining Eqs. 25.7 and 25.8 we obtain the desired result

$$\int_0^\infty \frac{x^{p-1}}{x^2+1}\,dx = \frac{\pi}{2}\csc\left(\frac{\pi p}{2}\right) \tag{25.9}$$

In evaluating I', we could also have chosen a different branch of the integrand, i.e., we could have adopted a convention different from that of 25.3. This would have led to a different value of the contour integral, since the contour would then have been located on a different sheet of the Riemann surface. But the relation between the contour integral I' and the definite integral I would also have changed, and of course the net result (Eq. 25.9) would not be modified.

26 · ANALYTIC CONTINUATION

We have seen by many examples that analytic functions possess rather unique properties. For example, Cauchy's integral formula shows that if a function is analytic in a certain region of the complex plane and on a curve delimiting that region, then the values of the function within the region are completely determined, once the values of the function on the boundary curve have been prescribed.

If we pursue this idea further it becomes natural to ask the following general question: Suppose that a function $f(z)$ is analytic within a certain region D. Which subsets of D have the property that, specifying the values of $f(z)$ over these subsets only, $f(z)$ is determined throughout the whole of D? In this connection a very important theorem can easily be proved.

Theorem. Let $f_1(z)$ and $f_2(z)$ be two functions of the complex variable z that are analytic within a region D. If the two functions coincide in the neighborhood of a point $z \in \mathrm{D}$, or on a segment of a curve lying in D, or (more generally) on a point set with an accumulation point belonging to D, then they coincide throughout D.

Proof. The validity of the theorem follows at once from the fact (see Sec. 21.1) that in a region where it is analytic, a function either has isolated zeros only (i.e., the zeros do not have an accumulation point) or it is identically equal to zero. Now, since $f_1(z)$ and $f_2(z)$ are assumed to be identical on a point set, this set is a set of zeros for the function $f_1(z) - f_2(z)$. Furthermore, since by hypothesis this set of zeros contains an accumulation point, and since $f_1(z) - f_2(z)$ is analytic throughout D, $f_1(z) - f_2(z)$ must be identically equal to zero in D. Hence,

$$f_1(z) - f_2(z) \equiv 0 \qquad z \in \mathrm{D}$$

i.e.,

$$f_1(z) \equiv f_2(z) \qquad z \in \mathrm{D}$$

The theorem that has just been proved states that two different analytic functions cannot coincide in the neighborhood of an arbitrary point.* Therefore, the behavior of a function in a region where it is analytic is uniquely determined by its behavior in the neighborhood of an arbitrary point of the region.

In a sense, the preceding theorem could be called a "uniqueness" theorem.

These results can also be obtained by using a different method. Consider a function $f(z)$ that is analytic in some region D. We can expand $f(z)$ in a Taylor series about an arbitrary point $z_0 \in \mathrm{D}$

$$f(z) = \sum_{n=0}^{\infty} a_n(z - z_0)^n \tag{26.1}$$

This series will be convergent within some circle γ_0. Let this circle be defined by

$$|z - z_0| = r_0 \tag{26.2}$$

Suppose that we know the behavior of $f(z)$ in the neighborhood of the point $z = z_0$ or, more precisely, that we know the coefficients a_n $(n = 0, 1, 2, \cdots)$ of the Taylor

* More generally, the functions cannot coincide throughout a point set that has this arbitrary point as an accumulation point.

expansion (26.1). We show that this knowledge is sufficient to determine the behavior of $f(z)$ in the neighborhood of an arbitrary point $z'_0 \in D$, which may be far removed from z_0.

By the definition of a region in the complex plane, it is possible to connect z_0 and z'_0 by a continuous path C lying entirely within D. Let us take a point $z_1 \in C$ such that

$$|z_1 - z_0| < r_0 \tag{26.3}$$

i.e., z_1 is within γ_0. Since the power series (26.1) converges uniformly for $|z - z_0| < r_0$, it can be differentiated term by term, and therefore one may find all derivatives of $f(z)$ from its Taylor expansion at all points within the circle γ_0 and in particular at the point z_1. Thus, we know the values of

$$f(z_1), \quad \left.\frac{df(z)}{dz}\right|_{z=z_1}, \cdots, \quad \left.\frac{d^nf(z)}{dz^n}\right|_{z=z_1}, \cdots \tag{26.4}$$

But these are, apart from a factor $1/n!$, just the values of the coefficients $a_n^{(1)}$ $(n = 0, 1, 2, \cdots)$ of the Taylor expansion of $f(z)$ about the point z_1 (Eq. 18.2). Hence, the coefficients in

$$f(z) = \sum_{n=0}^{\infty} a_n^{(1)}(z - z_1)^n \tag{26.5}$$

are known. Suppose that the sum in Eq. 26.5 converges within a circle γ_1 defined by

$$|z - z_1| = r_1 \tag{26.6}$$

Since $f(z)$ is analytic on $C \subset D$, r_1 is always nonzero, and it is always possible to choose the point z_1 (which can be anywhere within γ_0) in such a way that the circle γ_1 lies partly outside the circle γ_0 (Fig. 32). This is the crucial point, for from this fact it follows that now we are able to calculate $f(z)$ not only within the original region γ_0,

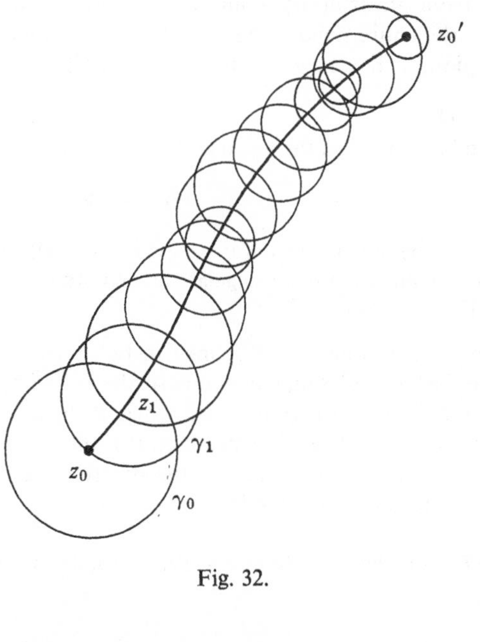

Fig. 32.

where it was assumed that $f(z)$ was known, but also within the larger region consisting of the union of the interiors of γ_0 and γ_1. This larger region will contain a segment of the curve C, which will be outside γ_0 but within γ_1. Thus, repeating over and over the argument, we cover the path C by overlapping circles $\gamma_0, \gamma_1, \gamma_2 \cdots$, which approach the point z'_0. After a certain number of steps, one of these circles will cover the point z'_0, and thus we shall be able to find the Taylor expansion of $f(z)$ about this point. Consequently, we can determine the behavior of $f(z)$ in the neighborhood of $z = z'_0$.

The process just described, which consists in determining the behavior of an analytic function outside the region where it was originally defined (in the present case, within the circle γ_0), is called **analytic continuation**.

It may appear strange to the reader that throughout this chapter we have insisted upon the fact that the results obtained should hold when the analytic function is defined **in a region**. Why not take a more arbitrary set of arguments? We cannot enter into a deep analysis of the foundations of the theory of analytic functions. However, the reasoning of the preceding discussion offers a good example to illustrate the role that the concept of a region plays in the theory. It is evident that if one knows the behavior of $f(z)$ at z_0, then one can deduce its behavior at z'_0 but only if one can connect z_0 and z'_0 by a path lying entirely within the domain of analyticity of $f(z)$. This means that each point of the path together with its neighborhood must belong to the domain of analyticity of $f(z)$, and this guarantees that each circle γ_j has a finite size. Hence, z_0 and z'_0 must belong to the same **region** of analyticity of $f(z)$.

It is important to realize that the technique of analytic continuation presented here determines also the location of the singular points of $f(z)$, since the radius of convergence of the Taylor expansion of $f(z)$ at a given point is equal to the distance from this point to the nearest singularity of the function. If we continue analytically a function along a path going through a singular point of the function, the radii of circles γ_j tend to zero as we approach the singularity. Hence, the singular point cannot be bypassed; the process of analytic continuation stops naturally there.

Suppose that the function we consider is single-valued. Then, continuing this function from an arbitrary point where it is analytic along all possible paths, we determine the entire region where the function is analytic. The "natural boundaries" of this region, if they exist, are the singular points of the function.

EXAMPLE

It can be shown that the function defined by the series

$$f(z) = \frac{1}{z^2} \sum_{k=0}^{\infty} z^{k!}$$

is analytic in the annular region $0 < |z| < 1$, while all the points lying on the unit circle $|z| = 1$ and the point $z = 0$ are singular* and constitute the boundary of the annular region $0 < |z| < 1$.

If an analytic function is defined in two regions that are separated by an "impenetrable barrier" of singular points so that it is impossible to perform an analytic continuation from one region to the other, then the values the function takes in the two regions are completely independent. It is therefore only a matter of convention to consider the two "components" of the function, corresponding to the two regions, as belonging to the same function.

* See E. C. Titchmarsh, *The Theory of Functions*, Sec. 4.7 (Oxford University Press, New York, 1939).

The analytic continuation along a closed path certainly restores the function to its initial value if the function is single-valued. This is not necessarily the case when the function is multivalued. In fact, a closed path that encircles a branch point leads to the next Riemann sheet and further analytic continuation yields values of the function proper to this Riemann sheet. Hence, by analytically continuing a multivalued function along all possible paths, we determine the geometry of its Riemann surface as well as the behavior of the function on the surface.

Suppose now that two functions $f_1(z)$ and $f_2(z)$, which have different functional forms, are analytic within regions D_1 and D_2, respectively, which overlap. If $f_1(z)$ and $f_2(z)$ are identical within the intersection $D_1 \cap D_2$ of the two regions, the result of the analytic continuation of $f_1(z)$ in D_2 (which is unique) must be identical with $f_2(z)$, and the result of the analytic continuation of $f_2(z)$ in D_1 must coincide with $f_1(z)$. Thus, in fact we may regard $f_1(z)$ and $f_2(z)$ as corresponding to a unique function $f(z)$.

$$f(z) = \begin{cases} f_1(z) & z \in D_1 \\ f_2(z) & z \in D_2 \end{cases}$$

which is analytic throughout the union $D_1 + D_2$ of the regions D_1 and D_2, and which is uniquely determined by $f_1(z)$, or $f_2(z)$, for $z \in D_1 \cap D_2$.

One says that $f_1(z)$ and $f_2(z)$ are **analytic continuations** of each other. Here the expression "analytic continuation" is used in a slightly different sense than before. It does not designate the action that tends to determine the behavior of an analytic function outside the subregion where it is known explicitly; rather, it signifies that this action applied to $f_1(z)$ yields $f_2(z)$, and vice versa. As an illustration we take

$$f_1(z) = 1 + z + z^2 + \cdots \tag{26.7}$$

which is defined for

$$|z| < 1 \tag{26.8}$$

where the series is convergent, and

$$f_2(z) = \tfrac{2}{3} + (\tfrac{2}{3})^2(z + \tfrac{1}{2}) + (\tfrac{2}{3})^3(z + \tfrac{1}{2})^2 + \cdots \tag{26.9}$$

which is defined for

$$|z + \tfrac{1}{2}| < 1 \tag{26.10}$$

The series can be summed as

$$f_1(z) = \frac{1}{1 - z} = 1 + z + z^2 + \cdots \qquad \text{for } |z| < 1$$

and

$$f_2(z) = \frac{1}{1 - z} \equiv \frac{1}{\frac{3}{2} - (z + \frac{1}{2})}$$

$$= \tfrac{2}{3} + (\tfrac{2}{3})^2(z + \tfrac{1}{2}) + \cdots \qquad \text{for } |z + \tfrac{1}{2}| < 1$$

We see that $f_1(z)$ and $f_2(z)$, which have the different functional forms (Eqs. 26.7 and 26.9) in the two overlapping regions defined by inequalities (26.8) and (26.10), respectively, represent in fact the same analytic function

$$f(z) = \frac{1}{1 - z} \tag{26.11}$$

In this example it was possible to find a unique expression (Eq. 26.11) for $f(z)$, which is valid in both circular regions defined by inequalities 26.8 and 26.10. It must be borne in mind, however, that it is in general impossible to find a unique mathematical expression which will hold within the entire region of analyticity of a function. This may appear paradoxical to the reader who learned from elementary calculus to consider the word "function" as synonymous with the idea of a unique and explicit mathematical expression. But in the theory of analytic functions, the functional forms that a function assumes in different regions of its domain of analyticity are merely considered as dissimilar "manifestations" of a really unique entity.

27 · THE SCHWARZ REFLECTION PRINCIPLE

We generalize one of the results of the preceding section by proving the following lemma.

Lemma. Consider two regions, D_f and D_g, that are nonoverlapping but that have in common a part R of their boundaries (see Fig. 33). Let $f(z)$ be analytic throughout D_f and continuous within $D_f + R$, and let $g(z)$ be analytic throughout D_g and continuous within $D_g + R$. If

$$f(z) = g(z) \qquad \text{for } z \in R$$

then $f(z)$ and $g(z)$ are analytic continuations of each other and together define a unique function

$$h(z) = \begin{cases} f(z) & z \in (D_f + R) \\ g(z) & z \in (D_g + R) \end{cases} \tag{27.1}$$

which is analytic throughout the entire region $D_f + D_g + R$.

The lemma generalizes the results previously obtained, since now we are supposing only that $f(z)$ and $g(z)$ are analytic within the **regions** (which are open sets) D_f and D_g but not necessarily **on** the boundary R.

Proof. Consider an arbitrary closed curve C entirely contained within $D_f + D_g + R$. When either $C \subset D_f$ or $C \subset D_g$, the integral

$$\int_C h(z')\,dz' \tag{27.2}$$

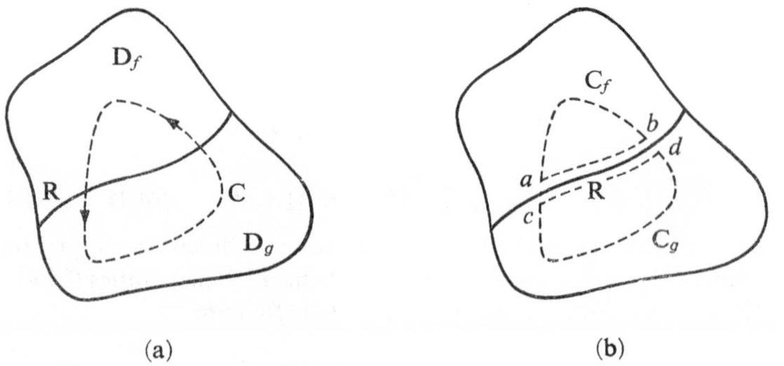

Fig. 33.

vanishes. To prove that $h(z)$ is analytic throughout $D_f + D_g + R$, we write (27.2) as the sum of two integrals and use Eq. 27.1, the definition of $h(z)$

$$\begin{aligned}\int_C h(z')\,dz' &= \int_{C_f} h(z')\,dz' + \int_{C_g} h(z')\,dz' \\ &= \int_{C_f} f(z')\,dz' + \int_{C_g} g(z')\,dz'\end{aligned} \tag{27.3}$$

In Eq. 27.3, $C_f \subset D_f$ and $C_g \subset D_g$ are closed contours, which follow in part (but in opposite directions) the boundary R shown in Fig. 33(b) and which are separated from it by an infinitesimal distance. On account of the analyticity of $f(z)$ and $g(z)$ in D_f and D_g, respectively, both integrals on the RHS of Eq. 27.3 vanish, which shows that

$$\int_C h(z')\,dz' = 0$$

for an arbitrary closed contour $C \subset (D_f + D_g + R)$. Therefore, by virtue of the theorem of Morera, $h(z)$ is analytic throughout the entire region $D_f + D_g + R$, including R, and $f(z)$ and $g(z)$ are analytic continuations of each other.

We are now in a position to prove the following theorem, known as the **Schwarz reflection principle**.

Theorem. Let $f(z)$ be a function analytic in a region D that has, as a part of its boundary, a segment R of the real axis. Then, provided $f(z)$ is real wherever z takes on real values, the analytic continuation of $f(z)$ into the region $\bar{D}$ (the mirror image of D with respect to the real axis) exists and is given by

$$g(z) = \bar{f}(\bar{z}) \qquad z \in \bar{D} \tag{27.4}$$

Proof. For any closed contour $C \subset D$, one has (Fig. 34)

$$\int_C f(z')\,dz' = 0 \tag{27.5}$$

Suppose that C is described by the parametric equation

$$z = \eta(t) \qquad (z \in D, \quad t_1 \le t \le t_2) \tag{27.6}$$

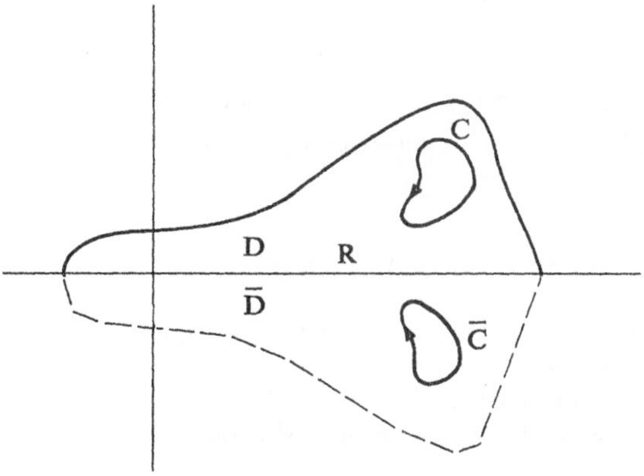

Fig. 34.

Then from Eq. 27.5 we have

$$\int_{t_1}^{t_2} f[\eta(t)] \frac{d\eta(t)}{dt}\, dt = 0 \tag{27.7}$$

We shall prove first that $g(z)$ is analytic in $\overline{\mathrm{D}}$. Let $\overline{\mathrm{C}}$ be the image of C in $\overline{\mathrm{D}}$. Its parametric equation is (compare Eq. 27.6)

$$z = \bar{\eta}(t) \qquad (z \in \overline{\mathrm{D}}, \quad t_1 \leq t \leq t) \tag{27.8}$$

Then, using Eqs. 27.4, 27.7, and 27.8, we have (the integral along $\overline{\mathrm{C}}$ is taken in the clockwise direction, since $\overline{\mathrm{C}}$ is the mirror image of C)

$$\begin{aligned}\int_{\overline{\mathrm{C}}} g(z')\, dz' &= \int_{t_1}^{t_2} g[\bar{\eta}(t)] \frac{d\bar{\eta}(t)}{dt}\, dt \\ &= \int_{t_1}^{t_2} \bar{f}[\eta(t)] \frac{d\bar{\eta}(t)}{dt}\, dt \\ &= \overline{\int_{t_1}^{t_2} f[\eta(t)] \frac{d\eta(t)}{dt}\, dt} = 0\end{aligned} \tag{27.9}$$

Hence, by the theorem of Morera, $g(z)$ is analytic in $\overline{\mathrm{D}}$.

Since $f(z)$ is real on the real axis, one has, because of Eq. 27.4

$$g(z) = f(z) \qquad \text{for real } z$$

Thus, $f(z)$ is analytic above the real axis (in D) and $g(z)$ is analytic below the real axis (in $\overline{\mathrm{D}}$), and these functions are equal to each other on the real axis (on R). Hence, by the preceding lemma, $f(z)$ and $g(z)$ or (what is the same) $f(z)$ and $\bar{f}(\bar{z})$ are analytic continuations of each other and together define a **unique** function

$$h(z) = \begin{cases} f(z), & z \in \mathrm{D} \\ g(z) \equiv \bar{f}(\bar{z}), & z \in \overline{\mathrm{D}} \end{cases} \tag{27.10}$$

which is analytic in the region $\mathrm{D} + \overline{\mathrm{D}} + \mathrm{R}$.

From Eq. 27.10 we immediately get

$$h(\bar{z}) = \bar{h}(z) \qquad z \in (\mathrm{D} + \overline{\mathrm{D}} + \mathrm{R}) \tag{27.11}$$

For example, when $z \in \mathrm{D}$, then $\bar{z} \in \overline{\mathrm{D}}$ and

$$h(\bar{z}) = g(\bar{z}) \equiv \bar{f}(z) = \bar{h}(z)$$

The relation (27.11) must clearly be satisfied by any function that is analytic throughout a region intersected by the real axis and which takes on real values when its argument is real.

28 · DISPERSION RELATIONS

We consider in this section a function $h(z)$ that is analytic throughout the entire complex plane except for a cut along the real axis extending from x_0 to infinity. We also suppose that $h(z)$ is real on the remainder of the real axis and that $|h(z)| \to 0$ faster than $1/|z|$ as $|z| \to \infty$.

For a point z not on the real axis, one has

$$h(z) = \frac{1}{2\pi i} \int_{\mathrm{C}} \frac{h(z')}{z' - z}\, dz' \tag{28.1}$$

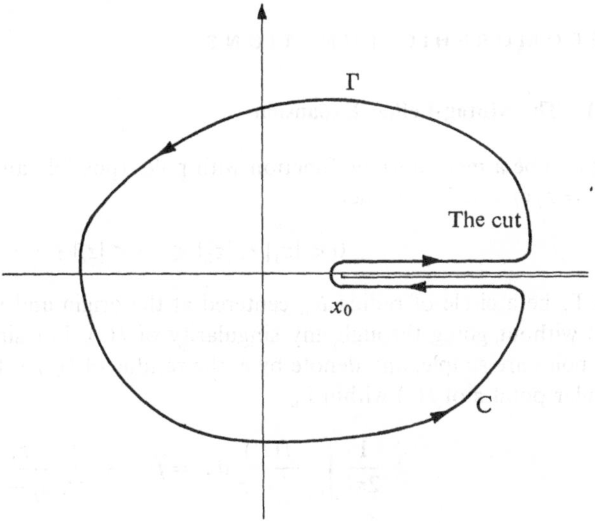

Fig. 35.

where C is the contour shown in Fig. 35. The contribution to Eq. 28.1 from the large circle Γ tends to zero as its radius tends to infinity, and we obtain

$$\begin{aligned} h(z) &= \frac{1}{2\pi i}\left\{\int_{x_0+i\varepsilon}^{\infty+i\varepsilon} \frac{h(z')}{z'-z}\,dz' - \int_{x_0-i\varepsilon}^{\infty-i\varepsilon} \frac{h(z')}{z'-z}\,dz'\right\} \\ &= \frac{1}{2\pi i}\left\{\int_{x_0}^{\infty} \frac{h(x'+i\varepsilon)}{x'-z+i\varepsilon}\,dx' - \int_{x_0}^{\infty} \frac{h(x'-i\varepsilon)}{x'-z-i\varepsilon}\,dx'\right\} \end{aligned} \tag{28.2}$$

As $\varepsilon \to +0$, we can neglect the $\pm i\varepsilon$ in the denominators above (remember that z is *not* on the real axis) and we get

$$h(z) = \frac{1}{2\pi i}\int_{x_0}^{\infty} \lim_{\varepsilon\to+0} \frac{[h(x'+i\varepsilon) - h(x'-i\varepsilon)]}{x'-z}\,dx' \tag{28.3}$$

The numerator of the integrand in Eq. 28.3 is the discontinuity of $h(z)$ across the cut. It can be evaluated if we note that $h(z)$ satisfies

$$\bar{h}(z) = h(\bar{z})$$

as explained at the end of Sec. 27 (Schwarz reflection principle). Hence,

$$\begin{aligned} \lim_{\varepsilon\to+0} [h(x+i\varepsilon) - h(x-i\varepsilon)] &= \lim_{\varepsilon\to+0} [h(x+i\varepsilon) - \bar{h}(x+i\varepsilon)] \\ &= \lim_{\varepsilon\to+0} 2i \operatorname{Im} h(x+i\varepsilon) = 2i \operatorname{Im} h(x+i0) \end{aligned} \tag{28.4}$$

Inserting Eq. 28.4 in Eq. 28.3, we get in the limit $\varepsilon \to +0$

$$h(z) = \frac{1}{\pi}\int_{x_0}^{\infty} \frac{\operatorname{Im} h(x'+i0)}{x'-z}\,dx' \tag{28.5}$$

This is a particular example of what physicists call a **dispersion relation.** It expresses the value of a function at any point of the complex plane in terms of an integral over its imaginary part on the upper lip of the cut.

29 · MEROMORPHIC FUNCTIONS

29.1 The Mittag-Leffler Expansion

Let $f(z)$ be a meromorphic function with poles (possibly an infinite number of them) at $z = z_j$ ($j = 1, 2, \cdots$) where

$$0 < |z_1| \leq |z_2| \leq \cdots \leq |z_n| \leq \cdots \tag{29.1}$$

Let Γ_n be a circle of radius R_n, centered at the origin and which encloses n poles of $f(z)$ without going through any singularity of $f(z)$. For simplicity, suppose that all the poles are simple, and denote by r_j the residue of $f(z)$ at the pole z_j. Then, for any regular point z of $f(z)$ within Γ_n

$$\frac{1}{2\pi i}\int_{\Gamma_n}\frac{f(z')}{z'-z}\,dz' = f(z) + \sum_{j=1}^{n}\frac{r_j}{z_j - z} \tag{29.2}$$

Writing the above equation for $z = 0$ and subtracting it from Eq. 29.2, we find

$$f(z) - f(0) = \frac{z}{2\pi i}\int_{\Gamma_n}\frac{f(z')}{z'(z'-z)}\,dz' + \sum_{j=1}^{n} r_j\left(\frac{1}{z-z_j} + \frac{1}{z_j}\right) \tag{29.3}$$

Let it be possible to choose the sequence of circles Γ_n such that on Γ_n

$$|f(z)| < A$$

where A is independent of n. Then, applying the Darboux inequality to the integral on the RHS of Eq. 29.3, we find

$$\left|\int_{\Gamma_n}\frac{f(z')}{z'(z'-z)}\,dz'\right| < \frac{A}{R_n(R_n - |z|)}\cdot 2\pi R_n = \frac{2\pi A}{R_n - |z|} \underset{R\to\infty}{\to} 0 \tag{29.4}$$

Hence, as $R_n \to \infty$, we obtain from Eq. 29.3

$$f(z) = f(0) + \sum_j r_j\left(\frac{1}{z-z_j} + \frac{1}{z_j}\right) \tag{29.5}$$

Equation 29.5 is a particular case of a general result due to Mittag-Leffler, which shows that any meromorphic function can be expressed as a sum of an entire function [in our case it is the constant $f(0)$] and a series (in general, infinite) of rational functions.

It can be shown that the infinite series in Eq. 29.5 can always be arranged so that it converges uniformly.

EXAMPLE

The residues of $1/\cos z$ at $z = (2n+1)(\pi/2)$ have already been calculated (see Example 2, Sec. 22.1). Using these results, we find

$$\frac{1}{\cos z} = 1 + 2\sum_{n=0}^{\infty}(-1)^{n+1}\left\{\frac{1}{2z-(2n+1)\pi} + \frac{1}{(2n+1)\pi}\right\}$$

Consider now an entire function $g(z)$ which has simple zeros only (not at the origin); then the function $\dfrac{dg(z)}{dz}\Big/ g(z)$ is meromorphic, and applying Eq. 29.5 to this function, one obtains

$$\frac{d}{dz}\operatorname{Ln} g(z) = \frac{\dfrac{dg(z)}{dz}}{g(z)}$$

$$= \frac{\left.\dfrac{dg(z)}{dz}\right|_{z=0}}{g(o)} + \sum_{j=1}^{\infty}\left(\frac{1}{z-z_j}+\frac{1}{z_j}\right) \tag{29.6}$$

since the residues r_j are equal to unity for $(dg/dz)/g(z)$. As the preceding series converges uniformly, Eq. 29.6 can be immediately integrated to give

$$g(z) = g(0)e^{cz}\prod_{j=1}^{\infty}\left(1-\frac{z}{z_j}\right)e^{z/z_j} \tag{29.7}$$

where

$$c = \frac{\left.\dfrac{dg(z)}{dz}\right|_{z=0}}{g(0)}$$

is a constant and $z_j \neq 0$.

29.2 A Theorem on Meromorphic Functions

Theorem. Let $f(z)$ be a meromorphic function in a region R and $g(z)$ an analytic function in R. Let C be a closed contour in R on which $f(z)$ is both analytic and nowhere zero. If $f(z)$ has, within C, Z zeros at $z = a_j\ (j = 1, 2, \cdots Z)$ of order $n_j\ (j = 1, 2, \cdots Z)$ and P poles at $z = b_j\ (j = 1, 2, \cdots, P)$ of order $m_j\ (j = 1, 2, \cdots, P)$, then

$$\frac{1}{2\pi i}\int_C g(z)\frac{\dfrac{df(z)}{dz}}{f(z)}\,dz = \sum_{j=1}^{Z} n_j g(a_j) - \sum_{j=1}^{P} m_j g(b_j) \tag{29.8}$$

Proof. Let z_j be either a pole or a zero of $f(z)$ of order l_j. We can expand $f(z)$ quite generally about z_j

$$f(z) = \alpha_0^{(j)}(z-z_j)^{l_j} + \alpha_1^{(j)}(z-z_j)^{l_j+1} + \cdots (\alpha_0^{(j)} \neq 0) \tag{29.9}$$

where $l_j = n_j$ if z_j is a zero and $l_j = -m_j$ if z_j is a pole. In the first case, Eq. 29.9 corresponds to a Taylor expansion about z_j, and in the second case, Eq. 29.9 corresponds to a Laurent expansion about that point.

From Eq. 29.9, by differentiation we have

$$\frac{df(z)}{dz} = \alpha_0^{(j)} l_j (z-z_j)^{l_j-1} + \alpha_1^{(j)}(l_j+1)(z-z_j)^{l_j} + \cdots \tag{29.10}$$

and therefore

$$\frac{\dfrac{df(z)}{dz}}{f(z)} = \frac{l_j}{z-z_j} + \text{(an analytic function)} \tag{29.11}$$

Since $g(z)$ is analytic throughout R, it can be expanded in a Taylor series about z_j

$$g(z) = g(z_j) + \left.\frac{dg(z)}{dz}\right|_{z=z_j} (z - z_j) + \cdots \tag{29.12}$$

and thus

$$g(z)\frac{\dfrac{df(z)}{dz}}{f(z)} = \frac{g(z_j)l_j}{z - z_j} + \text{(an analytic function)} \tag{29.13}$$

If C_j is a closed curve surrounding the point z_j only, we have

$$\operatorname{Res}\left[g\frac{\dfrac{df}{dz}}{f}\right]_{z=z_j} = \frac{1}{2\pi i}\int_{C_j} g(z)\frac{\dfrac{df(z)}{dz}}{f(z)}dz = g(z_j)l_j \tag{29.14}$$

Finally, for a curve C that surrounds all the poles and zeros z_j, we find

$$\frac{1}{2\pi i}\int_C g(z)\frac{\dfrac{df(z)}{dz}}{f(z)}\,dz = \sum_{j=1}^{Z} n_j g(a_j) - \sum_{j=1}^{P} m_j g(b_j) \tag{29.15}$$

In particular, for the case $g(z) \equiv 1$, we have

$$\frac{1}{2\pi i}\int_C \frac{\dfrac{df(z)}{dz}}{f(z)}\,dz = \sum_{j=1}^{Z} n_j - \sum_{j=1}^{P} m_j \tag{29.16}$$

In a sense, one can interpret the RHS of Eq. 29.16 as the difference between the number of zeros and the number of poles of $f(z)$ included within the contour C if one counts every zero of order n (pole of order m) as n zeros (m poles).

30 · THE FUNDAMENTAL THEOREM OF ALGEBRA

Using the results of the preceding section, we can easily prove the following theorem, known as the **fundamental theorem of algebra**.

Theorem. Every polynomial

$$p_n(z) = z^n + a_{n-1}z^{n-1} + \cdots + a_1 z + a_0 \tag{30.1}$$

of the nth degree can be written as

$$p_n(z) = \prod_j (z - z_j)^{n_j} \tag{30.2}$$

with

$$\sum_{j=1}^{Z} n_j = n \tag{30.3}$$

where Z is the number of zeros of $p_n(z)$ and n_j is the order of the jth zero.

Proof. $p_n(z)$ is an entire function, and all its zeros must be located at a finite distance from the origin because, as $|z| \to \infty$,

$$p_n(z) \approx z^n \tag{30.4}$$

Putting $f(z) = p_n(z)$ into Eq. 29.8 and choosing for C a circle C_R of radius R, which encloses all the zeros of $p_n(z)$, we obtain

$$\frac{1}{2\pi i}\int_{C_R} \frac{\dfrac{dp_n(z)}{dz}}{p_n(z)}\, dz = \sum_{j=1}^{Z} n_j \tag{30.5}$$

The integral on the LHS of Eq. 30.5 can be easily calculated for $R \to \infty$ using Eq. 30.4

$$\lim_{R\to\infty} \frac{1}{2\pi i}\int_{C_R} \frac{\dfrac{dp_n(z)}{dz}}{p_n(z)}\, dz = \lim_{R\to\infty} \frac{1}{2\pi i}\int_{C_R} \frac{n}{z}\, dz = n \tag{30.6}$$

From Eqs. 30.5 and 30.6 one obtains

$$\sum_{j=1}^{Z} n_j = n \tag{30.7}$$

This relation shows that if one counts a zero of order n_j as n_j zeros, a polynomial of the nth degree has exactly n zeros. Each zero contributes a factor $(z - z_j)$ to $p_n(z)$, and therefore

$$p_n(z) = \text{const} \prod_{j=1}^{Z} (z - z_j)^{n_j} \tag{30.8}$$

The constant is in fact equal to 1 because the coefficient of z^n in Eq. 30.1 was chosen to be equal to 1.

It is worthwhile to note that each zero of $p_n(z)$ is a root of the equation

$$p_n(z) = 0$$

and that the order of the zero is also called the **multiplicity** of the corresponding root.

31 · THE METHOD OF STEEPEST DESCENT; ASYMPTOTIC EXPANSIONS

In many applications it is required to evaluate integrals of the type

$$I(w) = \int_C e^{wf(z)} g(z)\, dz \tag{31.1}$$

when $|w|$ is very large. In Eq. 31.1, $f(z)$ and $g(z)$ are analytic functions in some region containing C. Without any loss of generality we can suppose that w is real and positive; otherwise, one could always absorb the constant factor $e^{i \arg w}$ into the function $f(z)$.

P. Debye has devised a method for evaluating approximately such integrals, called the **method of steepest descent** (sometimes called the **saddle point method**).

The method is based on the observation that the major contribution from the integrand in Eq. 31.1, when w is large, comes from the regions along C where Re $f(z)$

is largest or has a maximum. However, in these regions there would usually be very large oscillations, and consequently important cancellations due to the factor $e^{i\,\mathrm{Im} f(z)}$ in the integrand. These oscillations would make an evaluation of $I(w)$ very difficult.

The idea of Debye is to deform the path C in such a way that on a part C_0 of C, the following conditions are satisfied:

(a). Along C_0 $\mathrm{Im} f(z)$ is constant.

(b). C_0 goes through a point z_0 where

$$\left.\frac{df(z)}{dz}\right|_{z=z_0} = 0$$

(c). The path C_0 is so chosen that, at $z = z_0$, $\mathrm{Re} f(z)$ goes through a relative maximum.

The condition (a) will guarantee that there are no oscillations along C_0. Condition (a) and the fact that C_0 goes through the point z_0 determine the equation of the path C_0

$$\mathrm{Im} f(z) = \mathrm{Im} f(z_0) \tag{31.2}$$

The condition (c) ensures that the integrand in Eq. 31.1 has a peak at $z = z_0$, which in fact becomes more and more pronounced as w increases, so that one may hope that the main contribution to the integral will come from an environment of the point z_0.

We now consider in more detail the meaning of the conditions (b) and (c). We recall (see Sec. 14) that neither the real part nor the imaginary part of an analytic function can have an absolute maximum or minimum at a regular point z_0, although the first derivative of the function itself may vanish at that point

$$\left.\frac{df(z)}{dz}\right|_{z=z_0} = 0 \tag{31.3}$$

Such a point is called a **saddle point**. The reason for the name will become apparent. Let us study the behavior of $f(z)$ in the neighborhood of the saddle point z_0. Expanding $f(z)$ in a Taylor series about that point, we have due to Eq. 31.3

$$f(z) = f(z_0) + \frac{1}{2}(z - z_0)^2 \left.\frac{d^2f(z)}{dz^2}\right|_{z=z_0} + \cdots \tag{31.4}$$

For simplicity we suppose that

$$\left.\frac{d^2f(z)}{dz^2}\right|_{z=z_0} \neq 0 \tag{31.5}$$

Consider points z that are in an immediate neighborhood of z_0, so that the higher-order terms of the expansion (Eq. 31.4) can be neglected. Putting

$$z - z_0 = re^{i\varphi} \tag{31.6}$$

$$\frac{1}{2}\left.\frac{d^2f(z)}{dz^2}\right|_{z=z_0} = Re^{i\Phi} \tag{31.7}$$

from Eq. 31.4 we have

$$\mathrm{Re}[f(z) - f(z_0)] \approx r^2R\cos(2\varphi + \Phi) \tag{31.8}$$

$$\mathrm{Im}[f(z) - f(z_0)] \approx r^2R\sin(2\varphi + \Phi) \tag{31.9}$$

The condition (31.2) leads to

$$\sin(2\varphi + \Phi) = 0 \tag{31.10}$$

i.e., to

$$\varphi = -\frac{\Phi}{2} + n\frac{\pi}{2} \qquad (n = 0, 1, 2, 3) \tag{31.11}$$

Inserting Eq. 31.11 into Eq. 31.6, we get

$$z = z_0 \pm re^{-i\Phi/2} \qquad \begin{cases} + & \text{for } n = 0 \\ - & \text{for } n = 2 \end{cases} \tag{31.12a}$$

and

$$z = z_0 \pm re^{i[-(\Phi/2)+(\pi/2)]} \qquad \begin{cases} + & \text{for } n = 1 \\ - & \text{for } n = 3 \end{cases} \tag{31.12b}$$

Since Φ is a constant determined by the value of $d^2f(z)/dz^2$ at $z = z_0$, Eqs. 31.12a and 31.12b are the equations of two straight lines passing through the point z_0 and along which $\operatorname{Im} f(z)$ is constant.

Similarly, there are two lines of constant $\operatorname{Re} f(z)$ passing through z_0; they are determined by the condition

$$\cos(2\varphi + \Phi) = 0 \tag{31.13}$$

which leads to

$$z = z_0 \pm re^{i[-(\Phi/2)+(\pi/4)]} \qquad \begin{cases} + & \text{for } n = 0 \\ - & \text{for } n = 2 \end{cases} \tag{31.14a}$$

$$z = z_0 \pm re^{i[-(\Phi/2)+(3\pi/4)]} \qquad \begin{cases} + & \text{for } n = 1 \\ - & \text{for } n = 3 \end{cases} \tag{31.14b}$$

It can be seen from Eq. 31.8 that the lines defined by Eqs. 31.14a and 31.14b divide any neighborhood of z_0 into four sectors where alternatively

$$\operatorname{Re} f(z) > \operatorname{Re} f(z_0) \qquad (\text{since } \cos(2\varphi + \Phi) > 0) \tag{31.15}$$

and

$$\operatorname{Re} f(z) < \operatorname{Re} f(z_0) \qquad (\text{since } \cos(2\varphi + \Phi) < 0) \tag{31.16}$$

Furthermore, since between two zeros of the cosine function there is one zero of the sine function, there is in each of the four sectors, one and only one line of constant $\operatorname{Im} f(z)$.

Let us summarize the preceding discussion with a simple geometrical picture. $\operatorname{Re} f(z)$ is a function of two real variables $\operatorname{Re} z$ and $\operatorname{Im} z$ and, as is well known, it can be represented by a surface S in space, as shown in Fig. 36. The surface S is such that the points on S which project onto the sectors I and III of the z plane (i.e., those for which the condition (31.15) is satisfied) are higher than the point p_0 on S which is directly above z_0, whereas the points that project onto the sectors II and IV (i.e., those for which inequality 31.16 is satisfied) are lower than p_0. Thus, we can visualize S as having the shape of a horse's saddle (whence the name "saddle point").

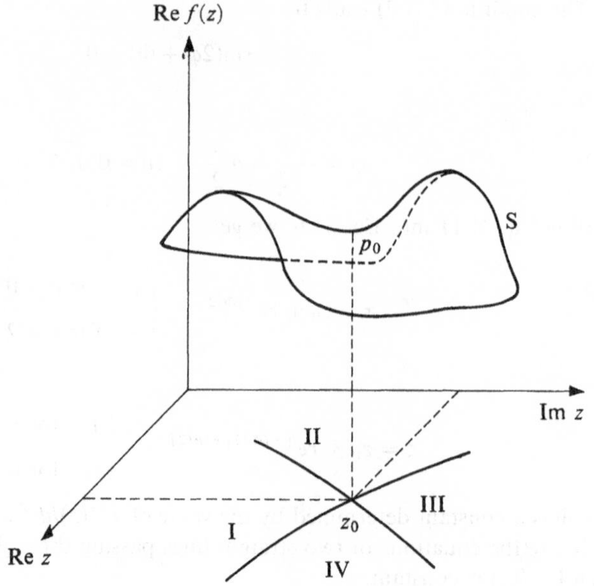

Fig. 36. The behavior of the real part of an analytic function in the neighborhood of a saddle point. When z lies in the sectors I and III, $\operatorname{Re} f(z) > \operatorname{Re} f(z_0)$; when z lies in the sectors II and IV, $\operatorname{Re} f(z) < \operatorname{Re} f(z_0)$.

In order for $\operatorname{Re} f(z)$ to go through a relative maximum at the saddle point, in accordance with condition (c), we must choose among the two paths of constant $\operatorname{Im} f(z)$ the one that lies in the sectors II and IV; along this line, $\operatorname{Re} f(z)$ goes through a relative maximum rather than a relative minimum at z_0. Furthermore, we can easily show that as z describes this line, $\operatorname{Re} f(z)$ varies as rapidly as possible. Using the geometrical picture, we can say that the curve on S that projects onto this line in the z plane has the property that it is as steep as possible (whence the name "method of steepest descent"). In fact, the directional derivative of $f(z)$ along any path in the complex plane is given by

$$\frac{\partial f}{\partial l} = \vec{\nabla} \operatorname{Re} f \cdot \vec{l} + i \vec{\nabla} \operatorname{Im} f \cdot \vec{l}$$

and therefore the modulus of $\partial f / \partial l$ is

$$\left| \frac{\partial f}{\partial l} \right| = \{[\vec{\nabla} \operatorname{Re} f \cdot \vec{l}]^2 + [\vec{\nabla} \operatorname{Im} f \cdot \vec{l}]^2\}^{1/2} \tag{31.17}$$

where ∂l is an element of arc length and $\vec{l}$ is the unit vector along the path. Since at a given point $|\partial f / \partial l|$ has a given value, (31.17) shows that $\vec{\nabla} \operatorname{Re} f \cdot \vec{l}$ is largest when $\vec{\nabla} \operatorname{Im} f \cdot \vec{l} = 0$. This last condition is clearly satisfied along a path of constant $\operatorname{Im} f(z)$.

The previous discussion can be generalized by taking into account higher-order terms in the expansion (31.4). Everything that has been said remains valid except that the paths of constant $\operatorname{Im} f(z)$ are no longer straight lines, but curves having these lines as tangents at the saddle point z_0.

The point to be retained from the foregoing discussion is that if $I(w)$ is evaluated along a path in the z plane satisfying Debye's conditions, then it is likely (and in fact it is so in most of the applications) that the main contribution to the integral will come from a neighborhood around the saddle point, since contributions from distant parts of the path are in general very quickly attenuated.

We evaluate the contribution $I_0(w)$ to $I(w)$ which comes from the integration along the part C_0 of the contour C where Debye's conditions are satisfied. Of course we need to assume that there exists at least one saddle point and that the contour C in Eq. 31.1 can be properly deformed. We write $f(z)$ as

$$f(z) = f(z_0) - \tau^2 \tag{31.18}$$

where τ is real along C_0 (see Eq. 31.2) and the negative sign appears because C_0 is a path of steepest descent; i.e., $\operatorname{Re} f(z)$ decreases as it leaves $z = z_0$. Then the Taylor expansion of $f(z)$ about z_0 gives

$$\tau^2 = -\frac{1}{2}(z - z_0)^2 \left.\frac{d^2 f(z)}{dz^2}\right|_{z=z_0} + \cdots \tag{31.19}$$

If we wish to approximate τ by the first nonvanishing term in the Taylor expansion, then we will have

$$z - z_0 = \frac{\sqrt{2}\,\tau e^{i\theta}}{\left|\left[\frac{d^2 f(z)}{dz^2}\right]_{z=z_0}\right|^{\frac{1}{2}}} \tag{31.20}$$

where θ is the phase angle of $z - z_0$. Putting Eq. 31.18 into Eq. 31.1, we find

$$I_0(w) \approx e^{wf(z_0)} \int_{C_0} e^{-w\tau^2(z)} g(z)\, dz \tag{31.21}$$

where C_0 is the path of steepest descent.

For large values of w, the dominant contribution to $I_0(w)$ comes from regions along C_0 where τ is small. For larger values of τ, the integrand falls off extremely rapidly. Therefore, we shall be making only a negligible error if we replace the integration in Eq. 31.21 over a region around $\tau \ll 1$ by an integration over the entire real axis (remember that τ is real). Hence we replace Eq. 31.21 by

$$I_0(w) \approx e^{wf(z_0)} \int_{-\infty}^{+\infty} e^{-w\tau^2} g[z(\tau)] \frac{dz(\tau)}{d\tau}\, d\tau \tag{31.22}$$

To evaluate (31.22), we replace $g[z(\tau)]$ and $dz(\tau)/d\tau$ by their power series expansions in τ about $\tau = 0$. Let us write for the product

$$g[z(\tau)] \frac{dz(\tau)}{d\tau} = \sum_{m=0}^{\infty} c_m \tau^m \tag{31.23}$$

The evaluation of $I_0(w)$, except the first term, is tedious in practical calculations because it involves expressing z as a power series in τ via Eq. 31.18, which is not always an easy task. Often, however, the first term is the only one required. Putting Eq. 31.23 into Eq. 31.22, we have

$$I_0(w) \approx e^{wf(z_0)} \sum_{m=0}^{\infty} c_m \int_{-\infty}^{+\infty} e^{-w\tau^2} \tau^m\, d\tau \tag{31.24}$$

To evaluate the integrals in Eq. 31.24, we write

$$J_m \equiv \int_{-\infty}^{\infty} e^{-w\tau^2}\tau^m \, d\tau \tag{31.25}$$

Then we note that all the J_m with odd m vanish because the integrand in (31.25) is an odd function of τ.

$$e^{-w\tau^2}\tau^m = -e^{-w(-\tau)^2}(-\tau)^m \qquad (m \text{ odd}) \tag{31.26}$$

Thus,

$$J_m = 0 \qquad m = 1, 3, 5, \cdots \tag{31.27}$$

On the other hand (see Eq. 22.20)

$$J_0 = \sqrt{\frac{\pi}{w}} \tag{31.28}$$

and since

$$J_{2m} = -\frac{\partial}{\partial w} J_{2m-2} \tag{31.29}$$

we easily obtain

$$J_{2m} = \frac{\sqrt{\pi} \cdot 1 \cdot 3 \cdot 5 \cdots (2m-1)}{2^m w^{(2m+1)/2}} \qquad m > 0 \tag{31.30}$$

Hence, combining Eqs. 31.24 and 31.30, we find

$$I_0(w) \approx e^{wf(z_0)}\left[c_0\sqrt{\frac{\pi}{w}} + \sum_{m=1}^{\infty} c_{2m} \frac{\sqrt{\pi} \cdot 1 \cdot 3 \cdot 5 \cdots (2m-1)}{2^m \cdot w^{(2m+1)/2}}\right] \tag{31.31}$$

The expression (31.31), valid for large values of w, is called an **asymptotic expansion** with respect to w.

Asymptotic series were first introduced by H. Poincaré. They are defined as follows. A series

$$\sum_{k=0}^{\infty} c_k z^{-k} \tag{31.32}$$

is said to be an asymptotic series for $f(z)$, and one writes

$$f(z) \sim \sum_{k=0}^{\infty} c_k z^{-k} \tag{31.33}$$

if, for any positive integer n

$$\lim_{|z| \to \infty} \{z^n[f(z) - \sum_{k=0}^{n} c_k z^{-k}]\} = 0 \tag{31.34}$$

The series (31.32) may or may not be convergent (in fact these series are usually divergent); nevertheless the series 31.32 may represent quite accurately the function $f(z)$ when $|z|$ is large. The reason is that, on account of Eq. 31.34, the difference between $f(z)$ and the first $n+1$ terms of the series (31.32)

$$\left| f(z) - c_0 - \frac{c_1}{z} - \frac{c_2}{z^2} - \cdots - \frac{c_n}{z^n} \right|$$

is of the order of $1/z^{n+1}$, and this can be made arbitrarily small when $|z|$ is sufficiently large. When an asymptotic series diverges, it is clear that for a fixed value of z, the inclusion of too many terms of the series renders the approximation worse, rather than better. In other words, for any fixed value of z, there is an optimum number of terms of the series that gives rise to the best approximation.

It is easy to show that a function has a unique asymptotic expansion, for if $f(z)$ also had the asymptotic expansion

$$f(z) \sim \sum_{k=0}^{\infty} d_k z^{-k} \tag{31.35}$$

one would have

$$\lim_{|z|\to\infty} \{z^n[f(z) - \sum_{k=0}^{n} d_k z^{-k}]\} = 0 \tag{31.36}$$

which, combined with Eq. 31.34, yields

$$\lim_{|z|\to\infty} \{z^n[\sum_{k=0}^{n} (d_k - c_k) z^{-k}]\} = 0 \tag{31.37}$$

for all n. Hence, we must have $c_k = d_k$ for all k. On the other hand, different functions may have the same asymptotic expansion. One can show that asymptotic series can be added, multiplied, and integrated term by term. However, one cannot differentiate an asymptotic series term by term.

EXAMPLE

Consider the function $I(w)$ defined by the integral

$$I(w) = \int_0^{\infty} e^{-z} z^w \, dz \tag{31.38}$$

where w is real. Put

$$\zeta = \frac{z}{w} \tag{31.39}$$

Then Eq. 31.38 becomes

$$I(w) = w^{w+1} \int_0^{\infty} e^{-\zeta w} \zeta^w \, d\zeta = w^{w+1} \int_0^{\infty} e^{w f(\zeta)} \, d\zeta \tag{31.40}$$

where

$$f(\zeta) = \ln \zeta - \zeta \tag{31.41}$$

The saddle point is at $df(\zeta)/d\zeta = 0$, i.e., at

$$\zeta = 1$$

In order to find the path of steepest descent, we expand $f(\zeta)$ about $\zeta = 1$

$$f(\zeta) = -1 - \tfrac{1}{2}(\zeta - 1)^2 + \cdots \tag{31.42}$$

Comparing Eq. 31.42 with Eqs. 31.9 and 31.12a, we see that, in the neighborhood of $\zeta = 1$, the lines of constant $\operatorname{Im} f(\zeta)$ are given by

$$\zeta = 1 \pm r \tag{31.43}$$

$$\zeta = 1 \mp ir \tag{31.44}$$

Equation 31.43 is the equation of a line along the real axis and Eq. 31.44 is the equation of a line parallel to the imaginary axis. It is easy to verify that along the line defined by Eq. 31.43, $f(\zeta)$ goes through a relative maximum, whereas along the line given by Eq. 31.44, $f(\zeta)$ goes through a relative minimum. Therefore, the path of steepest descent is along the real axis.

Putting

$$f(\zeta) = f(1) - \tau^2 = -1 - \tau^2 \tag{31.45}$$

and expanding $f(\zeta)$ about $\zeta = 1$, we find

$$\tau^2 = -1 + \zeta - \ln \zeta = (\zeta - 1) - [(\zeta - 1) - \tfrac{1}{2}(\zeta - 1)^2 + \cdots]$$

Hence, to lowest order

$$\frac{d\zeta(\tau)}{d\tau} \doteq \sqrt{2}$$

and comparing with Eq. 31.23, we see that

$$c_0 = \sqrt{2}$$

so that Eq. 31.31 yields

$$I(w) \sim \sqrt{2}\pi e^{-w} w^{w+1/2}$$

32 · THE GAMMA FUNCTION

The gamma function $\Gamma(z)$ is defined for all values of z by the integral

$$\frac{1}{\Gamma(z)} = \frac{1}{2\pi i} \int_C \frac{e^t}{t^z}\, dt \tag{32.1}$$

When z is not an integer, the integrand has a branch cut extending from the origin to infinity along the negative real axis. The convention

$$\arg t = \begin{cases} \pi & \text{above the cut} \\ -\pi & \text{below the cut} \end{cases} \tag{32.2}$$

will be adopted, and defines the branch of the integrand in Eq. 32.1. The contour C is then chosen as in Fig. 37. With this choice for C, $1/\Gamma(z)$ is single-valued and finite everywhere for all z real (integer or not) or complex. When z is a real integer n, C can be taken to be simply the circle γ around the origin, since in this case there is no branch

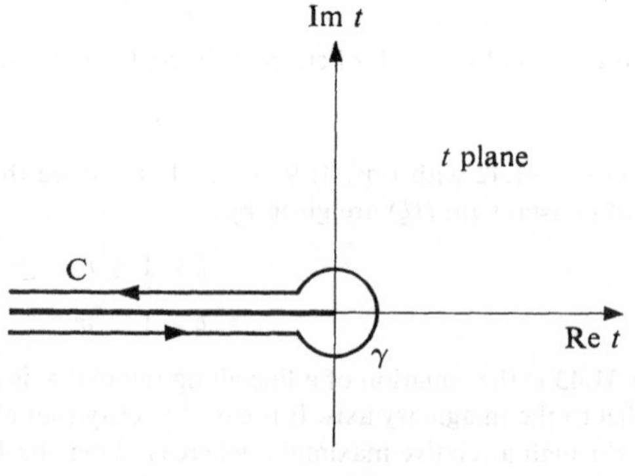

Fig. 37.

cut at all. Then the integral in Eq. 32.1 can be easily evaluated. When $n \leq 0$, the integrand is analytic within γ, and therefore $1/\Gamma(z) = 0$; when $n > 0$, we use

$$\frac{e^t}{t^n} = \sum_k \frac{t^{k-n}}{k!}$$

to find that

$$\operatorname{Res}\left\{\frac{e^t}{t^n}\right\}_{t=0} = \frac{1}{(n-1)!}$$

Hence

$$\Gamma(n) = \begin{cases} (n-1)! & n > 0 \\ \infty & n \leq 0 \end{cases} \tag{32.3}$$

where, as usual, $0! = 1$.

Another integral representation of the Γ function is

$$\Gamma(z) = \int_0^\infty e^{-t} t^{z-1}\, dt \qquad (\operatorname{Re} z > 0) \tag{32.4}$$

This integral is called **Euler's integral of the second kind**.

It is necessary to have $\operatorname{Re} z > 0$ in order for the integral to be convergent. We shall prove a number of useful properties of the Γ function and also demonstrate the equivalence of the integral representations 32.4 and 32.1.

(i) We first show that

$$\Gamma(z) = (z-1)\Gamma(z-1) \tag{32.5}$$

This follows by performing a single partial integration on Eq. 32.1.

$$\frac{1}{\Gamma(z)} = \frac{1}{2\pi i}\left\{\frac{-e^t}{(z-1)t^{z-1}}\Bigg|_{-\infty-i\varepsilon}^{-\infty+i\varepsilon} + \frac{1}{z-1}\int_C \frac{e^t}{t^{z-1}}\, dt\right\}$$

The first term vanishes and the second integral is simply $1/\Gamma(z-1)$. This proves the result.

Differentiating Eq. 32.4, we see that the derivative $d\Gamma(z)/dz$ exists whenever $\operatorname{Re} z > 0$. Hence, $\Gamma(z)$ is an analytic function of z for $\operatorname{Re} z > 0$. What happens for $\operatorname{Re} z < 0$? If n is a positive integer, from Eq. 32.5 we find

$$\Gamma(z) = \frac{\Gamma(z+n)}{z(z+1)(z+2)\cdots(z+n-1)}$$

The numerator is analytic for $\operatorname{Re} z > -n$. Hence, $\Gamma(z)$ is analytic for $\operatorname{Re} z > -n$ except for simple poles at

$$z = 0, -1, -2, \cdots, (-n+1)$$

Since n can be arbitrarily large, we deduce that $\Gamma(z)$ is analytic in the entire complex plane except when z is a negative integer or 0; at those points, $\Gamma(z)$ has simple poles.

(ii) We show next that for two complex numbers a and b, such that $\operatorname{Re} a$, $\operatorname{Re} b > 0$

$$\frac{\Gamma(a)\Gamma(b)}{\Gamma(a+b)} = \mathrm{B}(a,b) \tag{32.6}$$

where

$$\mathrm{B}(a,b) = \int_0^1 t^{a-1}(1-t)^{b-1}\,dt \qquad (\mathrm{Re}\,a,\ \mathrm{Re}\,b > 0) \tag{32.7}$$

or equivalently

$$\mathrm{B}(a,b) = \int_1^\infty t^{-a-b}(t-1)^{b-1}\,dt \tag{32.8}$$

in which $\mathrm{B}(a,b)$ is known as the **beta function**, or **Euler's integral of the first kind**. The second form (Eq. 32.8) is obtained from Eq. 32.7 by making the substitution $t \to 1/t$. By letting $t = \sin^2\theta$, Eq. 32.7 can also be expressed as

$$\mathrm{B}(a,b) = 2\int_0^{\pi/2} \sin^{2a-1}\theta\,\cos^{2b-1}\theta\,d\theta \tag{32.9}$$

We now prove Eq. 32.6. After making an obvious change of integration variable, from Eq. 32.4 we have

$$\begin{aligned}\Gamma(a)\Gamma(b) &= 4\int_0^\infty e^{-y^2}y^{2a-1}dy\cdot\int_0^\infty e^{-x^2}x^{2b-1}dx\\ &= 4\int_0^\infty\int_0^\infty e^{-(x^2+y^2)}x^{2b-1}y^{2a-1}dx\,dy\end{aligned} \tag{32.10}$$

Changing to polar coordinates, Eq. 32.10 leads to

$$\Gamma(a)\Gamma(b) = \left[\int_0^\infty dr^2 e^{-r^2}(r^2)^{a+b-1}\right]\cdot\left[2\int_0^{\pi/2}\sin^{2a-1}\theta\,\cos^{2b-1}\theta\,d\theta\right] \tag{32.11}$$

The first integral is equal to $\Gamma(a+b)$; compare Eq. 32.4. The second integral is $\mathrm{B}(a,b)$; compare Eq. 32.9. Thus, we obtain the desired result, and from this result the symmetry property of the beta function

$$\mathrm{B}(a,b) = \mathrm{B}(b,a) \tag{32.12}$$

(iii) The third property of the Γ function we wish to deduce is

$$\Gamma(z)\Gamma(1-z) = \pi\csc\pi z \tag{32.13}$$

Setting $b = z$, $a = 1 - z$ in Eq. 32.11, one has

$$\begin{aligned}\Gamma(z)\Gamma(1-z) &= \left[\int_0^\infty dr^2\,e^{-r^2}\right]\cdot\left[2\int_0^{\pi/2}\cot^{2z-1}\theta\,d\theta\right]\\ &= 2\int_0^{\pi/2}\cot^{2z-1}\theta\,d\theta \qquad (0 < \mathrm{Re}\,z < 1)\end{aligned} \tag{32.14}$$

Setting $\cot\theta \equiv \zeta$, Eq. 32.14 leads to

$$\Gamma(z)\Gamma(1-z) = 2\int_0^\infty \frac{\zeta^{2z-1}}{\zeta^2+1}\,d\zeta \qquad (0 < \mathrm{Re}\,z < 1) \tag{32.15}$$

The integral has been evaluated in Sec. 25 for real values of z. Thus, we obtain Eq. 32.13 for $0 < z < 1$, and by analytic continuation Eq. 32.13 holds for any z where both sides of the equation are analytic functions.

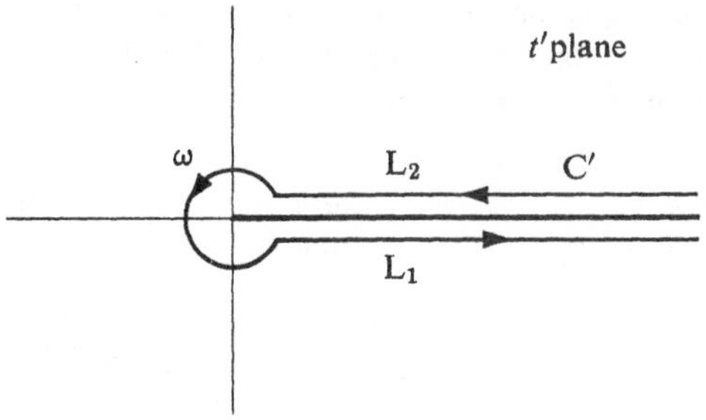

Fig. 38.

Setting $z = \frac{1}{2}$ in Eq. 32.13, we find

$$\Gamma(\tfrac{1}{2}) = \sqrt{\pi} \tag{32.16}$$

With the aid of Eq. 32.13, we can now demonstrate the equivalence of the two representations (32.1 and 32.4). We first transform Eq. 32.1 by the substitution

$$t = e^{-i\pi}t' \tag{32.17}$$

$$\frac{1}{\Gamma(1-z)} = -\frac{e^{-i\pi(z-1)}}{2\pi i}\int_{C'} e^{-t'}t'^{z-1}\,dt' \tag{32.18}$$

where C′ is the contour of Fig. 38. We assume in the beginning that z is not an integer.

From Eqs. 32.2 and 32.17 we have

$$\arg t' = \begin{cases} 0 & \text{above the cut} \\ 2\pi & \text{below the cut} \end{cases} \tag{32.19}$$

On the small circle ω of radius ρ, we set $t' = \rho e^{i\theta}$. Then we have

$$\frac{1}{\Gamma(1-z)} = -\frac{e^{-i\pi(z-1)}}{2\pi i}\left\{\int_{\infty}^{0} e^{-t'}t'^{z-1}\,dt' + e^{2\pi i(z-1)}\int_{0}^{\infty} e^{-t'}t'^{z-1}\,dt' \right.$$
$$\left. + i\rho^z\int_0^{2\pi} e^{\rho(\cos\theta + i\sin\theta) + iz\theta}\,d\theta\right\} \tag{32.20}$$

In the limit as $\rho \to 0$, the third integral vanishes when $\operatorname{Re} z > 0$ and we are left with

$$\frac{1}{\Gamma(1-z)} = -\frac{1}{2\pi i}\left\{\left[e^{i\pi(z-1)} - e^{-i\pi(z-1)}\right]\int_0^{\infty} e^{-t'}t'^{z-1}\,dt'\right\}$$
$$= -\frac{\sin\pi(z-1)}{\pi}\int_0^{\infty} e^{-t'}t'^{z-1}\,dt'$$
$$= \frac{\sin\pi z}{\pi}\int_0^{\infty} e^{-t'}t'^{z-1}\,dt' \qquad \operatorname{Re} z > 0 \tag{32.21}$$

Using Eqs. 32.4 and 32.13 (derived from 32.4) we get

$$\left[\frac{1}{\Gamma(1-z)}\right]\text{def. by 32.1} = \left[\frac{1}{\Gamma(1-z)}\right]\text{ def. by 32.4}$$

Although this equation has been derived for non-integer z and for Re $z > 0$, it holds by analytic continuation for all values of z. We have thus checked that Eq. 32.4 is also a proper representation of the Γ function.

(iv) From Eq. 32.5, one sees that $1/\Gamma(z)$ is an entire function of z with simple zeros at $z = 0, -1, -2, \cdots$. Therefore, using Eq. 29.7 for $1/\Gamma(z+1)$ and Eq. 32.5, $1/\Gamma(z)$ can be written in the product form as

$$\frac{1}{\Gamma(z)} = ze^{\gamma z} \prod_{k=1}^{\infty} \left(1 + \frac{z}{k}\right) e^{-z/k} \tag{32.22}$$

Since $\Gamma(1) = 1$, the constant γ can be determined from the relation

$$e^{-\gamma} = \prod_{k=1}^{\infty} \left(1 + \frac{1}{k}\right) e^{-1/k} \tag{32.23}$$

γ is called the Euler-Mascheroni constant.

A numerical evaluation of γ yields

$$\gamma = 0.57721566 \cdots$$

From Eq. 32.22 one immediately obtains a useful relation for the logarithmic derivative of the Γ function, usually denoted by $\psi(z)$

$$\psi(z) = \frac{d}{dz} \ln \Gamma(z) = -\gamma - \frac{1}{z} - \sum_{k=1}^{\infty} \left(\frac{1}{z+k} - \frac{1}{k}\right) \tag{32.24}$$

(v) In the example of Sec. 31 we derived an expression for the asymptotic behavior of the integral $I(w)$, which in fact is just the gamma function. (Compare Eqs. 31.38 and 32.4.)

Thus, relabeling the argument by putting $z + 1$ instead of w, we find

$$\Gamma(z+1) \sim \sqrt{2\pi} e^{-z} z^{z+1/2} \tag{32.25}$$

This formula is known as **Stirling's approximation for the gamma function**, and although it has been derived for real z, it is in fact also valid for complex z, provided $|\arg z| < \pi$.

33 · FUNCTIONS OF SEVERAL COMPLEX VARIABLES. ANALYTIC COMPLETION

It is possible to generalize some of the preceding results to the case where a function depends on several complex variables. To be specific, consider a function $f(z_1,z_2)$ of the two complex variables z_1 and z_2, which belong to two distinct complex planes. Suppose that there exists a simply-connected region R_1 in the z_1 plane and a simply-connected region R_2 in the z_2 plane such that

(i) $f(z_1,z_2)$ is an analytic function of z_1, for $z_1 \in R_1$ and for fixed $z_2 \in R_2$.
(ii) $f(z_1,z_2)$ is an analytic function of z_2 for $z_2 \in R_2$ and for fixed $z_1 \in R_1$.

If now C_1 is a closed contour belonging to R_1 and C_2 a closed contour belonging to R_2, we have

$$f(z_1,z_2) = \frac{1}{2\pi i} \int_{C_1} \frac{f(z'_1,z_2)}{z'_1 - z_1} dz'_1 \qquad \begin{matrix} z_1 \in R_1 \\ z_2 \in R_2 \end{matrix} \tag{33.1}$$

and

$$f(z_1,z_2) = \frac{1}{2\pi i} \int_{C_2} \frac{f(z_1,z'_2)}{z'_2 - z_2} dz'_2 \qquad \begin{matrix} z_1 \in R_1 \\ z_2 \in R_2 \end{matrix} \tag{33.2}$$

Combining Eqs. 33.1 and 33.2, we obtain the Cauchy integral formula for two complex variables

$$f(z_1,z_2) = \left(\frac{1}{2\pi i}\right)^2 \int_{C_1}\int_{C_2} \frac{f(z'_1,z'_2)}{(z'_1 - z_1)(z'_2 - z_2)} dz'_1\, dz'_2 \qquad \begin{matrix} z_1 \in \mathrm{R}_1 \\ z_2 \in \mathrm{R}_2 \end{matrix} \tag{33.3}$$

As in the case of a function of one complex variable, many consequences can be deduced from this generalized Cauchy integral representation. For example, every analytic function of two variables can be expanded in a (double) power series about any pair of regular points z_{10},z_{20}

$$f(z_1,z_2) = \sum_{m,n=0}^{\infty} a_{mn}(z_1 - z_{10})^m(z_2 - z_{20})^n \tag{33.4}$$

Similarly, all the partial derivatives

$$\frac{\partial^{m+n} f(z_1,z_2)}{\partial^m z_1\, \partial^n z_2}$$

exist and are analytic functions for $z_1 \in \mathrm{R}_1$, $z_2 \in \mathrm{R}_2$. One can also show, using the power series expansion, that two analytic functions which are equal in a common region are analytic continuations of each other.

It should be understood, however, that beyond these simple generalizations, functions of several complex variables display quite new features which have no analog in the one-variable case. This is similar to the situation one encounters in generalizing functions of a real variable to functions of a complex variable.

The following example will illustrate the type of unexpected result that one obtains.

Suppose that a function $f(z_1,z_2)$ is analytic throughout the open segment of the real z_2 axis

$$-b < z_2 < b \tag{33.5a}$$

provided $|z_1|$ satisfies the inequalities

$$|z_1| < \sqrt{b^2 - (z_2)^2} \qquad \text{for} \quad \begin{cases} a < z_2 < b \\ -b < z_2 < -a \end{cases} \tag{33.5b}$$

and

$$\sqrt{a^2 - (z_2)^2} < |z_1| < \sqrt{b^2 - (z_2)^2} \qquad \text{for } -a < z_2 < a \tag{33.5c}$$

Of course, when we say that $f(z_1,z_2)$ is analytic on a segment of the real z_2 axis, we mean that it is analytic in some neighborhood of any point of the segment and therefore that it is also analytic in some environment of the segment. We limit our attention to real values of z_2 only to make the discussion simpler and easier to visualize.

If we consider z_2, Re z_1, and Im z_1 as Cartesian coordinates of a point in space, then the region of analyticity of $f(z_1,z_2)$, as defined by inequalities 33.5, can be thought of as the domain between two spheres of radii a and b, respectively. The inequalities (33.5) determine nothing more than the intersection of this domain, with a plane perpendicular to the z_2 axis and crossing this axis at the point z_2 (Fig. 39). Such a plane will be denoted by P_{z_2}. It is clear that this intersection is either the annular region between two circles or the interior of a circle, depending upon whether P_{z_2} cuts the smaller sphere or not. Furthermore, let us denote by R_{z_2} the radius of an arbitrary circle centered on the z_2 axis, which lies in the plane P_{z_2} and which is entirely contained within the region between the two spheres [i.e., within the domain of analyticity of $f(z_1,z_2)$].

Let $g(z_1,z_2)$ be defined for $|z_1| < R_{z_2}$ by the integral

$$g(z_1,z_2) = \frac{1}{2\pi i} \int_{|z'_1| = R_{z_2}} \frac{f(z'_1,z_2)}{z'_1 - z_1} dz'_1 \tag{33.6}$$

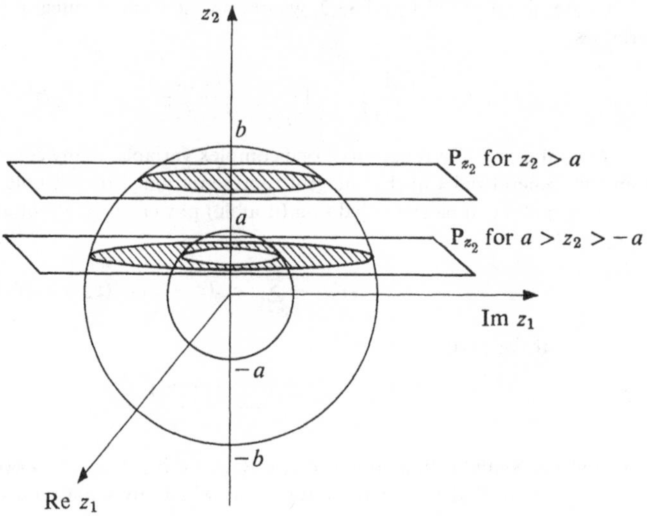

Fig. 39.

Of course the circle of integration can be enlarged without affecting the value of $g(z_1,z_2)$ for given z_1 and z_2. The enlargement of the circle simply yields the analytic continuation of the function $g(z_1,z_2)$ for larger $|z_1|$.

By virtue of Cauchy's integral formula

$$g(z_1,z_2) = f(z_1,z_2) \tag{33.7}$$

for

$$|z_1| < R_{z_2} \qquad \text{and} \qquad a < z_2 < b$$

Let us take two planes $P_{a+\delta}$ and $P_{a-\delta}$; the first is above the inner sphere and the second intersects it. Since the spacing between the two spheres is finite, we can always choose δ in such a way that it is possible to draw in the two planes, circles centered on the z_2 axis that are entirely within the domain of analyticity of $f(z_1,z_2)$, and which lie one above the other, so that

$$R_{a+\delta} = R_{a-\delta} \tag{33.8}$$

We now expand $f(z'_1,z_2)$ in a power series about a point z_{20} such that

$$a < z_{20} < a + \delta \tag{33.9}$$

for

$$|z'_1| = R_{a+\delta} = R_{a-\delta} \tag{33.10}$$

The expansion is

$$f(z'_1,z_2) = \sum_{n=0}^{\infty} \frac{1}{n!} \frac{\partial^n f(z'_1,z_{20})}{\partial^n z_{20}} (z_2 - z_{20})^n \tag{33.11}$$

Let us impose an additional condition on δ, namely, we demand that the series in Eq. 33.11 converges uniformly for $z_2 = a - \delta$. In spite of the fact that the radius of convergence of the series in Eq. 33.11 depends on z'_1, it has a finite lower limit for z'_1 satisfying Eq. 33.10, and our requirement can always be met by choosing δ small enough and z_{20} sufficiently close to a. From Eq. 33.11 one has

$$\frac{1}{2\pi i} \int_{|z'_1|=R_{a+\delta}} \frac{f(z'_1,z_2)}{z'_1 - z_1} dz'_1 = \frac{1}{2\pi i} \sum_{n=0}^{\infty} \frac{(z_2 - z_{20})^n}{n!} \int_{|z'_1|=R_{a+\delta}} \frac{\dfrac{\partial^n f(z'_1,z_{20})}{\partial^n z_{20}}}{z'_1 - z_1} dz'_1 \tag{33.12}$$

Putting in Eq. 33.6

$$R_{z_2} = R_{a+\delta} = R_{a-\delta} \tag{33.13}$$

for

$$a - \delta \leq z_2 \leq a + \delta \tag{33.14}$$

we obtain

$$g(z_1,z_2) = \frac{1}{2\pi i} \sum_{n=0}^{\infty} \frac{(z_2 - z_{20})^n}{n!} \int_{|z'_1| = R_{a+\delta}} \frac{\dfrac{\partial^n f(z'_1,z_{20})}{\partial^n z_{20}}}{z'_1 - z_1} dz'_1 \tag{33.15}$$

which proves that $g(z_1,z_2)$ is an analytic function of z_2 for

$$a - \delta < z_2 < a + \delta \tag{33.16}$$

when

$$|z_1| < R_{a+\delta} = R_{a-\delta} \tag{33.17}$$

On the other hand, from the result of Sec. 13 we know that $g(z_1,z_2)$ is an analytic function of z_1 wherever $f(z'_1,z_2)$ is continuous on the contour of integration and z_1 is not on the contour. Hence, $g(z_1,z_2)$ is an analytic function of z_1 for

$$|z_1| < R_{a+\delta} = R_{a-\delta}$$

when $a - \delta < z_2 < a + \delta$.

We have shown that the part of the inner sphere which is above the plane $P_{a-\delta}$ belongs to the region of analyticity of $g(z_1,z_2)$ and, therefore, also of $f(z_1,z_2)$ by analytic continuation because the two functions satisfy Eq. 33.7.

We can now enlarge the circle in a plane lying between P_a and $P_{a-\delta}$, project it onto a plane lying below $P_{a-\delta}$ and repeat the reasoning. In this manner one finds that $f(z_1,z_2)$ can be analytically continued throughout the whole interior of the inner sphere.

Summarizing: If a function $f(z_1,z_2)$ is analytic in a region between two spheres, it is also analytic throughout the inner sphere. This result is independent of the particular form of the function $f(z_1,z_2)$; it rests solely on the geometrical properties of the region between the two spheres.

A process whereby the extension of the region of analyticity of a function is based only on the geometrical properties of the region where the function was originally defined and which can be carried out for any analytic function defined in that region, is called **analytic completion.**

A consequence of the result we have just proved is that a function of two complex variables cannot have isolated singularities on the real z_2 axis, since such a singularity could always be enclosed within a sphere inside of which the function would be analytic except at the point itself.

By similar arguments, one can show that a function of two complex variables cannot have isolated singularities anywhere. The simplest set of singularities of a function of two complex variables is a trajectory, $z_2 = h(z_1)$, which must extend to infinity.

CHAPTER II

LINEAR VECTOR SPACES

1 · INTRODUCTION

The ideas presented in this chapter may be regarded as a generalization of elementary vector calculus. The way of presentation will be, however, quite different from the one that is usual in vector calculus. From the very beginning we emphasize those points that are really essential, and use a notation which, although it may seem strange at first, has a great many advantages, especially in physical applications. The rather abstract formulation of the theory will be supplemented by very simple examples in order to help to establish a link between the abstract notions introduced and the more familiar and intuitively understandable ideas of elementary mathematics.

The reader who is not yet accustomed to rather abstract reasoning should study all the more carefully the first few sections of this chapter. He will then undoubtedly be amply repaid by finding the rest of this chapter, and Chapter III, relatively easy reading.

2 · DEFINITION OF A LINEAR VECTOR SPACE

Let us consider a set S of certain abstract objects, represented by the symbol $|\rangle$*; in order to distinguish these objects, we provide them with labels.

EXAMPLE 1

For instance, some examples of $|\rangle$ are

$$|a\rangle, \quad |3\rangle, \quad |\alpha\beta\rangle$$

Having introduced these objects, we must define "rules of manipulation" or of "composition" as one calls them, of these objects, i.e., their algebra. This is similar to, say, the real numbers. One can introduce the set of real numbers, but unless one also specifies rules of addition and multiplication, one has done little more than distribute names. It is up to us to define the rules of algebra, but we must require that these rules be unambiguous.

The first of these operations to be defined and which in analogy to the case of real numbers we call addition of $|\rangle$, allows us to construct from any two $|\rangle$ a third $|\rangle$, which is called the sum of the first two $|\rangle$. In order to indicate that the particular object $|c\rangle$ is the sum of the particular objects $|a\rangle$ and $|b\rangle$, we write

$$|c\rangle = |a\rangle + |b\rangle \tag{2.1}$$

* The notation is due to P. A. M. Dirac.

The second operation to be defined, which we call "multiplication of a $|\rangle$ by a number", allows us to construct from any complex number and any $|\rangle$ another $|\rangle$. The equation

$$|c\rangle = \alpha \cdot |b\rangle \tag{2.2}$$

will mean that $|c\rangle$ is the product of $|b\rangle$ by the complex number α.*

We have now defined operations of addition and multiplication for a general set of objects $|\rangle$. This chapter, however, will deal only with special sets of objects that have the following properties:

A (a) If $|a\rangle, |b\rangle \in \mathrm{S}$, then $(|a\rangle + |b\rangle) \in \mathrm{S}$.
(b) If $|a\rangle \in \mathrm{S}$ and α is a complex number, then $(\alpha\,|a\rangle) \in \mathrm{S}$.
(c) There exists a null element $|0\rangle \in \mathrm{S}$ such that for any $|a\rangle \in \mathrm{S}$ one has $|a\rangle + |0\rangle = |a\rangle$
(d) For any $|a\rangle \in \mathrm{S}$ there exists an element $|a'\rangle \in \mathrm{S}$ such that $|a\rangle + |a'\rangle = |0\rangle$.

So far we have said that there exists certain operations, called addition and multiplication, but we have not yet specified the properties of these operations. The following properties B(a), B(b), B(c) will ensure that addition and multiplication are well-defined operations.

B For any $|a\rangle, |b\rangle, |c\rangle \in \mathrm{S}$ and for any complex numbers α and β one has:

(a) $|a\rangle + |b\rangle = |b\rangle + |a\rangle$; (commutative law of addition).
$(|a\rangle + |b\rangle) + |c\rangle = |a\rangle + (|b\rangle + |c\rangle)$; (associative law of addition).
(b) $1 \cdot |a\rangle = |a\rangle$.
(c) $\alpha \cdot (\beta \cdot |a\rangle) = (\alpha \cdot \beta) \cdot |a\rangle$; (associative law of multiplication).
$(\alpha + \beta)|a\rangle = \alpha\,|a\rangle + \beta\,|a\rangle$; (distributive law with respect to the addition of complex numbers).
$\alpha(|a\rangle + |b\rangle) = \alpha\,|a\rangle + \alpha\,|b\rangle$; (distributive law with respect to the addition of $|\rangle$).

A set S of $|\rangle$ that has the properties A and B is called a **linear vector space**. The elements of this set, $|\rangle$, are called **vectors**.

EXAMPLE 2

Suppose that the set S of objects $|\rangle$ is the set of all complex numbers. Then the list of B properties simply contains the well-known rules of arithmetic for complex numbers. This example justifies the naming of the abstract operations (2.1 and 2.2) as addition and multiplication.

EXAMPLE 3

Suppose that the set S consists of all the arrows lying in a plane, including the "arrow" of zero length. For the rule of addition of $|\rangle$, we take the familiar geometrical rule of the addition of arrows, as illustrated in Fig. 40(a). The reader can verify that this rule obeys all the conditions enumerated in B properties.

The multiplication of a $|\rangle$ by the number $z = re^{i\psi}$ (r and ψ being real) will be defined as the elongation of the arrow r times and its subsequent rotation by the angle ψ (as shown in Fig. 40(b)). When z is real, this rule reduces to the conventional one of multiplying arrows by numbers.

* The dot will often be omitted.

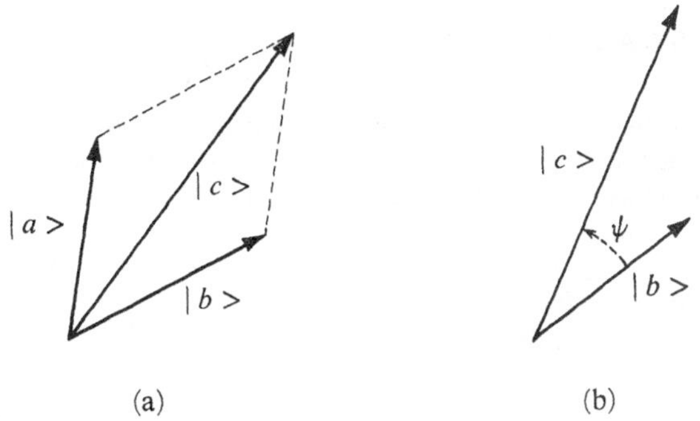

Fig. 40. Geometrical illustration of Eqs. 2.1 and 2.2 in the particular case of Example 3. (a) $|c\rangle = |a\rangle + |b\rangle$. (b) $|c\rangle = (re^{i\psi})|b\rangle$. The length of the arrow $|c\rangle$ is r times the length of $|b\rangle$.

Comparing these properties of the arrows with the properties A, we see that the set of all arrows constitutes a linear vector space. For example, the addition (as defined above) of two arrows is an arrow, and the multiplication of an arrow by a complex number is another arrow, etc.

Starting from A and B properties, one can easily demonstrate that a linear vector space contains only one null vector $|0\rangle$ and that to each vector $|a\rangle$ there corresponds one and only one vector $|a'\rangle$ satisfying $|a\rangle + |a'\rangle = |0\rangle$.

We now verify that any vector $|a\rangle$ multiplied by the number 0 gives $|0\rangle$. We have

$$|a\rangle = 1 \cdot |a\rangle = (0 + 1)\,|a\rangle = 0 \cdot |a\rangle + 1 \cdot |a\rangle$$
$$= 0 \cdot |a\rangle + |a\rangle$$

Hence, $|a\rangle = 0 \cdot |a\rangle + |a\rangle$.

Let $|a'\rangle$ be the vector satisfying

$$|a'\rangle + |a\rangle = |0\rangle$$

Then

$$|0\rangle = |a\rangle + |a'\rangle = (0 \cdot |a\rangle + |a\rangle) + |a'\rangle$$
$$= 0 \cdot |a\rangle + (|a\rangle + |a'\rangle)$$
$$= 0 \cdot |a\rangle + |0\rangle = 0\,.\,|a\rangle$$

or briefly

$$0 \cdot |a\rangle = |0\rangle \qquad \text{for any} \quad |a\rangle \in S \tag{2.3}$$

Because of Eq. 2.3, no ambiguity will result when, for simplicity, we write briefly 0 instead of $|0\rangle$.

It is easy to define subtraction of vectors

$$|a\rangle - |b\rangle \underset{\text{def}}{=} |a\rangle + (-1)\,|b\rangle$$

Of course

$$|a\rangle - |a\rangle = |a\rangle + (-1)\,|a\rangle = (1 - 1)\,|a\rangle = 0$$

as it should be.

3 · THE SCALAR PRODUCT

Suppose one has established a rule that associates with any pair of vectors $|b\rangle \in S$ and $|a\rangle \in S$ a certain complex number; we shall denote* it by $\langle b|a\rangle$ and call it the **scalar product** of $|b\rangle$ with $|a\rangle$. The properties of the scalar product will be, by definition, the following:

C (a) $\langle b|a\rangle = \overline{\langle a|b\rangle}$

(b) If

$$|d\rangle = \alpha\,|a\rangle + \beta\,|b\rangle$$

then

$$\langle c|d\rangle = \alpha\langle c|a\rangle + \beta\langle c|b\rangle$$

(c) $\langle a|a\rangle \geq 0$, the equality sign appears only when $|a\rangle = 0$.

Note that because of C(a), the number $\langle a|a\rangle$ is real. This is an important property, which will enable us to regard $\sqrt{\langle a|a\rangle}$ as the "length" of the vector $|a\rangle$.

From C(a) we see that in general the scalar product of $|b\rangle$ with $|a\rangle$ is not the same as the scalar product of $|a\rangle$ with $|b\rangle$, since

$$\langle b|a\rangle = \overline{\langle a|b\rangle} \neq \langle a|b\rangle$$

Two vectors $|a\rangle$ and $|b\rangle$ are said to be **orthogonal** to each other if their scalar product vanishes

$$\langle a|b\rangle = \langle b|a\rangle = 0$$

The definition C(c) implies that if a vector $|a\rangle \in S$ is orthogonal to every vector of S

$$\langle a|\,\rangle = 0 \qquad \text{for all} \quad |\,\rangle \in S \tag{3.1}$$

then $|a\rangle = 0$, since from Eq. 3.1 one has in particular $\langle a|a\rangle = 0$.

4 · DUAL VECTORS AND THE CAUCHY-SCHWARZ INEQUALITY

The form of the definition C(b) of the preceding section introduces an apparent asymmetry between the vectors $|c\rangle$ and $|d\rangle$ which enter into the scalar product $\langle c|d\rangle$. The meaning of C(b) is that the scalar product $\langle c|d\rangle$ depends linearly upon the vector $|d\rangle$ in the sense that if we set

$$|d\rangle = \alpha\,|a\rangle + \beta\,|b\rangle$$

then

$$\langle c|d\rangle = \alpha\,\langle c|a\rangle + \beta\,\langle c|b\rangle$$

is a linear function of α and β. However, if we set

$$|c\rangle = \alpha\,|a\rangle + \beta\,|b\rangle$$

* A "closed bracket" expression $\langle\,|\,\rangle$ will, by convention, always denote a number (complex in general) and not a vector.

then

$$\langle c|d\rangle = \overline{\langle d|c\rangle} = \overline{[\alpha\langle d|a\rangle + \beta\langle d|b\rangle]}$$
$$= \bar{\alpha}\langle a|d\rangle + \bar{\beta}\langle b|d\rangle \tag{4.1}$$

is no longer a linear function of α and β, since it depends linearly on $\bar{\alpha}$ and $\bar{\beta}$.*

To remove this assymetry, it is convenient to introduce, besides the vectors $|\rangle$, other vectors belonging to a **different** space and which will be denoted by the symbol $\langle|$. We shall assume that there is a one-to-one correspondence between vectors $|\rangle$ and vectors $\langle|$. A pair of vectors in which each is in correspondence with the other will be called a **pair of dual vectors**, and such pairs will always carry the same identification label. Thus, e.g., $\langle b|$ is the dual vector of $|b\rangle$.

We now define the multiplication of vectors $|\rangle$ by vectors $\langle|$ by requiring the following

D (a) The product of $\langle b|$ with $|a\rangle$ is identified with the scalar product $\langle b|a\rangle$.

$$\langle b|\cdot|a\rangle \equiv \langle b|a\rangle$$

(b) The scalar product $\langle c|d\rangle$ depends linearly on $\langle c|$.

$$[\langle a|\,\alpha + \langle b|\beta]\cdot|d\rangle = \alpha\langle a|d\rangle + \beta\langle b|d\rangle$$

From D(a) and C(b) we have

$$\langle c|\cdot[\alpha\,|a\rangle + \beta\,|b\rangle] = \alpha\langle c|a\rangle + \beta\langle c|d\rangle$$

Thus, the vectors $\langle|$ and $|\rangle$ play a symmetrical role in the scalar product.

Setting

$$\langle c| = \langle a|\,\bar{\alpha} + \langle b|\,\bar{\beta}$$

we have

$$\langle c|d\rangle = \bar{\alpha}\langle a|d\rangle + \bar{\beta}\langle b|d\rangle \tag{4.2}$$

Comparing Eq. 4.2 with Eq. 4.1, we see that $\langle a|\,\bar{\alpha} + \langle b|\,\bar{\beta}$ is the dual vector of $\alpha|a\rangle + \beta\,|b\rangle$; hence, the rule for obtaining a dual of a linear combination of vectors $|\rangle$ is to replace the vectors by their duals and the coefficients by their complex conjugates. The reason why the scalar product is symmetric with respect to vectors $\langle|$ and $|\rangle$ is now apparent; we have, so to say, included the complex conjugation in the definition of the vectors $\langle|$.

The advantage of considering $\langle|\rangle$ as a product of $\langle|$ with $|\rangle$ is that now a simple distributive law of multiplication holds for the vectors $\langle|$ as well as for the vectors $|\rangle$.

The manner in which we introduced dual vectors, namely, as a device for simplifying the notation, is neither very rigorous nor the most general, although it is quite sufficient for our purpose. The interested reader can find the general definition sketched below.

Let f be a function defined in S by a rule that associates with every **vector** $|x\rangle \in$ S a complex number $f(|x\rangle)$; such a function is usually called a **functional**. The functional f is linear if

$$f(\alpha\,|x\rangle + \beta\,|y\rangle) = \alpha f(|x\rangle) + \beta f(|y\rangle)$$

* One says that $\langle c|d\rangle$ is linear with respect to $|d\rangle$ but antilinear with respect to $|c\rangle$.

The set of all linear functionals in S forms a linear vector space, for adding two linear functionals and multiplying a linear functional by a number results again in a linear functional. The space of all linear functionals in S is called the **dual space** of S.

Suppose that the scalar product has been defined in S and consider all the functionals of the particular type

$$f(|x\rangle) \equiv \langle f|x\rangle, \qquad |f\rangle \in \mathrm{S}$$

Owing to the linearity of the scalar product, these are linear functionals, and it is clear that they form a linear vector space. This space is just the space of vectors $\langle|$; the use of vectors $\langle|$ corresponds to using the notation

$$f \equiv \langle f|$$

which should be understood in the sense that attaching an argument $|x\rangle$ to f is equivalent to "multiplying" $\langle f|$ by $|x\rangle$. The dual vector of $|f\rangle$ is $\langle f|$.

Consider now the vector

$$|c\rangle = |a\rangle - x\,\langle b|a\rangle\,|b\rangle$$

with real x. Since

$$\langle c|c\rangle \geq 0$$

we have

$$x^2\,\langle b|a\rangle\langle a|b\rangle\langle b|b\rangle - 2x\,\langle b|a\rangle\langle a|b\rangle + \langle a|a\rangle \geq 0 \tag{4.3}$$

Inequality 4.3 implies that the above quadratic equation in x with real coefficients has either a double real root or no real roots. Therefore

$$\langle a|a\rangle\langle b|b\rangle \geq \langle b|a\rangle\langle a|b\rangle = |\langle b|a\rangle|^2$$

or

$$\sqrt{\langle a|a\rangle}\cdot\sqrt{\langle b|b\rangle} \geq |\langle b|a\rangle| \tag{4.4}$$

Inequality 4.4 is known as the **Cauchy-Schwarz inequality**.

5 · REAL AND COMPLEX VECTOR SPACES

Until now the word "number" has been understood in the sense of a complex number. A real number is, however, a special case of a complex number. It is obvious, therefore, that one can repeat all the considerations of the previous sections, restricting ourselves to real numbers exclusively. The only difference would be that complex conjugation would become a redundant operation and consequently would never appear. For instance, the scalar product would be symmetric, since we would have the relation

$$\langle a|b\rangle = \langle b|a\rangle$$

instead of the more general relation

$$\langle a|b\rangle = \overline{\langle b|a\rangle}$$

In the light of these remarks, one can speak of "real" and "complex" vector spaces.

A simple example of a real vector space is the set of arrows lying in a plane. Here the addition of arrows is to be understood in the usual sense of the parallelogram rule, and the multiplication of an arrow by a number x is to be understood as the elongation of the vector x times. This is not the same as assuming that the arrows belong to a complex vector space, as in Example 3 of Sec. 2, where the multiplication of an arrow by a complex number was defined; it not only elongated the arrow, but also rotated it and brought it onto another arrow. This double operation was possible because a complex number contains two real parameters. The importance of this difference will be properly understood after we have introduced the notion of a dimension of the space.

The reader who feels ill at ease with some of the abstract notions of complex vector spaces introduced in this text is advised to think in terms of arrows in a plane or in space. This should help his intuitive understanding of the more abstract case.

EXAMPLE

Consider arrows in a plane. As we mentioned above, they form a real vector space. The scalar product of two vectors $|a\rangle$ and $|b\rangle$ will be defined conventionally as the product of the lengths of the corresponding arrows by the cosine of the angle between them

$$\langle a|b\rangle = \langle b|a\rangle \underset{\text{def}}{=} \vec{a}\cdot\vec{b} = |\vec{a}|\,|\vec{b}|\cos\psi_{a,b} \tag{5.1}$$

The reader can verify that this definition of the scalar product is consistent with C. For example

$$\langle a|[|b\rangle + |c\rangle] = \vec{a}\cdot(\vec{b}+\vec{c}) = |\vec{a}|\,|\vec{b}+\vec{c}|\cos\psi_{a,b+c}$$

On the other hand, using the well-known trigonometrical relations between the sines of angles and the sides of a triangle, after some calculations we get

$$\begin{aligned}
\langle a|b\rangle + \langle a|c\rangle &= |\vec{a}|\,|\vec{b}|\cos\psi_{a,b} + |\vec{a}|\,|\vec{c}|\cos\psi_{a,c} \\
&= |\vec{a}|\,|\vec{b}+\vec{c}|\left\{\frac{|\vec{b}|}{|\vec{b}+\vec{c}|}\cos\psi_{a,b} + \frac{|\vec{c}|}{|\vec{b}+\vec{c}|}\cos\psi_{a,c}\right\} \\
&= |\vec{a}|\,|\vec{b}+\vec{c}|\left\{\frac{\sin\psi_{c,b+c}}{\sin\psi_{b,c}}\cos\psi_{a,b} + \frac{\sin\psi_{b,b+c}}{\sin\psi_{b,c}}\cos\psi_{a,c}\right\} \\
&= |\vec{a}|\,|\vec{b}+\vec{c}|\cos\psi_{a,b+c}
\end{aligned}$$

Thus

$$\langle a|[|b\rangle + |c\rangle] = \langle a|b\rangle + \langle a|c\rangle$$

as required by C(b) in Sec. 3.

From the definition (5.1) it is evident that the orthogonality of two vectors means that the corresponding arrows are perpendicular.

6 · METRIC SPACES

E A set R is called a metric space if a real, positive number $\rho(a,b)$ is associated with any pair of its elements $a,b \in$ R and if

$$\begin{aligned}
&\text{(a)}\quad \rho(a,b) = \rho(b,a) \\
&\text{(b)}\quad \rho(a,b) = 0 \text{ only when } a \equiv b \\
&\text{(c)}\quad \rho(a,b) + \rho(b,c) \geq \rho(a,c)
\end{aligned} \tag{6.1}$$

The number $\rho(a,b)$ is called the distance between a and b. Conditions E(a) and E(b) simply mean that the distance from a to b is the same as that from b to a, and that the distance vanishes only when two elements coincide. The condition E(c) is known as the **triangle inequality**.

EXAMPLE 1

Any set of points on a plane is a metric space if $\rho(a,b)$ is identified with the "ordinary" distance between the points a and b. The condition E(c) is then the familiar statement that the sum of the lengths of two sides of a triangle is not smaller than the length of the third side of this triangle. (Cf. the relations 1.8 of the preceding chapter.)

This notion of a distance between elements of a set is now extended to the case where the set constitutes a linear vector space. First, a few comments are in order.

The scalar product of $|a\rangle$ with its dual vector $\langle a|$ is, by the very definition of the scalar product, a positive number. $\sqrt{\langle a|a\rangle}$ is called the **norm**, or the **length**, of the vector $|a\rangle$.

EXAMPLE 2

In the elementary case of arrows in a plane (example of Sec. 5), the norm is simply the length of the arrow

$$\sqrt{\langle a|a\rangle} = |\vec{a}|$$

In elementary vector calculus one considers a vector as an arrow that joins two points of the space; each vector has its origin and its end. In the general theory of linear vector spaces it is also often very helpful to consider vectors as having a common origin and extending out from that origin. Each vector may then be considered as a "radius vector" which defines a point (the "end" of the vector) in the space. It must, however, be kept very clearly in mind that this notion of a point in space is introduced only as a pictorial representation of a vector, and that it never enters in a fundamental way into the theory of linear vector spaces. In fact, in defining a linear vector space (Sec. 2), the idea of a point was not even mentioned.

Let us define the distance between two vectors $|a\rangle$ and $|b\rangle$ (or, if we wish, the distance between the points that they determine) as the norm of the vector $\{|a\rangle - |b\rangle\}$. We do this in analogy to elementary vector calculus, where the distance between two points is defined as the length of the vector joining the ends of the respective radius vectors or, equivalently (according to the rule of vectors addition) the length of the difference between two vectors.

We show that the distance so defined satisfies the triangle inequality. Let

$$|3\rangle = |1\rangle + |2\rangle$$

We have

$$\begin{aligned}\langle 3|3\rangle &= \langle 1|1\rangle + \langle 2|2\rangle + 2\,\mathrm{Re}\langle 1|2\rangle \\ &\leq \langle 1|1\rangle + \langle 2|2\rangle + 2|\langle 1|2\rangle|\end{aligned}$$

Using the Cauchy-Schwarz inequality, we get

$$\begin{aligned}\langle 3|3\rangle &\leq \langle 1|1\rangle + \langle 2|2\rangle + 2\sqrt{\langle 1|1\rangle\langle 2|2\rangle} \\ &= (\sqrt{\langle 1|1\rangle} + \sqrt{\langle 2|2\rangle})^2\end{aligned}$$

Thus

$$\sqrt{\langle 3|3\rangle} \leq \sqrt{\langle 1|1\rangle} + \sqrt{\langle 2|2\rangle} \tag{6.2}$$

Putting

$$|1\rangle = |a\rangle - |b\rangle$$

$$|2\rangle = |b\rangle - |c\rangle$$

$$|3\rangle = |a\rangle - |c\rangle$$

we recognize in 6.2 the triangle inequality.

It is evident that the norm of $\{|a\rangle - |b\rangle\}$ satisfies conditions E(a) and E(b). **Therefore, a linear vector space in which there is defined a scalar product is a metric space.**

It should be borne in mind, however, that a linear vector space is not necessarily a metric space. The reader will notice in what follows that there exist properties of linear vector spaces which can be discussed whether or not a scalar product in the space has been defined.

Consider now an infinite sequence of elements of a metric space: $a_{(1)}, a_{(2)}, \cdots, a_{(k)} \cdots$. Suppose there exists an element of the space such that the distances $\rho(a, a_{(k)})$ $(k = 1, 2, \cdots n, \cdots)$ between the members of the sequence become smaller and smaller as k increases and in the limit as $k \to \infty$ tend to 0

$$\lim_{k \to \infty} \rho(a, a_{(k)}) = 0 \tag{6.3}$$

We prove that a is unique. In fact, suppose that besides Eq. 6.3, one also has

$$\lim_{k \to \infty} \rho(b, a_{(k)}) = 0 \tag{6.4}$$

Then by virtue of the triangle inequality

$$\rho(a,b) \leq \rho(a, a_{(k)}) + \rho(b, a_{(k)}) \tag{6.5}$$

and since both members of the RHS of (6.5) tend to zero as $k \to \infty$, one must have

$$\rho(a,b) = 0$$

This result is based only on the fundamental properties of metric spaces. Therefore, it remains valid in the case of a linear vector space in which there is defined a scalar product. Thus, if a sequence of vectors $|a_{(1)}\rangle, |a_{(2)}\rangle, \cdots, |a_{(k)}\rangle, \cdots$, converges to some vector $|a\rangle$ in the sense that

$$\rho^2(|a\rangle, |a_{(k)}\rangle) = [\langle a| - \langle a_{(k)}|][|a\rangle - |a_{(k)}\rangle] \to 0$$

then $|a\rangle$ is unique.

7 · LINEAR OPERATORS

A function was generally defined in Chapter I, Sec. 1.5. An example of a function is a rule that associates with a number x another number, say y. Let the rule of association be represented by the symbol $f(\ )$. Thus $f(\ )$ associates the number $y = f(x)$ in a particular way with the number x. One can also define a function of a **vector argument** $|x\rangle$. In this case, one writes for the function $f(|x\rangle)$. As an example, if one lets

$$f(|x\rangle) \equiv \langle a|x\rangle \tag{7.1}$$

then, Eq. 7.1 defines a rule, $f(\,)$, which associates with a vector $|x\rangle$ a number $\langle a|x\rangle$. We can generalize still further the notion of a function and introduce the notion of a **vector function** of a **vector argument**. Thus

$$|f(|x\rangle)\rangle$$

defines a rule that associates with the vector $|x\rangle$ the vector function $|f(|x\rangle)\rangle$. The simplest example of this rule is provided by the multiplication of $|x\rangle$ by a number c.

$$|f(|x\rangle)\rangle \equiv c\,|x\rangle$$

This example suggests a particular notation. We shall say that $|f(|x\rangle)\rangle$ results from the multiplication of $|x\rangle$ by an object called an **operator**. Accordingly, we write $F|x\rangle$ instead of $|f(|x\rangle)\rangle$, where F is an operator. Then F defines a rule that associates with a vector $|x\rangle$ another vector $F|x\rangle$. Of course, when we say that $F|x\rangle$ is a vector, we tacitly assume that $F|x\rangle$ belongs to some linear vector space, provided $|x\rangle$ is such a vector that $F|x\rangle$ is meaningful. In the following discussion, we shall use capital italic letters to denote operators.

We shall be interested in a rather special class of operators, the **linear** ones, which are defined as follows:

F The operator A is a linear* operator if

$$A\{\alpha\,|a\rangle + \beta\,|b\rangle\} = \alpha\{A\,|a\rangle\} + \beta\{A\,|b\rangle\}$$

To simplify the writing, we shall assume in this chapter and in the following one that, given an operator A, the expression $A\,|\rangle$ is meaningful for any $|\rangle \in$ S and, moreover, that $\{A\,|\rangle\} \in$ S. In Chapter IV we shall drop the first of these assumptions.

It is well known that a function $f(x)$ may not be defined for all values of its argument x. Similarly, the vector function $|f(|x\rangle)\rangle$ of the vector argument $|x\rangle$ may not be defined for all vectors $|x\rangle$. The set of vectors $|x\rangle$, for which $|f(|x\rangle)\rangle = F|x\rangle$ is defined, is called the **domain** of the operator F.

In general, the vector $F|x\rangle$ will not belong to S (compare p. 156), but to some other vector space. The totality of vectors $F|x\rangle$ obtained by letting F operate on all vectors of its domain is called the **range** of F.

Thus, our assumption means that the domain of any operator that we consider is identical to the space S itself and that the range of the operator is included in S.

In Chapter IV we shall abandon the first assumption when we discuss the so-called linear differential operators. As for the second condition, it will always be possible to sufficiently enlarge the space S so as to satisfy it.

The operator associated with the function

$$|f(|x\rangle)\rangle = |x\rangle$$

is called the **identity**, or **unit**, operator, and will be denoted by E.

$$E\,|\rangle = |\rangle \qquad \text{for any } |\rangle$$

---

* Sometimes one also defines **antilinear** operators. They satisfy

$$A\{\alpha\,|a\rangle + \beta\,|b\rangle\} = \bar{\alpha}\{A\,|a\rangle\} + \bar{\beta}\{A\,|b\rangle\}$$

8 · THE ALGEBRA OF LINEAR OPERATORS

Let A and B be two linear operators defined in a linear space S of vectors $|\rangle$. The equation $A = B$ will be understood in the sense that

$$A\,|\rangle = B\,|\rangle \qquad \text{for any } |\rangle \in \text{S}$$

We define the addition and multiplication of linear operators as

G $C = A + B$ and $D = A \cdot B$ if for any $|\rangle \in \text{S}$

$$C\,|\rangle = (A + B)\,|\rangle = A\,|\rangle + B\,|\rangle$$

$$D\,|\rangle = (A \cdot B)\,|\rangle = A \cdot (B\,|\rangle)$$

Using the linearity properties of vector spaces together with the definitions F and G, it is easy to show that $A + B$ and $A \cdot B$ are themselves linear operators and that the addition and the multiplication of operators satisfy all the rules of the addition and of the multiplication of numbers, with the exception of the commutative law for multiplication.

The reader may verify the preceding statements by using the methods of the examples below.

EXAMPLE 1

Verification that $A \cdot B$ is itself a linear operator is

$$(A \cdot B)(\alpha\,|a\rangle + \beta\,|b\rangle) = A\{\alpha(B\,|a\rangle) + \beta(B\,|b\rangle)\} = \alpha(AB)\,|a\rangle + \beta(AB)\,|b\rangle$$

EXAMPLE 2

Verification that $C(A + B) = CA + CB$ is

$$C(A + B)\,|\rangle = C(A\,|\rangle + B\,|\rangle) = CA\,|\rangle + CB\,|\rangle$$

We have mentioned that in general

$$AB - BA \neq 0 \tag{8.1}$$

The quantity of the LHS of (8.1) is called the **commutator** of A and B and is denoted by the symbol $[A,B]$

$$[A,B] \underset{\text{def}}{=} AB - BA$$

Operators whose commutator vanishes are called **commuting** operators. Of course any operator commutes with the unit operator, since

$$(AE)\,|\rangle = A\,|\rangle$$

$$(EA)\,|\rangle = A\,|\rangle$$

We shall consider as meaningful the multiplication of operators by numbers, treating the operator equation

$$B = \alpha A = A\alpha$$

as equivalent to the vector equation

$$B\,|\rangle = \alpha(A\,|\rangle) \qquad \text{for any } |\rangle$$

In order to preserve a consistent notation, however, we interpret the vector equation

$$A\,|\rangle = \alpha\,|\rangle \qquad \text{for any } |\rangle$$

as equivalent to the operator equation

$$A = \alpha E,$$

while the equation $A = \alpha$ is meaningless.

Having defined the product of two operators, we can, of course, also define an operator raised to a certain power. For example, by $A^m\,|\rangle$ we mean that

$$A^m\,|\rangle = \underbrace{A \cdot A \cdots A}_{m \text{ factors}}\,|\rangle$$

Similarly, one can define functions of operators by their (formal) power series expansions. Thus, for example, e^A formally means

$$e^A \underset{\text{def}}{=} 1 + A + \frac{A^2}{2!} + \frac{A^3}{3!} + \cdots$$

Given an operator A that acts on vectors $|\rangle$, one can define the action of the same operator on vectors $\langle|$. The action of A on a vector $\langle|$ is defined by requiring* that for any $|a\rangle$ and $\langle b|$, one has

$$\{\langle b|\,A\}\,|a\rangle \underset{\text{def}}{=} \langle b|\,\{A\,|a\rangle\} \equiv \langle b|\,A\,|a\rangle \tag{8.2}$$

The preceding definition maintains the symmetry between the vectors $\langle|$ and $|\rangle$.

It should, however, be stressed that $\langle|\,A$ is in general **not** the dual vector of $A|\rangle$, as the following example shows.

EXAMPLE 3

The dual vector of $E\,|b\rangle = |b\rangle$ is $\langle b|\,E = \langle b|$, but the dual vector of $(\alpha E)\,|b\rangle = \alpha\,|b\rangle$ is $\langle b|\,(E\bar{\alpha}) = \langle b|\,\bar{\alpha}$ and not $\langle b|\,(E\alpha) = \langle b|\,\alpha$. ($\alpha$ is an arbitrary complex number.)

9 · SOME SPECIAL OPERATORS

Certain operators with rather special properties play particularly important roles in theory and its applications. We shall consider some of them below.

The operator X satisfying $XA = E$ is called the **left inverse** of A and will be denoted by A_l^{-1}. Thus,

$$A_l^{-1}A \underset{\text{def}}{=} E \tag{9.1}$$

Similarly, the **right inverse** operator of A is defined by the equation

$$AA_r^{-1} \underset{\text{def}}{=} E \tag{9.2}$$

It is worth mentioning that, in general, $AA_l^{-1} \neq E$ and $A_r^{-1}A \neq E$. Also, A_l^{-1} or A_r^{-1}, or both, may not be unique and even may not exist at all. One has, however, the following important theorem:

* We shall use the convention of operating on $\langle|$ from the right.

Theorem. If, for a given A, both operators A_l^{-1} and A_r^{-1} exist, they then are unique and

$$A_l^{-1} = A_r^{-1}$$

If A_l^{-1} is unique, then

$$AA_l^{-1} = E$$

and A_l^{-1} is also a unique right inverse of A. Similarly, if A_r^{-1} is unique, then

$$A_r^{-1}A = E$$

and A_r^{-1} is a unique left inverse of A.

Proof. Multiplying Eq. 9.1 from the right by A_r^{-1} and Eq. 9.2 from the left by A_l^{-1}, we get

$$A_l^{-1}AA_r^{-1} = A_r^{-1}$$
$$A_l^{-1}AA_r^{-1} = A_l^{-1} \tag{9.3}$$

Hence,

$$A_l^{-1} = A_r^{-1}$$

The proof holds for any pair of operators A_l^{-1} and A_r^{-1} and Eq. 9.3 ensures that there exists only one such pair.

Multiplying Eq. 9.1 from the left by A, we have

$$AA_l^{-1}A = A \tag{9.4}$$

Thus, adding Eqs. 9.1 and 9.4, we get

$$AA_l^{-1}A + A_l^{-1}A = A + E$$

or

$$(AA_l^{-1} + A_l^{-1} - E)A = E$$

Assuming that A_l^{-1} is unique, we obtain

$$AA_l^{-1} + A_l^{-1} - E = A_l^{-1}$$

and therefore

$$AA_l^{-1} = E$$

Hence, A_l^{-1} is also a right inverse of A, and from the first part of the theorem it follows that it is a unique right inverse of A. Similarly, one proves an analogous result for A_r^{-1}.

When both A_l^{-1} and A_r^{-1} exist, then the unique operator (see the preceding theorem) A^{-1} defined by the equation

$$A^{-1} = A_l^{-1} = A_r^{-1} \tag{9.5}$$

is called the operator **inverse** to A.

Using the rules of operator multiplication one easily obtains*

$$(AB)^{-1} = B^{-1}A^{-1} \tag{9.6}$$

* The analogous relation holds for right and left inverses of product of operators.

provided B^{-1} and A^{-1} exist, since then we have

$$(AB)^{-1}AB = B^{-1}(A^{-1}A)B = B^{-1}EB = B^{-1}B = E$$
$$AB(AB)^{-1} = A(BB^{-1})A^{-1} = AEA^{-1} = AA^{-1} = E$$

Suppose now that the scalar product is defined in S. Then the operator X satisfying

$$\langle a| X |b\rangle = \overline{\langle b| A |a\rangle}$$

for any $|a\rangle$, $|b\rangle \in$ S is called the **adjoint** operator of A and is denoted by A^+. Hence

$$\langle a| A^+ |b\rangle \underset{\text{def}}{=} \overline{\langle b| A |a\rangle} \qquad \text{for any } |a\rangle, |b\rangle \in \text{S} \tag{9.7}$$

By inspection of Eq. 8.2 we see that $\langle| A^+$ is a dual vector of $A|\rangle$. From Eq. 9.7 we find

$$\langle b| (A^+)^+ |a\rangle = \langle b| A |a\rangle$$

for any $|a\rangle$ and $|b\rangle$. Thus

$$(A^+)^+ = A \tag{9.8}$$

Since, for any $|a\rangle$, $|b\rangle$, $\langle b| B^+$ and $B|b\rangle$ and $\langle a| A^+$ and $A|a\rangle$ are pairs of dual vectors, one has

$$\begin{aligned}\langle b| B^+A^+ |a\rangle &= [\langle b| B^+][A^+ |a\rangle] \\ &= \overline{[\langle a| A][B |b\rangle]} \\ &= \overline{\langle a| AB |b\rangle} \\ &= \langle b| (AB)^+ |a\rangle\end{aligned}$$

and therefore

$$(AB)^+ = B^+A^+ \tag{9.9}$$

We leave to the reader the verification that

$$(A + B)^+ = A^+ + B^+ \tag{9.10}$$

An operator H that is equal to its adjoint, i.e., which obeys the relation

$$H = H^+ \tag{9.11}$$

is called **Hermitian**.

An operator U that satisfies the condition

$$U^+ = U^{-1} \tag{9.12}$$

is called **unitary**.

Unitary operators have the remarkable property that their action on a vector preserves the length of that vector. In fact, the length of $|a\rangle$ is $\sqrt{\langle a|a\rangle}$, which is the same as the length of $U|a\rangle$, for we have

$$\{\langle a| U^+\}\{U |a\rangle\} = \langle a| U^{-1}U |a\rangle = \langle a|a\rangle$$

A particular notation turns out to be quite useful. A symbol of the type of $|a\rangle\langle b|$ has all the properties of a linear operator; multiplied from the right by a

$|\rangle$, it gives another $|\rangle$; multiplied from the left by a $\langle|$ it gives a $\langle|$. The linearity of $|a\rangle\langle b|$ results from the linear properties of the scalar product. One also has

$$\{|a\rangle\langle b|\}^+ = |b\rangle\langle a| \tag{9.13}$$

since

$$\begin{aligned}\langle x|\{|b\rangle\langle a|\}|y\rangle &\equiv \langle x|b\rangle\langle a|y\rangle \\ &= \overline{\langle y|a\rangle\langle b|x\rangle} \equiv \overline{\langle y|\{|a\rangle\langle b|\}|x\rangle}\end{aligned}$$

for arbitrary $|x\rangle$ and $|y\rangle$.
Note that since quantities of the type $\langle|\rangle$ are pure numbers, they can be placed either to the left or to the right of vectors $|\rangle$ or $\langle|$.

We need to assume

$$|a\rangle\{\langle b| + \langle c|\} \underset{\text{def}}{=} |a\rangle\langle b| + |a\rangle\langle c| \tag{9.14}$$

From Eqs. 9.13 and 9.14 it follows then that

$$\{|b\rangle + |c\rangle\}\langle a| = |b\rangle\langle a| + |c\rangle\langle a|$$

since

$$\begin{aligned}[|b\rangle + |c\rangle]\langle a| &= \{|a\rangle[\langle b| + \langle c|]\}^+ = [|a\rangle\langle b| + |a\rangle\langle c|]^+ \\ &= [|a\rangle\langle b|]^+ + [|a\rangle\langle c|]^+ = |b\rangle\langle a| + |c\rangle\langle a|\end{aligned}$$

Consider the set S_e of all vectors that can be obtained by multiplying the vector $|e\rangle$ of unit length, $\langle e|e\rangle = 1$, by a complex number. Evidently this set constitutes a linear vector space, and moreover $|a\rangle \in S_e$ implies $|a\rangle \in S$ so that $S_e \subset S$. A space that is a subset of a larger space is called a **subspace** of this space.

The operator $P_e = |e\rangle\langle e|$ has the property that if any $|\rangle$ is multiplied by it, one gets a vector proportional to $|e\rangle$, and therefore belonging to S_e.

$$P_e|\rangle = \langle e|\rangle|e\rangle \in S_e \qquad \text{for any } |\rangle$$

since $\langle e|\rangle$ is simply a complex number. Also

$$P_e|\rangle = |\rangle \qquad \text{for any } |\rangle \in S_e$$

We say that P_e projects $|\rangle$ on the subspace S_e. P_e is a very particular example of a **projection operator**.

A linear operator P is called a projection operator if it is Hermitian and if

$$P^2 = P \tag{9.15}$$

If P had an inverse, then by multiplying both sides of Eq. 9.15 by P^{-1}, one would have

$$P = E$$

Therefore, the only projection operator that has an inverse is the identity operator. (E is obviously a projection operator.)

If P_1 and P_2 are two projection operators, then $P_1 + P_2$ is also a projection operator if and only if

$$P_1P_2 = P_2P_1 = 0 \tag{9.16}$$

To see this, we note that $P_1 + P_2$ is a projection operator if

$$(P_1 + P_2)^2 = P_1 + P_2 \tag{9.17}$$

i.e., since $P_1^2 = P_1$ and $P_2^2 = P_2$, if

$$P_1P_2 + P_2P_1 = 0 \tag{9.18}$$

Multiplying Eq. 9.18 by P_1 from either the right or left, we have

$$P_1P_2P_1 + P_2P_1 = 0 \tag{9.19}$$

$$P_1P_2 + P_1P_2P_1 = 0 \tag{9.20}$$

Hence

$$P_1P_2 - P_2P_1 = 0 \tag{9.21}$$

Equations 9.18 and 9.21 yield Eq. 9.16. Conversely, if Eq. 9.16 is satisfied, then so is Eq. 9.17.

Operators that satisfy the conditions (9.16) are called **orthogonal operators.** More generally if the P_i $(i = 1, 2, \cdots, N)$ are a set of N orthogonal projection operators satisfying

$$P_iP_j = \begin{cases} P_i & i = j \\ 0 & i \neq j \end{cases}$$

then $P = \sum_{i=1}^{N} P_i$ is also a projection operator.

EXAMPLE

Take a real space of vectors represented by arrows in a plane, as discussed in Example 3, Sec. 2. Let $|e_1\rangle$ and $|e_2\rangle$ be two orthogonal (i.e., perpendicular) unit vectors

$$\begin{aligned} \langle e_1|e_1\rangle &= \langle e_2|e_2\rangle = 1 \\ \langle e_1|e_2\rangle &= \langle e_2|e_1\rangle = 0 \end{aligned} \tag{9.22}$$

Then

$$P_1 = |e_1\rangle\langle e_1| \quad \text{and} \quad P_2 = |e_2\rangle\langle e_2|$$

are projection operators, since

$$P_1^2 = |e_1\rangle\langle e_1|e_1\rangle\langle e_1| = |e_1\rangle\langle e_1| = P_1$$

and similarly for P_2. Furthermore, P_1 and P_2 are orthogonal projection operators, since for any vector $|a\rangle$, we have (on account of Eq. 9.22)

$$P_1P_2|a\rangle = |e_1\rangle\langle e_1|e_2\rangle\langle e_2|a\rangle = 0$$

and similarly

$$P_2P_1|a\rangle = 0$$

Applying P_1 to an arbitrary vector $|a\rangle$, we have

$$P_1|a\rangle = |e_1\rangle\langle e_1|a\rangle \equiv |e_1\rangle|\vec{a}|\cos\psi_{a,e_1}$$

Thus, $P_1|a\rangle$ is a vector directed along $|e_1\rangle$ and has a length reduced, as compared to $|\vec{a}|$, by the usual cosine factor.

Analogously, P_2 projects an arbitrary vector along the direction of $|e_2\rangle$.

10·LINEAR INDEPENDENCE OF VECTORS

The vectors $|1\rangle, \cdots, |N\rangle, \cdots$ are said to be **linearly independent** if the relation

$$\sum_i a^i|i\rangle = 0 \qquad (|i\rangle \neq 0) \tag{10.1}$$

necessarily implies that all $a^i = 0$.

On the contrary, if one had a relation like that of Eq. 10.1, with at least two nonvanishing a^i, we would say that the vectors $|1\rangle, \cdots, |N\rangle, \cdots$ are linearly dependent.

The maximum number of linearly independent vectors in a space, if it is finite, is called the **dimension** of this space. In the case where the number of linearly independent vectors is not bounded, the space has an infinite number of dimensions.

EXAMPLE 1

Notice that arrows in a plane may be regarded as forming a two-dimensional real vector space. However, where the multiplication of arrows by complex numbers has been defined, as in Example 3 of Sec. 2, the arrows in a plane form a complex vector space that is one-dimensional. In fact, the multiplication of an arrow by a complex number, as defined in Example 3, involves a rotation of the arrow. Hence, any arrow may be brought into any other one when we multiply it by a proper complex number.

The set of vectors $|i\rangle$ that are linearly independent and have the property that each vector $|a\rangle \in S$ can be expressed as a linear combination of the vectors $|i\rangle$.

$$|a\rangle = \sum_i a^i \, |i\rangle \tag{10.2}$$

is called a **basis** of the space S. One also says that the set $|i\rangle$ **spans** the space S.

The numbers a^i in Eq. 10.2 are called the **components** of $|a\rangle$ with respect to the basis vectors $|i\rangle$.

EXAMPLE 2

We again return to the example of arrows in a plane. It is well known that any arrow in a plane can be uniquely decomposed along any two directions. The discussion in this section aims at generalizing such decompositions.

Given a basis, there corresponds to any vector $|a\rangle \in S$ a **unique** set of components. If this were not so, one could write

$$|a\rangle = \sum_i a^i \, |i\rangle \tag{10.3}$$

and also

$$|a\rangle = \sum_i a'^i \, |i\rangle \tag{10.4}$$

Subtracting Eq. 10.3 from Eq. 10.4, one gets

$$\sum_i (a^i - a'^i) \, |i\rangle = 0$$

and this implies that $a^i = a'^i$, since by definition, the $|i\rangle$ are linearly independent.

11 · EIGENVALUES AND EIGENVECTORS

11.1 Ordinary Eigenvectors

An operator B operating on a vector $|b\rangle$ may have no other effect on $|b\rangle$ than to change the length of that vector while preserving its original "direction." In that case we would have

$$B \, |b\rangle = b \, |b\rangle \tag{11.1}$$

where b is in general a complex number.

Equations of this type are very important and occur often in applications. They are called **eigenvalue equations**. The vector $|b\rangle$ is called an **eigenvector** of the operator B and the number b is an **eigenvalue** of that operator.

EXAMPLE 1

All vectors $|\rangle \in S$ are eigenvectors of the unit operator with the eigenvalue 1, since

$$E|\rangle = 1 \cdot |\rangle$$

EXAMPLE 2

The vector $|a\rangle$ is an eigenvector of the operator $|a\rangle\langle b|$ with eigenvalue $\langle b|a\rangle$.

$$\{|a\rangle\langle b|\}|a\rangle = \langle b|a\rangle|a\rangle$$

However, it is in general not an eigenvector of the operator $|b\rangle\langle a|$.

Hermitian operators play a particularly important role in physics, especially in wave phenomena and in quantum mechanics. This is the reason that compels us to examine in greater detail the properties of Hermitian operators. Two important properties of Hermitian operators should be noted.

Theorem

(i) The eigenvalues of a Hermitian operator are all real.
(ii) Eigenvectors corresponding to two different eigenvalues of a Hermitian operator are orthogonal.

Proof. Let

$$H|h_1\rangle = h_1|h_1\rangle \tag{11.2}$$

$$H|h_2\rangle = h_2|h_2\rangle \tag{11.3}$$

where $H = H^+$ and $h_1 \neq h_2$.

Without loss of generality we may suppose that

$$|h_i\rangle \neq 0 \qquad (\mathrm{i} = 1, 2)$$

since otherwise the theorem becomes trivial.

(i) We multiply Eq. 11.2 by $\langle h_1|$.

$$\langle h_1|H|h_1\rangle = h_1\langle h_1|h_1\rangle$$

However, using Eq. 9.8

$$\langle h_1|H|h_1\rangle = \overline{\langle h_1|H^+|h_1\rangle} = \overline{\langle h_1|H|h_1\rangle} = \overline{h_1}\langle h_1|h_1\rangle$$

Therefore

$$h_1 = \overline{h_1} \tag{11.4}$$

and h_1 is real.

(ii) We multiply Eq. 11.2 by $\langle h_2|$ and Eq. 11.3 by $\langle h_1|$.

$$\langle h_2|H|h_1\rangle = h_1\langle h_2|h_1\rangle \tag{11.5}$$

$$\langle h_1|H|h_2\rangle = h_2\langle h_1|h_2\rangle \tag{11.6}$$

Taking the complex conjugate of Eq. 11.6 (see Eq. 9.8), subtracting it from Eq. 11.5, using Eq. 11.4, and the fact that H is Hermitian, we get

$$(h_1 - h_2)\langle h_2|h_1\rangle = 0$$

Since $h_1 \neq h_2$, it follows that

$$\langle h_2|h_1\rangle = 0$$

and so $|h_2\rangle$ is orthogonal to $|h_1\rangle$.

11.2 Generalized Eigenvectors

The material presented in this section will be utilized only in Secs. 22 through 24. The reader may, if he wishes, skip this section now and return to it later on.

The equation defining an eigenvector (Eq. 11.1) may also be written in the form

$$(B - bE)|b\rangle = 0$$

$$|b\rangle \neq 0$$

This suggests the following generalization of the notion of an eigenvector: The vector $|\beta\rangle$ satisfying

$$\begin{aligned}(B - bE)^m|\beta\rangle &= 0\\ (B - bE)^{m-1}|\beta\rangle &\neq 0\end{aligned} \tag{11.7}$$

is called a **generalized eigenvector of rank** m of the operator B, and the number b is called a **generalized eigenvalue** of B. Equation 11.7 implies that if $k > m$, then evidently

$$(B - bE)^k|\beta\rangle = (B - bE)^{k-m}(B - bE)^m|\beta\rangle = 0 \tag{11.8}$$

If $j < m$, then

$$|\alpha\rangle = (B - bE)^j|\beta\rangle$$

is also a generalized eigenvector of B, but has rank $m - j$

$$(B - bE)^{m-j}|\alpha\rangle = (B - bE)^{m-j}(B - bE)^j|\beta\rangle = (B - bE)^m|\beta\rangle = 0$$

Lemma 1. If $|\beta\rangle$ is a generalized eigenvector of rank m, then

$$\begin{aligned}&(B - bE)^{m-1}|\beta\rangle\\ &(B - bE)^{m-2}|\beta\rangle\\ &\cdots\cdots\cdots\cdots\\ &(B - bE)|\beta\rangle\\ &|\beta\rangle\end{aligned}$$

are linearly independent.

Proof. Let us suppose that the lemma is false and that there exists a set of numbers c^i not all zero, such that

$$\sum_{i=0}^{m-1} c^i(B - bE)^i|\beta\rangle = 0$$

Multiplying the preceding equation by $(B - bE)^{m-1}$, and using Eq. 11.8, we get

$$c^0(B - bE)^{m-1}|\beta\rangle = 0$$

which implies that $c^0 = 0$. Thus, we are left with

$$\sum_{i=1}^{m-1} c^i(B - bE)^i|\beta\rangle = 0$$

Multiplying now by the operator $(B - bE)^{m-2}$, we get

$$c^1(B - bE)^{m-1}|\beta\rangle = 0$$

which implies that $c^1 = 0$.

Continuing this procedure, we finally obtain

$$c^i = 0 \qquad \text{for } i = 0, 1, \cdots, m - 1$$

and this proves the lemma.

Lemma 2. Generalized eigenvectors corresponding to different generalized eigenvalues are linearly independent.

Proof. The proof is by induction. We shall first prove the linear independence of two generalized eigenvectors. Let

$$\begin{aligned} (B - b_1E)^{m_1}|\beta_1\rangle &= 0 \\ (B - b_1E)^{m_1-1}|\beta_1\rangle &\neq 0 \end{aligned} \tag{11.9}$$

and

$$\begin{aligned} (B - b_2E)^{m_2}|\beta_2\rangle &= 0 \\ (B - b_2E)^{m_2-1}|\beta_2\rangle &\neq 0 \end{aligned} \tag{11.10}$$

with $b_1 \neq b_2$ and $m_1 \geq m_2$.

Suppose that there exists a linear relation between $|\beta_1\rangle$ and $|\beta_2\rangle$

$$c_1|\beta_1\rangle + c_2|\beta_2\rangle = 0 \qquad (c_1, c_2 \neq 0)$$

Multiplying the preceding equation by $(B - b_1E)^{m_1}$, we get

$$(B - b_1E)^{m_1}|\beta_2\rangle = 0 \tag{11.11}$$

On the other hand, one has*

$$\begin{aligned} (B - bE_1)^{m_1}|\beta_2\rangle &= [(B - b_2E) + (b_2 - b_1)E]^{m_1}|\beta_2\rangle \\ &= \sum_{k=0}^{m_1}\binom{m_1}{k}(b_2 - b_1)^{m_1-k}(B - b_2E)^k|\beta_2\rangle \\ &= \sum_{k=0}^{m_2-1}\binom{m_1}{k}(b_2 - b_1)^{m_1-k}(B - b_2E)^k|\beta_2\rangle \end{aligned}$$

The last equality follows from the fact that when $k \geq m_2$, $(B - bE)^k|\beta_2\rangle$ vanishes because of Eq. 11.8. Hence, due to Eq. 11.11 we have

$$\sum_{k=0}^{m_2-1}\binom{m_1}{k}(b_2 - b_1)^{m_1-k}(B - b_2E)^k|\beta_2\rangle = 0$$

---

* The binomial coefficient is $\binom{m}{k} = \dfrac{m!}{(m-k)!k!}$

But such a relation cannot exist because, according to the preceding lemma, the vectors $(B - b_2E)^k|\beta_2\rangle$ $(b = 0, 1, \cdots, m_2 - 1)$ are linearly independent. Thus, $|\beta_1\rangle$ and $|\beta_2\rangle$ must also be linearly independent.

Suppose now that the lemma is true for any set of $(n - 1)$ generalized eigenvectors. Consider n generalized eigenvectors $|\beta_i\rangle$ $(i = 1, 2, \cdots, n)$. Suppose that there also exists a linear relation between the $|\beta_i\rangle$

$$\sum_{i=1}^{n} c^i|\beta_i\rangle = 0$$

Without any loss of generality, we can assume that the rank m_n of the nth eigenvector is the highest. Multiplying the preceding equation by $(B - b_nE)^{m_n}$ and proceeding as before, we get

$$\sum_{i=1}^{n-1} c^i(B - b_nE)^{m_n}|\beta_i\rangle$$
$$= \sum_{i=1}^{n-1} c^i\left\{\sum_{k=0}^{m_i-1}\binom{m_n}{k}(b_i - b_n)^{m_n-k}(B - b_iE)^k|\beta_i\rangle\right\} = 0$$

Since the expression in brackets is a generalized eigenvector of rank m_i corresponding to the eigenvalue b_i, each c^i must vanish, since we have assumed that the lemma is true for any set of $(n - 1)$ eigenvectors. Therefore, the lemma is proved by induction.

Similarly, one can show that eigenvectors having different rank but corresponding to the same eigenvalue are linearly independent. In fact, from

$$(B - b_iE)^m|\beta_i,1\rangle = 0 \qquad |\beta_i,1\rangle \neq 0$$
$$(B - b_iE)^n|\beta_i,2\rangle = 0 \qquad |\beta_i,2\rangle \neq 0$$
$$(B - b_iE)^{n-1}|\beta_i,2\rangle \neq 0$$

with $m < n$, it follows that a relation of the kind

$$c_1|\beta_i,1\rangle + c_2|\beta_i,2\rangle = 0$$

implies that $c_1 = c_2 = 0$. This can be shown by multiplying the preceding equation by $(B - b_iE)^m$. Then

$$c_2(B - b_iE)^m|\beta_i,2\rangle = 0$$

which means that $c_2 = 0$, since $m \leq n - 1$. If $c_2 = 0$, then necessarily $c_1 = 0$. The proof for an arbitrary number of eigenvectors of different rank is completely analogous.

On the other hand, eigenvectors of the same rank and corresponding to the same eigenvalue may be linearly dependent. Suppose, for instance, that $|\beta_i,1\rangle$ and $|\beta_i,2\rangle$ are generalized eigenvectors of rank n corresponding to the eigenvalue β_i. A vector

$$|\beta_i,3\rangle = c_1|\beta_i,1\rangle + c_2|\beta_i,2\rangle$$

satisfies

$$(B - b_iE)^n|\beta_i,3\rangle = c_1(B - b_iE)^n|\beta_i,1\rangle + c_2(B - b_iE)^n|\beta_i,2\rangle = 0$$

for arbitrary c_1 and c_2, while

$$(B - b_iE)^{n-1}|\beta_i,3\rangle = c_1(B - b_iE)^{n-1}|\beta_i,1\rangle + c_2(B - b_iE)^{n-1}|\beta_i,2\rangle$$

cannot be zero for arbitrary c_1 and c_2. Therefore, there exists an infinite number of vectors $|\beta_i,3\rangle$ which are generalized eigenvectors of rank n with eigenvalue b_i and which are at the same time linear combinations of $|\beta_i,1\rangle$ and $|\beta_i,2\rangle$.

The Hermitian operators have an important property, the consequences of which will be seen in Sec. 24; namely, a Hermitian operator cannot have generalized eigenvectors of rank higher than one. In other words, eigenvectors of Hermitian operators are always the "ordinary" ones. This can be easily proved. Let

$$H^+ = H$$

and

$$\begin{aligned} (H - hE)^n |h\rangle &= 0 \\ (H - hE)^{n-1} |h\rangle &\neq 0 \end{aligned} \tag{11.12}$$

From the second of the preceding relations we have

$$\langle h| \{(H - hE)^{n-1}\}^+(H - hE)^{n-1} |h\rangle \neq 0$$

But the generalized eigenvalues of a Hermitian operator are real just as are the "ordinary" eigenvalues. This can be immediately seen by reducing the generalized eigenvalue equation to the "ordinary" one. Putting

$$(H - hE)|h'\rangle = 0$$

where

$$|h'\rangle \equiv (H - hE)^{n-1} |h\rangle \neq 0$$

it follows from the Theorem of Sec. 11.1 that h is real. Then, since

$$\{(H - hE)^{n-1}\}^+ = (H^+ - \bar{h}E)^{n-1} = (H - hE)^{n-1}$$

we have

$$\langle h| (H - hE)^{2n-2} |h\rangle \neq 0$$

Hence, due to Eq. 11.12, we have $2n - 2 < n$, and therefore $n < 2$.

12 · ORTHOGONALIZATION THEOREM

Suppose now that a scalar product has been defined in the vector space we are considering.

Before proving an important theorem, let us introduce the symbol δ_{ij} defined by

$$\delta_{ij} = \delta_{ji} = \begin{cases} 1 & i = j \\ 0 & i \neq j \end{cases} \tag{12.1}$$

δ_{ij} is called the **Kronecker delta**.

Theorem. From any set of linearly independent vectors $|i\rangle$ $(i = 1, 2, \cdots, N)$ one can always construct a set $|e_i\rangle$ $(i = 1, 2, \cdots, N)$ of mutually orthogonal and normalized vectors

$$\langle e_i|e_j\rangle = \delta_{ij} \tag{12.2}$$

such that each vector $|e_i\rangle$ is a linear combination of the vectors $|i\rangle$.

Proof. In Sec. 9 we introduced a projection operator $P_e = |e\rangle\langle e|$ which projected any vector $|\rangle$ onto a "direction" parallel to $|e\rangle$. Hence

$$P_{e_j}|i\rangle = \langle e_j|i\rangle\,|e_j\rangle$$

is the projection of $|i\rangle$ in the direction of $|e_j\rangle$ and the quantity

$$\left\{|i\rangle - \sum_{j=1}^{i-1}\langle e_j|i\rangle\,|e_j\rangle\right\} \tag{12.3}$$

is a quantity in which one has removed from $|i\rangle$ all its projections along the mutually orthogonal directions $|e_1\rangle, |e_2\rangle, \cdots |e_{i-1}\rangle$. Therefore it is natural to write

$$|e_i\rangle = \frac{1}{L_i}\left\{|i\rangle - \sum_{j=1}^{i-1}\langle e_j|i\rangle\,|e_j\rangle\right\} \tag{12.4}$$

where L_i is the norm of the vector expression in brackets and which can be calculated if one knows $|i\rangle$ and $|e_j\rangle$ for $j < i$.

The proof of the theorem proceeds by induction. Suppose that vectors $|e_j\rangle$, $(j < i)$ have been found such that

$$\langle e_j|e_k\rangle = \delta_{jk} \qquad \text{for } j, k < i \tag{12.5}$$

Multiplying Eq. 12.4 by $\langle e_k|$, as a result of Eq. 12.5 we evidently get

$$\langle e_k|e_i\rangle = 0 \qquad k < i$$

The factor $1/L_i$ in Eq. 12.4 guarantees that $\langle e_i|e_i\rangle = 1$. Thus

$$\langle e_k|e_i\rangle = \delta_{ki} \qquad \text{for } k \le i$$

Equation 12.5 is certainly true for $i = 2$, since

$$|e_1\rangle = \frac{1}{\sqrt{\langle 1|1\rangle}}|1\rangle$$

satisfies this condition.

Of course $|e_i\rangle$ as given by Eq. 12.4 is a linear combination of the vectors $|j\rangle$, for $j \le i$, and hence cannot be a null vector due to the linear independence of the vectors $|j\rangle$. This completes the proof.

A system of mutually orthogonal and normalized vectors (i.e., normalized to unity) is called briefly an **orthonormal** system. The method used for constructing an orthonormal system, starting from a set of linearly independent vectors, is known as the **Schmidt method**. A very simple lemma, which is the converse of the theorem, will be of use later on.

Lemma. Any two vectors, $|1\rangle, |2\rangle \neq 0$, that are orthogonal to each other are linearly independent.

The proof is almost immediate. Suppose that the lemma is not true, i.e., that $|1\rangle$ and $|2\rangle$ are linearly dependent. Then one can find numbers $a^i \neq 0$ such that

$$a^1|1\rangle + a^2|2\rangle = 0 \tag{12.6}$$

Multiplying Eq. 12.6 from the left by $\langle 1|$, one obtains $a^1\langle 1|1\rangle = 0$. Since $\langle 1|1\rangle > 0$, we have $a^1 = 0$. Similarly, one also finds $a^2 = 0$, which proves the linear independence of $|1\rangle$ and $|2\rangle$.

13 · N-DIMENSIONAL VECTOR SPACE

13.1 Preliminaries

From this point to the end of the chapter we shall consider only finite dimensional spaces. Infinite dimensional spaces will be discussed in the next chapter. By definition, an N-dimensional space S_N contains N (and no more!) linearly independent vectors.

In Sec. 10 we defined the basis of a space as a set of linearly independent vectors which is such that any vector of the space can be expressed as a linear combination of the vectors of this set. However, we have not yet discussed the question of the existence of a basis. For an N-dimensional space, this problem is settled by the following theorem.

Theorem. Any set $|1\rangle, \cdots, |N\rangle$ of N linearly independent vectors in an N-dimensional space S_N forms a basis for this space.

Proof. Consider the expression

$$c^0 |a\rangle + \sum_{i=1}^{N} c^i |i\rangle$$

with arbitrary c^i $(i = 0, 1, \cdots, N)$ and $|a\rangle \in S_N$. The equation

$$c^0 |a\rangle + \sum_{i=1}^{N} c^i |i\rangle = 0 \tag{13.1}$$

cannot imply

$$c^i = 0 \qquad i = 0, 1, \cdots, N$$

since that would mean that there exist $N + 1$ independent vectors in S_N.

$$|a\rangle, |1\rangle, \cdots, |N\rangle$$

and the dimension of S_N would not be N, but would be at least $N + 1$. Therefore, a set c^i exists, with at least two nonzero members, such that Eq. 13.1 is satisfied. One cannot have $c^0 = 0$, since it would imply that the vectors $|i\rangle$ are linearly dependent. Multiplying Eq. 13.1 by $1/c^0$, one gets

$$|a\rangle + \sum_{i=1}^{N} \left(\frac{c^i}{c^0}\right) |i\rangle = 0$$

or

$$|a\rangle = \sum_{i=1}^{N} -\left(\frac{c^i}{c_0}\right) |i\rangle$$

Therefore, an arbitrary vector $|a\rangle$ has been expressed as a linear combination of the vectors $|i\rangle$, and this proves the theorem.

We have proved that any N linearly independent vectors $|i\rangle \in S_N$ $(i = 1, 2, \cdots, N)$ span the space S_N (i.e., form a basis in S_N). Previously (Sec. 10) we showed that the decomposition

$$|a\rangle = \sum_{i=1}^{N} a^i |i\rangle$$

is unique, provided the vectors $|i\rangle$ are linearly independent.

Thus, once we have chosen a basis in S_N, any set of complex numbers determines a vector and, conversely, any vector determines uniquely N complex numbers, which are its components with respect to the basis.

13.2 Representations

In the foregoing text, we introduced certain objects called vectors and demanded that these vectors obey a number of laws of composition. We may regard these vectors as abstract objects in the sense that no specific properties, aside from the laws of composition, have been attributed to them. On the other hand, we may decompose a vector with respect to some basis vectors

$$|a\rangle = \sum_{i=1}^{N} a^i \, |i\rangle \tag{13.2}$$

and then we may regard the set of the N numbers a^i as **representing** the vector $|a\rangle$; for we have seen that, given a set of basis vectors, the decomposition (13.2) is unique. In that case, manipulations with abstract vectors can be replaced by manipulations with the components a^i, that is by the familiar operations of arithmetics, since the components are well-known pure numbers. It is therefore not astonishing that in the applications of mathematics to physics, we encounter very often the problem of finding representations of abstract objects.

Notice that there is a one-to-one correspondence between vectors in an N-dimensional complex space and vectors in a $2N$-dimensional real space. In fact, N complex numbers determine $2N$ real numbers, and vice versa.

As an example, the equations

$$|a\rangle + |b\rangle = \sum_{i=1}^{N} a^i |i\rangle + \sum_{i=1}^{N} b^i |i\rangle = \sum_{i=1}^{N} (a^i + b^i) |i\rangle$$

$$x\,|a\rangle = x \sum_{i=1}^{N} a^i |i\rangle = \sum_{i=1}^{N} (xa^i) |i\rangle$$

mean simply that the addition of two vectors is represented by the addition of their components and that the multiplication of a vector by a number is represented by the multiplication of its components by this number. Here, the addition and multiplication of components are understood in the usual sense as addition and multiplication of complex numbers. This results because the rules that define abstract operations in a linear vector space (Sec. 2,A) are modeled upon the rules of arithmetic of complex numbers. For instance, the commutative rule of addition of vectors

$$|c\rangle = |a\rangle + |b\rangle = |b\rangle + |a\rangle$$

becomes, in the language of components, the familiar commutative law of the addition of numbers

$$c^i = a^i + b^i = b^i + c^i \qquad (i = 1, 2, \cdots, N)$$

Clearly, an N-dimensional vector space is represented by all sets of N complex numbers. We shall show how all abstract operations previously defined can be reduced to the well-known manipulations with finite sets of numbers.

We shall assume first that some fixed basis has been chosen in S_N. The question of what happens to sets of numbers that represent abstract objects when one changes the basis will be discussed later.

13.3 The Representation of a Linear Operator in an N-dimensional Space

In Sec. 13.2 we discussed the representation of a vector by a set of numbers; we now discuss the representation of a linear operator. This will lead us to the concept of a matrix.

As before, let $|i\rangle$ $(i = 1, 2, \cdots, N)$ denote the basis vectors of S_N. Consider a linear operator A. $A\,|i\rangle$ is a vector that also belongs to S_N and therefore can be decomposed.

$$A\,|i\rangle = \sum_{j=1}^{N} A_i^j\,|j\rangle \tag{13.3}$$

The components of $A\,|i\rangle$ have two indices. One, the superscript, identifies as before the **component** of the vector that is being decomposed. The other, the subscript, identifies the **vector** that is decomposed. Thus A_i^j is the jth component of the ith vector $A\,|i\rangle$.

Using Eq. 13.3, let us find the result of the multiplication by A of an arbitrary vector $|a\rangle$, i.e., one that is not necessarily a basis vector. Let

$$|a'\rangle = A\,|a\rangle$$

and

$$|a\rangle = \sum_{i=1}^{N} a^i\,|i\rangle \tag{13.4}$$

$$|a'\rangle = \sum_{i=1}^{N} a'^i\,|i\rangle \tag{13.5}$$

From Eqs. 13.3 and 13.4 we get

$$|a'\rangle = A\,|a\rangle = \sum_{i=1}^{N}\sum_{j=1}^{N} A_i^j a^i\,|j\rangle \tag{13.6}$$

Hence, because of the uniqueness of the decomposition, by comparing Eqs. 13.5 and 13.6, we have

$$a'^j = \sum_{i=1}^{N} A_i^j a^i \qquad (j = 1, 2, \cdots, N) \tag{13.7}$$

Before going further, let us introduce a useful rule, known as the **Einstein convention.** Each time an index appears twice, once as an upper index and once as a lower index, the summation over the whole range of this index will be understood, and the summation symbol will be omitted. Wherever a repeated index is not summed over, we shall place that index in parenthesis. Thus, instead of Eq. 13.7, we write

$$a'^j = A_i^j a^i \qquad (j = 1, 2, \cdots, N) \tag{13.8}$$

where a summation over the index i is implied. Of course it does not matter what letter is used to denote the indices being summed, i.e., the "dummy indices," as they are called. Thus, Eq. 13.8 could well be written as

$$a'^j = A_m^j a^m \qquad (j = 1, 2, \cdots, N)$$

The set of numbers A_i^j **represents** the abstract operator A in the sense that these numbers completely determine (by Eq. 13.8) the effect of A on an arbitrary vector of S_N. In other words, once a basis has been chosen in an N-dimensional space, the multiplication of a vector by a linear operator is **represented** by a linear transformation of the components of this vector.

The numbers A_i^j can be arranged in a table

$$\begin{pmatrix} A_1^1 & A_2^1 & \cdots & A_N^1 \\ A_1^2 & A_2^2 & \cdots & A_N^2 \\ \vdots & \vdots & \vdots & \vdots \\ A_1^N & A_2^N & \cdots & A_N^N \end{pmatrix} \tag{13.9}$$

if we agree to consider the lower index as the column number and the upper index as the row number that determine the position of A_i^j within the table. Such a table, which represents a linear operator, is called a **matrix**. In contrast to A_i^j, which is a number, we shall denote the matrix, i.e., the set of all A_i^j by **A**. In the following, matrices will be denoted by capital boldface letters.

The decompositions (13.3) of vectors $A|i\rangle$ into the basis vectors $|j\rangle$ are unique. Therefore, given a basis in an N-dimensional space, the correspondence between linear operators and matrices is one-to-one. Two matrices are said to be equal if they represent the same operator with respect to the same basis. Of course equality of two matrices means that all their elements lying in the intersection of the same row and the same column are equal, hence

$$\mathbf{A} = \mathbf{A}'$$

means that

$$A_i^j = A'^j_i \qquad (i, j = 1, 2, \cdots, N)$$

14 · MATRIX ALGEBRA

In the preceding section the notion of a matrix was introduced to describe the set of numbers that represents a linear operator. The addition and multiplication of operators will therefore be represented by some operations of "addition" and "multiplication" of matrices. In this section we discuss these operations.

Let matrices **A** and **B** represent, respectively, the operators A and B. A matrix is called the **sum** of **A** and **B** if it represents the operator $A + B$; similarly, a matrix is called the **product** of **A** and **B** if it represents the operator $A \cdot B$.

The uniqueness of the following decompositions of vectors

$$A|i\rangle = \sum_{j=1}^{N} A_i^j |j\rangle \tag{14.1}$$

$$B|i\rangle = \sum_{j=1}^{N} B_i^j |j\rangle \tag{14.2}$$

$$(A + B)|i\rangle = \sum_{j=1}^{N} (A + B)_i^j |j\rangle \tag{14.3}$$

$$(A \cdot B)|i\rangle = \sum_{j=1}^{N} (A \cdot B)_i^j |j\rangle \tag{14.4}$$

will guarantee the uniqueness of the result of matrix addition and multiplication.

Adding Eqs. 14.1 and 14.2, one gets

$$(A + B)|i\rangle = \sum_{j=1}^{N} (A_i^j + B_i^j)|j\rangle \tag{14.5}$$

and comparing Eq. 14.5 with Eq. 14.3, we find

$$(A + B)_i^j = A_i^j + B_i^j \qquad (i,j = 1, 2, \cdots, N) \tag{14.6}$$

Multiplying Eq. 14.2 by A and using Eq. 14.1 with an obvious change of summation index, one obtains

$$(A \cdot B)|i\rangle = \sum_{k=1}^{N} B_i^k A|k\rangle = \sum_{j=1}^{N}\sum_{k=1}^{N} A_k^j B_i^k |j\rangle \tag{14.7}$$

Comparing Eq. 14.7 with Eq. 14.4, we get

$$(A \cdot B)_i^j = A_k^j B_i^k \qquad (i,j = 1, 2, \cdots, N) \tag{14.8}$$

where the Einstein summation convention has been used. Equations 14.6 and 14.8 mean that a matrix with elements $(A_i^j + B_i^j)$ represents $A + B$, and a matrix with elements $A_k^j \cdot B_i^k$ represents $A \cdot B$. Therefore, according to what has been said at the beginning of this section:

(i) **The addition of matrices consists in adding their corresponding elements.** Thus, $\mathbf{C} = \mathbf{A} + \mathbf{B}$ means

$$C_i^j = A_i^j + B_i^j \qquad (i,j = 1, \cdots, N)$$

EXAMPLE 1

An example of matrix addition is

$$\begin{pmatrix} 1 & 3 \\ 0 & 2 \end{pmatrix} + \begin{pmatrix} 2 & 5 \\ 0 & 0 \end{pmatrix} = \begin{pmatrix} 1+2 & 3+5 \\ 0+0 & 2+0 \end{pmatrix} = \begin{pmatrix} 3 & 8 \\ 0 & 2 \end{pmatrix}$$

(ii) **The multiplication of two matrices consists in multiplying term by term the elements of the row of the first matrix by the elements of the column of the second matrix and adding the result, to get the element of the product matrix that lies in the intersection of the row and the column that have been multiplied by each other.** Thus, $\mathbf{C} = \mathbf{A} \cdot \mathbf{B}$ means

$$C_i^j = A_k^j B_i^k \qquad (i,j = 1, 2, \cdots, N)$$

EXAMPLE 2

An example of matrix multiplication is

$$\begin{pmatrix} 1 & 3 \\ 0 & 2 \end{pmatrix} \cdot \begin{pmatrix} 2 & 5 \\ 0 & 4 \end{pmatrix} = \begin{pmatrix} 1\cdot 2+3\cdot 0 & 1\cdot 5+3\cdot 4 \\ 0\cdot 2+2\cdot 0 & 0\cdot 5+2\cdot 4 \end{pmatrix} = \begin{pmatrix} 2 & 17 \\ 0 & 8 \end{pmatrix}$$

The reader should verify that

$$\begin{pmatrix} 2 & 3 & 1 \\ 8 & 5 & 0 \\ 1 & 4 & 0 \end{pmatrix} \begin{pmatrix} 1 & 8 & 2 \\ 0 & 7 & 5 \\ 3 & 0 & 4 \end{pmatrix} = \begin{pmatrix} 5 & 37 & 23 \\ 8 & 99 & 41 \\ 1 & 36 & 22 \end{pmatrix}$$

The addition and multiplication of matrices have, of course, the same properties as the addition and multiplication of linear operators, as discussed in Sec. 8, where we stressed that multiplication of operators does not obey the commutative law. Since each matrix determines some operator, it may be worthwhile to illustrate the non-commutation of operators by the noncommutation of matrices. This is done in the following example.

EXAMPLE 3

Consider two matrices

$$\mathbf{A} = \begin{pmatrix} 0 & 1 \\ 1 & 0 \end{pmatrix} \quad \text{and} \quad \mathbf{B} = \begin{pmatrix} 1 & 0 \\ 1 & -1 \end{pmatrix}$$

We have

$$\mathbf{A} \cdot \mathbf{B} = \begin{pmatrix} 0 & 1 \\ 1 & 0 \end{pmatrix} \cdot \begin{pmatrix} 1 & 0 \\ 0 & -1 \end{pmatrix} = \begin{pmatrix} 0 & -1 \\ 1 & 0 \end{pmatrix}$$

and

$$\mathbf{B} \cdot \mathbf{A} = \begin{pmatrix} 1 & 0 \\ 0 & -1 \end{pmatrix} \begin{pmatrix} 0 & 1 \\ 1 & 0 \end{pmatrix} = \begin{pmatrix} 0 & 1 \\ -1 & 0 \end{pmatrix}$$

so that

$$\mathbf{A} \cdot \mathbf{B} \neq \mathbf{B} \cdot \mathbf{A}$$

In this particular case, we have $\mathbf{A} \cdot \mathbf{B} = -\mathbf{B} \cdot \mathbf{A}$. Two matrices that obey such a relation are said to **anticommute.**

Consider now the representation of the unit operator E.

$$E\,|i\rangle = \sum_{j=1}^{N} E_i^j\,|j\rangle = |i\rangle$$

Hence

$$E_i^j = \begin{cases} 1 & i = j \\ 0 & i \neq j \end{cases}$$

This result is completely independent of the chosen basis. Thus, for any basis, the unit operator is represented by the matrix

$$\mathbf{E} = \begin{pmatrix} 1 & 0 & 0 & 0 & \cdots & 0 \\ 0 & 1 & 0 & 0 & \cdots & 0 \\ 0 & 0 & 1 & 0 & \cdots & \\ \cdots & \cdots & \cdots & \cdots & \cdots & \cdots \\ 0 & 0 & 0 & 0 & \cdots & 1 \end{pmatrix}$$

in which all elements lying on the diagonal of the table are equal to unity and all others equal zero.

The multiplication of an operator by a number is represented by the multiplication of the corresponding matrix by this number. Consider the operator

$$A' = \alpha A \tag{14.9}$$

where α is some number. Multiplying Eq. 14.1 by α and comparing with Eq. 14.9, we get

$$A'\,|i\rangle = \sum_{j=1}^{N} (\alpha A_i^j)\,|j\rangle$$

which shows that the matrix with elements αA_i^j represents the operator $A' = \alpha A$. Therefore $\mathbf{A}' = \alpha \cdot \mathbf{A}$ means

$$A'^j_i = \alpha \cdot A_i^j \qquad (i,j = 1, 2, \cdots, N)$$

(iii) **The multiplication of a matrix by a number consists in multiplying by that number all the elements of the matrix.**

Let us notice that the components a^i of a vector $|a\rangle$ can also be arranged in a "table" as

$$\begin{pmatrix} a^1 \\ a^2 \\ \vdots \\ a^N \end{pmatrix} \tag{14.10}$$

This is consistent with matrix convention; a^i has an upper index that distinguishes rows, but has no lower index to distinguish columns, so that the a^i form a "table" with N rows and only one column. Such a table representation of a vector will be denoted by a small boldface letter; for instance, the table in (14.10) will be symbolized by **a**. The equation

$$a'^j = A^j_i a^i$$

can then be written in the matrix form

$$\mathbf{a}' = \mathbf{A}\mathbf{a} \tag{14.11}$$

because the rules of matrix multiplication require only that the number of columns of the first matrix equals the number of rows of the second matrix. (Of course **aA** is meaningless.) When written in full, Eq. 14.11 reads as

$$\begin{pmatrix} a'^1 \\ a'^2 \\ \vdots \\ a'^N \end{pmatrix} = \begin{pmatrix} A^1_1 A^1_2 & \cdots & A^1_N \\ A^2_1 A^2_2 & \cdots & A^2_N \\ \vdots \;\; \vdots & \cdots & \vdots \\ A^N_1 A^N_2 & \cdots & A^N_N \end{pmatrix} \begin{pmatrix} a^1 \\ a^2 \\ \vdots \\ a^N \end{pmatrix} \tag{14.12}$$

Multiplying out the RHS of Eq. 14.12 and using the Einstein convention, we have

$$\begin{pmatrix} a'^1 \\ a'^2 \\ \vdots \\ a'^N \end{pmatrix} = \begin{pmatrix} A^1_j a^j \\ A^2_j a^j \\ \vdots \\ A^N_j a^j \end{pmatrix}$$

or, remembering what is meant by the equality of two matrices

$$a'^i = A^i_j a^j$$

15 · THE INVERSE OF A MATRIX

The inverse matrix to **A** is defined by the equation

$$\mathbf{A}^{-1}\mathbf{A} = \mathbf{E} \tag{15.1}$$

When **A** represents the operator A, $\mathbf{A}^{-1}$ will represent the left inverse operator A_l^{-1}. However, we shall prove below that $\mathbf{A}^{-1}$ either is unique or does not exist at all. Hence, $\mathbf{A}^{-1}$, if it exists, represents the unique left inverse operator A_l^{-1}, and therefore (in accord with the notation we have introduced) $\mathbf{A}^{-1}$ represents the operator A^{-1} inverse to A. (Refer to the theorem of Sec. 9.)

The value of the determinant det $[A^i_j]$, to be denoted briefly by det **A**, is crucial for the existence of $\mathbf{A}^{-1}$, as can be seen from the following theorem.

Theorem. The necessary and sufficient condition for the matrix $\mathbf{A}^{-1}$, the inverse to the matrix **A**, to exist is that det $\mathbf{A} \neq 0$. In this case $\mathbf{A}^{-1}$ is unique and is given by

$$(A^{-1})^j_i \equiv (-1)^{i+j} \frac{M^i_j}{\det \mathbf{A}} \tag{15.2}$$

where M^i_j is the minor of det**A** corresponding to the element A^i_j.

Proof. Write Eq. 15.1 in the form

$$(A^{-1})^i_k A^k_j = E^i_j \qquad (i, j = 1, 2, \cdots, N) \tag{15.3}$$

For any fixed i, Eqs. 15.3 can be considered as a system of linear equations with N unknowns $(A^{-1})^i_k$, $(k = 1, 2, \cdots, N)$.

Suppose first that det $\mathbf{A} = 0$. From the properties of determinants we know that this implies that one of the columns of **A** (the mth, say) can be expressed as a linear combination of the other columns of **A**

$$A^k_m = \sum_{j \neq m} A^k_j c^j \tag{15.4}$$

Putting $i = m$ in Eq. 15.3, multiplying by c_j, and summing over $j \neq m$, we get

$$(A^{-1})^m_k \sum_{j \neq m} A^k_j c^j = \sum_{j \neq m} E^m_j c^j = 0 \tag{15.5}$$

On the other hand, from Eqs. 15.4 and 15.3, we obtain

$$(A^{-1})^m_k \sum_{j \neq m} A^k_j c^j = \{(A^{-1})^i_k A^k_m\}_{i=m} = 1 \tag{15.6}$$

Equations 15.5 and 15.6 show that Eqs. 15.3 are inconsistent. Thus, $\mathbf{A}^{-1}$ does not exist.

Suppose now that det $\mathbf{A} \neq 0$. In this case, as is well known, Eqs. 15.3 have a unique solution, given by (15.2).

With respect to a given basis a matrix determines uniquely an operator and vice versa. Therefore the condition that det $\mathbf{A} \neq 0$, which is a necessary and sufficient condition for the existence of a unique left inverse* of A, is also a necessary and sufficient condition for the existence of the inverse operator A^{-1}. In other words, in an N-dimensional space, if one has an operator X that satisfies

$$XA = E \tag{15.7}$$

it must also satisfy

$$AX = E \tag{15.8}$$

i.e., $X = A^{-1}$. To arrive at this result, we have made use of the fact that A can be represented by a matrix with a finite number of rows and columns, so that simple properties of systems of linear algebraic equations could be used to prove the uniqueness of $\mathbf{A}^{-1}$ and thus of A^{-1}. The finite dimensionality of the space was an essential assumption. For a space with an infinite number of dimensions, Eq. 15.7 no longer implies Eq. 15.8 for an arbitrary linear operator A, since A_l^{-1} (or A_r^{-1}) is not necessarily unique even if it exists.

* Equation 15.3 when written in matrix form reads $\mathbf{A}^{-1}\mathbf{A} = \mathbf{E}$ and represents the operator equation that defines a left inverse of A (see Eq. 9.1).

16 · CHANGE OF BASIS IN AN N-DIMENSIONAL SPACE

In the past few sections we have examined the representations of vectors and linear operators with respect to a given basis in S_N. A basis is, however, by no means unique. On the contrary, there exists an infinite number of sets of N linearly independent vectors in S_N, and each one of these sets of vectors can be equally well chosen as a basis of the space. In fact, let R be a linear operator represented in the basis $|i\rangle$ $(i = 1, 2, \cdots, N)$ by the matrix $\mathbf{R}$ with nonvanishing determinant

$$\det \mathbf{R} \neq 0 \tag{16.1}$$

Consider a set of vectors

$$|i'\rangle = R\,|i\rangle = \sum_{j=1}^{N} R_i^j\,|j\rangle \qquad (i = 1, \cdots, N) \tag{16.2}$$

Because of (16.1), R has an inverse which (according to Eq. 15.8) satisfies the relation

$$R_i^j(R^{-1})_k^i = E_k^j \tag{16.3}$$

Hence, Eq. 16.2 can be inverted. Multiplying both sides of Eq. 16.2 by $(R^{-1})_k^i$ and using Eq. 16.3, after making an obvious change of indices we find

$$|i\rangle = \sum_{j=1}^{N} (R^{-1})_i^j\,|j'\rangle \tag{16.4}$$

We can show that the $|i'\rangle$ are linearly independent, for if this were not so, there would exist a set of N numbers c^i not all vanishing such that

$$\sum_{i=1}^{N} c^i\,|i'\rangle = 0$$

or

$$\sum_{j=1}^{N} c^i R_i^j\,|j\rangle = 0$$

However, the linear independence of vectors $|j\rangle$ implies that

$$R_i^j c^i = 0 \qquad (j = 1, 2, \cdots, N)$$

Because of Eq. 16.1, this set of equations has the unique solution

$$c^i = 0 \qquad (i = 1, 2, \cdots, N)$$

and this proves the linear independence of $|i'\rangle$.

Suppose now that we want to switch basis in S_N from the set $|i\rangle$ to the set $|i'\rangle$. The question that immediately arises is, "what is the relation between the representations of a given vector or of a given operator in the new and in the old bases?"

Consider a vector $|a\rangle$

$$|a\rangle = \sum_{i=1}^{N} a^i\,|i\rangle \tag{16.5}$$

In the new basis, $|a\rangle$ is decomposed as

$$|a\rangle = \sum_{i=1}^{N} a'^i|i'\rangle \tag{16.6}$$

Thus, using Eq. 16.4, Eq. 16.5 can be written as

$$|a\rangle = \sum_{j=1}^{N} a^i (R^{-1})^j_i |j'\rangle \tag{16.7}$$

By comparing Eq. 16.6 with Eq. 16.7, we get

$$a'^j = (R^{-1})^j_i a^i \tag{16.8}$$

or in matrix form

$$\mathbf{a}' = \mathbf{R}^{-1}\mathbf{a} \tag{16.9}$$

Using a similar technique, we obtain the transformation law for the elements of a matrix **A**, which represents in the old basis the linear operator A. Again using Eqs. 16.2 and 16.4, we have

$$\begin{aligned} A|i'\rangle &= A \sum_{j=1}^{N} R^j_i |j\rangle \\ &= \sum_{k=1}^{N} R^j_i A^k_j |k\rangle \\ &= \sum_{m=1}^{N} (R^{-1})^m_k R^j_i A^k_j |m'\rangle \end{aligned} \tag{16.10}$$

In the new basis, A is represented by the matrix $\mathbf{A}'$ defined by

$$A|i'\rangle = \sum_{m=1}^{N} A'^m_i |m'\rangle \tag{16.11}$$

Therefore, comparing Eqs. 16.10 and 16.11, we find

$$A'^m_i = (R^{-1})^m_k R^j_i A^k_j \tag{16.12}$$

or in matrix form

$$\mathbf{A}' = \mathbf{R}^{-1}\mathbf{A}\mathbf{R} \tag{16.13}$$

(Note that the order of the matrices in Eq. 16.13 follows from Eq. 16.12.) Equations 16.8 and 16.12 are the transformation laws for the components of a vector and for the elements of a matrix, respectively, when the basis has been changed according to the transformation

$$|i'\rangle = \sum_{j=1}^{N} R^j_i |j\rangle \qquad (i = 1, 2, \cdots, N)$$

17 · SCALARS AND TENSORS

The choice of a basis in an abstract vector space usually corresponds in physical applications to the choice of what one calls a reference frame associated with an observer. Two observers may be in motion, one with respect to the other, or may use instruments of a different kind, but the meaning of their observations should be essentially the same. Only those observations are meaningful whose results can somehow be formulated independently of the choice of the reference frame. Stated differently, the laws of physics, i.e., the equations of physics, should not depend upon a particular reference frame. If physical laws were to depend upon the particular frame of reference in which the measurements were carried out, we would not simply have a single set of laws of physics, but rather an infinite set of such laws, one set corresponding to each reference frame and each observer having his own laws. Therefore, even though each physical quantity involved in an equation may vary from one reference frame to another, the variation must be such as to "cancel" the effect of the changes brought upon the other quantities involved in the equation, and to make the law globally invariant. (The understanding that physical laws must remain invariant with respect to changes of reference frames led to the formulation of the theory of relativity.)

Since the result of a measurement is always a set of numbers, the manner in which this set of numbers transforms as we change the reference frame is of utmost importance. Mathematically, the problem is one of determining the abstract object associated with the physical quantity that is to be measured and to find the transformation law for the representation of this mathematical object as we change the reference frame. For these reasons, it is of interest to study the transformation properties of various mathematical objects. We shall see that a vector, with the transformation property (16.8) is a particular case of a more general class of objects called tensors.

The simplest behavior with the change of basis is that of a quantity which does not vary at all. Such a quantity is called a **scalar**. For instance, the sum of the diagonal elements of a matrix is a scalar. In fact from Eq. 16.12 and using Eq. 16.3, we have

$$A_i^{\prime i} = (R^{-1})_j^k \, A_m^j R_k^m = E_j^m A_m^j = A_j^j \tag{17.1}$$

A_i^i is called the **trace** of the matrix **A** and is denoted by

$$\operatorname{Tr} \mathbf{A} \underset{\text{def}}{=} A_i^i \qquad \text{(summation implied)}$$

Another example of a scalar is the determinant of a matrix. Comparing the rule of multiplication of determinants with the rule of matrix multiplication, we get immediately

$$\begin{aligned} \det \mathbf{A}' &= \det(\mathbf{R}^{-1}\mathbf{A}\mathbf{R}) = \det \mathbf{R}^{-1} \det \mathbf{A} \det \mathbf{R} \\ &= \det \mathbf{A} \end{aligned}$$

Now let us examine sets of numbers with more complicated transformation properties. This will be a generalization, from a slightly different point of view, of what was done in the preceding section.

The representation of a vector $|a\rangle$ by the set of components $a^i (i = 1, 2, \cdots, N)$ that determine its decomposition

$$|a\rangle = \sum_{i=1}^{N} a^i \, |i\rangle$$

into basis vectors is not the only possible way of associating a set of N numbers with this vector. When a scalar product has been defined we can consider, for instance, the set of numbers

$$a_i = \sum_{j=1}^{N} \bar{a}^j \langle j|i\rangle \qquad (i = 1, 2, \cdots, N) \tag{17.2}$$

The reason for using a subscript here instead of a superscript is to distinguish the a_i from the a^i. The transformation (17.2) is not a linear one, since it contains a complex conjugation. The set of a^j determines uniquely the set of a_i. The converse is also true, since

$$\det(\langle j|i\rangle) \neq 0 \tag{17.3}$$

and the system of equations (17.2) can be solved for the $\bar{a}^j$, given the a_i. The condition 17.3 follows from the linear independence of the basis vectors. In fact, $\det(\langle j|i\rangle) = 0$ would imply the existence of numbers c^i, not all zero, satisfying

$$\sum_{i=1}^{N} c^i \langle j|i\rangle = 0 \qquad \text{for all } j$$

The vector

$$|c\rangle = \sum_{i=1}^{N} c^i\,|i\rangle \tag{17.4}$$

would therefore either be equal to zero, in contradiction to the supposition that the vectors $|i\rangle$ are linearly independent, or would be orthogonal to all basis vectors $|j\rangle$, which is impossible because the vector $|c\rangle$ is a linear combination of the vectors $|i\rangle$. (Remember that orthogonal vectors are linearly independent.)

Let us find the transformation law for the numbers a_i. The equation

$$|i'\rangle = \sum_{m=1}^{N} R_i^m |m\rangle \tag{17.5}$$

is equivalent to

$$\langle j'| = \sum_{k=1}^{N} \bar{R}_j^k \langle k|$$

Therefore

$$\langle j'|i'\rangle = \sum_{k,m=1}^{N} \bar{R}_j^k R_i^m \langle k|m\rangle \tag{17.6}$$

Taking the complex conjugate of

$$a'^j = (R^{-1})_i^j a^i$$

we have

$$\bar{a}'^j = \overline{(R^{-1})}{}_n^j \bar{a}^n \tag{17.7}$$

From Eqs. 17.6 and 17.7 we obtain

$$\sum_{j=1}^{N} \bar{a}'^j \langle j'|i'\rangle = \sum_{j,k,m=1}^{N} \overline{(R^{-1})}{}_k^j \bar{R}_j^k R_i^m \bar{a}^n \langle k|m\rangle \tag{17.8}$$

However

$$\overline{(R^{-1})}{}_n^j \bar{R}_j^k = E_n^k \tag{17.9}$$

since all elements of **E** are real. Inserting Eq. 17.9 in Eq. 17.8 and using the definition of the a_i we finally have

$$a'_i = R_i^m a_m \tag{17.10}$$

This can be written in the matrix form

$$\tilde{\mathbf{a}}' = \tilde{\mathbf{a}}\mathbf{R} \tag{17.11}$$

if we agree to arrange the numbers a_i in a table with only **one row** and N columns, and denote this table by the symbol $\tilde{\mathbf{a}}$. Note that Eq. 17.11 is different from the arrangement of Eq. 14.11, where the vector was written as a matrix with N rows and **one column**. Written out in full, Eq. 17.11 is

$$(a'_1, a'_2, \cdots, a'_N) = (a_1, a_2, \cdots, a_N)\begin{pmatrix} R_1^1 & R_2^1 & \cdots & R_N^1 \\ R_1^2 & R_2^2 & \cdots & R_N^2 \\ \vdots & \vdots & \vdots & \vdots \\ R_1^N & R_2^N & \cdots & R_N^N \end{pmatrix}$$

and the "row vector" **premultiplies** the matrix **R**. In Eq. 14.11, **a postmultiplied** the transformation matrix. This is consistent with the rules of matrix multiplication.

Compare now Eqs. 17.5 and 17.10. We see that a_i transforms according to the same linear transformation as that used to go from the old to the new basis. The numbers a_i are called the **covariant** components of the vector $|a\rangle$, to stress the fact that they transform in the same way as basis vectors. On the other hand the numbers a^i, which are transformed by a matrix that is the inverse of the matrix which transforms the basis vectors, are called the **contravariant** components of $|a\rangle$, since, roughly speaking, they transform in a manner "opposite" to that of the basis.

Take a vector $|b\rangle$. Its contravariant components transform as

$$b'^j = (R^{-1})^j_k b^k \tag{17.12}$$

Multiplying Eq. 17.12 by Eq. 17.10, we have

$$a'_i b'^j = R_i^m (\mathrm{R}^{-1})^j_k a_m b^k \tag{17.13}$$

Comparing Eq. 17.13 with the transformation law for the elements of a matrix, as given by Eq. 16.12 of the preceding section, we find that A^k_m transforms exactly in the same way as the set of products $a_m b^k$. This example leads us in a natural way to the concept of a **tensor**:

H Sets of numbers that transform like products of components (covariant or contravariant, or both) of vectors are called tensors. Thus, the tensor $A^{ij\cdots}_{kl\cdots}$ transforms like $a^i b^j \cdots c_k d_l \cdots$; namely,

$$A'^{mn\cdots}_{rs\cdots} = (R^{-1})^m_i (R^{-1})^n_j \cdots R^k_r R^l_s \cdots A^{ij\cdots}_{kl\cdots}$$

The lower (upper) indices of a tensor are called covariant (contravariant) indices. A tensor that has only covariant (contravariant) indices is called a **covariant (contravariant) tensor**. If a tensor has both upper and lower indices, it is called a **mixed tensor**. A tensor that has M indices (this includes covariant and contravariant indices) is called a tensor of the Mth rank. In this sense, components of a vector form a tensor of the first rank, while the elements of a matrix representing a linear operator constitute a mixed tensor of the second rank.

In defining matrices we have represented vectors and linear operators by sets of numbers, and found that, from the point of view of their transformation properties, these sets of numbers are particular examples of more general objects called tensors.

In introducing tensors, we no longer attempted to define them abstractly, i.e., independently of a basis, as we did in the case of linear operators. On the contrary, our attention was focused entirely on transformation properties, and in fact we defined tensors by characterizing their behavior under a change of basis. The difference in approach should be evident.

18 · ORTHOGONAL BASES AND SOME SPECIAL MATRICES

Until now in our discussion of representations of vectors and linear operators, no use was made of the scalar product (except in defining the covariant components of a vector). Certain new aspects of the problem appear, however, if the scalar product is introduced. According to the orthogonalization theorem, one can obtain a basis of orthonormal vectors $|e_i\rangle (i = 1, \cdots, N)$ by a suitable linear transformation. Consider then the decomposition

$$|a\rangle = \sum_{j=1}^{N} a^j |e_j\rangle \tag{18.1}$$

Multiplying the preceding equation by $\langle e_k|$ and using the orthonormality of the basis vectors (Eq. 12.2), we get

$$a^k = \langle e_k|a\rangle \qquad (k = 1, 2, \cdots, N) \tag{18.2}$$

We see from Eq. 18.2 that a characteristic of the decomposition of a vector with respect to an orthonormal basis is that it allows us to express any component of the decomposed vector in terms of a simple scalar product. The importance of this quite general feature will become more apparent in Chapter III.

With respect to an orthonormal basis, co- and contravariant components of a vector are simply complex conjugates of each other. This follows from the definition of covariant components as given by the formula 17.2

$$a_i \equiv \sum_{j=1}^{N} \bar{a}^j \langle e_j|e_i\rangle = \bar{a}^i \tag{18.3}$$

Writing

$$|b\rangle = \sum_{i=1}^{N} b^i |i\rangle$$

$$\langle a| = \sum_{j=1}^{N} \bar{a}^j \langle j|$$

and using Eq. 17.2, one finds that

$$\langle a|b\rangle = \sum_{i,j=1}^{N} b^i \bar{a}^j \langle j|i\rangle = a_i b^i \tag{18.4}$$

In an orthonormal basis, because of Eq. 18.3, this becomes

$$\langle a|b\rangle = \bar{a}^1 b^1 + \bar{a}^2 b^2 + \cdots + a^N b^N \tag{18.5}$$

Apart from the complex conjugations resulting from taking vectors with complex components, Eq. 18.5 is the usual expression used in elementary vector calculus to define the scalar product in orthogonal coordinates.

Using the same argument as that which led to Eq. 18.2, we get from

$$A\,|e_i\rangle = \sum_{j=1}^{N} A^j_i\,|e_j\rangle$$

a simple expression for the elements of the matrix representing the operator A in an orthonormal basis

$$A^k_i = \langle e_k|\,A\,|e_i\rangle \tag{18.6}$$

Let us examine in more detail the structure of matrices representing in an orthonormal basis the adjoint, Hermitian, and unitary operators defined in Sec. 8. The corresponding matrices are also called adjoint, Hermitian, and unitary, respectively.

The operator A^+ adjoint to A is represented by the matrix $\mathbf{A}^+$

$$(A^+)^k_i = \langle e_k|\,A^+\,|e_i\rangle \tag{18.7}$$

Taking the complex conjugate of Eq. 18.6 and remembering the definition of an adjoint operator (Eq. 9.8), we get

$$\bar{A}^k_i = \langle e_i|\,A^+\,|e_k\rangle$$

Thus, comparing with Eq. 18.7, we have

$$(A^+)^k_i \equiv \bar{A}^i_k \tag{18.8}$$

This identity defines the **adjoint** matrix to $\mathbf{A}$. It is seen that in order to obtain the elements of $\mathbf{A}^+$, knowing the elements of $\mathbf{A}$, we have to perform two operations. First, we replace all the elements of $\mathbf{A}$ by their complex conjugates. The operation

$$A^k_i \to \bar{A}^k_i \tag{18.9}$$

transforms $\mathbf{A}$ into the matrix $\mathbf{A}^*$ called the **conjugate** matrix to $\mathbf{A}$. Secondly, we interchange the rows and columns of $\mathbf{A}^*$

$$(A^*)^k_i \to (A^*)^i_k \tag{18.10}$$

This operation is called a **transposition**. Therefore, (18.8) can be described by saying that the adjoint matrix to $\mathbf{A}$ is its transposed conjugate matrix. One writes

$$\mathbf{A}^+ = (\mathbf{A}^*)^{\mathrm{T}}$$

in which "T" represents a transposition.

The relation

$$(\mathbf{A}\cdot\mathbf{B})^+ = \mathbf{B}^+\mathbf{A}^+ \tag{18.11}$$

which is the analog of the corresponding expression for the adjoint of the product of two linear operators, holds for matrices. That this should be so is a direct consequence of the manner in which matrix multiplication and the notion of an adjoint matrix have been introduced. Equation 18.11 can also be verified by an explicit calculation.

$$[(AB)^+]^i_j \equiv \bar{A}^j_k\bar{B}^k_i \equiv (B^+)^i_k(A^+)^k_j$$

Consider now a Hermitian operator H

$$H = H^+ \tag{18.12}$$

From Eqs. 18.6 and 18.7, we see that in an orthogonal basis, Eq. 18.12 implies for the matrices that

$$\mathbf{H} = \mathbf{H}^+ \tag{18.13}$$

A matrix equal to its adjoint is called a **Hermitian** matrix. Equation 18.13 is equivalent to

$$H_i^k \equiv \overline{H}_k^i \tag{18.14}$$

Thus, a Hermitian matrix is characterized by the fact that those of its elements that are symmetrical with respect to the principal diagonal are complex conjugates of each other. Consequently, the diagonal elements of a Hermitian matrix are all real. This can also be seen by setting $i = k$ in Eq. 18.14.

EXAMPLE

An example of a Hermitian matrix is

$$\begin{pmatrix} a & c+id \\ c-id & b \end{pmatrix} \tag{18.15}$$

with a,b,c, and d real.

The transpose of (18.15) is

$$\begin{pmatrix} a & c-id \\ c+id & b \end{pmatrix} \tag{18.16}$$

and the complex conjugate of Eq. 18.16 is

$$\begin{pmatrix} a & c+id \\ c-id & b \end{pmatrix}$$

which is the same as (18.15).

An unitary operator U has been defined by the equation

$$U^+ = U^{-1}$$

U^{-1} is represented in any basis by the matrix $\mathbf{U}^{-1}$ inverse to $\mathbf{U}$. We have seen above that U^+ is represented in an orthogonal basis by the matrix $\mathbf{U}^+$ adjoint to $\mathbf{U}$. Therefore, in an orthonormal basis, a unitary operator U is represented by a matrix $\mathbf{U}$ satisfying

$$\mathbf{U}^+\mathbf{U} = \mathbf{E} \tag{18.17}$$

and also called **unitary**. It has already been shown that in a space with a finite number of dimensions, Eq. 18.17 implies

$$\mathbf{U}\mathbf{U}^+ = \mathbf{E}$$

Let us perform a transformation of the orthonormal basis $|e_i\rangle (i = 1, 2, \cdots, N)$ generated by a unitary operator U.

$$|e'_i\rangle = U\,|e_i\rangle$$

The vectors $|e'_i\rangle$ again form an orthonormal system

$$\langle e'_i|e'_k\rangle = \langle e_i|\, U^+U\, |e_k\rangle = \langle e_i|e_k\rangle$$

Since the vectors $|e'_i\rangle$ form an orthonormal set, they are obviously linearly independent (see lemma, Sec. 11) and may be chosen as a basis.

Consider now two different orthonormal bases $|e_i\rangle$ and $|e'_i\rangle (i = 1, \cdots, N)$. The vectors $|e'_i\rangle$ can be decomposed in terms of the set $|e_i\rangle$.

$$|e'_i\rangle = \sum_{k=1}^{N} e'^k_{(i)} |e_k\rangle \tag{18.18}$$

Let us define a matrix **U** by the identity

$$U^k_i \equiv e'^k_{(i)}$$

so that the $\mathbf{e}'_{(i)}$ occupies the ith column of the matrix **U**.

$$\mathbf{U} = \begin{pmatrix} e'^1_{(1)} & e'^1_{(2)} & \cdots & e'^1_{(N)} \\ e'^2_{(1)} & e'^2_{(2)} & \cdots & e'^2_{(N)} \\ \vdots & \vdots & \vdots & \vdots \\ e'^N_{(1)} & e'^N_{(2)} & \cdots & e'^N_{(N)} \end{pmatrix}$$

$\mathbf{U}^+$ is, according to the definition of the adjoint matrix, given by (compare Eq. 18.3)

$$(U^+)^i_k \equiv \bar{e}'^k_{(i)} \equiv e'_{(i)k}$$

which means that the $\tilde{\mathbf{e}}'_{(i)}$ occupies the ith row of $\mathbf{U}^+$. Equation 18.18 can be understood as the basis transformation generated by the matrix **U**. We can easily verify that **U** is unitary

$$(U^+)^i_k U^k_j = e'_{(i)k} e'^k_{(j)} = \langle e'_i | e'_j \rangle$$

where we have made use of Eq. 18.4. Thus

$$(\mathbf{U}^+\mathbf{U})^i_j = \begin{cases} 0 & i \neq j \\ 1 & i = j \end{cases}$$

which means that

$$\mathbf{U}^+\mathbf{U} = \mathbf{E}$$

The foregoing discussion proves the following theorem.

Theorem. The necessary and sufficient condition for an orthonormal basis to be transformed into another orthonormal basis is that the transformation matrix is unitary.

We have defined Hermitian and unitary matrices as the matrices that represent in an orthonormal basis Hermitian and unitary operators, respectively. Although we referred the matrices to some orthonormal basis, no particular orthonormal basis was chosen. Therefore, a transformation leading from one orthonormal basis to another orthonormal basis should preserve the hermiticity or unitarity of a matrix, as will be shown now. According to the theorem just proved, such a transformation is generated by a unitary matrix; we denote it by **U**.

$$\mathbf{U}^+ = \mathbf{U}^{-1}$$

Consider two matrices, one Hermitian and the other unitary.

$$\mathbf{H} = \mathbf{H}^+$$
$$\mathbf{V}^{-1} = \mathbf{V}^+ \tag{18.19}$$

Remembering the transformation law for matrices (Eq. 16.13), we have, in the new basis

$$\mathbf{H}' = \mathbf{U}^{-1}\mathbf{H}\mathbf{U} = \mathbf{U}^{+}\mathbf{H}\mathbf{U}$$

$$\mathbf{V}' = \mathbf{U}^{-1}\mathbf{V}\mathbf{U} = \mathbf{U}^{+}\mathbf{V}\mathbf{U}$$

Using Eqs. 18.11 and 18.19, we obtain

$$\begin{aligned} \mathbf{H}'^{+} &= [(\mathbf{U}^{+}\mathbf{H})\mathbf{U}]^{+} = \mathbf{U}^{+}(\mathbf{U}^{+}\mathbf{H})^{+} \\ &= \mathbf{U}^{+}\mathbf{H}^{+}\mathbf{U} = \mathbf{H}' \end{aligned} \tag{18.20}$$

and

$$\mathbf{V}'^{+} = \mathbf{U}^{+}\mathbf{V}^{+}\mathbf{U} = \mathbf{U}^{-1}\mathbf{V}^{-1}\mathbf{U}$$

Thus

$$\mathbf{V}'^{+}\mathbf{V}' = \mathbf{U}^{-1}\mathbf{V}^{-1}\mathbf{U}\mathbf{U}^{-1}\mathbf{V}\mathbf{U} = 1$$

Hence

$$\mathbf{V}'^{+} = \mathbf{V}'^{-1} \tag{18.21}$$

Equations 18.20 and 18.21 show that the transformed matrices $\mathbf{H}'$ and $\mathbf{V}'$ are, respectively, Hermitian and unitary.

It is evident that an arbitrary basis transformation would not preserve the hermiticity or the unitarity of a matrix since, in the preceding demonstration the unitary of the transformation matrix was essential. Thus, the property of hermiticity or unitarity of a matrix is invariant with respect to a restricted class of basis transformations. This is in contrast, for example, to the notion of an inverse matrix, since from the equation

$$\mathbf{A}^{-1}\mathbf{A} = \mathbf{E}$$

follows

$$\mathbf{R}^{-1}\mathbf{A}^{-1}\mathbf{R}\mathbf{R}^{-1}\mathbf{A}\mathbf{R} = \mathbf{R}^{-1}\mathbf{E}\mathbf{R} = \mathbf{E}$$

which means that

$$\mathbf{A}'^{-1}\mathbf{A}' = \mathbf{E}$$

without any restriction on the transformation matrix **R**.

19 · INTRODUCTION TO TENSOR CALCULUS

19.1 Tensors in a Real Vector Space

In Sec. 17 we defined tensors by considering general transformations of the basis in an N-dimensional complex space. One could study further the properties of such tensors, but it would necessitate a rather complicated notation. We believe that from a didactic point of view, it is reasonable to limit considerably the scope of the discussion. In this section we limit ourselves to **real** vector spaces, in which a scalar product has been defined.

Consider the transformation of the basis vectors

$$|i'\rangle = \sum_{m=1}^{N} R_i^m |m\rangle \qquad (i = 1, \cdots, N) \tag{19.1}$$

Since all numbers are now real, this is equivalent to

$$\langle j'| = \sum_{n=1}^{N} R_j^n \langle n| \qquad (j = 1, \cdots, N) \tag{19.2}$$

Multiplying Eq. 19.1 by Eq. 19.2, we get

$$\langle j'|i'\rangle = \sum_{n,m=1}^{N} R_j^n R_i^m \langle n|m\rangle \tag{19.3}$$

This shows that the set of numbers $\langle n|m\rangle$ transforms like a covariant tensor of the second rank. We shall call this tensor the **metric tensor,** and denote it by a special symbol g_{nm}.* Thus, Eq. 19.3 can be written as

$$g'_{ji} = R_j^n R^m_i g_{nm} \tag{19.4}$$

In a real vector space, g_{ij} is a symmetric tensor, i.e., $g_{ij} = g_{ji}$.

The definition of the covariant components of a vector $|a\rangle$ (Eq. 17.2) now reads

$$a_i = g_{ij} a^j \tag{19.5}$$

Since det $(g_{ij}) \neq 0$, (compare 17.3) the preceding equation can be uniquely solved for a^j. We write this solution as

$$a^j = g^{ji} a_i \tag{19.6}$$

Inserting Eq. 19.5 in Eq. 19.6, we obtain

$$a_i = g_{ij} g^{jk} a_k$$

Since the vector $|a\rangle$ is arbitrary, we must have

$$g_{ij} g^{jk} = E_i^k \tag{19.7}$$

Analogously we can also obtain the relation

$$g^{kj} g_{ji} = E_i^k \tag{19.8}$$

The notation in Eq. 19.6 is justified by the fact that the numbers g^{ij} really form a contravariant tensor of the second rank. In fact, in another basis in analogy to Eq. 19.5 we have

$$a'_i = g'_{ij} a'^j \tag{19.9}$$

with

$$\begin{aligned} a'_i &= R_i^k a_k \\ a'^j &= (R^{-1})_k^j a^k \end{aligned} \tag{19.10}$$

As before, Eq. 19.9 can be inverted

$$a'^j = g'^{ji} a'_i \tag{19.11}$$

---

* Notice that when we speak about a tensor $a_{k \cdots l}^{i \cdots j}$ we mean the entire set of numbers $a_{k \cdots l}^{i \cdots j}$ $(i,j,k,l, \cdots = 1, 2, \cdots, N)$ and not some particular number belonging to this set.

Using Eqs. 19.10, we get

$$(R^{-1})^j_k a^k = g'^{ji} R^m_i a_m$$

Therefore

$$R^s_j (R^{-1})^j_k a^k = a^s = g'^{ji} R^s_j R^m_i a_m$$

Comparing with Eq. 19.6, we have

$$g^{sm} = R^s_j R^m_i g'^{ji}$$

or

$$g'^{ji} = (R^{-1})^j_s (R^{-1})^i_m g^{sm} \qquad (19.12)$$

which is the transformation law for a contravariant tensor of the second rank.

Equation 19.8 shows that

$$g^{kj} \equiv (-1)^{k+j} \frac{M_{jk}}{\det(g_{mn})} \qquad (19.13)$$

where M_{jk} is the minor of $\det(g_{mn})$ corresponding to the element g_{jk} (compare Eq. 15.2). We leave it to the reader to show that if g_{ij} is symmetric, then so is g^{ij}.

Before going further, let us observe that given a tensor $a^{i\cdots j}_{k\cdots l}$ with m_1 contravariant and m_2 covariant indices (thus, of rank $m_1 + m_2$), and given another tensor $b^{m\cdots n}_{r\cdots s}$ with n_1 contravariant and n_2 covariant indices (thus of rank $n_1 + n_2$), the set of all products

$$a^{i\cdots j}_{k\cdots l} \cdot b^{m\cdots n}_{r\cdots s} \qquad (i,j,k,l,m,n,r,s,\cdots = 1,\cdots, N) \qquad (19.14)$$

forms a tensor with $m_1 + n_1$ contravariant and $m_2 + n_2$ covariant indices (thus, of rank $m_1 + n_1 + m_2 + n_2$). This follows immediately from the definition of a tensor, as can be seen by multiplying

$$a'^{i\cdots j}_{k\cdots l} = (R^{-1})^i_{i'} \cdots (R^{-1})^j_{j'} R^{k'}_k \cdots R^{l'}_l a^{i'\cdots j'}_{k'\cdots l'}$$

with

$$b'^{m\cdots n}_{r\cdots s} = (R^{-1})^m_{m'} \cdots (R^{-1})^{n'}_n R^{r'}_r \cdots R^{s'}_s b^{m'\cdots n'}_{r'\cdots s'}$$

to get the transformation law for the tensor 19.14.

Consider now an arbitrary mixed tensor $a^{i\cdots j\cdots k}_{r\cdots s\cdots t}$ and let us examine the transformation properties of the set of numbers* $a^{i\cdots j\cdots k}_{r\cdots j\cdots t}$, where we have replaced the subscript s by j. The operation, which consists in assigning to an upper and a lower index of a tensor the same numerical value and then summing over all possible assignments, is called the **contraction** of this tensor with respect to these indices.

The index over which the summation is performed is called a "dummy" index, since the summation over it destroys its individuality. Thus $a^{i\cdots j\cdots k}_{r\cdots j\cdots t}$ is obtained by contraction of $a^{i\cdots j\cdots k}_{r\cdots s\cdots t}$ with respect to the indices j and s. Under a change of basis, $a^{i\cdots j\cdots k}_{r\cdots j\cdots t}$ transforms as

$$a'^{i\cdots j\cdots k}_{r\cdots j\cdots t} = (R^{-1})^i_l \cdots (R^{-1})^j_m \cdots (R^{-1})^k_n R^p_r \cdots R^q_j \cdots R^s_t \cdots a^{l\cdots m\cdots n}_{p\cdots q\cdots s} \qquad (19.15)$$

However

$$(R^{-1})^j_m R^q_j = E^q_m \qquad (19.16)$$

* Remember that the summation over j is understood according to the Einstein convention.

Therefore

$$a'^{i\cdots j\cdots k}_{r\cdots j\cdots t} = (R^{-1})^i_l \cdots (R^{-1})^k_n R^p_r \cdots R^s_t a^{l\cdots q\cdots n}_{p\cdots q\cdots s} \tag{19.17}$$

where two of the transformation matrices that appeared in Eq. 19.15 have now been eliminated. The dummy index q on the RHS of Eq. 19.17 can, of course, be replaced by j. The transformation law 19.17 is that of a tensor, but because of the relation 19.16 which has eliminated two transformation matrices, the rank of the contracted tensor is two less than the rank of the original tensor. In other words we have demonstrated the following lemma.

Lemma 1. Dummy indices do not participate in the transformation of a tensor.

We now illustrate the previous general result by a particular example. Let a_i and b^j denote the covariant and contravariant components of the vectors $|a\rangle$ and $|b\rangle$, respectively. The array of numbers $a_i b^j$ forms a tensor of second rank. Contracting the indices i and j, we obtain a tensor of rank zero, i.e., a scalar

$$a'_j b'^j = (R^{-1})^j_m R^n_j a_n b^m = a_m b^m$$

It should not be surprising that $a_i b^i$ is an invariant, since

$$a_i b^i = \langle a|b\rangle$$

and the scalar product $\langle a|b\rangle$ has been introduced without any reference to a particular basis.

Using Eq. 19.5, we can rewrite the scalar product in the form

$$\langle a|b\rangle = g_{ij} a^i b^j = g^{ij} a_i b_i \tag{19.18}$$

Thus the metric properties of the space are determined by the tensor g_{ij}, which justifies its naming.

Let us consider now a tensor $a^{i\cdots j}_{k\cdots l}$ whose elements are in some reference frame (i.e., basis) equal, element by element, to the corresponding elements of another tensor $b^{i\cdots j}_{k\cdots l}$

$$a^{i\cdots j}_{k\cdots l} = b^{i\cdots j}_{k\cdots l}$$

for a particular choice of the basis. Multiplying both sides of the preceding equation by the tensor $(R^{-1})^m_i \cdots (R^{-1})^n_j (R)^k_p \cdots R^l_q$, and contracting the resulting tensor

$$(R^{-1})^m_i \cdots (R^{-1})^n_j R^k_p \cdots R^l_q a^{i\cdots j}_{k\cdots l} = (R^{-1})^m_i \cdots (R^{-1})^n_j R^k_p \cdots R^l_q b^{i\cdots j}_{k\cdots l}$$

we see that

$$a'^{m\cdots n}_{p\cdots q} = b'^{m\cdots n}_{p\cdots q}$$

in the new basis. This proves

Lemma 2. The equality, element by element, of two tensors in a particular reference frame implies their equality in any reference frame.

Lemma 2 makes clear that the validity of a tensor equation is independent of the choice of a basis, provided all the terms entering such an equation transform in the same way (i.e., have the same covariant and contravariant indices, although not necessarily in the same order).

EXAMPLE 1

A tensor $a_{i\cdots j\cdots k\cdots l}$ is called symmetric or antisymmetric with respect to the indices j and k if it satisfies the equation

$$a_{i\ldots j\ldots k\ldots l} = \pm a_{i\ldots k\ldots j\ldots l} \tag{19.19}$$

with the upper sign for the symmetric tensor and the lower sign for the antisymmetric tensor. Equation 19.19 is a meaningful tensor equation, since (as may easily be verified) both sides transform in the same way and thus the property that a tensor be symmetric or antisymmetric with respect to a pair of indices is independent of the choice of a basis.

It can easily be verified that in an N-dimensional space, a symmetric tensor of the second rank has $N(N+1)/2$ independent components, whereas an antisymmetric tensor of the second rank has $N(N-1)/2$ independent components, instead of the usual number of components N^2.

We have seen that a vector can be represented either by a set of covariant or by a set of contravariant components. Similarly, the representation of an operator A by a matrix whose elements transform like a mixed tensor of rank 2 is not the only possible representation of the operator. Consider the equation

$$A\,|i\rangle = \sum_{j=1}^{N} A_i^j\,|j\rangle \tag{19.20}$$

Multiplying Eq. 19.20 by $\langle k|$, we get

$$\langle k|\,A\,|i\rangle = \sum_{j=1}^{N} A_i^j\,\langle k|\,j\rangle = g_{kj}A_i^j \tag{19.21}$$

The numbers $\langle k|\,A\,|i\rangle$, which result from the contraction with respect to indices j and l of the tensor $g_{kl}A_i^j$, form a covariant tensor of the second rank

$$A_{ki} = g_{kj}A_i^j \tag{19.22}$$

The set of N^2 numbers A_{ij} $(i,j = 1, \cdots, N)$ is uniquely determined by the operator A once a basis has been chosen, and provides a representation of A, which may in some applications be more useful than the representation by a matrix **A** (also determined, of course, by N^2 parameters). Although the transformation properties of the tensors A_{ij} and A_i^j are, in general, quite different so that the equation $A_{ij} = A_i^j$ is meaningless, one can say that they have their common origin in the operator A. Moreover, they are related by Eq. 19.22 which makes it possible, once the metric properties of the space (more precisely the metric tensor) are known, to find A_{ij} when A_i^j is given, and vice versa. In fact, Eq. 19.22 can easily be inverted. Multiplying Eq. 19.22 by g^{pq} and contracting the indices q and k, we get

$$g^{pk}A_{ki} = g^{pk}g_{kj}A_i^j$$

Using Eq. 19.8, we obtain

$$A_i^p = g^{pk}A_{ki}$$

The operation of the multiplication of a tensor by the metric tensor with a subsequent contraction is called the operation of **raising** or **lowering** of an index. For instance, in Eq. 19.22, one has lowered an index of A_i^j, while in Eq. 19.23, one has raised an index in A_{ij}. By raising both indices in A_{ij}, one gets a contravariant tensor related to the operator A

$$A^{ij} = g^{ip}g^{jq}A_{pq} = g^{ip}A_p^j$$

Since the tensors A_{ij}, A_i^j and A^{ij} are all related through operations involving the metric tensor, it is usual in physical applications to regard these tensors as corresponding to the same physical quantity.

If we take, instead of an arbitrary operator A, the unit operator E, from Eq. 19.22 we get

$$E_{ki} = g_{kj}E_i^j = g_{ki}$$

Thus, the metric tensor provides an alternative representation of the unit operator. We can introduce the notation

$$g_i^k = E_i^k$$

which is the one commonly used in the literature when one wants to stress the tensor aspect of the problem.

19.2 Tensor Functions

Consider a function

$$w_{m\cdots n}^{k\cdots l}(|x\rangle) \tag{19.24}$$

In a given reference frame, $|x\rangle$ is determined by its components $x^1, x^2, \cdots, x^N$ and $w_{m\cdots n}^{k\cdots l}$ can be regarded as a function of N independent variables $x^i (i = 1, 2, \cdots, N)$.

$$w_{m\cdots n}^{k\cdots l}(x^1, x^2, \cdots, x^N) \tag{19.25}$$

or, more briefly,

$$w_{m\cdots n}^{k\cdots l}(x^i)$$

Under the basis transformation

$$|i'\rangle = D\,|i\rangle \tag{19.26}$$

the contravariant components of $|x\rangle$ transform as

$$x'^j = (D^{-1})_i^j x^i \tag{19.27}$$

Inverting Eq. 19.27, we have

$$x^i = D_j^i x'^j$$

The function (19.25) is called a **tensor function of rank** N if under the basis transformation (19.26) it transforms like a tensor of rank N

$$w'^{p\cdots q}_{r\cdots s}(x'^i) = (D^{-1})_k^p \cdots (D^{-1})_l^q D_r^m \cdots D_s^k w_{m\cdots n}^{k\cdots l}(D_j^i x'^j) \tag{19.28}$$

Notice that the components of the tensor function and their arguments participate in the transformation induced by the change of the basis.

In particular, a **scalar function** $w(x^i)$ transforms as

$$w'(x'^i) = w(D_j^i x'^j) \tag{19.29}$$

In general, the function $w'(x'^1, x'^2, \cdots, x'^N)$ will be a different function of its arguments than will $w(x^1, x^2, \cdots, x^N)$. However, it may happen that

$$w'(x'^i) = w(x'^i)$$

i.e.,

$$w(x^i) = w[(D^{-1})_j^i x^j] \tag{19.30}$$

In this case, w retains its functional form under a change of coordinates and is called an **invariant function.**

EXAMPLE 2

The function $w = x_i x^i = g_{ij} x^i x^j$ is an example of an invariant function. We have

$$w'(x'^i) = (D^{-1})^m_j D^j_k x'_m x'^k = x'_k x'^k$$

and w' has the same functional form as w.

On the other hand, $w = a_i x^i$, where $|a\rangle$ is an arbitrary constant vector, is a scalar function, but is not an invariant function

$$w'(x'^i) = (a_j D^j_i) x'^i$$

Differentiating both sides of Eq. 19.29, we get

$$\frac{\partial w'}{\partial x'^j} = \sum_k \frac{\partial w}{\partial x^k} \frac{\partial x^k}{\partial x'^j} = \sum_k D^k_j \frac{\partial w}{\partial x^k} \tag{19.31}$$

Therefore, the derivatives $\partial w / \partial x^k$ of the scalar function $w(x^i)$, transform like the covariant components of a vector.*

The calculations that have been carried out for the scalar function w can be easily generalized to include higher derivatives of tensor functions. This leads to the

Lemma 3. Provided one considers only linear transformations on the space, the mth order derivative

$$\frac{\partial^m w^{i \cdots j}_{k \cdots l}}{\partial x^r \cdots \partial x^s}$$

of a mixed tensor function $w^{i \cdots j}_{k \cdots l}(x^1, \cdots, x^N)$ is itself a mixed tensor, but has the number of covariant indices increased by m, as compared to $w^{i \cdots j}_{k \cdots l}$.

EXAMPLE 3

(i) The electrostatic potential $V(x^1, x^2, x^3)$ is an example of a scalar function. Its derivatives $\partial V / \partial x^i (i = 1, 2, 3)$ transform like the ith covariant component of a vector. $\partial^2 V / \partial x^i \partial x_j$ is a mixed second-rank tensor, which upon contraction yields the scalar $\partial^2 V / \partial x^i \partial x_i$.

(ii) The electric field $\vec{E}(x^1, x^2, x^3)$ is an example of a vector function. Its ith covariant component $E_i (i = 1, 2, 3)$ is related to the scalar potential V by the equation

$$E_i = -\frac{\partial V}{\partial x^i}$$

(iii) The vector potential $\vec{A}$ is also an example of a vector function. The quantities $\partial A^i / \partial x^j$ form a mixed tensor of the second rank, while upon contraction, $\partial A^i / \partial x^j$ becomes the scalar $\partial A^i / \partial x^i$ which is the divergence of $\vec{A}$.

Let us introduce a very particular mathematical object, which turns out to be of great use because it transforms like a tensor under a wide class of transformations.

The symbol $\varepsilon_{ij \ldots k} (i, j, \cdots, k = 1, \cdots, N)$, with the number of indices equal to the dimension of the space, is defined as

$$\varepsilon_{ij \cdots k} = \begin{cases} +1 & \text{if } (i, j, \cdots, k) \text{ is an even permutation of } (1, 2, \cdots, N) \\ -1 & \text{if } (i, j, \cdots, k) \text{ is an odd permutation of } (1, 2, \cdots, N) \\ 0 & \text{otherwise} \end{cases}$$

* Note that since we are considering only linear transformations, the elements of the transformation matrix **D** are simply numbers. The generalization of tensor calculus to the case of arbitrary transformations of the space, i.e., when the x'^i are arbitrary functions of the x^i, can be found in any textbook on general relativity. For a clear introduction, see, for instance, H. C. Corben and P. Stehle, *Classical Mechanics*, John Wiley & Sons, Inc., New York, 1950, chap. I.

Therefore, a particular element $\varepsilon_{ij\cdots k}$ is zero if the sequence of numbers $(i,j,\cdots,k)$ is not a permutation of $(1, 2, \cdots, N)$ which means that at least two of the indices are equal to each other.

Let us transform the set of numbers $\varepsilon_{ij\cdots k}$ $(i,j,\cdots,k = 1,\cdots,N)$ as if they constituted a covariant tensor

$$\varepsilon'_{ij\cdots k} = D_i^r D_j^s \cdots D_k^t \varepsilon_{rs\cdots t} \tag{19.32}$$

What are the properties of the symbol $\varepsilon'_{ij\cdots k}$? This question can be easily answered by noticing that the RHS of Eq. 19.32 can be written in the determinantal form

$$D_i^r D_j^s \cdots D_k^t \varepsilon_{rs\cdots t} = \begin{vmatrix} D_i^1 & D_j^1 & \cdots & D_k^1 \\ D_i^2 & D_j^2 & \cdots & D_k^2 \\ \vdots & \vdots & \vdots & \vdots \\ D_i^N & D_j^N & \cdots & D_k^N \end{vmatrix}$$

Thus

$$\varepsilon'_{ij\cdots k} = \begin{cases} \det \mathbf{D} & \text{if } (1, 2, \cdots, N) \text{ is an even permutation of } (1, 2, \cdots, N) \\ -\det \mathbf{D} & \text{if } (1, 2, \cdots, N) \text{ is an odd permutation of } (1, 2, \cdots, N) \\ 0 & \text{otherwise} \end{cases}$$

and we see that $\varepsilon_{ij\cdots k}$ transforms effectively like a tensor, provided we restrict ourselves to transformations of the basis satisfying the condition

$$\det \mathbf{D} = 1 \tag{19.33}$$

A set of numbers transforming like a tensor with respect to a restricted class of transformations is called a **pseudotensor**.

The condition 19.33 is satisfied by all the transformations that generate a rotation in the space, so that $\varepsilon_{ij\cdots k}$ behaves like a tensor with respect to rotations.

We turn now briefly to a consideration of some of the properties of rotations.

19.3 Rotations

A rotation can be defined as a transformation whose matrix $\mathbf{D}$ is determined by a set of parameters $\alpha,\beta,\cdots,\gamma$ and satisfying the following conditions:

(i) $\mathbf{D}(\alpha,\beta,\cdots,\gamma)$ is a continuous function of the parameters $\alpha,\beta,\cdots,\gamma$.
(ii) $\mathbf{D}(0, 0, \cdots, 0) = \mathbf{E}$.
(iii) The transformation generated by $\mathbf{D}$ does not change the scalar product of any two vectors belonging to the space.

Condition (iii) can be visualized by saying that the angles between vectors are left unchanged, since in analogy to elementary vector calculus, one may define the angle θ between the vectors $|a\rangle$ and $|b\rangle$ by the relation

$$\cos\theta = \frac{\langle a|b\rangle}{\sqrt{\langle a|a\rangle\,\langle b|b\rangle}} \tag{19.34}$$

Because of the Cauchy-Schwarz inequality applied to the RHS of Eq. 19.34, we always have

$$-1 \leq \cos\theta \leq 1$$

From condition (iii) it follows that a rotation transforms an orthogonal basis into another orthogonal basis. Assume first that our basis is an orthonormal one; then, as a consequence of the theorem of Sec. 18, **D** is a unitary matrix. Since in a real vector space the adjoint of a matrix is the same as the transpose of a matrix, we have

$$\mathbf{D}^{\mathrm{T}}\mathbf{D} = \mathbf{E} \tag{19.35}$$

where the superscript "T" means transposed. Equation 19.35 gives

$$\det \mathbf{D}^{\mathrm{T}} \det \mathbf{D} = (\det \mathbf{D})^2 = 1$$

Thus

$$\det \mathbf{D} = \pm 1$$

However, since a rotation is a continuous transformation, $\det \mathbf{D}$ must be a continuous function of $\alpha, \beta, \cdots, \gamma$, and the condition

$$\det \mathbf{D}(0, 0, \cdots, 0) = 1$$

implies that **D** must satisfy Eq. 19.33. This result has been obtained by assuming that the basis in which **D** has been defined is an orthonormal basis. However, an arbitrary change of the basis does not change the value of the determinant of a matrix, which, as we have seen, is a scalar. Thus, a rotation is always generated by a matrix with determinant equal to unity.

EXAMPLE 4

The matrix for rotations in a plane about an axis perpendicular to the plane is given by

$$\mathbf{D}(\theta) = \begin{pmatrix} \cos\theta & \sin\theta \\ -\sin\theta & \cos\theta \end{pmatrix}$$

where θ is the angle of rotation.

We notice that $\mathbf{D}(\theta)$ is a continuous function of θ, satisfying $\det \mathbf{D} = +1$, that

$$\mathbf{D}(\theta = 0) = \begin{pmatrix} 1 & 0 \\ 0 & 1 \end{pmatrix}$$

is the unit matrix and that

$$\mathbf{D}^{\mathrm{T}}\mathbf{D} = \begin{pmatrix} \cos\theta & -\sin\theta \\ \sin\theta & \cos\theta \end{pmatrix}\begin{pmatrix} \cos\theta & \sin\theta \\ -\sin\theta & \cos\theta \end{pmatrix} = \begin{pmatrix} 1 & 0 \\ 0 & 1 \end{pmatrix}$$

is also the unit matrix.

A two-dimensional vector with components (x,y) is transformed by $\mathbf{D}(\theta)$ into a vector with components (x',y').

$$\begin{pmatrix} x' \\ y' \end{pmatrix} = \begin{pmatrix} \cos\theta & \sin\theta \\ -\sin\theta & \cos\theta \end{pmatrix}\begin{pmatrix} x \\ y \end{pmatrix}$$

and therefore

$$\begin{aligned} x' &= x\cos\theta + y\sin\theta \\ y' &= -x\sin\theta + y\cos\theta \end{aligned}$$

The matrices for rotation in three dimensions are given in many textbooks (see e.g., M. E. Rose, *Elementary Theory of Angular Momentum*, John Wiley & Sons, Inc., New York, 1961).

19.4 Vector Analysis in a Three-dimensional Real Space

In this section we consider a three-dimensional real space in which an orthonormal basis has been chosen, and we limit our attention to the unitary transformations on the basis. The condition for a matrix $\mathbf{U}$ to be unitary

$$\mathbf{U}^{+} = \mathbf{U}^{-1}$$

now reads*

$$\mathbf{U}^{\mathrm{T}} = \mathbf{U}^{-1} \tag{19.36}$$

(where, as before, "T" means transposed)

$$(\mathbf{U}^{\mathrm{T}})^{i}_{k} \equiv \mathbf{U}^{k}_{i} \tag{19.37}$$

It is easy to see that, if we restrict ourselves to unitary transformations, then in a real vector space the covariant and contravariant indices of a tensor transform in the same way. For instance, the transformation law

$$a'_i = U^k_i a_k$$

is the same as the transformation law

$$a'^i = (U^{-1})^i_k a^k$$

since, using Eq. 19.36 and 19.37, we get

$$a'^i = (U^{\mathrm{T}})^i_k a^k = \sum_k U^k_i a^k$$

Due to the orthonormality of the basis, we also have

$$g_{ik} = \delta_{ik}$$

where δ_{ik} is the Kronecker symbol defined in Sec. 12. From Eq. 19.7, we get immediately

$$g^{ik} \equiv \delta_{ik}$$

Thus, the raising or lowering of indices does not alter in any way the properties of a tensor. For instance, raising indices in ε_{ijk}, we get a pseudotensor

$$\varepsilon^{ijk} = g^{ir}g^{js}g^{kt}\varepsilon_{rst} \equiv \varepsilon_{ijk}$$

with the same properties as ε_{ijk}. Although with these restrictions there is no difference between co- and contravariant indices, we shall use both kinds of indices in order to still be able to use the Einstein convention, as formulated in Sec. 13.

By inspection, one can verify the following relation, which will be useful in further discussions

$$\varepsilon_{ijk}\varepsilon^{lmk} = g^l_i g^m_j - g^l_j g^m_i \tag{19.38}$$

Take the vectors $|a\rangle$ and $|b\rangle$. From their components one may construct objects with various transformation properties. We know already that the set of products of components of two vectors (for instance, the numbers $a_i b^j$) forms a tensor. Contracting indices i and j in $a_i b^j$, we get a scalar, the scalar product $\langle a|b\rangle$. Using ε^{ijk}, one can

---

* A matrix satisfying Eq. 19.36 is called an **orthogonal matrix.** Hence, a unitary transformation in a real vector space is an orthogonal transformation.

construct a vector, or rather a pseudovector, from the numbers a_i, b_j $(i,j = 1, 2, 3)$

$$c^i \underset{\text{def}}{=} \varepsilon^{ijk} a_j b_k \tag{19.39}$$

Written in terms of components, Eq. 19.39 becomes

$$\begin{aligned} c^1 &= a^2b^3 - a^3b^2 \\ c^2 &= a^3b^1 - a^1b^3 \\ c^3 &= a^1b^2 - a^2b^1 \end{aligned} \tag{19.40}$$

and we see that the c^i are components of the familiar vector product of two vectors.

We also use the conventional notation $|a\rangle \equiv \vec{a}$, $|b\rangle \equiv \vec{b}$ for vectors and

$$\langle a|b\rangle = (\vec{a} \cdot \vec{b})$$

for the scalar product so that Eq. 19.39 can be written in the usual form as

$$\vec{c} = \vec{a} \times \vec{b}$$

where $\vec{c}$ stands for the vector with components c^i. The c^i transform indeed as components of a vector under rotations. However, the transformation that changes the directions of all basic vectors (i.e., the **reflection**)

$$\vec{e'}_i = -\vec{e}_i \qquad (i = 1, 2, 3)$$

changes the sign of all components of any "true" vector, but leaves the components of $\vec{c}$ unchanged, as may be seen from Eq. 19.40. This is a manifestation of the pseudovectorial character of $\vec{c}$. In fact, a reflection in three dimensions is generated by a matrix with determinant equal to -1.

Relations involving the vector product can immediately be obtained with the help of the formalism just presented. Let us take the well-known relation

$$\vec{a} \times (\vec{b} \times \vec{c}) = \vec{b}(\vec{a} \cdot \vec{c}) - \vec{c}(\vec{a} \cdot \vec{b}) \tag{19.41}$$

To derive it, we use Eq. 19.38. In tensor notation we have

$$\begin{aligned} [\vec{a} \times (\vec{b} \times \vec{c})]^i &= \varepsilon^{ijk} a_j \varepsilon_{klm} b^l c^m \\ &= \varepsilon^{ijk} \varepsilon_{lmk} a_j b^l c^m \\ &= (g^i_l g^j_m - g^i_m g^j_l) a_j b^l c^m \\ &= b^i a_j c^j - c^i a_j b^j \end{aligned}$$

The last line is clearly the RHS of Eq. 19.41 written in tensor notation.

Let us now focus our attention on functions of a vector argument. In particular, let $w = w(\vec{x})$ denote a scalar function and let $\vec{A} = \vec{A}(\vec{x})$ be a vector function of the vector argument $\vec{x}$. According to the last lemma of this section, the application of the derivative operator $\partial/\partial x^i$ adds an index i to a tensor function of $\vec{x}$. Thus, we can formally consider the operators $\partial/\partial x^i$ as the components of a vector $\vec{\nabla}$. In vector analysis one defines the gradient, the divergence, and the curl as

$$\begin{aligned} \text{grad } w &= \vec{\nabla} w \\ \text{div } \vec{A} &= (\vec{\nabla} \cdot \vec{A}) \\ \text{curl } \vec{A} &= (\vec{\nabla} \times A) \end{aligned}$$

The results of repeated applications of grad, div, and curl operators can easily be found with the help of the tensor formalism. Let us calculate, for instance, div(curl $\vec{A}$).

$$[\text{div}(\text{curl}\,\vec{A})] = \vec{\nabla}\cdot(\vec{\nabla}\times\vec{A}) = \nabla^i\varepsilon_{ijk}\nabla^j A^k = \varepsilon_{ijk}\nabla^i\nabla^j A^k \tag{19.42}$$

The dummy indices i and j may be interchanged

$$[\text{div}(\text{curl}\,\vec{A})] = \varepsilon_{ijk}\nabla^j\nabla^i A^k = -\,\varepsilon_{ijk}\nabla^i\nabla^j A^k \tag{19.43}$$

In Eq. 19.43 we used the facts that interchanging two indices in ε_{ijk} changes only its sign and that $\nabla^i\nabla^j A^k = \nabla^j\nabla^i A^k$, which is the usual property of partial derivatives. Comparing Eqs. 19.42 and 19.43 yields

$$\text{div}(\text{curl}\,\vec{A}) = 0 \tag{19.44}$$

Using the same arguments as those employed in the derivation of Eq. 19.41, we also get

$$\text{curl}(\text{curl}\,\vec{A}) = \text{grad}(\text{div}\,\vec{A}) - \Delta\vec{A}$$

where

$$\Delta \equiv \vec{\nabla}\cdot\vec{\nabla} = \frac{\partial^2}{\partial x_1^2} + \frac{\partial^2}{\partial x_2^2} + \frac{\partial^2}{\partial x_3^2}$$

is the Laplace operator. We leave to the reader as an exercise the derivation of the relations

$$\begin{aligned} \text{div}(w\vec{A}) &= w\,\text{div}\,\vec{A} + (\text{grad}\,w)\cdot\vec{A} \\ \text{div}(\vec{A}\times\vec{B}) &= \vec{B}\cdot\text{curl}\,\vec{A} - \vec{A}\cdot\text{curl}\,\vec{B} \\ \text{curl}(w\vec{A}) &= (\text{grad}\,w)\times\vec{A} + w\,\text{curl}\,\vec{A} \end{aligned} \tag{19.45}$$

20 · INVARIANT SUBSPACES

We mentioned in Sec. 9 that a vector space whose vectors belong to some larger vector space is called a **subspace** of the larger space. Any set $|i_1\rangle, \cdots, |i_M\rangle$ of M $(M < N)$ linearly independent vectors of an N-dimensional linear vector space S_N may be considered as a basis for some M-dimensional subspace S_M of S_N; one says that the vectors $|i_1\rangle, \cdots, |i_M\rangle$ span the subspace S_M.

Consider a linear operator A. In the preceding sections we assumed that the multiplication of an arbitrary vector by an arbitrary operator resulted in a vector that belonged to the same space as the original vector. This is not necessarily true, as we shall see later, but we again assume for the moment that it is the case for A.

Multiplying an arbitrary vector $|x\rangle$ of S_N by A, we get some other vector belonging to S_N, since A operates only "inside" S_N. Repeating this multiplication, we obtain a series of vectors

$$|x\rangle, \quad A|x\rangle, \quad A^2|x\rangle, \quad \cdots, \quad A^N|x\rangle, \cdots \tag{20.1}$$

Let M be the maximum number of linearly **independent** vectors in the sequence (20.1).* These vectors span some subspace $\mathrm{S}_M^{(A)} \subset \mathrm{S}_N$. The subspace $\mathrm{S}_M^{(A)}$ has the

* Of course $M \leq N$, since the full space is N-dimensional.

same property as S_N; thus

$$A\,|\rangle \in S_M^{(A)} \qquad \text{for any } |\rangle \in S_M^{(A)}$$

i.e., every vector of $S_M^{(A)}$ is transformed by A into another vector of $S_M^{(A)}$. One says that $S_M^{(A)}$ is transformed or mapped into a subspace of itself.

A subspace that has this property is called an **invariant subspace** of the operator A.

EXAMPLE 1

Let $|a\rangle$ be an eigenvector of A.

$$A\,|a\rangle = a\,|a\rangle$$

The set of vectors

$$|a_x\rangle = x\,|a\rangle$$

obtained by multiplying $|a\rangle$ by an arbitrary number x, forms a one-dimensional invariant subspace of S_N, since

$$A\,|a_x\rangle = xA\,|a\rangle = (ax)\,|a\rangle$$

belongs to this set

The set $N^{(A)}$ of vectors satisfying the equation

$$A\,|x\rangle = 0$$

forms a linear vector space called the **null space** of the operator A. This can be easily shown by using the properties of linear operators and remembering the definition A of Sec. 2, where the notion of a linear vector space was introduced. For instance if,

$$|x_1\rangle, |x_2\rangle \in N^{(A)}$$

then

$$[|x_1\rangle + |x_2\rangle] \in N^{(A)}$$

since

$$A\,[|x_1\rangle + |x_2\rangle] = A\,|x_1\rangle + A\,|x_2\rangle = 0$$

We leave to the reader the verification that the remaining conditions of the definition A are also satisfied. $N^{(A)}$ is clearly an invariant subspace of the full space.

Let us choose as the first M basis vectors $|i\rangle$ $(i = 1, 2, \cdots, M)$ of S_N a set of M vectors that span the invariant subspace $S_M^{(A)}$. As the vectors $|i\rangle$ are basis vectors of $S_M^{(A)}$, we have

$$A\,|i\rangle = \sum_{j=1}^{M} A_i^j\,|j\rangle \qquad (i = 1, 2, \cdots, M)$$

The operator A is therefore represented by a matrix having the structure

$$\mathbf{A} = \left(\begin{array}{cccc|c} \multicolumn{4}{c|}{\mathbf{A}_1} & \\ 0 & 0 & \cdots & 0 & \mathbf{A}_2 \\ 0 & 0 & \cdots & 0 & \\ \vdots & \vdots & & \vdots & \\ 0 & 0 & \cdots & 0 & \end{array}\right)$$

$\mathbf{A}_1$ is a $M \times M$ matrix representing the operator A in the M-dimensional subspace $S_M^{(A)}$

$$\mathbf{A}_1 = \begin{pmatrix} A_1^1 & A_2^1 & & A_M^1 \\ A_1^2 & A_2^2 & \cdots & A_M^2 \\ \vdots & \vdots & \vdots & \vdots \\ A_1^M & A_2^M & \cdots & A_M^M \end{pmatrix}$$

$\mathbf{A}_2$ is a matrix with N rows and $(N - M)$ columns.

$$\mathbf{A}_2 = \begin{pmatrix} A_{M+1}^1 & A_{M+2}^1 & \cdots & A_N^1 \\ A_{M+1}^2 & A_{M+2}^2 & \cdots & A_N^2 \\ \vdots & \vdots & \vdots & \vdots \\ A_{M+1}^N & A_{M+2}^N & \cdots & A_N^N \end{pmatrix}$$

$\mathbf{A}_2$ is an example of a rectangular matrix. The array of numbers that forms $\mathbf{A}_2$ represents the operator A in the subspace $C_{(N-M)}^{(A)}$ spanned by the $(N - M)$ vectors $|i\rangle$ $(i = M + 1, M + 2, \cdots, N)$. In fact, the usual relation

$$A\,|i\rangle = \sum_{j=1}^{N} A_i^j\,|j\rangle \qquad (i = M + 1, \cdots, N)$$

again defines the matrix elements A_i^j. However, the ranges of the upper and lower indices are no longer the same since $A|\rangle$ does not in general belong to $C_{(N-M)}^{(A)}$ when $|\rangle \in C_{(N-M)}^{(A)}$.

Instead of operators that transform a space into its own subspace (or into itself), one can consider a more general class of operators transforming some space S_1 into a distinct space S_2. This amounts to considering vector functions $|f(|x\rangle)\rangle$ of a vector argument $|x\rangle$ such that

$$|f(|x\rangle)\rangle \in S_2 \qquad \text{whenever} \quad |x\rangle \in S_1$$

or using the operator notation (compare Sec. 7)

$$F|x\rangle \in S_2 \qquad \text{whenever } |x\rangle \in S_1$$

In the case when the dimension of S_2 differs from the dimension of S_1, F will be represented by the elements of a rectangular matrix $\mathbf{F}$.

$$F|i_1\rangle = \sum_{j=1}^{N_2} F_i^j\,|j_2\rangle \qquad (i = 1, \cdots, N_2)$$

where

$$|i_1\rangle \qquad (i = 1, \cdots, N_1)$$

$$|i_2\rangle \qquad (i = 1, \cdots, N_2)$$

are the bases of the spaces S_1 and S_2, respectively.

We defined the addition of linear operators by the relation

$$(A + B)\,|x\rangle = A\,|x\rangle + B\,|x\rangle \qquad \text{for any } |x\rangle.$$

This requires that $A\,|x\rangle$ and $B\,|x\rangle$ belong to the same space, since only the addition of vectors within a space is meaningful. In matrix language it means that one can add only matrices that have the same number of rows and the same numbers of columns.

As far as matrix multiplication is concerned, we have already noticed (Sec. 14) that the usual rule of multiplication requires only the equality of the number of rows of the multiplied matrix and the number of columns of the multiplying matrix. Let

$$A|\rangle \in S_2 \qquad \text{when} \quad |\rangle \in S_1$$
$$B|\rangle \in S_3 \qquad \text{when} \quad |\rangle \in S_2$$

One can, as usual, define the operator $C = BA$ as

$$C|\rangle = B(A|\rangle) \in S_3 \qquad \text{when} \quad |\rangle \in S_1$$

C connects the spaces S_1 and S_3 and is represented by the matrix $\mathbf{C}$, given by

$$\mathbf{C} = \mathbf{BA}$$

where the above equation has the usual meaning

$$C^i_j = \sum_{k=1}^{N_2} B^i_k A^k_j \qquad \begin{pmatrix} i = 1, \cdots, N_3 \\ j = 1, \cdots, N_1 \end{pmatrix}$$

In general, however, the operator $A \cdot B$ is not defined at all, even though the operator $B \cdot A$ is, since the equation

$$(AB)|\rangle = A(B|\rangle)$$

is meaningless; for the operator A is defined in S_1 and not in the space S_3 to which the vector $B|\rangle$ belongs.

EXAMPLE 3

An example of multiplication of rectangular matrices is

$$\begin{pmatrix} 1 & 3 & 2 \\ 0 & 1 & 4 \end{pmatrix} \begin{pmatrix} 2 & 1 & 3 & 1 \\ 0 & 0 & 3 & 1 \\ 5 & 0 & 2 & 1 \end{pmatrix}$$
$$= \begin{pmatrix} 1\cdot2 + 3\cdot0 + 2\cdot5 & 1\cdot1 + 3\cdot0 + 2\cdot0 & 1\cdot3 + 3\cdot3 + 2\cdot2 & 1\cdot1 + 3\cdot1 + 2\cdot1 \\ 0\cdot2 + 1\cdot0 + 4\cdot5 & 0\cdot1 + 1\cdot0 + 4\cdot0 & 0\cdot3 + 1\cdot3 + 4\cdot2 & 0\cdot1 + 1\cdot1 + 4\cdot1 \end{pmatrix}$$
$$= \begin{pmatrix} 12 & 1 & 16 & 6 \\ 20 & 0 & 11 & 5 \end{pmatrix}$$

After this lengthy digression, let us come back to the main subject of this section. Suppose now that the subspace $C^{(A)}_{(N-M)}$ is, as is the subspace $S^{(A)}_M$, an invariant subspace of A. Then the matrix $\mathbf{A}_2$ has the structure

$$\mathbf{A}_2 = \begin{pmatrix} 0 & 0 & \cdots & 0 \\ 0 & 0 & \cdots & 0 \\ \vdots & \vdots & & \vdots \\ 0 & 0 & \cdots & 0 \\ \hline & \boxed{\mathbf{A}_3} & & \end{pmatrix} \tag{20.2}$$

where $\mathbf{A}_3$ is an $(N - M) \times (N - M)$ matrix.

$$\mathbf{A}_3 = \begin{pmatrix} A^{M+1}_{M+1} & A^{M+1}_{M+2} & \cdots & A^{M+1}_N \\ A^{M+2}_{M+1} & A^{M+2}_{M+2} & \cdots & A^{M+2}_N \\ \vdots & \vdots & \vdots & \vdots \\ A^N_{M+1} & A^N_{M+2} & \cdots & A^N_N \end{pmatrix}$$

The matrix $\mathbf{A}$ now assumes the quasidiagonal form

$$\mathbf{A} = \begin{pmatrix} \mathbf{A}_1 & 0 \\ 0 & \mathbf{A}_3 \end{pmatrix}$$

where the zeros represent matrices with all elements equal to zero. A matrix that can be brought into the preceding form by a suitable choice of the basis vectors is called **reducible**. A matrix that cannot be brought into this form is called **irreducible**. The reducibility of a matrix means that the full space can be split into subspaces, which transform only into themselves under the transformation induced by the corresponding operator. In other words, the space S_N is split into a number l, say, of subspaces $\mathrm{S}^{(i)}$ $(i = 1, 2, \cdots, l)$ such that if $|\rangle \in \mathrm{S}^{(i)}$, then also

$$A\,|\rangle \in S^{(i)}$$

In the subsequent sections we shall examine the question of the complete decomposition of an N-dimensional space into invariant irreducible subspaces of an arbitrary linear operator.

21 · THE CHARACTERISTIC EQUATION AND THE HAMILTON-CAYLEY THEOREM

Let $\mathbf{A}$ be a square $N \times N$ matrix representing in some basis the operator A, and let λ be a parameter. The equation

$$\varphi(\lambda) \equiv \det(\lambda\mathbf{E} - \mathbf{A}) = 0 \tag{21.1}$$

is called the **characteristic equation** of the operator A (or of the matrix $\mathbf{A}$). It is evident that $\varphi(\lambda)$ is a polynomial of the Nth degree in λ with numerical coefficients, the leading coefficient (that of λ^N) being equal to 1

$$\varphi(\lambda) = \varphi_0 + \varphi_1\lambda + \cdots + \varphi_{N-1}\lambda^{N-1} + \lambda^N \tag{21.2}$$

The form of the characteristic equation (and thus the numerical values of the coefficients, φ_j) does not depend on the choice of the basis, since the determinant of a matrix, in our case the matrix $(\lambda\mathbf{E} - \mathbf{A})$, is a scalar.

Replacing λ by the operator A, we get the operator

$$\varphi(A) = \varphi_0 E + \varphi_1 A + \cdots + \varphi_{N-1}A^{N-1} + A^N \tag{21.3}$$

We shall now prove the important Hamilton-Cayley theorem.

Theorem. Let $\varphi(\lambda)$ be the characteristic polynomial of an operator A defined in an N-dimensional linear vector space

$$A\,|\rangle \in \mathrm{S}_N \qquad \text{when } |\rangle \in \mathrm{S}_N$$

Then

$$\varphi(A)\,|\rangle = 0 \qquad \text{for any } |\rangle \in \mathrm{S}_N$$

or in matrix notation

$$\varphi(\mathbf{A}) \equiv \varphi_0\mathbf{E} + \varphi_1\mathbf{A} + \cdots + \varphi_{N-1}\mathbf{A}^{N-1} + \mathbf{A}^N \equiv 0$$

In other words, all vectors of S_N belong to the null space of $\varphi(A)$.

Proof. Let $\mathbf{G}(\lambda)$ be the matrix defined as

$$(-1)^{i+k} G_k^i(\lambda) \qquad \text{is the minor of} \quad (\lambda E - A)_i^k$$

Since each element of $\mathbf{G}(\lambda)$ is a polynomial of degree $\leq N-1$, one can write $\mathbf{G}(\lambda)$ in the form of a polynomial with matricial coefficients

$$\mathbf{G}(\lambda) = \sum_{\nu=0}^{N-1} \mathbf{G}^{(\nu)} \lambda^\nu$$

Consider now the matrix

$$\mathbf{H}(\lambda) = \mathbf{G}(\lambda)(\lambda \mathbf{E} - \mathbf{A})$$

Using in the second step the well-known property of determinants, we have

$$H_j^i(\lambda) = G_k^i(\lambda)(\lambda E_j^k - A_j^k) = \det(\lambda \mathbf{E} - \mathbf{A}) E_j^i$$

Thus

$$\mathbf{G}(\lambda)(\lambda \mathbf{E} - \mathbf{A}) = \varphi(\lambda)\mathbf{E}$$

or

$$\sum_{\nu=0}^{N-1} \mathbf{G}^{(\nu)}(\lambda \mathbf{E} - \mathbf{A})\lambda^\nu = \varphi(\lambda)\mathbf{E}$$

If we now replace λ by $\mathbf{A}$, we obtain $\varphi(\mathbf{A}) = 0$.

22 · THE DECOMPOSITION OF AN N-DIMENSIONAL SPACE

We denote by λ_i $(i = 1, 2, \cdots, L)$ the roots of the characteristic equation of the operator A and by r_i the multiplicity of the ith root

$$\varphi(\lambda) \equiv \prod_{i=1}^{L} (\lambda - \lambda_i)^{r_i} \tag{22.1}$$

Since $\varphi(\lambda)$ is a polynomial of the Nth degree, one has

$$\sum_{i=1}^{L} r_i = N \tag{22.2}$$

It follows directly from Eq. 22.1 that the inverse of the characteristic polynomial can be decomposed as

$$\frac{1}{\varphi(\lambda)} = \sum_{i=1}^{L} \frac{f_i(\lambda)}{(\lambda - \lambda_i)^{r_i}} \tag{22.3}$$

where $f_i(\lambda)$ is a polynomial of degree $\leq r_i - 1$. Multiplying both sides of Eq. 22.3 by $\varphi(\lambda)$, we get

$$1 \equiv \sum_{i=1}^{L} f_i(\lambda) \prod_{k \neq i} (\lambda - \lambda_k)^{r_k} \tag{22.4}$$

Introducing the notation

$$\varphi_i(\lambda) \equiv f_i(\lambda) \prod_{k \neq i} (\lambda - \lambda_k)^{r_k} \tag{22.5}$$

and replacing in Eq. 22.4 the parameter λ by the operator A, we obtain

$$E = \sum_{i=1}^{L} \varphi_i(A)$$

Multiplying both sides of the preceding operator equation by an arbitrary vector $|\rangle \in S_N$, one obtains the decomposition of this vector

$$|\rangle = \sum_{i=1}^{L} \varphi_i(A)\,|\rangle \tag{22.6}$$

We now prove an important property of the operators $\varphi_i(A)$.

Lemma 1. The operators $\varphi_i(A)$ satisfy

$$\varphi_i(A)\varphi_k(A)\,|\rangle = \delta_{ik}\varphi_k(A)\,|\rangle \qquad \text{for any } |\rangle \in S_N$$

Proof. Suppose first that $k \neq i$. Then, from Eq. 22.5, we obtain

$$\begin{aligned}\varphi_i(A)\varphi_k(A)\,|\rangle &= \{f_i(A)f_k(A) \prod_{l \neq i} (A - \lambda_l E)^{r_l} \prod_{m \neq k} (A - \lambda_m E)^{r_m}\}|\rangle \\ &= \{f_i(A)f_k(A) \prod_{\substack{l \neq i \\ l \neq k}} (A - \lambda_l E)^{r_l}\}\varphi(A)\,|\rangle = 0 \end{aligned} \tag{22.7}$$

The last step follows from the Hamilton-Cayley theorem.

To prove the lemma when $i = k$, in Eq. 22.6 we replace $|\rangle$ by $\varphi_k(A)|\rangle$. Then, using Eq. 22.7, we get

$$\begin{aligned}\varphi_k(A)\,|\rangle &= \sum_{i=1}^{L} \varphi_i(A)\varphi_k(A)\,|\rangle \\ &= \sum_{i \neq k} \varphi_i(A)\varphi_k(A)\,|\rangle + \varphi_k(A)\varphi_k(A)\,|\rangle = \varphi_k(A)\varphi_k(A)|\rangle\end{aligned}$$

which proves the lemma.

Let us denote by $S^{(i)}$ the subspace containing all vectors that can be expressed as $\varphi_i(A)\,|\rangle$, where $|\rangle$ is some arbitrary vector. Clearly $S^{(i)}$ is an invariant subspace of the operator A, as can be seen from

$$A[\varphi_i(A)\,|\rangle] = \varphi_i(A) \cdot A\,|\rangle = \varphi_i(A)[A\,|\rangle]$$

Thus

$$A[\varphi_i(A)\,|\rangle] \in S^{(i)}$$

With the help of the last lemma, we can show that vectors belonging to two different subspaces $S^{(i)}$ and $S^{(k)}$ are necessarily linearly independent. In fact, multiplying the equation

$$\sum_k c_k\varphi_k(A)\,|k\rangle = 0$$

by $\varphi_i(A)$, we get

$$c_i\varphi_i(A)\,|i\rangle = 0$$

which means, since i is arbitrary, that

$$c_i = 0 \qquad \text{for all } i$$

It follows that the decomposition (22.6) of an arbitrary vector $|\rangle$ into vectors $[\varphi_i(A)|\rangle]$ is unique. Therefore, an arbitrary vector $|\rangle \in S_N$ either belongs entirely to one of the subspaces $S^{(i)}$ ($i = 1, 2, \cdots, L$) or can be decomposed uniquely into vectors belonging to different subspaces $S^{(i)}$. One says that S_N is the **direct sum** of the subspaces $S^{(i)}$, and one writes

$$S_N \doteq S^{(1)} \oplus S^{(2)} \oplus \cdots \oplus S^{(L)} \tag{22.8}$$

In other words, to the decomposition (22.6) of an arbitrary vector corresponds the decomposition (22.8) of the space. We still need to examine in greater detail the nature of the subspaces $S^{(i)}$.

Lemma 2. The subspace $S^{(i)}$ is identical to the null space of the operator $(A - \lambda_i E)^{r_i}$, where r_i is the multiplicity of the root λ_i of the characteristic equation.

Proof. First we prove that

$$(A - \lambda_i E)^{r_i} |\lambda_i\rangle = 0 \tag{22.9}$$

implies

$$|\lambda_i\rangle \in S^{(i)}$$

We decompose the vector $|\lambda_i\rangle$, using (22.6)

$$|\lambda_i\rangle = \sum_{k=1}^{L} \varphi_k(A) |\lambda_i\rangle$$

Inserting Eq. 22.5 into the preceding equation, we obtain

$$|\lambda_i\rangle = \sum_{k=1}^{L} \left\{ f_k(A) \prod_{l \neq k} (A - \lambda_l E)^{r_l} \right\} |\lambda_i\rangle$$

All terms in the sum except the term $k = i$ contain in the product a factor $(A - \lambda_i E)^{r_i}$. Therefore, because of Eq. 22.9, these terms vanish and we are left with

$$\begin{aligned} |\lambda_i\rangle &= f_i(A) \prod_{l \neq i} (A - \lambda_l E)^{r_l} |\lambda_i\rangle \\ &= \varphi_i(A) |\lambda_i\rangle \in S^{(i)} \end{aligned}$$

Conversely, each vector $\varphi_i(A)|\rangle \in S^{(i)}$ satisfies Eq. 22.9. To show this, we multiply $\varphi_i(A)|\rangle$ by the operator $(A - \lambda_i E)^{r_i}$ and use Eqs. 22.5 and 22.1

$$\begin{aligned} (A - \lambda_i E)^{r_i} \varphi_i(A)|\rangle &= f_i(A) \prod_{l} (A - \lambda_l E)^{r_l} |\rangle \\ &= f_i(A) \varphi(A)|\rangle = 0 \end{aligned}$$

The Hamilton-Cayley theorem has been used in the last step. Hence the lemma is established.

It is evident that a vector $|\lambda_i\rangle$ which belongs to the null space of the operator $(A - \lambda_i E)^{r_i}$ and which therefore satisfies the equation

$$(A - \lambda_i E)^{r_i} |\lambda_i\rangle = 0$$

is either a null vector (this is the trivial possibility) or a generalized eigenvector of the operator A with eigenvalue λ_i. If $|\lambda_i\rangle \neq 0$, there must exist an integer $0 \leq j < r_i$ such that

$$(A - \lambda_i E)^j |\lambda_i\rangle \neq 0 \qquad \text{while } (A - \lambda_i E)^{j+1} |\lambda_i\rangle = 0$$

This is precisely the definition of a generalized eigenvector.

Thus, we see that the roots of the characteristic equation of an operator A are generalized eigenvalues of this operator. It is not difficult to show that A cannot have other eigenvalues, for we have seen that two generalized eigenvectors corresponding to different eigenvalues are always linearly independent (see the last lemma of Sec. 11). Therefore, if there existed generalized eigenvalues distinct from all the roots of the characteristic equation, the corresponding eigenvector would not belong to any of the subspaces $S^{(i)}$, which is impossible.

Using the same argument, it is found also that one cannot have generalized eigenvectors of rank higher than the multiplicity of the corresponding root of the characteristic equation.

We have proved that an arbitrary vector can be expressed as a linear combination of vectors belonging to subspaces $S^{(i)}$ $(i = 1, 2, \cdots, L)$ which, according to the last lemma, are the null spaces of the operators $(A - \lambda_i E)^{r_i}$ $(i = 1, 2, \cdots, L)$. On the other hand, the vectors belonging to the null space of the operator $(A - \lambda_i E)^{r_i}$ are just the generalized eigenvectors of this operator. Thus, an arbitrary vector can be expressed as a linear combination of generalized eigenvectors of the operator A.

We can now summarize the results of this section:

The set of all linearly independent generalized eigenvectors of an arbitrary linear operator* A forms a basis of the space.

23 · THE CANONICAL FORM OF A MATRIX

We shall see in this section that there exists a basis of the space, with respect to which the matrix $\mathbf{A}$ representing the operator A takes a particularly simple form. It turns out that the basis needed is the one whose basis vectors are the generalized eigenvectors of the operator A.

Before going further, let us recall some of the results of Sec. 11: Two generalized eigenvectors that correspond to different eigenvalues are necessarily linearly independent. The same is true for eigenvectors of different ranks corresponding to the same eigenvalue. However, eigenvectors of the same rank and corresponding to the same eigenvalue may be linearly dependent. This complicates somewhat the explicit construction of the "proper" basis we seek.

It was shown in Sec. 11 that with every generalized eigenvector $|\lambda_i\rangle$ of rank m, say, corresponding to the eigenvalue λ_i, one can associate a **chain** of m linearly independent vectors

$$\begin{gathered} |\lambda_i\rangle \\ (A - \lambda_i E)|\lambda_i\rangle \\ (A - \lambda_i E)^2|\lambda_i\rangle \\ \cdots\cdots\cdots\cdots \\ (A - \lambda_i E)^{m-1}|\lambda_i\rangle \end{gathered}$$

which are themselves generalized eigenvectors, but of lower rank. Given two linearly independent generalized eigenvectors corresponding to the same eigenvalue, one can

* Remember, however, that we limit ourselves to operators that can be represented in S_N by a square matrix (compare Sec. 20), i.e., we assumed that $\mathbf{A}|\rangle \in S_N$ if $|\rangle \in S_N$.

associate a chain of eigenvectors with each of them. Those members of the two chains that are eigenvectors of the same rank are not necessarily linearly independent. One can, however, prove the following lemma.

Lemma.* For an arbitrary** linear operator A and for an arbitrary eigenvalue λ_i, there always exists a set of linearly independent generalized eigenvectors

$$|\lambda_i,1\rangle, \quad |\lambda_i,2\rangle \quad \cdots \quad |\lambda_i,I_i\rangle \tag{23.1}$$

such that the members of the chains associated with each of these eigenvectors are linearly independent; the totality of members of the chains form a basis of the null space of the operator $(A - \lambda_i E)^{r_i}$, r_i being the multiplicity of the root λ_i of the characteristic equation.

Hence each eigenvector of A with eigenvalue λ_i can be represented as a linear combination of vectors that are members of chains generated by the eigenvectors of the sequence 23.1.

The proof of the preceding lemma is easy but cumbersome, and we shall skip it here.

Let us denote by $m_{1i}, m_{2i}, \cdots, m_{I_i i}$ the ranks of the eigenvectors of the sequence 23.1. Such a sequence can be found for any of the eigenvalues of A. For the eigenvalue λ_j we have the following sequences of chains

$$\begin{array}{llll} |\lambda_j,1\rangle, & (A - \lambda_j E)|\lambda_j,1\rangle, & \cdots, & (A - \lambda_j E)^{m_{1j}-1}|\lambda_j,1\rangle \\ |\lambda_j,2\rangle, & (A - \lambda_j E)|\lambda_j,2\rangle, & \cdots, & (A - \lambda_j E)^{m_{2j}-1}|\lambda_j,2\rangle \\ \cdots & \cdots & \cdots & \cdots \\ |\lambda_j,I_j\rangle, & (A - \lambda_j E)|\lambda_j,I_j\rangle, & \cdots, & (A - \lambda_j E)^{m_{I_j j}-1}|\lambda_j,I_j\rangle \end{array} \tag{23.2}$$

In sequence 23.2, there are $m_{1j} + m_{2j} + \cdots + m_{I_j j}$ eigenvectors, and these span the null space of the operator $(A - \lambda_j E)^{r_j}$. Since the full space is a direct sum of the null spaces of all the operators $(A - \lambda_j E)^{r_j}$, $(j = 1, 2, \cdots, L)$ the basis vectors of S_N will consist of the totality of the eigenvectors (23.2), with $j = 1, 2, \cdots, L$. It is convenient to make a table of these basis vectors and to label them in a certain order. We shall order them as follows: The chains generated from the first eigenvalue λ_1 will precede those generated from the second eigenvalue λ_2, and so on. Within each chain the order will be as in sequence 23.2 and will be read horizontally. We are then led to the following choice for the basis vectors of S_N.

Eigenvectors corresponding to the eigenvalue λ_1

FIRST CHAIN:

$$\begin{aligned} |1,\lambda_1\rangle &= |\lambda_1,1\rangle \\ |2,\lambda_1\rangle &= (A - \lambda_1 E)|\lambda_1,1\rangle \\ &\cdots\cdots \\ |m_{11},\lambda_1\rangle &= (A - \lambda_1 E)^{m_{11}-1}|\lambda_1,1\rangle \end{aligned}$$

---

* See V. I. Smirnov, *A Course of Mathematical Analysis*, Pergamon Press, New York, 1964, Vol. 3, part II.

** Arbitrary in the same sense as in the preceding section; i.e., c.erators that can be represented in S_N by a square matrix.

SECOND CHAIN:

$$|m_{11}+1,\lambda_1\rangle = |\lambda_1,2\rangle$$
$$|m_{11}+2,\lambda_1\rangle = (A-\lambda_1 E)|\lambda_1,2\rangle$$
$$|m_{11}+m_{21},\lambda_1\rangle = (A-\lambda_1 E)^{m_{21}-1}|\lambda_1,2\rangle$$
$$\cdots\cdots\cdots$$

(I_1)st CHAIN:

$$\left|\sum_{i=1}^{I_1-1} m_{i1}+1,\lambda_1\right\rangle = |\lambda_1,I_1\rangle$$
$$\left|\sum_{i=1}^{I_1-1} m_{i1}+2,\lambda_1\right\rangle = (A-\lambda_1 E)\,|\lambda_1,I_1\rangle$$
$$\cdots\cdots\cdots$$
$$\left|\sum_{i=1}^{I_1} m_{i1},\lambda_1\right\rangle = (A-\lambda_1 E)^{m_{I_1 1}-1}|\lambda_1,I_1\rangle$$
$$\cdots\cdots\cdots$$
$$\cdots\cdots\cdots$$

Eigenvectors corresponding to the eigenvalue λ_k

FIRST CHAIN:

$$\left|\sum_{j=1}^{k-1}\sum_{i=1}^{I_j} m_{ij}+1,\lambda_k\right\rangle = |\lambda_k,1\rangle$$
$$\left|\sum_{j=1}^{k-1}\sum_{i=1}^{I_j} m_{ij}+2,\lambda_k\right\rangle = (A-\lambda_k E)\,|\lambda_k,1\rangle$$
$$\left|\sum_{j=1}^{k-1}\sum_{i=1}^{I_j} m_{ij}+m_{1k},\lambda_k\right\rangle = (A-\lambda_k E)^{m_{1k}-1}\,|\lambda_k,1\rangle$$
$$\cdots\cdots\cdots$$

(I_k)th CHAIN:

$$\left|\sum_{j=1}^{k-1}\sum_{i=1}^{I_j} m_{ij}+\sum_{j=1}^{I_k-1} m_{jk}+1,\lambda_k\right\rangle = |\lambda_k,I_k\rangle$$
$$\left|\sum_{j=1}^{k-1}\sum_{i=1}^{I_j} m_{ij}+\sum_{j=1}^{I_k-1} m_{jk}+2,\lambda_k\right\rangle = (A-\lambda_k E)\,|\lambda_k,I_k\rangle$$
$$\cdots\cdots\cdots$$
$$\left|\sum_{j=1}^{k}\sum_{i=1}^{I_k} m_{ij},\lambda_k\right\rangle = (A-\lambda_k E)^{m_{I_k k}-1}\,|\lambda_k,I_k\rangle$$
$$\cdots\cdots\cdots$$
$$\cdots\cdots\cdots$$

We use the notation $|i,\lambda_k\rangle$ for basis vectors where $i = 1, 2, \cdots, N$ stands for the reference number and λ_k indicates that $|i,\lambda_k\rangle$ is a generalized eigenvector with eigenvalue λ_k.

Let us examine now the action of the operator A on the basis vectors $|i,\lambda_k\rangle$.

$$\begin{aligned} A\,|i,\lambda_k\rangle &= (A-\lambda_k+\lambda_k)\,|i,\lambda_k\rangle \\ &= \lambda_k\,|i,\lambda_k\rangle + (A-\lambda_k E)\,|i,\lambda_k\rangle \\ &= \begin{cases} \lambda_k\,|i,\lambda_k\rangle & \text{if } |i,\lambda_k\rangle \text{ is the last vector of a chain} \\ \lambda_k\,|i,\lambda_k\rangle + |i+1,\lambda_k\rangle & \text{otherwise} \end{cases} \end{aligned} \tag{23.3}$$

From Eq. 23.3 one sees that in the new basis the operator $(A - \lambda_k E)$ acts as a kind of "raising operator"; i.e., it "raises" an eigenvector $|i,\lambda_k\rangle$ to the next higher one $|i + 1,\lambda_k\rangle$ unless the former eigenvector $|i,\lambda_k\rangle$ happens to be the last of a chain: in which case, since it cannot be raised, it is annihilated.

Remembering the definition of the elements of the matrix $\mathbf{A}$ representing the operator A (formula 13.3), we see from Eq. 23.3 that $\mathbf{A}$ has, in the chosen basis, the quasidiagonal form

$$\mathbf{A} = \begin{pmatrix} \boxed{\mathbf{A}_{11}} & & & & & & & & & \\ & \boxed{\mathbf{A}_{21}} & & & & & & & & \\ & & \boxed{\mathbf{A}_{I_1 1}} & & & & & & & \\ & & & \boxed{\mathbf{A}_{1k}} & & & & & & \\ & & & & \boxed{\mathbf{A}_{2k}} & & & & & \\ & & & & & \boxed{\mathbf{A}_{I_k k}} & & & & \\ & & & & & & \boxed{\mathbf{A}_{1L}} & & & \\ & & & & & & & \boxed{\mathbf{A}_{2L}} & & \\ & & & & & & & & \boxed{\mathbf{A}_{I_L L}} \end{pmatrix} \tag{23.4}$$

where L is, as before, the number of distinct roots of the characteristic polynomial and $\mathbf{A}_{ik}$ stands for the square $m_{ik} \times m_{ik}$ matrix

$$\mathbf{A}_{ik} = \begin{pmatrix} \lambda_k & & & & & & \\ 1 & \lambda_k & & & & & \\ & 1 & \lambda_k & & & & \\ & & 1 & \cdot & & & \\ & & & & \cdot & & \\ & & & & & \cdot & \\ & & & & & \lambda_k & \\ & & & & & 1 & \lambda_k \end{pmatrix} \tag{23.5}$$

representing A in the invariant irreducible subspace, spanned by the m_{ik} generalized eigenvectors forming the ith chain corresponding to the eigenvalue λ_k. In the case when a chain contains one eigenvector only, the corresponding matrix $\mathbf{A}_{ik}$ reduces to a number, the eigenvalue of this eigenvector.

The form 23.4 is called the **Jordan canonical form** of a matrix.

Consider an arbitrary square matrix $\mathbf{A}'$. It represents some operator A in some basis. We have seen, however, that for any such linear operator, there exists a basis in which this operator is represented by a matrix in Jordan canonical form. Since to the change of basis

$$|i\rangle = R|i'\rangle$$

corresponds the transformation of matrices (see Sec. 16)

$$\mathbf{A} = \mathbf{R}^{-1}\mathbf{A}'\mathbf{R}$$

we have the following theorem.

Theorem. Every square matrix $\mathbf{A}'$ can, by a suitable transformation

$$\mathbf{A} = \mathbf{R}^{-1}\mathbf{A}'\mathbf{R}$$

be brought to the Jordan canonical form.

When the characteristic polynomial has only simple roots λ_k $(k = 1, 2, \cdots, N)$, each of the matrices $\mathbf{A}_{ik}$ reduces to the number λ_k.

Corollary. Every square matrix $\mathbf{A}'$, whose characteristic polynomial $\det(\lambda\mathbf{E} - \mathbf{A}')$ has only simple roots, can, by a suitable transformation

$$\mathbf{A} = \mathbf{R}^{-1}\mathbf{A}'\mathbf{R}$$

be brought to diagonal form.

We have already noticed that the characteristic equation has an invariant form, which is independent of the choice of a basis. In particular, one can find the characteristic polynomial by calculating the determinant

$$\varphi(\lambda) = \det(\lambda\mathbf{E} - \mathbf{A})$$

with $\mathbf{A}$ in Jordan canonical form.

From 23.4 and 23.5 one has

$$\varphi(\lambda) = \prod_{i=1}^{L} \prod_{j=1}^{I_i} (\lambda - \lambda_i)^{m_{ji}}$$

$$= \prod_{i=1}^{L} (\lambda - \lambda_i)^{\sum_{j=1}^{I_i} m_{ji}}$$

Comparing this with the formula 22.2, we get

$$\sum_{j=1}^{I_i} m_{ji} = r_i \tag{23.6}$$

The sum on the left of Eq. 23.6 is the number of linearly independent generalized eigenvectors with eigenvalue λ_i, and therefore determines the dimensions of the null space of the operator $(A - \lambda_i E)^{r_i}$. Thus, the multiplicity of a root of the characteristic equation has the meaning of the dimension of the invariant subspace spanned by the eigenvectors for which this root plays the role of an eigenvalue.

Notice that the structure of the characteristic polynomial of a matrix does not, in general, determine completely the canonical form of this matrix; the numbers m_{ik} remain undetermined and have to be found by other methods.

EXAMPLE

Consider the matrix

$$\mathbf{A} = \begin{pmatrix} 1 & 3 & -2 & 2 \\ -1 & 4 & -1 & 1 \\ -1 & 2 & 1 & 1 \\ 0 & 1 & -1 & 3 \end{pmatrix} \tag{23.7}$$

We wish to transform **A** to canonical form.
Setting the determinant

$$\det(\lambda\mathbf{E} - \mathbf{A}) = \begin{vmatrix} \lambda - 1 & -3 & 2 & -2 \\ 1 & \lambda - 4 & 1 & -1 \\ 1 & -2 & \lambda - 1 & -1 \\ 0 & -1 & 1 & \lambda - 3 \end{vmatrix}$$

equal to zero, the reader may verify that the characteristic equation has the form

$$(\lambda - 3)(\lambda - 2)^3 = 0$$

Therefore, the space splits into two invariant subspaces: The first is the one-dimensional null space of the operator $(A - 3E)$ and the other is the three-dimensional null space of the operator $(A - 2E)^3$. We must now find vectors that span these invariant subspaces. A vector that belongs to the first invariant subspace must satisfy an eigenvalue equation, which in matrix form reads

$$(\mathbf{A} - 3\mathbf{E})\mathbf{a} = 0$$

Hence,

$$\begin{pmatrix} -2 & 3 & -2 & 2 \\ -1 & 1 & -1 & 1 \\ -1 & 2 & -2 & 1 \\ 0 & 1 & -1 & 0 \end{pmatrix} \begin{pmatrix} a^1 \\ a^2 \\ a^3 \\ a^4 \end{pmatrix} = 0$$

or

$$\begin{aligned} -2a^1 + 3a^2 - 2a^3 + 2a^4 &= 0 \\ -a^1 + a^2 - a^3 + a^4 &= 0 \\ -a^1 + 2a^2 - 2a^3 + a^4 &= 0 \\ a^2 - a^3 &= 0 \end{aligned}$$

The preceding equations are equivalent to

$$a^1 = a^4, \qquad a^2 = a^3 = 0$$

These are the conditions that the components of a vector must satisfy in order for that vector to belong to the null space of $(A - 3E)$. For example, one such vector is

$$\mathbf{b}_1 = \begin{pmatrix} 1 \\ 0 \\ 0 \\ 1 \end{pmatrix} \tag{23.8}$$

We now proceed to find the conditions that must be satisfied by the components of a vector which belongs to the null space of $(A - 2E)^3$. The corresponding (generalized) eigenvalue equation is

$$(\mathbf{A} - 2\mathbf{E})^3 a = 0 \tag{23.9}$$

The reader will verify that

$$(\mathbf{A}-2\mathbf{E})^3 = \begin{pmatrix} 0 & 1 & -1 & 1 \\ 0 & 0 & 0 & 0 \\ 0 & 0 & 0 & 0 \\ 0 & 1 & -1 & 1 \end{pmatrix}$$

and that Eq. 23.9 yields the relation

$$a^2 - a^3 + a^4 = 0$$

Three linearly independent vectors satisfying this condition are

$$\mathbf{b}_2 = \begin{pmatrix} 1 \\ 0 \\ 0 \\ 0 \end{pmatrix}, \qquad \mathbf{b}_3 = \begin{pmatrix} 0 \\ 0 \\ 1 \\ 1 \end{pmatrix}, \qquad \mathbf{b}_4 = \begin{pmatrix} 0 \\ 1 \\ 1 \\ 0 \end{pmatrix} \tag{23.10}$$

The vectors $|b_i\rangle$ $(i = 1, 2, 3, 4)$ are chosen as basis vectors. This corresponds to the basis transformation

$$|b_i\rangle = B|i\rangle = \sum_j B_i^j |j\rangle$$

The elements of the matrix **B** are immediately obtained from Eqs. 23.8 and 23.10, since B_j^i is the jth component of $|b_i\rangle$. Hence,

$$\mathbf{B} = \begin{pmatrix} 1 & 1 & 0 & 0 \\ 0 & 0 & 0 & 1 \\ 0 & 0 & 1 & 1 \\ 1 & 0 & 1 & 0 \end{pmatrix} \tag{23.11}$$

The matrix $\mathbf{B}^{-1}$ is obtained according to the rules given in Sec. 15.

$$\mathbf{B}^{-1} = \begin{pmatrix} 0 & 1 & -1 & 1 \\ 1 & -1 & 1 & -1 \\ 0 & -1 & 1 & 0 \\ 0 & 1 & 0 & 0 \end{pmatrix} \tag{23.12}$$

From Eqs. 23.7, 23.11, and 23.12, we get

$$\mathbf{A}' = \mathbf{B}^{-1}\mathbf{A}\mathbf{B} = \begin{pmatrix} 3 & 0 & 0 & 0 \\ 0 & 1 & 0 & 1 \\ 0 & 0 & 2 & 0 \\ 0 & -1 & 0 & 3 \end{pmatrix}$$

The matrix $\mathbf{A}'$ represents the operator A in the new basis. It is already in quasidiagonal form. We now examine the structure of the null space of $(A - 2E)^3$. One easily finds that

$$(\mathbf{A}' - 2\mathbf{E})^2 = \begin{pmatrix} 1 & 0 & 0 & 0 \\ 0 & 0 & 0 & 0 \\ 0 & 0 & 0 & 0 \\ 0 & 0 & 0 & 0 \end{pmatrix}$$

Therefore, any vector that belongs to the null space of $(A - 2E)^3$

$$\mathbf{a} = \begin{pmatrix} 0 \\ a^2 \\ a^3 \\ a^4 \end{pmatrix}$$

satisfies the equation

$$(\mathbf{A}' - 2\mathbf{E})^2\mathbf{a} = 0$$

This means that A has no generalized eigenvector of rank 3, but rather generalized eigenvectors of rank 2 and 1. The "ordinary" eigenvector of A must satisfy the equation

$$(\mathbf{A}' - 2\mathbf{E})\mathbf{a} = 0$$

that is

$$\begin{pmatrix} 1 & 0 & 0 & 0 \\ 0 & -1 & 0 & 1 \\ 0 & 0 & 0 & 0 \\ 0 & -1 & 0 & 1 \end{pmatrix} \begin{pmatrix} a^1 \\ a^2 \\ a^3 \\ a^4 \end{pmatrix} = 0$$

This yields the condition

$$a^2 = a^4, \qquad a^1 = 0 \tag{23.13}$$

which is satisfied, for example, by the vector

$$\mathbf{c}_2 = \begin{pmatrix} 0 \\ 0 \\ 1 \\ 0 \end{pmatrix}$$

To find a generalized eigenvector of second rank, we must remember that for such a vector

$$\begin{pmatrix} 1 & 0 & 0 & 0 \\ 0 & -1 & 0 & 1 \\ 0 & 0 & 0 & 0 \\ 0 & -1 & 0 & 1 \end{pmatrix} \begin{pmatrix} a^1 \\ a^2 \\ a^3 \\ a^4 \end{pmatrix} \neq 0$$

which is equivalent to the condition

$$a^2 \neq a^4, \qquad a^1 = 0$$

One such vector is

$$\mathbf{c}_3 = \begin{pmatrix} 0 \\ 1 \\ 0 \\ 0 \end{pmatrix}$$

$|c_3\rangle$ generates a chain to which belongs $|c_3\rangle$, and the vector $|c_4\rangle = (A - 2E)\,|c_3\rangle$. Its components are readily found as

$$\mathbf{c}_4 = \begin{pmatrix} 1 & 0 & 0 & 0 \\ 0 & -1 & 0 & 1 \\ 0 & 0 & 0 & 0 \\ 0 & -1 & 0 & 1 \end{pmatrix} \begin{pmatrix} 0 \\ 1 \\ 0 \\ 0 \end{pmatrix} = \begin{pmatrix} 0 \\ -1 \\ 0 \\ -1 \end{pmatrix}$$

The vectors consisting of $|c_2\rangle$ and the chain generated by $|c_3\rangle$ (the chain generated by $|c_2\rangle$ reduces to $|c_2\rangle$ itself, since $|c_2\rangle$ is an eigenvector of rank 1) are linearly independent because $|c_2\rangle$ and $|c_3\rangle$ have been properly chosen. For other possible choices for $|c_2\rangle$ and $|c_3\rangle$, this linear independence would not necessarily be ensured. However, the very existence of proper vectors $|c_2\rangle$ and $|c_3\rangle$ is guaranteed by the lemma of this section.

We again change the basis by choosing now the vectors $|c_i\rangle$ $(i = 2, 3, 4)$ as basis vectors instead of $|b_i\rangle$ $(i = 2,3,4)$. The matrix $\mathbf{C}$, which effects this transformation, is once again easily found as

$$\mathbf{C} = \begin{pmatrix} 0 & 0 & 0 & 0 \\ 0 & 0 & 1 & -1 \\ 0 & 1 & 0 & 0 \\ 0 & 0 & 0 & -1 \end{pmatrix}$$

One also has for the inverse of **C**

$$\mathbf{C}^{-1} = \begin{pmatrix} 1 & 0 & 0 & 0 \\ 0 & 0 & 1 & 0 \\ 0 & 1 & 0 & -1 \\ 0 & 0 & 0 & -1 \end{pmatrix}$$

Therefore,

$$\mathbf{A}'' = \mathbf{C}^{-1}\mathbf{A}'\mathbf{C} = \begin{pmatrix} 3 & 0 & 0 & 0 \\ 0 & 2 & 0 & 0 \\ 0 & 0 & 2 & 0 \\ 0 & 0 & 1 & 2 \end{pmatrix}$$

This is the canonical matrix representing the operator A. Since

$$\mathbf{A}'' = \mathbf{C}^{-1}\mathbf{B}^{-1}\mathbf{ABC} = (\mathbf{BC})^{-1}\mathbf{A}(\mathbf{BC})$$

the matrix **R**, which transforms **A** into **A**″, is

$$\mathbf{R} = \mathbf{BC} = \begin{pmatrix} 1 & 0 & 1 & -1 \\ 0 & 0 & 0 & -1 \\ 0 & 1 & 0 & -1 \\ 1 & 1 & 0 & 0 \end{pmatrix}$$

24 · HERMITIAN MATRICES AND QUADRATIC FORMS

24.1 Diagonalization of Hermitian Matrices

The theory of the preceding few sections becomes very much simplified if one considers only Hermitian operators. We showed in Sec. 11 that Hermitian operators may not have eigenvectors of rank higher than 1. Therefore, each one of the chains of eigenvectors introduced in the preceding sections reduces trivially to only one eigenvector. Thus

$$m_{ik} = 1 \qquad \text{for any } i \text{ and } k$$

and from Eq. 23.6 we obtain the result that the number of linearly independent eigenvectors, corresponding to some eigenvalue is equal to the multiplicity r_k of the root λ_k of the characteristic polynomial. Consequently, a matrix that represents a Hermitian operator, when brought to the Jordan canonical form, is simply diagonal. The canonical form of this matrix is completely determined by the structure of its characteristic polynomial. The diagonal elements are equal to the eigenvalues of the Hermitian operator, or what is the same, to the roots of the characteristic equation, and each root appears along the diagonal a number of times equal to its multiplicity.

We now consider anew the problem of diagonalizing a Hermitian matrix using a more elementary and explicit approach. Consider the eigenvalue equation

$$H\,|h\rangle = h\,|h\rangle \tag{24.1}$$

Writing Eq. 24.1 in terms of components referred to some particular basis, we have

$$H^j_k h^k = h h^j \qquad (j = 1, 2, \cdots, N)$$

or

$$[hE^j_k - H^j_k]h^k = 0 \qquad (j = 1, 2, \cdots, N) \tag{24.2}$$

Equations 24.2 are a system of N linear homogeneous equations which have a non-trivial solution, provided

$$\det(h\mathbf{E} - \mathbf{H}) = 0 \tag{24.3}$$

We recognize in Eq. 24.3 the characteristic equation with respect to h. When this characteristic equation has only simple roots, there corresponds to each root a different solution of the system of equations 24.2. Since Eqs. 24.2 are homogeneous, these solutions can be normalized as*

$$h_{(i)k}h^{k}_{(i)} = 1 \qquad (i = 1, 2, \cdots, N) \tag{24.4}$$

or in vector notation

$$\langle h_{(i)}|h_{(i)}\rangle = 1$$

Furthermore, since the eigenvectors of a Hermitian operator corresponding to different eigenvalues are orthogonal (see Sec. 11), and since the numbers h_i^k ($k = 1, 2, \cdots, N$) have the meaning of the components of the ith eigenvector, we have

$$h_{(i)k}h^{k}_{(j)} = 0 \qquad \text{for } i \neq j \tag{24.5}$$

which reads in vector notation

$$\langle h_{(i)}|h_{(j)}\rangle = 0 \qquad \text{for } i \neq j \tag{24.6}$$

In the case when the characteristic equation has multiple roots, the situation is slightly more complicated. Suppose that $h_{(j)}$ is a root of Eq. 24.3, and as usual let us denote by r_j the multiplicity of this root. The null space of the operator $[H - h_{(j)}E]^{r_j}$, i.e., the manifold of all vectors $|\rangle$ satisfying

$$[H - h_{(j)}E]^{r_j}|\rangle = 0$$

is r_j-dimensional, as we saw in the preceding section. This manifold is spanned by the eigenvectors of H which are all of rank 1, since H is Hermitian. Thus, there exist r_j linearly independent vectors satisfying Eq. 24.1 with $h = h_{(j)}$. Consequently, the set of equations

$$(h_{(j)}E^i_k - H^i_k)h^k_{(j)} = 0 \qquad (i = 1, 2, \cdots, N) \tag{24.7}$$

has r_j linearly independent solutions. Again, each of these solutions determines an eigenvector. Since, however, these eigenvectors correspond to the same eigenvalue $h_{(j)}$, they are in general not orthogonal. Notice, however, that Eqs. 24.1 are linear, and therefore any linear combination of eigenvectors corresponding to the same eigenvalue is itself an eigenvector with the same eigenvalue

$$\begin{aligned} H\left\{\sum_k c^k\,|h_{(j)},k\rangle\right\} &= \sum_k c^k H\,|h_{(j)},k\rangle \\ &= \sum_k c^k h_{(j)}\,|h_{(j)},k\rangle \\ &= h_{(j)}\left\{\sum_k c^k\,|h_{(j)},k\rangle\right\} \end{aligned}$$

Any set of linearly independent vectors may be orthogonalized, as shown in Sec. 12, by taking suitable linear combinations of these vectors. It follows, that Eqs. 24.7 have r_j solutions, which are not only linearly independent but also orthogonal.

---

* i corresponds to the different solutions of Eq. 24.2; it is not an index identifying the component of a vector.

The set of all linearly independent generalized eigenvectors of a linear operator forms a basis, as has been shown in Sec. 22. This general result remains, of course, true in the particular case of Hermitian operators that have only "ordinary" eigenvectors.

The components of those eigenvectors are found by solving Eqs. 24.2; it is important that one can obtain explicitly N eigenvectors that form an orthonormal set, either because they are eigenvectors of a Hermitian operator corresponding to distinct eigenvalues or because they have been orthogonalized, if they correspond to the same eigenvalue. Let us denote these eigenvectors by $|k,h_{(i)}\rangle$; the first argument specifies the eigenvector, the second specifies the eigenvalue (these two arguments are clearly not independent, since the label determining the eigenvector determines also the eigenvalue). We have

$$H\,|k,h_{(i)}\rangle = h_{(i)}\,|k,h_{(i)}\rangle \tag{24.8}$$

and

$$\langle h_{(j)},l\,|k,h_{(i)}\rangle = \delta_{lk} \tag{24.9}$$

Of course, when $i \neq j$, the labels of the two eigenvectors are necessarily different, and Eq. 24.9 is equivalent to Eq. 24.6.

Choosing $|k,h_{(i)}\rangle$ as basis vectors, we get the matrix $\mathbf{H}$ representing H in the diagonal form

$$\mathbf{H} = \begin{pmatrix} h_1 & & & & & & & & & & & \\ & h_1 & & & & & & & & & & \\ & & \ddots & & & & & & & & & \\ & & & h_1 & & & & & & & & \\ & & & & h_2 & & & & & & & \\ & & & & & \ddots & & & & & & \\ & & & & & & h_2 & & & & & \\ & & & & & & & h_3 & & & & \\ & & & & & & & & \ddots & & & \\ & & & & & & & & & & & h_L \end{pmatrix} \tag{24.10}$$

Suppose that in some arbitrary basis $|i\rangle$, $(i = 1, 2, \cdots, N)$, the Hermitian operator H is represented by a matrix $\mathbf{H}'$. We shall explicitly construct the transformation that will bring $\mathbf{H}'$ into the diagonal form. Solving the equations

$$H'^{n}_{m}h^{m} = hh^{n} \qquad (n = 1, 2, \cdots, N) \tag{24.11}$$

we obtain the components $h^{m}_{(i)}(l)$ $(m,l = 1, 2, \cdots, N)$ of eigenvectors $|l,h_{(i)}\rangle$. Equation 24.9 now reads

$$h_{(j)m}(l)h^{m}_{(i)}(k) = \delta_{lk} \tag{24.12}$$

Let us define the matrix $\mathbf{R}$ by*

$$R^{m}_{k} \equiv h^{m}_{(i)}(k) \qquad (m,k = 1, 2, \cdots, N) \tag{24.13}$$

The elements of the matrix $\mathbf{R}^{-1}$ are evidently given by

$$(R^{-1})^{l}_{m} \equiv h_{(j)m}(l) \tag{24.14}$$

since, using Eq. 24.12, we have

$$(R^{-1})^{l}_{m}R^{m}_{k} \equiv h_{(j)m}(l)h^{m}_{(i)}(k) = E^{l}_{k}$$

Let us consider the matrix

$$\mathbf{H} = \mathbf{R}^{-1}\mathbf{H}'\mathbf{R}$$

which represents the operator H in the basis formed out of the eigenvectors

$$|k,h_{(i)}\rangle = \sum_{m} h^{m}_{(i)}(k)\,|m\rangle = R\,|k\rangle$$

Using Eqs. 24.12, 24.13, and 24.14 one has

$$\begin{aligned} H^{l}_{k} &= (R^{-1})^{l}_{n}H'^{n}_{m}R^{m}_{k} \\ &\equiv h_{(j)n}(l)H'^{n}_{m}h^{m}_{(i)}(k) \\ &\equiv h_{(i)}h_{(j)n}(l)h^{n}_{(i)}(k) \\ &\equiv \begin{cases} h_{(i)} & l = k \\ 0 & l \neq k \end{cases} \end{aligned}$$

or briefly

$$H^{l}_{k} = h_{(i)}E^{l}_{k}$$

Therefore, $\mathbf{H}$ has the diagonal form 24.10.

In the case when the original basis vectors $|i\rangle$ $(i = 1, 2, \cdots, N)$ form an orthonormal set so that the matrix $\mathbf{H}'$ is Hermitian, the matrix $\mathbf{R}$ is unitary. Indeed if

$$\langle i|k\rangle = \delta_{ik}$$

then

$$(R^{-1})^{l}_{m} \equiv h_{(j)m}(l) \equiv \sum_{n} \bar{h}^{n}_{(j)}(l)\langle n|m\rangle \equiv \bar{h}^{m}_{(j)}(l)$$

or

$$(R^{-1})^{l}_{m} \equiv \bar{R}^{m}_{l}$$

---

* Notice that we use the same techniques as in proving the theorem of Sec. 18.

Thus

$$\mathbf{R}^{-1}=\mathbf{R}^{+}$$

This proves the following

Theorem 1. Every Hermitian matrix $\mathbf{H}'$ can be brought to diagonal form by the transformation

$$\mathbf{H}=\mathbf{U}^{-1}\mathbf{H}'\mathbf{U}$$

where $\mathbf{U}$ is unitary

$$\mathbf{U}^{-1}=\mathbf{U}^{+}$$

EXAMPLE

Consider the matrix

$$\mathbf{H}=\begin{pmatrix} 3 & i \\ -i & 3 \end{pmatrix}$$

Let us diagonalize $\mathbf{H}$.

One has

$$\det(\lambda\mathbf{E}-\mathbf{H})=\begin{vmatrix} \lambda-3 & i \\ i & \lambda-3 \end{vmatrix}=(\lambda-4)(\lambda-2)$$

The characteristic equation is

$$(\lambda-4)(\lambda-2)=0$$

The eigenvector corresponding to the root $\lambda=4$ can be determined from the equation

$$(\mathbf{H}-4\mathbf{E})\mathbf{a}=0$$

or

$$\begin{pmatrix} -1 & i \\ -i & -1 \end{pmatrix}\begin{pmatrix} a^1 \\ a^2 \end{pmatrix}=0$$

which leads to

$$-a^1+ia^2=0$$

$$-ia^1-a^2=0$$

The components of $\mathbf{a}$ must therefore satisfy the condition

$$a^1=ia^2$$

An eigenvector satisfying this condition and normalized to unity $((a^1)^2+(a^2)^2=1)$ is

$$\mathbf{a}=\begin{pmatrix} \dfrac{i}{\sqrt{2}} \\ \dfrac{1}{\sqrt{2}} \end{pmatrix}$$

Similarly, the equation for the other root

$$(\mathbf{H}-2\mathbf{E})\mathbf{b}=0$$

yields the condition

$$b^1=-ib^2$$

The corresponding normalized eigenvector is

$$\mathbf{b} = \begin{pmatrix} \dfrac{-i}{\sqrt{2}} \\ \dfrac{1}{\sqrt{2}} \end{pmatrix}$$

The vector **b** could in fact have been written down by inspection, since **b** must be orthogonal to **a**.

According to the discussion of this section, the matrix that diagonalizes **H** is

$$\mathbf{U} = \begin{pmatrix} i/\sqrt{2} & -i/\sqrt{2} \\ 1/\sqrt{2} & 1/\sqrt{2} \end{pmatrix}$$

The adjoint matrix $\mathbf{U}^+$ is given by

$$\mathbf{U}^+ = \begin{pmatrix} -i/\sqrt{2} & 1/\sqrt{2} \\ i/\sqrt{2} & 1/\sqrt{2} \end{pmatrix}$$

The reader can verify that

$$\mathbf{U}^+\mathbf{U} = \mathbf{E}$$

and therefore that

$$\mathbf{U}^{-1} = \mathbf{U}^+$$

Hence

$$\mathbf{H}' = \mathbf{U}^{-1}\mathbf{H}\mathbf{U} = \begin{pmatrix} 4 & 0 \\ 0 & 2 \end{pmatrix}$$

24.2 Quadratic Forms

We shall now give an immediate application of the results obtained above. Let H be some Hermitian operator and let $|x\rangle$ be some vector. In an arbitrary orthonormal basis, we have

$$\begin{aligned} \langle x| H |x\rangle &= x_i H^i_j x^j \\ &\equiv \sum_{i,j} H^i_j \bar{x}^i x^j \end{aligned}$$

where H^i_j are elements of a Hermitian matrix **H**. The last expression is called a **Hermitian quadratic form.*** Let **U** be the unitary matrix that diagonalizes **H**; then the matrix

$$\mathbf{G} = \mathbf{U}^+\mathbf{H}\mathbf{U} \tag{24.15}$$

is diagonal and its diagonal elements $g_{(i)}$ $(i = 1, 2, \cdots, N)$ are equal to the eigenvalues of the operator H. Choosing as basis vectors the eigenvectors of H, one gets

$$\langle x| H |x\rangle \equiv \sum_i g_{(i)} |\xi^i|^2 \tag{24.16}$$

where ξ^i $(i = 1, 2, \cdots, N)$ denote the components of $|x\rangle$ in this particular basis. The linear transformation of x^j, which brings $\sum_{i,j} H^i_j \bar{x}^i x^j$ into the simplified form 24.16,

* Hermitian, because the matrix **H** is Hermitian; quadratic because it is a quadratic expression with respect to x^j.

is generated by the unitary matrix **U**, which diagonalizes the matrix **H**. This can be verified explicitly. Let us perform the transformation

$$x^j = U^j_m \xi^m \qquad j = 1, 2, \cdots, N \tag{24.17}$$

Taking the complex conjugates of both sides of the preceding equation and making the substitution of indices $j \to i$, $m \to n$, we obtain

$$\bar{x}^i = \bar{U}^i_n \bar{\xi}^n \equiv \sum_n (U^+)^n_i \bar{\xi}^n \tag{24.18}$$

The matrix equation (24.15) can be written as

$$g_{(m)} E^n_m = (U^+)^n_i H^i_j U^j_m \tag{24.19}$$

Collecting Eqs. 24.17, 24.18, and 24.19, we have

$$\begin{aligned}\sum_{i,j} H^i_j \bar{x}^i x^j &= \sum_{i,j,m,n} H^i_j U^j_m \xi^m (U^+)^n_i \bar{\xi}^n \\ &= \sum_{m,n} g_{(m)} E^n_m \xi^m \bar{\xi}^n \\ &= \sum_m g_{(m)} |\xi^m|^2\end{aligned}$$

The foregoing discussion can be summarized in the following theorem.

Theorem 2. Every Hermitian quadratic form

$$\sum_{i,j} H^i_j \bar{x}^i x^j$$

can be brought to the form

$$\sum_i g_{(i)} |\xi^i|^2$$

by an unitary transformation of the x^j.

The real symmetric matrix is a particular case of a Hermitian matrix; the two conditions

$$S^i_j \equiv S^j_i \qquad (i,j = 1, 2, \cdots, N)$$

and

$$\operatorname{Im} S^i_j = 0 \qquad (i,j = 1, 2, \cdots, N)$$

guarantee that the relation

$$S^i_j \equiv \bar{S}^j_i \qquad (i,j = 1, 2, \cdots, N)$$

which defines a Hermitian matrix, is trivially satisfied. One can construct a matrix **O** which diagonalizes **S** by the method described in this section. It may be easily seen that since all S^i_j are real and since the eigenvalues of **S** are also real, as it is a Hermitian matrix, all the elements of **O** will also be real. A unitary real matrix is called an **orthogonal** matrix (cf. footnote on p. 152). We have therefore the following corollaries:

Corollary 1. Every real, symmetric matrix **S**′ can be diagonalized by a transformation

$$\mathbf{S} = \mathbf{O}^{\mathrm{T}} \mathbf{S}' \mathbf{O}$$

where **O** is orthogonal and "T" means "transposed."

Corollary 2. Every expression

$$\sum_{i,j} S^j_i x^i x^j$$

where S is a real, symmetric matrix can be transformed to the form

$$\sum_i g_{(i)}(\xi^i)^2$$

by an orthogonal transformation of the x^j

$$x^j = O^j_m \xi^m$$

24.3 Simultaneous Diagonalization of Two Hermitian Matrices

We have seen that with the proper choice of a basis, we can considerably simplify the study of a matrix representing a Hermitian operator, since we can always bring it to diagonal form. If we encounter two Hermitian operators in a problem, then it is in general impossible to find a single basis in which both operators are represented by diagonal matrices. One has, however, the following theorem.

Theorem 3. Let A and B be two Hermitian operators

$$A^+ = A; \qquad B^+ = B$$

The necessary and sufficient condition for the existence of a basis with respect to which the matrices representing A and B are both diagonal is that the operators A and B commute.

$$[A,B] \equiv AB - BA = 0$$

Proof. Suppose that

$$AB - BA = 0 \tag{24.20}$$

i.e., for the corresponding matrix elements

$$A^i_k B^k_j - B^i_k A^k_j = 0 \tag{24.21}$$

We choose a basis such that one of the operators (A, say) is represented by a diagonal matrix. As we have seen, this is always possible. Thus

$$A^i_j = a_{(j)} E^i_j \tag{24.22}$$

Putting Eq. 24.22 in Eq. 24.21, we find

$$[a_{(i)} - a_{(j)}]B^i_j = 0 \qquad \text{(no summation!)} \tag{24.23}$$

Hence, when

$$a_{(i)} \neq a_{(j)} \qquad (i \neq j)$$

one has

$$B^i_j = 0 \qquad (i \neq j)$$

On the other hand, when $a_{(i)} = a_{(j)}$, one may have $B^i_j \neq 0$.

However, in this case, it is still possible to change the basis so as to get **B** in diagonal form, without disturbing the diagonality of **A**. In fact, the eigenvectors of A, which correspond to the same eigenvalue, span a subspace where A, being already in diagonal form, is represented by a unit matrix multiplied by this eigenvalue. In that subspace, B is represented by a Hermitian matrix. We now change the basis of this subspace to have B represented by a diagonal matrix in that subspace. This is always possible and, moreover, such a change of basis cannot alter the structure of the matrix that represents A in that subspace, since the unit matrix is always transformed into itself. Thus, we have shown that two commuting operators can be both represented by diagonal matrices when a basis has been properly chosen.

The proof of the converse is immediate. If, in a given basis, A and B are represented by diagonal matrices

$$\left.\begin{aligned} A^i_j &= a_{(j)}E^i_j = a_{(i)}E^i_j \\ B^i_j &= b_{(j)}E^i_j = b_{(i)}E^i_j \end{aligned}\right\} \quad \text{(no summation!)}$$

then one has

$$A^i_k B^k_j - B^i_k A^k_j = [a_{(i)}b_{(j)} - a_{(j)}b_{(i)}]E^i_j = 0 \tag{24.24}$$

or in matrix form

$$\mathbf{AB} - \mathbf{BA} = 0$$

Thus, if two operators are represented in a particular basis by diagonal matrices, these matrices commute. But by changing from that basis to an arbitrary basis, one has

$$\mathbf{A}' = \mathbf{R}^{-1}\mathbf{A}\mathbf{R}$$

$$\mathbf{B}' = \mathbf{R}^{-1}\mathbf{B}\mathbf{R}$$

and so

$$\mathbf{A}'\mathbf{B}' - \mathbf{B}'\mathbf{A}' = \mathbf{R}^{-1}[\mathbf{AB} - \mathbf{BA}]\mathbf{R} = 0 \tag{24.25}$$

Hence, if the matrices representing A and B commute in a particular basis, they commute in any basis. This means that the corresponding operators commute

$$AB - BA = 0$$

We can now better understand why an arbitrary operator C cannot be represented by a diagonal matrix. It is evidently always possible to write C as

$$C = A + iB \tag{24.26}$$

where

$$A \equiv \frac{1}{2}(C + C^+) \qquad \text{and} \qquad B \equiv \frac{-i}{2}(C - C^+) \tag{24.27}$$

are Hermitian operators. In general, A and B will not commute. However, if

$$C^+C = CC^+ \tag{24.28}$$

then

$$AB - BA = 0$$

Thus, when Eq. 24.28 is satisfied, both A and B, and therefore both C and C^+, can be represented by diagonal matrices. In particular, any unitary operator can be represented by a diagonal matrix.

CHAPTER III

FUNCTION SPACE, ORTHOGONAL POLYNOMIALS, AND FOURIER ANALYSIS

1 · INTRODUCTION

In the preceding chapter we focused our attention mainly on finite-dimensional linear vector spaces. Without having, of course, exhausted the subject, we presented those results that are the most important for physical applications. It is clear, however, that there exist important linear vector spaces other than the finite-dimensional ones.

In this chapter we consider infinite-dimensional linear vector spaces which, by definition, are those spaces whose number of linearly independent vectors is not bounded; i.e., given an arbitrary finite sequence of vectors in such a space, one can always find a vector that also belongs to the same space and which is linearly independent of all the vectors of this sequence. Fortunately, this branch of mathematics has received much attention, and its development has led to many subtle and beautiful results. We are obliged, however, to make a still more drastic selection of subjects than we did in the preceding chapter, and our discussion will be limited to those results which have already been found to be of much use in physics.

2 · SPACE OF CONTINUOUS FUNCTIONS

Consider the set of all complex functions (i.e., functions taking in general complex values) that are continuous on some interval of finite length of the real axis

$$a \leq x \leq b$$

Two such functions, $f(x)$ and $g(x)$, can be added together to construct the function

$$h(x) = f(x) + g(x), \qquad a \leq x \leq b$$

where the plus symbol has the usual operational meaning of "add the value of f at the point x to the value of g at the same point."

A function $f(x)$ can also be multiplied by a number c, to give the function

$$p(x) = c \cdot f(x) \qquad a \leq x \leq b$$

The centered dot, the multiplication symbol, is again understood in the conventional sense: "Multiply by c the value of the function f at the point x."

These rules for the addition of functions and the multiplication of functions by numbers, which are known to the reader from elementary analysis, reduce for each point $x \in [a,b]$ to the usual arithmetic manipulations, and therefore satisfy the conditions B of Sec. 2, of the preceding chapter. These rules, in fact, were fashioned after the rules of arithmetic.

It is evident that the following conditions are satisfied:

(a) By adding two continuous functions, one obtains a continuous function.
(b) The multiplication by a number of a continuous function yields again a continuous function.
(c) The function that is identically zero for $a \leq x \leq b$ is continuous, and its addition to any other function does not alter this function.
(d) For any function $f(x)$ there exists a function $(-1)f(x)$, which satisfies

$$f(x) + [(-1)f(x)] = 0$$

Comparing these four statements with the conditions A (Chapter II, Sec. 2), we see that the set of all continuous functions defined on some interval $a \leq x \leq b$ forms a linear vector space.

We shall, however, be slightly more sophisticated, and we shall not simply identify each continuous function with some vector $|\rangle$. Instead, we shall consider the entire set of values of a function $f(x)$ as **representing** a vector $|f\rangle$ belonging to some abstract linear vector space F, which we shall call **function space**. In other words, we shall treat the number $f(x)$ as the component with "index x" of an abstract vector $|f\rangle$. This is quite similar to what we did in the case of finite-dimensional spaces when we associated a component a^i of a vector $|a\rangle$ with each value of the index i. The only difference is that this index assumed a discrete set of values 1, 2, etc., up to N, whereas the argument x of a function $f(x)$ is a continuous variable. This resemblance would in fact be even more pronounced if one wrote f^x instead of $f(x)$ but this is merely a matter of notation. However, the objection may be raised that the components of a vector are defined with respect to some basis and we do not know what basis has been chosen in the function space F. Unfortunately, we are obliged to postpone the answer to this objection. Let us merely note that, once a basis has been chosen, we work only with the numbers that represent the vectors. Therefore, provided we do not change to other basis vectors, we need not be concerned about the particular basis that has been chosen.

Let us summarize the preceding discussion as follows:

The abstract function space F is defined as the linear vector space whose vectors are represented by functions defined on an interval $[a,b]$. Let the functions $f(x)$ and $g(x)$ represent the vectors $|f\rangle, |g\rangle \in$ F. Then, by definition, $|f\rangle + |g\rangle$ is represented by $f(x) + g(x)$ and $c \cdot |f\rangle$ is represented by $c \cdot f(x)$.

In the subsequent discussions, we shall at times consider certain properties of the vectors belonging to the function space; at other times, certain properties of the functions that represent these vectors will be considered. This dualism in our presentation is justifiable because many of the results of analysis have in fact an algebraic foundation, and it is often clearer and more convenient if these results are formulated in the language of algebra. On the other hand, there are problems, in applications, which, although they may be formulated algebraically, appear more naturally in their analytic form, and therefore it is important not to lose sight of the interconnection between the analytic and algebraic aspects of the problem.

3 · METRIC PROPERTIES OF THE SPACE OF CONTINUOUS FUNCTIONS

In Chapter II, Sec. 6, we introduced the notion of a metric space. This is a space in which a distance between its elements can be defined. This abstract distance is a straightforward generalization of the intuitive notion of a distance; one demands that it be a positive number, that the distance from a to b be the same as that from b to a, that the distance vanish only if two elements of the space are identical, and that the triangle inequality be satisfied.

We also noted in Chapter II that one can introduce the notion of a "point" in a linear vector space by considering vectors as "radius vectors," and it was shown that, given two vectors $|a\rangle$ and $|b\rangle$ which determine two points a and b, the length of the vector $\{|a\rangle - |b\rangle\}$ has all the required properties of the distance $\rho(|a\rangle,|b\rangle)$ as in elementary vector calculus.

We did not discuss separately the properties of the distance in the case of arbitrary finite-dimensional spaces, since they do not differ essentially from the corresponding properties of a three-dimensional physical manifold; the only difference is that components of vectors are, in general, complex numbers.

Suppose that an orthonormal basis $|e_i\rangle$ $(i = 1, 2, \cdots, N)$ has been chosen in an N-dimensional space, which we know is always possible. Take two vectors

$$|a\rangle = \sum_{i=1}^{N} a^i |e_i\rangle$$

$$|b\rangle = \sum_{i=1}^{N} b^i |e_i\rangle$$

The distance $\rho(|a\rangle,|b\rangle)$ between the points a and b is

$$\begin{aligned}\rho(|a\rangle, |b\rangle) &= \{[\langle a| - \langle b|][|a\rangle - |b\rangle]\}^{1/2} \\ &= \sum_{i,k=1}^{N} \{[\langle e_k|(\bar{a}^k - \bar{b}^k)][(a^i - b^i)|e_i\rangle]\}^{1/2} \\ &= \left[\sum_{i=1}^{N} |a^i - b^i|^2\right]^{1/2} \end{aligned} \tag{3.1}$$

This is the usual expression for the distance between two points in orthonormal coordinates, except that here one takes the sum of squares of the moduli of the differences between the coordinates instead of taking the sum of squares of these differences. Therefore, for real vector spaces (as, for example, the physical three-dimensional manifold), Eq. 3.1 becomes exactly the conventional definition of the distance.

We define the scalar product in the space of functions that are continuous in the interval $[a,b]$ by the expression

$$\langle f | g\rangle = \int_a^b \bar{f}(x)g(x)w(x)\,dx \tag{3.2}$$

where $w(x)$ is some real, positive function

$$w(x) > 0$$

called the density, or **weight function.**

The definition 3.2 has all the required properties of the scalar product. $\langle f|g\rangle$ is linear with respect to $|g\rangle$, since the function under the integral is linear with respect to $g(x)$; also

$$\langle f|g\rangle = \overline{\langle g|f\rangle}$$

since $w(x)$ is real, etc.

Equation 3.2 can be regarded as a direct generalization of the expression

$$\langle a|b\rangle = \sum_{i=1}^{N} \bar{a}^i b^i \tag{3.3}$$

for the scalar product in an orthonormal basis (see Chapter II, Sec. 18). We make the replacement

$$\sum_i \to \int_a^b dx\, w(x)$$

which is common in mathematical physics. The meaning of this replacement is evident. Given a function $h(x_{(i)})$ defined over an enumerable set of points $x_{(i)}$ in some interval $[a,b]$ and distributed with a density $w(x)$, the sum $\sum_{i=1}^{N} h(x_{(i)})\,(b-a)/N$ goes over into the integral $\int_a^b h(x)w(x)\,dx$ when the number of points $x_{(i)}$ increases indefinitely.

The Cauchy-Schwarz inequality, which was derived in complete generality (Chapter II, Sec. 4) holds, of course, also for the particular definition 3.2 of the scalar product. One immediately sees that

$$|\langle f|g\rangle|^2 \leq \langle f|f\rangle\langle g|g\rangle$$

now reads

$$\left|\int_a^b \bar{f}(x)g(x)w(x)\,dx\right|^2 \leq \left[\int_a^b \bar{f}(x)f(x)w(x)\,dx\right]\left[\int_a^b \bar{g}(x)g(x)w(x)\,dx\right] \tag{3.4}$$

The metric properties of a space are determined when the scalar product has been defined. For a function space, the definition 3.2 yields the following expression for the distance between the "points" defined by the vectors $|f\rangle, |g\rangle \in \text{F}$:

$$\rho(|f\rangle, |g\rangle) = \left\{\int_a^b |f(x) - g(x)|^2 w(x)\,dx\right\}^{\frac{1}{2}} \tag{3.5}$$

The formal resemblance of Eq. 3.5 to Eq. 3.1 is to be noted. However, an N-dimensional space, with the distance between two vectors $|a\rangle$ and $|b\rangle$ given by Eq. 3.1, has an important property which, as we shall see, is not shared by all linear vector spaces, and in particular by the space of continuous functions.

Let us consider an infinite sequence of vectors

$$|a_1\rangle,\quad a|_2\rangle,\quad \cdots,\quad |a_n\rangle,\quad \cdots\quad \in S_N$$

satisfying the condition

$$\lim_{k,l\to\infty} \rho(|a_k\rangle, |a_l\rangle) = 0 \tag{3.6}$$

It is not difficult to prove that Eq. 3.6 implies that there exists a vector $|a\rangle \in S_N$ to which the sequence $|a_1\rangle, |a_2\rangle, \cdots, |a_n\rangle \cdots$ converges. In fact, Eq. 3.6 means that for an arbitrary $\varepsilon > 0$, there exists a number L such that, provided $k,l > L$,

$$\sum_{i=1}^{N} |a^i_{(k)} - a^i_{(l)}|^2 < \varepsilon$$

Since this inequality is satisfied by the whole sum, it is necessarily satisfied by each term individually.

$$|a^i_{(k)} - a^i_{(l)}|^2 < \varepsilon \qquad \text{for any } i$$

Therefore, according to the Cauchy criterion for the convergence of a series, the sequence of components $a^i_{(1)}, a^i_{(2)}, \cdots, a^i_{(n)} \cdots$ converges to some number a^i for any $i = 1, 2, \cdots, N$. The N numbers a^i $(i = 1, 2, \cdots, N)$ determine a vector $|a\rangle$, which evidently belongs to the space in question, namely, S_N.

The intuitive meaning of Eq. 3.6 is simple: The distances between the points, determined by the sequence $|a_k\rangle$ $(k = 1, 2, \cdots, n, \cdots)$, become arbitrarily small for sufficiently large k. In other words, these points accumulate in some domain of the space and, however small the "dimension" of this domain may be, only a finite number of points remains outside of it. Moreover, as we have just shown in the case of a finite-dimensional space, the accumulation point itself belongs to the space.

This result, which may appear evident, is not true, however, for an arbitrary linear vector space. That is, the condition

$$\lim_{k,l \to \infty} \rho(|a_k\rangle, |a_l\rangle) = 0$$

does not necessarily imply, in general, that there exists a vector $|a\rangle$ such that

$$\lim_{k \to \infty} \rho(|a\rangle, |a_k\rangle) = 0$$

Spaces for which this implication is true are called **complete**. Any finite dimensional space is complete. A simple example is sufficient to demonstrate that the space of functions which are continuous on some interval is not complete.

Let the interval in question be $[-1, 1]$ and, for simplicity, let us put $w(x) = 1$. Define the sequence of functions $f_{(k)}(x)$ as

$$f_{(k)}(x) = \begin{cases} 1 \;, & \dfrac{1}{k} < x \le 1 \\[2ex] \dfrac{kx+1}{2}, & -\dfrac{1}{k} < x < \dfrac{1}{k} \\[2ex] 0 \;, & -1 \le x < -\dfrac{1}{k} \end{cases} \tag{3.7}$$

We have

$$\rho^2(|f_k\rangle, |f_l\rangle) = \int_{-1}^{1} |f_{(k)}(x) - f_{(l)}(x)|^2 \, dx \xrightarrow[k,l \to \infty]{} 0$$

Thus, the condition 3.6 is satisfied. However, the sequence of functions $f_{(k)}(x)$ converges to the function

$$f(x) = \begin{cases} 1 & 0 < x < 1 \\ 0 & -1 < x < 0 \end{cases}$$

which is discontinuous at $x = 0$ and therefore does not belong to the space of continuous functions.

It can easily be proved that a continuous function $g(x)$ satisfying

$$\lim_{k \to \infty} \rho(|g\rangle, |f_k\rangle) = 0 \tag{3.8}$$

does not exist. Take the interval [0, 1]. The function $f(x)$ is continuous in this interval. Thus, $f(x)$ and the $f_{(k)}(x)$ belong to the same space of continuous functions defined in the interval [0, 1]. Since

$$\lim_{k\to\infty} \int_0^1 |f(x) - f_{(k)}(x)|^2\, dx = 0 \tag{3.9}$$

the function $f(x) = 1$ for $x \in [0, 1]$ is unique, as we showed in Sec. 5 of Chapter II. Similarly, one gets the result that for $x \in [-1, 0]$, the function $f(x) = 0$ is the unique function satisfying

$$\lim_{k\to\infty} \int_{-1}^0 |f(x) - f_{(k)}(x)|^2\, dx = 0 \tag{3.10}$$

A function $g(x)$ that would be continuous for $x \in [-1, 1]$, and which would satisfy Eq. 3.8, would be continuous in the intervals $[-1, 0]$ and $[0, 1]$ separately and would also have to satisfy Eqs. 3.9 and 3.10. This would be incompatible with the fact that it is continuous, since one should have $g(x) = 0$ for $x \in [-1, 0]$ and $g(x) = 1$ for $x \in [0, 1]$.

4 · ELEMENTARY INTRODUCTION TO THE LEBESGUE INTEGRAL

There exists a complete space of functions which contains as a subspace the space of those functions that are continuous on an interval. The functions of this space may have a strange behavior; not only discontinuous functions with finite jumps are included but also, for instance, the function that is equal to 1 or 0, according as its argument is a rational or an irrational number. It is important to generalize the familiar notion of the Riemann integral so as to be able to integrate such rebellious functions. This generalization is achieved by introducing the Lebesgue integral, which is equal to the Riemann integral for functions that are integrable in the conventional sense.

We pass now to a more detailed discussion, which, however, is meant to give to the reader only a very general idea of what a Lebesgue integral is. The reader who likes subtle mathematical considerations will probably not be satisfied by this section. We advise him to consult more complete textbooks on the subject.*

One says that a **measure** has been defined on a set E if one has associated in a unique manner some nonnegative number with subsets of E

$$m(\mathrm{E}_i) \geq 0, \qquad \mathrm{E}_i \subset \mathrm{E}$$

where $m(E_i)$ is, by definition, zero for an empty set.**

There may exist subsets of E for which the measure does not exist. One demands, however, that within the class of measurable subsets, the measure be an additive number. This means that the measure of the sum of two nonoverlapping sets exists and is equal to the sum of the measures of these sets

$$m(\mathrm{E}_1 + \mathrm{E}_2) = m(\mathrm{E}_1) + m(\mathrm{E}_2) \qquad \text{when } \mathrm{E}_1 \cap \mathrm{E}_2 = 0 \tag{4.1}$$

---

* For instance, E. C. Titchmarsh *The Theory of Functions*, Oxford University Press, 1964.

** The converse is not true; there exist nonempty sets of measure zero.

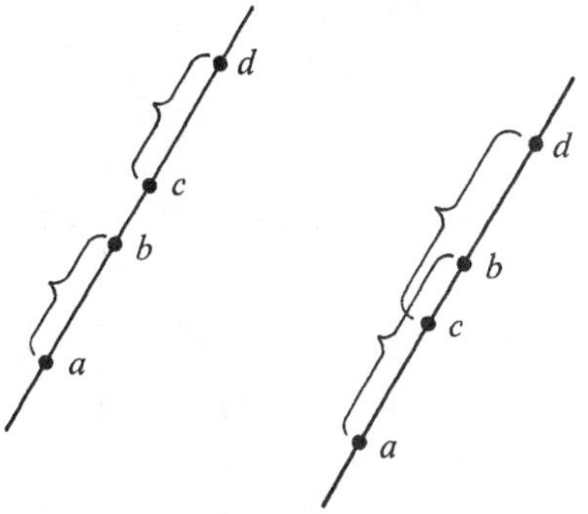

Fig. 41.

We recall that $E_1 + E_2$ denotes the set that contains both elements of E_1 and E_2, each element counted only once; $E_1 \cap E_2$ denotes the set that contains the common elements of E_1 and E_2. In the case where E_1 and E_2 overlap, one replaces Eq. 4.1 by

$$m(E_1 + E_2) = m(E_1) + m(E_2) - m(E_1 \cap E_2)$$

in order to count only once the points common to E_1 and E_2.

EXAMPLE 1

Consider the set of points on a line. The measure of an interval (a,b) will be taken to be the length of this interval. Given two intervals, one has an alternative: Either they overlap or they do not. In the first case, the set of points that belongs either to (a,b) or to (c,d) has a measure which is the sum of the lengths of (a,b) and of (c,d). In the latter case (see Fig. 41), it is the length of the interval (a,d).

The conventional manner of defining the Riemann integral of a function $f(x)$ over a closed interval $[a,b]$ is to divide the interval into a number of nonoverlapping subintervals*

$$[x_0,x_1), [x_1,x_2), \cdots, [x_k,x_{k+1}), \cdots, [x_{N-1},x_N] \tag{4.2}$$

where

$$a \equiv x_0 < x_1 < \quad \cdots \quad < x_N \equiv b$$

and to form the sum

$$\sum_{k=0}^{N-1} f(\xi_k)(x_{k+1} - x_k) \tag{4.3}$$

in which $\xi_{(k)}$ denotes a point of the subinterval $[x_k,x_{k+1})$

$$x_k < \xi_k < x_{k+1} \tag{4.4}$$

Now the number of subintervals 4.2 is increased indefinitely and in such a manner that

$$|x_{k+1} - x_k| \to 0 \qquad \text{for any } k$$

Then, provided the limit of the sum 4.3 exists and is independent of the manner in which the subdivision of $[a,b]$ was made, it is called the definite integral of $f(x)$ over the interval $[a,b]$.

---

* $[x_k,x_{k+1})$ denotes an interval that is closed from the left but open from the right.

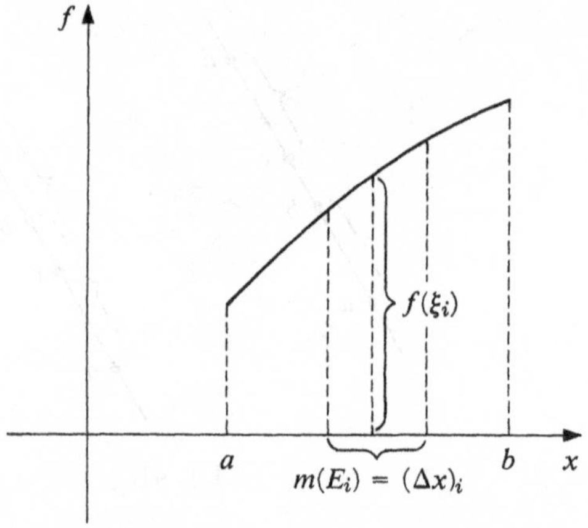

Fig. 42. The integration of a continuous function in the sense of Riemann.

Translated into the language of measure, the Riemann integral of a function $f(x)$ defined over a set of arguments $x \in \mathrm{E}$ is obtained as follows: One subdivides E into nonoverlapping subsets E_i

$$\mathrm{E} = \mathrm{E}_1 + \mathrm{E}_2 + \cdots + \mathrm{E}_N, \quad \mathrm{E}_i \cap \mathrm{E}_j = 0 \qquad \text{for any } i,j$$

and one forms the sum

$$\sum_{i=1}^{N} f(\xi_i) m(\mathrm{E}_i) \tag{4.5}$$

where $m(\mathrm{E}_i)$ is the measure of the subset E_i and ξ_i is any point that belongs to E_i. We now increase the number of subsets indefinitely and in such a way that

$$m(\mathrm{E}_i) \to 0 \qquad \text{for any } \mathrm{E}_i$$

Then, provided the limit of the sum 4.5 exists and is independent of the subdivision process, it is called the (Riemann) integral of $f(x)$ over E.

It is clear that a Riemann integral can be defined if, as the measure of each subset E_i tends to zero, all the values of the function defined over that subset tend to a common (well-defined) limit.* Such a requirement, of course, excludes defining a Riemann integral for violently discontinuous functions.

EXAMPLE 2

Consider a continuous function $f(x)$ defined in the interval $[a,b]$ as plotted in Fig. 42. The Riemann integral is obtained if one divides the interval $[a,b]$ into subintervals E_i of length (measure) $(\Delta x)_i$, such that $\sum_i (\Delta x)_i$ is equal to the length of $[a,b]$. When all the $(\Delta x)_i$ are small enough, the function $f(x)$ does not vary appreciably within each of these intervals and the limit

$$\lim_{\max(\Delta x)_i \to 0} \sum_i f(\xi_i)(\Delta x)_i = \int_a^b f(x)\,dx$$

exists and is equal to the area under the curve.

* There may, however, be a finite number of points where this condition is not satisfied, but which give in the limit a vanishing contribution to the integral.

We are now in a position to explain the main idea of the Lebesgue integral. Let the function $f(x)$ defined on a set E be bounded

$$f^{(\min)} \leq f(x) \leq f^{(\max)} \qquad \text{for any } x \in E$$

To each interval of values of f around some given value f_i

$$f_i - (\Delta f)_i \leq f \leq f_i + (\Delta f)_i \tag{4.6}$$

there corresponds a set E_{f_i} of values x such that

$$f_i - (\Delta f)_i \leq f(x) \leq f_i + (\Delta f)_i \qquad \text{for any } x \in E_{f_i} \tag{4.7}$$

We form the sum of products

$$f_i \cdot m(E_{f_i}) \tag{4.8}$$

of all possible values of f by the measure of the corresponding set of arguments, and the limit

$$\lim_{\max(\Delta f)i \to 0} \sum_i f_i m(E_{f_i}) = \int_E f(x) \tag{4.9}$$

if it exists, is called the **Lebesgue integral** of $f(x)$ over the set E.

Notice that the sum in Eq. 4.9, although apparently similar to the Riemann sum 4.5, is in fact quite different from it, and one could say that the two definitions are almost complementary. In the sum 4.5, the set E was subdivided into a number of subsets E_i and $f(\xi_i)$ was the value of the function $f(x)$ at **any** point $\xi_i \in E_i$. On the other hand, in Eq. 4.9 the function $f(x)$ in each subset E_{f_i} has a **definite** value f_i, and we must then find the measure $m(E_{f_i})$ of the set where inequality 4.7 is satisfied.

For the existence of the Lebesgue integral, we require the existence of a measure of E_{f_i} for any f_i, but the conditions imposed on the integrated function are very weak, since we no longer need the "local smoothness" of $f(x)$. The reason, of course, is that in Eq. 4.9, $f(x)$ has a definite value f_i and is not allowed to vary within each subset as in the case of a Riemann integral. Therefore, the question of smoothness does not enter.

Thus the problem of constructing an integral reduces essentially to that of finding a nontrivial measure for the sets of arguments. In the case of sets of points on an interval, such a measure can be defined for a wide variety of sets and is called the **Lebesgue measure**. If the set contains all points of an interval, the Lebesgue measure of this set is simply equal to the length of this interval.

EXAMPLE 3

Let us consider again the problem of the integration of a continuous function defined on an interval $[a,b]$ from the point of view of the Lebesgue integral. For each f_i the horizontal lines (Fig. 43)

$$f = f_i + (\Delta f)_i$$
$$f = f_i - (\Delta f)_i$$

cut out a segment of the curve $f(x)$. Thus, the whole curve is divided into small segments. Projecting the ith segment on the x axis, we get some interval with measure (length) $m(E_{f_i})$. The sum $\sum_i f_i \cdot m(E_{f_i})$ clearly also tends to a quantity equal to the area under the curve $f(x)$, as in the case of the Riemann integral.

The preceding example demonstrates that when the Riemann integral exists, the Lebesgue integral also exists, and both integrals are equal.

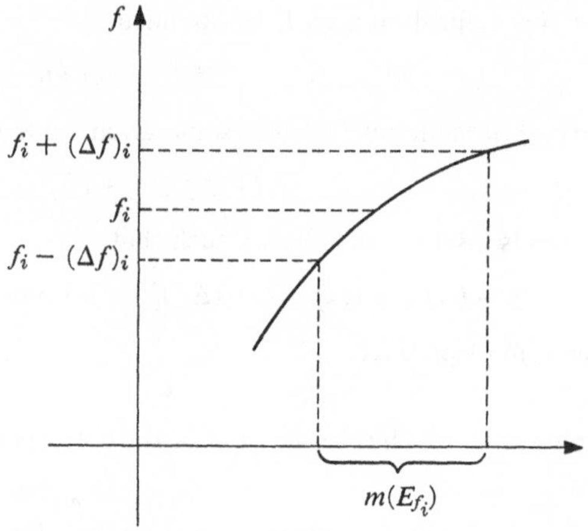

Fig. 43. The integration of a continuous function in the sense of Lebesgue.

We briefly indicate how one constructs the Lebesgue measure of a point set. Let us take a finite interval (a,b) of length L. Consider a set E of points $x \in (a,b)$. We denote by E′ the set of all points $x \in (a,b)$ which do not belong to E. Let us denote by $l_1, l_2 \cdots$ the lengths of non-overlapping intervals $\Lambda_i \subset (a,b)$ such that

$$\mathrm{E} \subset (\Lambda_1 + \Lambda_2 + \cdots)$$

Thus

$$0 \leq \sum_i l_i \leq L$$

The smallest such sum, $\sum_i l_i$ is called the **exterior measure** of E

$$m_{\mathrm{ext}}(\mathrm{E}) = \min(\sum_i l_i)$$

In the same manner, one can find intervals $\Lambda'_i \subset (a,b)$, of lengths $l'_1, l'_2 \cdots$ which cover the set of points of E′, i.e., such that

$$\mathrm{E}' \subset (\Lambda'_1 + \Lambda'_2 + \cdots)$$
$$0 \leq \sum_i l'_i \leq L$$

The number

$$m_{\mathrm{int}}(\mathrm{E}) = L - \min(\sum_i l'_i)$$

is called the **interior measure** of E. It can be proved that in general

$$0 \leq m_{\mathrm{int}}(\mathrm{E}) \leq m_{\mathrm{ext}}(\mathrm{E}) \tag{4.10}$$

In the case when

$$m_{\mathrm{int}}(\mathrm{E}) = m_{\mathrm{ext}}(\mathrm{E}) = m(\mathrm{E})$$

the number $m(\mathrm{E})$ is called the **Lebesgue measure** of the point set E. One can show that it has all the required properties of a measure. Clearly, when E contains all the points of (a,b), the smallest interval that covers (a,b) is (a,b) itself, and $m(\mathrm{E}) = L$.

As an example, we show that any enumerable set has a measure equal to zero, for let x_k $(i = 1, 2, \cdots, n, \cdots)$ denote the points of an enumerable set E which are contained in an interval of length L. We cover each point x_n by an open interval of length $\varepsilon/2^n$, where ε is an arbitrary positive number. Since these intervals may overlap, the entire set can be covered by an open set of measure not greater than

$$\sum_{n=0}^{\infty} \frac{\varepsilon}{2^n} = 2\varepsilon \tag{4.11}$$

Since ε can be made arbitrarily small, we find that

$$m_{\text{ext}}(\text{E}) = 0 \tag{4.12}$$

From inequality 4.10 we immediately get

$$m_{\text{int}}(\text{E}) = 0$$

and thus

$$m(\text{E}) = 0$$

The set of rational numbers is enumerable; for example, the rational numbers within the interval [0, 1] can be arranged in a sequence of proper fractions as

$$0, +1, +\tfrac{1}{2}, +\tfrac{1}{3}, +\tfrac{2}{3}, +\tfrac{1}{4}, +\tfrac{3}{4}, +\tfrac{1}{5}, +\tfrac{2}{5}, +\tfrac{3}{5}, \cdots$$

Hence this set has zero measure.

Consider now the so-called **Dirichlet function**, which is defined in the interval $0 \leq x \leq 1$ as

$$\chi(x) = \begin{cases} 1 & \text{if } x \text{ is a rational number} \\ 0 & \text{if } x \text{ is an irrational number} \end{cases}$$

This highly discontinuous function, which cannot be integrated in the sense of Riemann, nevertheless possesses a Lebesgue integral. Since the set of all rational numbers within the interval [0, 1] has zero measure, the Lebesgue integral of $\chi(x)$ over this interval is well defined and equal to zero.

The Lebesgue integral (which is, just as is the Riemann integral, a limit of a sum) has the usual properties of a sum

$$\int [f(x) + g(x)] = \int f(x) + \int g(x)$$

$$\int [cf(x)] = c \int f(x)$$

Because of the formal resemblance between Lebesgue and Riemann integrals, we use the same notation for them; there will be no confusion, since for functions that are integrable in the conventional sense, the Lebesgue and the Riemann integrals give the same result.

5 · THE RIESZ-FISCHER THEOREM

We ended Section 3 with an example showing that the space of continuous functions is not complete. The reason in this particular example, as well as in general, is that a sequence of continuous functions may converge to a function that has quite an odd behavior. This suggests that if we wish to have a complete function space, we must considerably enlarge the class of admissible functions; for instance, if we include discontinuous functions with finite jumps, the difficulty with the example in Sec. 3 would disappear.

Let us consider the function space whose vectors are represented by functions that are defined on a finite interval $[a,b]$ and for which the Lebesgue integral

$$\int_a^b |f(x)|^2 w(x)\,dx \tag{5.1}$$

exists and is finite. This space is called the $L_w^2(a,b)$ space; L represents the name Lebesgue, the superscript 2 indicates the integrability of the square of the modulus of each function representing a vector belonging to $L_w^2(a,b)$, and $w(x)$ is the weight function. It can be proved that, provided

$$\int_a^b |f(x)|^2 w(x)\,dx \quad . \quad \text{and} \qquad \int_a^b |g(x)|^2 w(x)\,dx$$

exist, which is the necessary condition that $|f\rangle, |g\rangle \in L_w^2(a,b)$, the integral $\int_b^a \bar{f}(x)g(x)w(x)\,dx$ and consequently the integral $\int_a^b |f(x)+g(x)|^2 w(x)\,dx$ exists and therefore $\{f(x)+g(x)\}$ represents a vector $\{|f\rangle + |g\rangle\} \in L_w^2(a,b)$ as required by the linearity of this space. The scalar product is defined in $L_w^2(a,b)$ as

$$\langle f|g\rangle = \int_a^b \bar{f}(x)g(x)w(x)\,dx \tag{5.2}$$

which is the same as Eq. 3.2 except that the integral has to be understood in the sense of Lebesgue. From the formal resemblance between the properties of the Lebesgue and the Riemann integrals, it follows that the expression 3.5 for the distance remains valid also in $L_w^2(a,b)$.

It is possible to prove (but the proof does not fall within the scope of this book) the following very important theorem due to F. Riesz and E. Fischer:

Theorem. The space $L_w^2(a,b)$ is complete; i.e., provided there exists a sequence of functions $f_{(1)}(x), f_{(2)}(x), \cdots$ that represent vectors of $L_w^2(a,b)$ and which satisfy the condition

$$\lim_{k,l\to\infty} \int_a^b |f_{(k)}(x) - f_{(l)}(x)|^2 w(x)\,dx = 0$$

then there always exists a function $f(x)$ which also represents a vector belonging to $L_w^2(a,b)$ and such that

$$\lim_{k\to\infty} \int_a^b |f(x) - f_{(k)}(x)|^2 w(x)\,dx = 0 \tag{5.3}$$

A few comments are needed in order to understand properly the meaning of the Riesz-Fischer theorem, and particularly the meaning of the relation 5.3. The integral in Eq. 5.3 is a Lebesgue integral; its value will not be changed by the modification of the integrand over a set of x of total measure zero. Therefore, the function $f(x)$ is not uniquely determined by the condition 5.3, since if we alter its value on a set of arguments of total measure zero, Eq. 5.3 will still be satisfied. However, we have agreed to interpret the integral

$$\int_a^b |f(x) - g(x)|^2 w(x)\, dx$$

as the square of the distance $\rho^2(|f\rangle, |g\rangle)$, and we have proved (Chapter II, Sec. 6) the uniqueness of the limit vector of a sequence of vectors. If there are two functions $f(x)$ and $g(x)$ satisfying Eq. 5.3, we must have

$$\rho(|f\rangle, |g\rangle) = 0$$

or

$$\int_a^b |f(x) - g(x)|^2 w(x)\, dx = 0$$

This condition is indeed satisfied, since $f(x)$ and $g(x)$ differ only on a set of measure zero. But two elements of a metric space must be identical if the distance between them is zero. Therefore

$$|f\rangle = |g\rangle$$

and we are led to consider two functions that differ only on a set of measure zero, or, as one says, two functions that are **equal almost everywhere** (or **equivalent**), as representing the same vector in $L^2_w(a,b)$.

The Riesz-Fischer theorem states that there exists a class of equivalent functions rather than a unique function satisfying Eq. 5.3. This nonuniqueness is expressed by saying that when Eq. 5.3 is satisfied, the sequence of functions $f_{(k)}(x)$ $(k = 1, 2, \cdots)$ **converges in the mean** to $f(x)$; this allows us to distinguish between this type of convergence and the usual convergence

$$\lim_{k \to \infty} f_{(k)}(x) = f(x), \qquad x \in [a,b]$$

6 · EXPANSIONS IN ORTHOGONAL FUNCTIONS

We have already noted in Chapter II, Sec. 19, the importance of the decomposition of a vector into a set of orthonormal vectors, since the coefficients of such a decomposition can be directly expressed as the scalar products of the decomposed vector and the vectors of the orthonormal set.

Suppose first that a vector $|f\rangle \in L^2_w(a,b)$ can be represented as a finite sum of vectors $|e_i\rangle \in L^2_w(a,b)$.

$$|f\rangle = \sum_{i=1}^{M} f^i |e_i\rangle \tag{6.1}$$

where the vectors $|e_i\rangle$ satisfy

$$\langle e_k | e_i \rangle = \delta_{ik} \qquad (i,k = 1, 2, \cdots, M)$$

Multiplying Eq. 6.1 from the left by $\langle e_k|$, we get the expansion coefficients

$$f^k \equiv \langle e_k | f \rangle \qquad (k = 1, 2, \cdots, M)$$

i.e.,

$$f^k \equiv \int_a^b \bar{e}_{(k)}(x) f(x) w(x)\, dx \qquad (k = 1, 2, \cdots, M)$$

where $e_{(k)}(x)$ is the function that represents $|e_k\rangle$. Thus, the calculation of the expansion coefficients f^i reduces to the evaluation of a series of integrals, which is very often feasible, at least numerically. If the functions $e_{(i)}(x)$ have, from some point of view, simple properties (for example, they may be elementary functions, or they may satisfy a simple differential equation, or their appearance in the expansion may have some interesting physical interpretation) the expansions of functions into functions $e_{(i)}(x)$ have a very wide range of applications. This is why we devote much time to the study of such particular expansions and the corresponding orthonormal sets of functions.

Let us now drop the assumption (which can be satisfied only accidentally) that the vector we are considering, $|f\rangle \in L^2_w(a,b)$, can be represented as a linear combination of a finite number of given orthonormal vectors.

For an infinite sequence of orthonormal vectors

$$|e_1\rangle, \quad |e_2\rangle, \quad \cdots, \quad |e_n\rangle, \quad \cdots \quad \in L^2_w(a,b)$$

we can again construct the scalar products

$$f^i \equiv \langle e_i|f\rangle \equiv \int_a^b \bar{e}_{(i)}(x)f(x)w(x)\,dx \tag{6.2}$$

although we cannot yet interpret them as the coefficients of an expansion. Let us construct a vector

$$|f_k\rangle = \sum_{i=1}^{k} \langle e_i|f\rangle\, |e_i\rangle \equiv \sum_{i=1}^{k} f^i\, |e_i\rangle$$

Using the Cauchy-Schwarz inequality, we get

$$|\langle f|f_k\rangle|^2 \leq \langle f|f\rangle\langle f_k|f_k\rangle = \langle f|f\rangle \left\{\sum_{i=1}^{k} |f^i|^2\right\} \tag{6.3}$$

From the definition 6.2 we have

$$\langle f|f_k\rangle = \sum_{i=1}^{k} f^i\langle f|e_i\rangle = \sum_{i=1}^{k} |f_i|^2 \tag{6.4}$$

Combining Eqs. 6.3 and 6.4, we get

$$\sum_{i=1}^{k} |f^i|^2 \leq \langle f|f\rangle$$

Since k is completely arbitrary, we can let $k \to \infty$ and obtain the so-called **Bessel inequality**

$$\sum_{i=1}^{\infty} |f^i|^2 \leq \langle f|f\rangle \tag{6.5}$$

Notice that the series on the LHS of the preceding inequality is always convergent, since $|f\rangle \in L^2_w(a,b)$, and hence has a finite norm

$$\langle f|f\rangle = \int_a^b |f(x)|^2 w(x)\,dx < \infty$$

To explain the meaning of the Bessel inequality, let us reconsider finite-dimensional spaces. In an N-dimensional space one can always find an orthonormal basis. Expanding an arbitrary vector $|a\rangle$

$$|a\rangle = \sum_{i=1}^{N} a^i\,|e_i\rangle \tag{6.6}$$

we obtain the already derived expression

$$\langle a|a\rangle = \sum_{i=1}^{N} |a^i|^2$$

Since each term in the sum on the RHS above is a non-negative number, $\langle a|a\rangle$ is always greater than, or equal to, any partial sum of these terms:

$$\sum_{1 \le i \le N} |a^i|^2 \le \langle a|a\rangle$$

This is the finite-dimensional analog of the Bessel inequality; the equality sign appears in general only when one has taken the components of $|a\rangle$ with respect to all the N orthonormal basis vectors. The inequality sign signifies that some of these basis vectors have been "forgotten."

We now need to generalize the notion of an orthonormal basis in order that it also be applicable to a function space. We shall then see under what conditions the equality sign in inequality 6.5 will appear.

A sequence (finite or infinite) of orthonormal vectors

$$|e_1\rangle, \quad |e_2\rangle, \cdots \tag{6.7}$$

is called a basis* of the space if the only vector orthogonal to all vectors of the sequence 6.7 is the trivial null vector.

This condition is clearly satisfied by a set of N orthonormal vectors in an N-dimensional space; if there existed a vector orthogonal to all N vectors of the set, it would necessarily be linearly independent of all vectors of the set, and one would have $N + 1$ linearly independent vectors in the N-dimensional space, which is self-contradictory.

Given an orthonormal basis in a space with an infinite number of dimensions, one is tempted to write, as in the finite-dimensional case

$$|f\rangle = \sum_{i=1}^{\infty} f^i |e_i\rangle \tag{6.8}$$

with

$$f^i \equiv \langle e_i|f\rangle$$

For the moment, Eq. 6.8 is, however, only a formal expansion without any clear-cut meaning, since the conditions for the summability of an infinite sum of vectors have not yet been specified. The convergence of an infinite sum means that the partial sums tend to some limit. In our case, this limit should be a vector belonging to the same space as the vectors that form the sequence of partial sums, and it should not be surprising that the condition for the completeness of the space will ensure that Eq. 6.8 is a meaningful expression.

We now prove a theorem that holds for any complete space; for definiteness, however, we state it explicitly for the space $L_w^2(a,b)$.

* Or a complete set of vectors. In order to avoid confusion between the concepts of a complete set of vectors and that of a complete space, we prefer to use the word "basis."

Theorem. Assume that there exists an orthonormal basis* $|e_i\rangle (i = 1, 2 \cdots, n, \cdots)$ in $L_w^2(a,b)$. Then, for any $|f\rangle \in L_w^2(a,b)$, the sequence of vectors

$$|f_k\rangle = \sum_{i=1}^{k} f^i |e_i\rangle$$

with

$$f^i \equiv \langle e_i | f\rangle$$

has $|f\rangle$ as the limit vector in the sense that

$$\lim_{k\to\infty} \rho(|f\rangle, |f_k\rangle) = 0$$

Proof. Let us calculate the limit of $\rho(|f_k\rangle, |f_l\rangle)$ when $k,l \to \infty$. Without any loss of generality, we can assume $k > l$. One has

$$\begin{aligned}\rho^2(|f_k\rangle, |f_l\rangle) &\equiv (\langle f_{(k)}| - \langle f_{(l)}|)(|f_{(k)}\rangle - |f_{(l)}\rangle)\\ &= \left\{\sum_{j=l+1}^{k} \langle e_j | \bar{f}^j\right\}\left\{\sum_{i=l+1}^{k} f^i |e_i\rangle\right\}\\ &= \sum_{i=l+1}^{k} |f^i|^2 \end{aligned} \tag{6.9}$$

According to the Bessel inequality, the sum

$$\sum_{i=1}^{\infty} |f^i|^2 \le \langle f | f\rangle < \infty$$

converges, and therefore the Cauchy criterion for the summability of a numerical series gives

$$\lim_{k,l\to\infty} \sum_{i=l+1}^{k} |f^i|^2 = 0$$

or according to Eq. 6.9

$$\lim_{k,l\to\infty} \rho(|f_k\rangle, |f_l\rangle) = 0$$

Therefore, because of the Riesz-Fischer theorem, which states that $L_w^2(a,b)$ is a complete vector space, the series of vectors $|f_k\rangle$ has some unique limit vector $|g\rangle$

$$\lim_{k\to\infty} \rho(|g\rangle, |f_k\rangle) = 0$$

It is not difficult to see that this limit vector $|g\rangle$ is identical to $|f\rangle$. Using the Cauchy-Schwarz inequality, we have

$$\left|\langle g|e_j\rangle - \langle f_k|e_j\rangle\right| \le \rho(|g\rangle, |f_k\rangle)\langle e_j|e_j\rangle \xrightarrow[k\to\infty]{} 0 \tag{6.10}$$

Thus

$$\begin{aligned}\langle g|e_j\rangle &= \lim_{k\to\infty} \langle f_k|e_j\rangle\\ &= \lim_{k\to\infty}\left\{\sum_{i=1}^{k} \langle e_i| \langle f|e_i\rangle\right\}|e_j\rangle\\ &= \langle f|e_j\rangle\end{aligned} \tag{6.11}$$

* Explicit examples of such bases will be given later on in this chapter.

Since the index j is arbitrary, Eq. 6.11 shows that the vector $\{|g\rangle - |f\rangle\}$ is orthogonal to each of the basis vectors and therefore, by the very definition of an orthonormal basis, we must have

$$|g\rangle - |f\rangle = 0$$

or

$$|g\rangle = |f\rangle$$

The numbers

$$f^i \equiv \langle e_i|f\rangle$$

appearing in the expansion

$$|f\rangle = \sum_{i=1}^{\infty} f^i |e_i\rangle \tag{6.12}$$

are called the **Fourier coefficients** of $|f\rangle$ with respect to the basis $|e_1\rangle, |e_2\rangle, \cdots$. The statement

$$\lim_{k\to\infty} \rho^2(|f\rangle, |f_k\rangle) = 0$$

becomes, when the explicit expression for $|f_k\rangle$ is used

$$\lim_{k\to\infty}\left\{\left[\langle f| - \sum_{i=1}^{k}\langle e_i|\bar{f}^i\right]\left[|f\rangle - \sum_{j=1}^{k} f^j|e_j\rangle\right]\right\} = \lim_{k\to\infty}\left\{\langle f|f\rangle - \sum_{i=1}^{k}|f^i|^2\right\} = 0$$

Thus, we see that in the case where the vectors $|e_i\rangle$ form a basis of $L_w^2(a,b)$, one has the equality sign in 6.5 for any $|f\rangle \in L_w^2(a,b)$

$$\langle f|f\rangle = \sum_{i=1}^{\infty} |f^i|^2 \tag{6.13}$$

The converse is also true. When Eq. 6.13, which is called **Parseval's relation**, is satisfied for any $|f\rangle \in L_w^2(a,b)$, the vectors $|e_i\rangle$ $(i = 1, 2, \cdots)$, which have been used to construct the Fourier coefficients f^i, form a basis of $L_w^2(a,b)$. In fact, suppose that a vector $|a\rangle$, $\langle a|a\rangle = 1$, is orthogonal to all vectors $|e_i\rangle$, $(i = 1, 2, \cdots)$, so that the set $|e_i\rangle$ does not form a basis. Then, if we assume that

$$\sum_{i=1}^{\infty} |\langle e_i|f\rangle|^2 = \langle f|f\rangle \qquad \text{for any } |f\rangle \in L_w^2(a,b) \tag{6.14}$$

since $|a\rangle \neq 0$ we get

$$\sum_{i=1}^{\infty} |f^i|^2 + |\langle a|f\rangle|^2 > \langle f|f\rangle$$

for some vector $|f\rangle$, in contradiction with the Bessel inequality, which is valid for any sequence of orthonormal vectors and in particular for the sequence $|a\rangle, |e_1\rangle, |e_2\rangle, \cdots$. Therefore, for $L_w^2(a,b)$, the definition J and Eq. 6.13 are equivalent definitions of an orthonormal basis; this is essentially due to the completeness of $L_w^2(a,b)$.

In the language of analysis the vector equation 6.12 reads

$$f(x) = \sum_{i=1}^{\infty} f^i e_{(i)}(x)$$

where the expansion coefficients are given by the integral formula (6.2). The preceding equality sign merely expresses the fact that both sides of the equation are functions that are equal almost everywhere; i.e., they differ at most on a set of arguments of total measure zero.

The theorem of this section may be reformulated in the following way: Given a set $e_{(i)}(x)$ $(i = 1, 2, \cdots)$ of orthonormal functions representing a set of basis vectors of $L^2_w(a,b)$, any function $f(x)$, for which the Lebesgue integral

$$\int_a^b |f(x)|^2 w(x)\,dx$$

exists and is finite, can be expanded in the infinite sum

$$f(x) = \sum_{i=1}^{\infty} f^i e_{(i)}(x) \tag{6.15}$$

with

$$f^i = \int_a^b \bar{e}_i(x) f(x) w(x)\,dx$$

The equality sign in Eq. 6.15 means that the partial sums $\sum_{i=1}^{k} f^i e_{(i)}(x)$ converge in the mean to $f(x)$

$$\lim_{k\to\infty} \int_a^b |f(x) - \sum_{i=1}^{k} f^i e_{(i)}(x)|^2 w(x)\,dx = 0$$

7 · HILBERT SPACE

In the N-dimensional space, it was possible to associate with any vector a set of N complex numbers, its components

$$|a\rangle \leftrightarrow (a^1, a^2, \cdots, a^n) \tag{7.1}$$

With respect to an orthonormal basis, the scalar product was given by

$$\langle b|a\rangle = \sum_{i=1}^{N} \bar{b}^i a^i \tag{7.2}$$

and the Cauchy-Schwarz inequality became

$$\left|\sum_{i=1}^{N} \bar{b}^i a^i\right| \le \sqrt{\sum_{i=1}^{N} |a^i|^2} \cdot \sqrt{\sum_{i=1}^{N} |b^i|^2}$$

Since the vectors $|a\rangle$ and $|b\rangle$ are completely arbitrary, this relation is valid for any two sets $a^i(i = L, \cdots, L+N)$ and $b^i(i = L, \cdots, L+N)$ of complex numbers

$$\left|\sum_{i=L}^{L+N} \bar{b}^i a^i\right| \le \sqrt{\sum_{i=L}^{L+N} |a^i|^2} \cdot \sqrt{\sum_{i=L}^{L+N} |b^i|^2} \tag{7.3}$$

The most straightforward generalization of the concept of an N-dimensional space consists in introducing an infinite-dimensional space whose vectors, instead of being represented by finite sequences of numbers as in 7.1, are represented by infinite sequences of complex numbers

$$|a\rangle \leftrightarrow (a^1, a^2, \cdots, a^n, \cdots) \tag{7.4}$$

with the condition that the infinite sum

$$\sum_{i=1}^{\infty} |a^i|^2$$

always converges.

By definition we take

$$\begin{aligned} \{|a\rangle + |b\rangle\} &\leftrightarrow (a^1 + b^1, a^2 + b^2, \cdots, a^N + b^N, \cdots) \\ c\,|a\rangle &\leftrightarrow (ca^1, ca^2, \cdots, ca^N, \cdots) \\ \langle a| &\leftrightarrow (\bar{a}^1, \bar{a}^2, \cdots, \bar{a}^N, \cdots) \end{aligned} \tag{7.5}$$

and we treat the numbers of the sequences in brackets as if they were components of the corresponding vectors with respect to an orthonormal basis, the only difference being that their number is infinite. Thus, letting $N \to \infty$, we obtain for the scalar product

$$\langle b|a\rangle = \sum_{i=1}^{\infty} \bar{b}^i a^i \tag{7.6}$$

This equation is meaningful because, on account of 7.3 and the assumed convergence of the series

$$\sum_{i=1}^{\infty} |b^i|^2 \quad \text{and} \quad \sum_{i=1}^{\infty} |a^i|^2$$

the series on the RHS of Eq. 7.6 converges.

Relations 7.5 are also meaningful, since the sums

$$\sum_{i=1}^{\infty} |a^i + b^i|^2 = \sum_{i=1}^{\infty} |a^i|^2 + \sum_{i=1}^{\infty} |b^i|^2 + 2\,\mathrm{Re}\left\{\sum_{i=1}^{\infty} \bar{b}^i a^i\right\}$$

and

$$\sum_{i=1}^{\infty} |ca^i|^2 = |c|^2 \sum_{i=1}^{\infty} |a^i|^2$$

are clearly convergent.

The infinite dimensional space we have just constructed is called a **Hilbert space.**

The discussion of the preceding section has shown that with each vector $|f\rangle$ of the function space $L_w^2(a,b)$, one can associate an infinite series of numbers, its Fourier coefficients f^i. These Fourier coefficients satisfy the convergence condition

$$\sum_{i=1}^{\infty} |f^i|^2 < \infty$$

and therefore determine a vector in Hilbert space. Conversely, each vector of Hilbert space whose components are treated as Fourier coefficients determines some vector in $L_w^2(a,b)$. Thus, there is a one-to-one correspondence between the elements of the Hilbert space and the elements of the function space $L_w^2(a,b)$. One says that the two spaces are **isomorphic.**

8 · THE GENERALIZATION OF THE NOTION OF A BASIS

We have shown (Chapter II, Sec. 11) that any finite set of linearly independent vectors can be orthogonalized by the Schmidt method; by taking suitable linear combinations of these vectors, one can construct an orthonormal set. Continuing the Schmidt orthogonalization procedure, one can orthogonalize an arbitrary enumerable set of linearly independent vectors. Given any sequence

$$|g_1\rangle, \quad |g_2\rangle, \quad \cdots, \quad |g_n\rangle, \quad \cdots \tag{8.1}$$

of linearly independent vectors (i.e., a sequence such that for any n, $|g_{n+1}\rangle$ is always linearly independent of the set of vectors $|g_i\rangle$ $(i \leq n)$), we can construct a sequence of orthonormal vectors

$$|e_1\rangle, \quad |e_2\rangle, \quad \cdots, \quad |e_n\rangle, \quad \cdots \tag{8.2}$$

where $|e_n\rangle$ is a linear combination of all $|g_i\rangle$ for which $i \leq n$.

In general the set 8.2 will not be a basis of $L_w^2(a,b)$. This is completely analogous to the case of an N-dimensional space; in order to get an orthonormal basis for this space, one has to orthogonalize a set of N (and not less!) linearly independent vectors, i.e., a set which itself is a basis of the space. A corresponding condition exists for $L_w^2(a,b)$.

Lemma. The sequence of orthonormal vectors 8.2 obtained by the orthogonalization of the linearly independent vectors 8.1 is a basis of the space if and only if each vector $|f\rangle \in L_w^2(a,b)$ is a limit vector of a sequence of linear combinations of vectors $|g_i\rangle$.

Proof. Suppose first that an arbitrary vector $|f\rangle$ is a limit vector of some sequence of linear combinations of vectors $|g_k\rangle$. Each vector $|e_k\rangle$ is a linear combination of a finite number of $|g_i\rangle$ (those for which $i \leq k$) and, vice versa, each vector $|g_k\rangle$ may also be written as a linear combination of vectors $|e_i\rangle$ (again those with $i \leq k$). Thus, $|f\rangle$ is a limit vector of a sequence of vectors

$$|a_k\rangle = \sum_{i=1}^{k} a^i_{(k)} |e_i\rangle \qquad (k = 1, 2, \cdots, n, \cdots)$$

This means that

$$\lim_{k \to \infty} \rho(|f\rangle, |a_k\rangle) = 0$$

However, as can easily be verified

$$\rho^2(|f\rangle, |a_k\rangle) = \left[\langle f|f\rangle - \sum_{i=1}^{k} \left|\langle e_i|f\rangle\right|^2\right] + \sum_{i=1}^{k} \left|a^i_{(k)} - \langle e_i|f\rangle\right|^2$$

The second term on the RHS above is evidently ≥ 0 and so is the term in brackets, owing to the Bessel inequality, 6.5; therefore, both terms tend separately to zero, and as $k \to \infty$, we get

$$\langle f|f\rangle = \sum_{i=1}^{\infty} \left|\langle e_i|f\rangle\right|^2$$

This proves that the vectors $|e_i\rangle$ $(i = 1, 2, \cdots)$ form a basis of the space.

If the vectors $|e_i\rangle$ $(i = 1, 2, \cdots)$ form a basis of the space, then according to the theorem of Sec. 6, any vector $|f\rangle \in L_w^2(a,b)$ is a limit vector of the sequence

$$|f_k\rangle = \sum_{i=1}^{k} \langle e_i|f\rangle |e_i\rangle \qquad (k = 1, 2, \cdots, n, \cdots)$$

and therefore it may also be regarded as a limit vector of a sequence of linear combinations of the vectors $|g_i\rangle$.

The definition J of Sec. 6 generalized the notion of an orthonormal basis to the case of $L_w^2(a,b)$. We are now in a position to generalize the notion of an arbitrary basis.

In a finite-dimensional space, a basis is a set of vectors such that each vector of the space can be expressed as a linear combination of basis vectors. In $L_w^2(a,b)$ the infinite set of linearly independent vectors 8.1 may be considered to be a basis of the space if an arbitrary vector of $L_w^2(a,b)$ can be expressed as a limit vector of linear combinations of the vectors 8.1.

The basis vectors are represented by functions having the remarkable property that each function whose modulus is square integrable can be arbitrarily well approximated in the mean by linear combinations of these functions. A set of functions that represent a basis in $L_w^2(a,b)$ is called a **complete** set of functions.

Having completed the general discussion, we must now find proper orthogonal functions that can serve for orthogonal expansions.

9 · THE WEIERSTRASS THEOREM

We now prove the important fact that the sequence of vectors

$$|1\rangle, \quad |2\rangle, \quad |3\rangle, \quad \cdots, \quad |n\rangle, \quad \cdots \quad \in L_w^2(a,b) \tag{9.1}$$

defined by the requirement that they are represented by the functions

$$1, \quad x, \quad x^2, \quad x^3, \quad \cdots, \quad x^{n-1}, \quad \cdots$$

forms a basis of $L_w^2(a,b)$.

First we prove a lemma.

Lemma. For any $1 > \delta > 0$ and with

$$A_n(\delta) = \int_\delta^1 (1 - y^2)^n \, dy$$

one has

$$\lim_{n\to\infty} \left\{\frac{A_n(\delta)}{A_n(0)}\right\} = 0$$

Proof. We have

$$A_n(\delta) = \int_\delta^1 (1 - y^2)^n \, dy < (1 - \delta^2)^n(1 - \delta) < (1 - \delta^2)^n$$

and (for $n > 0$)

$$A_n(0) = \int_0^1 (1 - y^2)^n \, dy > \int_0^1 (1 - y)^n \, dy = \frac{1}{n + 1}$$

Thus

$$\lim_{n\to\infty} \left\{\frac{A_n(\delta)}{A_n(0)}\right\} \le \lim_{n\to\infty} \{(n + 1)(1 - \delta^2)^n\} = 0$$

This proves the lemma, since $A_n(\delta)/A_n(0) > 0$.

We can now prove the following theorem due to Weierstrass, on the approximation of continuous functions by polynomials.

Theorem (Weierstrass). Let the function $f(x)$ be continuous on the finite, closed interval $[a,b]$. For any $\varepsilon > 0$, there exists a positive integer n and a corresponding polynomial $p_n(x)$ of the nth degree such that

$$|f(x) - p_n(x)| < \varepsilon \qquad \text{for any } x \in [a,b]$$

In other words, since ε is independent of x, an arbitrary continuous function defined on a closed, finite interval can be approximated uniformly by suitable polynomials.

Proof. The transformation of argument

$$\frac{1}{b-a}[(x-a)(1-\sigma) + (b-x)\sigma] \to x$$

with $\sigma < \frac{1}{2}$ brings the interval $[a,b]$ into the interval $[\sigma, 1-\sigma] \subset [0, 1]$. The function $f(x)$ is then determined for $\sigma \leq x \leq 1 - \sigma$. We complete this function by defining it in the entire interval $[0, 1]$; the values of $f(x)$ for $x < \sigma$ and $x > 1 - \sigma$ may be chosen quite arbitrarily apart from the requirements of boundedness and continuity. For instance, one can put

$$f(x) = \begin{cases} f(\sigma) & 0 \leq x < \sigma \\ f(x) & \sigma \leq x \leq 1 - \sigma \\ f(1-\sigma) & 1 - \sigma < x \leq 1 \end{cases}$$

Without any loss of generality we can also suppose that $f(x)$ is a real function; in the case of a complex $f(x)$, the separate validity of the theorem for $\operatorname{Re} f(x)$ and $\operatorname{Im} f(x)$ will ensure its validity for $f(x)$.

In the interval

$$\sigma \leq x \leq 1 - \sigma \tag{9.2}$$

we consider the following polynomials of degree $2n$

$$p_{(2n)}(x) = \frac{1}{2A_n(0)} \int_0^1 f(z)[1 - (x-z)^2]^n \, dz \tag{9.3}$$

Let ε be an arbitrary positive number. One can write

$$\begin{aligned} \int_0^1 f(z)[1 - (x-z)^2]^n dz &= \int_{-x}^{1-x} f(x+y)(1-y^2)^n \, dy \\ &= \int_{-x}^{-\delta} f(x+y)(1-y^2)^n \, dy + \int_{-\delta}^{+\delta} f(x+y)(1-y^2)^n \, dy \\ &\quad + \int_{+\delta}^{1-x} f(x+y)(1-y^2)^n \, dy \end{aligned} \tag{9.4}$$

where $0 < \delta < 1$ has been chosen in order to have

$$|f(x+y) - f(x)| < \frac{\varepsilon}{2} \tag{9.5}$$

for $|y| < \delta$, which is always possible because f is continuous in the whole interval [0, 1]. Therefore

$$\begin{aligned}\frac{1}{2A_n(0)}\int_{-\delta}^{\delta} f(x+y)(1-y^2)^n\,dy &\equiv \frac{f(x)}{2A_n(0)}\int_{-\delta}^{\delta}(1-y^2)^n\,dy \\ &\quad + \frac{1}{2A_n(0)}\int_{-\delta}^{\delta}[f(x+y)-f(x)](1-y^2)^n\,dy \\ &= f(x) - \frac{A_n(\delta)}{A_n(0)}f(x) \\ &\quad + \frac{1}{2A_n(0)}\int_{-\delta}^{\delta}[f(x+y)-f(x)](1-y^2)^n\,dy \end{aligned} \tag{9.6}$$

According to inequality 9.5, the last term satisfies the inequality

$$\frac{1}{2A_n(0)}\left|\int_{-\delta}^{\delta}[f(x+y)-f(x)](1-y^2)^n\,dy\right| < \frac{1}{2}\varepsilon\left[1-\frac{A_n(\delta)}{A_n(0)}\right] < \frac{1}{2}\varepsilon \tag{9.7}$$

Because $f(x)$ is continuous in a finite interval, it is necessarily bounded, and therefore

$$\begin{aligned}&\frac{1}{2A_n(0)}\left|\left\{\int_{-x}^{-\delta} f(x+y)(1-y^2)^n\,dy + \int_{+\delta}^{1-x} f(x+y)(1-y^2)^n\,dy\right\}\right| \\ &\quad < \frac{\max|f|}{2A_n(0)}\left\{\int_{-x}^{-\delta}(1-y^2)^n\,dy + \int_{+\delta}^{1-x}(1-y^2)^n\,dy\right\} \\ &\quad < \frac{\max|f|}{2A_n(0)}\left\{\int_{-1}^{-\delta}(1-y^2)^n\,dy + \int_{+\delta}^{1}(1-y^2)^n\,dy\right\} \\ &\quad = \max|f|\,\frac{A_n(\delta)}{A_n(0)}\end{aligned} \tag{9.8}$$

Collecting Eq. 9.6 and inequalities 9.7 and 9.8, we obtain

$$|p_{(2n)}(x) - f(x)| < \frac{1}{2}\varepsilon + 2\max|f|\,\frac{A_n(\delta)}{A_n(0)}$$

For sufficiently large n, $A_n(\delta)/A_n(0)$ can be made arbitrarily small according to the preceding lemma. Therefore, one has

$$2\max|f|\,\frac{A_n(\delta)}{A_n(0)} < \frac{\varepsilon}{2}$$

for n large enough and

$$|p_{(2n)}(x) - f(x)| < \varepsilon \qquad \text{for any } x \in [\sigma, 1-\sigma]$$

This proves the theorem.

Hence, any continuous function can be approximated uniformly and with any desired accuracy by linear combinations of the functions

$$1,\quad x,\quad x^2,\quad x^3,\quad \cdots$$

A Generalization of the Weierstrass Theorem

The preceding theorem can be extended to the case of continuous functions of several real variables. Thus, if the function $f(x_1, \cdots, x_s)$ is continuous in the domain

$$a_i \leq x_i \leq b_i \qquad (i = 1, 2, \cdots, s)$$

it can be approximated uniformly with any desired accuracy by linear combinations of monomials

$$x_1^{m_1} \cdot \ x_2^{m_2} \cdot \ \cdots \ x_s^{m_s} \qquad m_i \geq 0$$

The fact that for any continuous function, there exists a sequence of polynomials that converges uniformly to it is a very strong result. It implies, of course, the convergence in the mean, which is a much weaker requirement. Thus, if $f(x)$ represents a vector $|f\rangle \in L_w^2(a,b)$ and if the polynomials $p_{(n)}(x)$ $(n = 0, 1, \cdots)$, represented by the vectors $|p_n\rangle$ $(n = 0, 1, \cdots)$ (which are linear combinations of the first $(n + 1)$ vectors 9.1) converge uniformly to $f(x)$, one has for an $\varepsilon > 0$ and provided n is large enough

$$\int_a^b |f(x) - p_{(n)}(x)|^2 w(x)\, dx < \varepsilon^2 \int_a^b w(x)\, dx$$

Therefore

$$\lim_{n\to\infty} \rho^2(|f\rangle,|p_n\rangle) \equiv \lim_{n\to\infty} \int_a^b |f(x) - p_n(x)|^2 w(x)\, dx$$
$$= 0$$

This means that $\rho(|f\rangle,|p_n\rangle)$ can be made arbitrarily small by the proper choice of $|p_n\rangle$.

As for the vector $|f\rangle$, it can be shown (but we shall skip the proof) that **for any $|f\rangle \in L_w^2(a,b)$, there exists a sequence of vectors that can be represented by continuous functions, and which converges to** $|f\rangle$. The result we stated at the beginning of this section, that vectors $|1\rangle,|2\rangle, \cdots,|n\rangle, \cdots$ form a basis of $L_w^2(a,b)$, follows immediately. For, let the vector $|f\rangle$ be the limit vector of the sequence $|f_1\rangle,|f_2\rangle, \cdots$. Then there exists a positive number ε such that

$$\rho(|f\rangle,|f_n\rangle) \leq \frac{\varepsilon}{2}$$

provided n is sufficiently large. On the other hand, if the vectors $|f_i\rangle$ can be represented by continuous functions, there exists a polynomial of degree m, say, which approximates the function $f_n(x)$ with any desired accuracy; and consequently there exists a vector

$$|p_m\rangle = \sum_{i=1}^{m+1} a^i \,|i\rangle$$

such that

$$\rho(|f_n\rangle,|p_m\rangle) \leq \frac{\varepsilon}{2}$$

Then, owing to the triangle inequality

$$\rho(|f\rangle,|p_m\rangle) \leq \rho(|f\rangle,|f_n\rangle) + \rho(|f_n\rangle,|p_m\rangle) \leq \varepsilon$$

This proves that $|f\rangle$ is a limit vector of a sequence of linear combinations of vectors belonging to the set 9.1.

10 · THE CLASSICAL ORTHOGONAL POLYNOMIALS

10.1 Introductory Remarks

In this section we introduce a class of orthogonal polynomials, the so-called **classical** ones, which are of particular importance in physical applications. In the present chapter we focus our attention on the "algebraic" properties of these functions; in the next chapter, we shall consider them again from a different point of view.

The Schmidt method will not be used to orthogonalize the series of functions

$$1, \quad x, \quad x^2, \quad x^3, \quad \cdots, \quad x^n, \quad \cdots$$

but rather a much more elegant, although less general, approach due essentially to Tricomi.

It is not our aim to give a complete description of the classical polynomials. The detailed enumeration of their properties as well as of the properties of other higher transcendental functions, can be found in the excellent and exhaustive book by Erdelyi *et al.** We feel that it makes no sense to overload this book with numerous particular results that are sure to be forgotten by the reader. We hope only that after a careful study of this book, the reader will be able to understand and to use these particular results, without being terrified by the apparent enormous complexity of the topics. Of course a deep understanding of all the basic problems requires a quite separate study.

10.2 The Generalized Rodrigues Formula

Let us consider the functions

$$C_{(n)}(x) = \frac{1}{w}\frac{d^n}{dx^n}(ws^n) \qquad (n = 0, 1, 2, \cdots) \tag{10.1}$$

where $C_{(1)}(x)$, $w = w(x)$, and $s = s(x)$ satisfy the following conditions:

(i) $C_{(1)}(x)$ is a first degree polynomial in x.

(ii) $s(x)$ is a polynomial in x of degree ≤ 2, with real roots.

(iii) $w(x)$ is real, positive, and integrable in the interval $[a,b]$ and satisfies the boundary conditions

$$w(a)s(a) = w(b)s(b) = 0$$

These three requirements are, as we shall see, very restrictive. In this subsection, we examine the properties of the functions $C_{(n)}(x)$. We show that $C_{(n)}(x)$ is a polynomial in x of the nth degree and that the polynomials $C_{(k)}(x)$ $(k = 0, 1, \cdots)$ form a set of functions that is orthogonal with weight $w(x)$ on the interval $[a,b]$. In the next subsection we determine the class of weight functions $w(x)$, which satisfy the conditions (i) to (iii).

Lemma 1. We denote by the symbol $p_{(\leq k)}(x)$ an arbitrary polynomial in x of degree $\leq k$. The following identity holds

$$\frac{d^m}{dx^m}(ws^n p_{(\leq k)}) \equiv ws^{(n-m)} p_{(\leq k+m)} \tag{10.2}$$

* For the reader's convenience, we follow Erdelyi *et al.* notation and conventions.

Proof. From Eq. 10.1 with $n = 1$, we get

$$s\frac{dw}{dx} = w\left(C_{(1)} - \frac{ds}{dx}\right)$$

Thus

$$\frac{d}{dx}(ws^n p_{(\le k)}) = \frac{dw}{dx}s^n p_{(\le k)}$$

$$+ nws^{n-1}\frac{ds}{dx}p_{(\le k)} + ws^n\frac{dp_{(\le k)}}{dx}$$

$$= ws^{n-1}\left\{\left[C_{(1)}(x) + (n-1)\frac{ds}{dx}\right]p_{(\le k)} + s\frac{dp_{(\le k)}}{dx}\right\}$$

$$\equiv ws^{n-1}p_{(\le k+1)}$$

In the last step, we used (i) and (ii). Repeating the differentiation, we obtain the required identity.

Lemma 2. All the derivatives $(d^m/dx^m)(ws^n)$ with $m < n$ vanish at $x = a$ and $x = b$.

Proof. This result is a direct consequence of the preceding lemma. Putting in 10.2 $k = 0$ and $p_{(\le 0)} = 1$, we get

$$\frac{d^m}{dx^m}(ws^n) = ws^{(n-m)}p_{(\le m)}$$

The RHS of the preceding equation vanishes at $x = a$ and $x = b$ when $n > m$, according to assumption (iii); in the case of an infinite interval, it will be shown in the next subsection that from the conditions (i) to (iii) it follows that ws vanishes at infinity faster than any polynomial.

Theorem. $C_{(n)}(x)$ is a polynomial in x of the nth degree, which is orthogonal on the interval $[a,b]$, with weight $w(x)$, to any polynomial $p_{(m)}(x)$ of degree m less than n

$$\int_a^b p_{(m)}(x)C_{(n)}(x)w(x)\,dx = 0, \qquad (m < n)$$

Proof. We first prove the second part of the theorem. Using Eq. 10.1, integrating by parts n times, and remembering Lemma 2, we obtain for $m < n$

$$\int_a^b p_{(m)}(x)C_{(n)}(x)w(x)\,dx = \int_a^b p_{(m)}\frac{d^n}{dx^n}(ws^n)\,dx$$

$$= \int_a^b ws^n\frac{d^n}{dx^n}(p_{(m)})\,dx = 0 \tag{10.3}$$

The first part of the theorem follows immediately from Lemma 1:

$$C_{(n)}(x) = \frac{1}{w}\frac{d^n}{dx^n}(ws^n) \equiv p_{(\le n)}$$

which means that $C_{(n)}(x)$ is a polynomial of degree $\le n$. We can thus write

$$C_n(x) = p_{(\le n-1)}(x) + a_{(n)}x^n$$

where $a_{(n)}$ is some number that (as we shall show) is not zero. Multiplying both parts of the preceding equation by $C_n(x)w(x)$, integrating, and using Eq. 10.3, we get

$$\int_a^b \left[C_{(n)}(x)\right]^2 w(x)\,dx = \int_a^b p_{(\le n-1)}C_{(n)}(x)w(x)\,dx + a_{(n)}\int_a^b x^n C_n(x)w(x)\,dx$$

$$= a_{(n)}\int_a^b x^n C_{(n)}(x)w(x)\,dx$$

Since the LHS is certainly >0, we must have $a_{(n)} \neq 0$. This proves that $C_{(n)}(x)$ is a polynomial of the nth degree. Incorporating a normalization constant for latter convenience, from Eq. 10.1 we get the so-called **generalized Rodrigues formula**

$$C_{(n)}(x) = \frac{1}{K_n w}\frac{d^n}{dx^n}(ws^n) \tag{10.4}$$

From the foregoing theorem it is found that the sequence of functions

$$C_{(0)}(x),\quad C_{(1)}(x),\quad \cdots,\quad C_{(n)}(x),\quad \cdots$$

forms an orthogonal set of polynomials (on the interval $[a,b]$ and with weight $w(x)$), which can, if one wishes, be normalized by a suitable choice of constants K_n.

10.3 Classification of the Classical Polynomials

The aim of this subsection is to give a classification of the orthogonal polynomials that were just introduced.

The normalization constants K_n remain arbitrary; they will be fixed by convention later on. Thus, we are not concerned for the present about multiplicative numerical factors that can be absorbed into K_n. Notice also that a simultaneous linear transformation of the argument in $w(x)$ and $s(x)$ does not modify either the degree of the corresponding polynomials $C_{(n)}(x)$ or their orthogonality property, since (although it changes the limits a and b of the interval) the conditions (i) to (iii) remain satisfied. For this reason, we choose the argument in such a way that a and b will have some standard values.

According to (i), $C_{(1)}(x)$ is a first-degree polynomial, and we can always, by a suitable choice of the argument x, define it as

$$C_{(1)}(x) = -\frac{x}{K_1} \tag{10.5}$$

Equation 10.5 together with the Rodrigues formula gives

$$\frac{1}{w}\frac{dw}{dx} = -\frac{\left(x + \dfrac{ds}{dx}\right)}{s} \tag{10.6}$$

For $s(x)$, we take successively the zeroth-, first-, and second-degree polynomials, and we examine the possibility of finding a $w(x)$ that satisfies the differential equation 10.6 as well as the boundary condition (iii)

$$w(a)s(a) = w(b)s(b) = 0 \tag{10.7}$$

(a) Let us take

$$s(x) = \alpha$$

Equation 10.6 takes the form

$$\frac{1}{w}\frac{dw}{dx} = -\frac{x}{\alpha}$$

and has the solution

$$w = \text{const } e^{-x^2/2\alpha} \tag{10.8}$$

$s(x)w(x)$ vanishes only at $x = \pm\infty$, provided $\alpha > 0$. To satisfy the conditions 10.7, we have to put

$$a = -\infty, \qquad b = +\infty$$

We make the change of argument $x/\sqrt{2\alpha} \to x$, and we forget about α and the constant in Eq. 10.8, which affect only the multiplicative factor in front of each polynomial. Thus, without loss of generality, we can take

$$s = 1$$

$$w = e^{-x^2}$$

$$a = -\infty, \qquad b = +\infty$$

(b) Let us now take

$$s(x) = \beta(x - \alpha)$$

Equation 10.6 becomes

$$\frac{1}{w}\frac{dw}{dx} = \frac{(x + \beta)}{\beta(x - \alpha)}$$

and has the solution

$$w(x) = \text{const}(x - \alpha)^{\left[-\frac{\alpha+\beta}{\beta}\right]} e^{-\frac{x}{\beta}}$$

If

$$\beta > 0$$

$$\nu \equiv \left[-\frac{\alpha + \beta}{\beta}\right] > -1$$

then $s(x)w(x)$ vanishes at $x = \alpha$ and $x = +\infty$, and $w(x)$ is integrable in the interval $(\alpha, +\infty)$. We therefore identify $(\alpha, +\infty)$ with our interval (a,b). Making the substitution $(x - \alpha)/\beta \to x$, and again forgetting about unimportant multiplicative factors, we obtain

$$s = x$$

$$w(x) = x^\nu e^{-x} \qquad \nu > -1$$

$$a = 0, \qquad b = +\infty$$

(c) Finally, let us take

$$s(x) = \gamma(x - \alpha)(\beta - x), \qquad \beta > \alpha$$

Equation 10.6 now reads as

$$\frac{1}{w}\frac{dw}{dx} = -\left[\frac{x + \gamma(\beta - x) - \gamma(x - \alpha)}{\gamma(x - \alpha)(\beta - x)}\right]$$

and has the solution

$$w = \text{const}(x - \alpha)^{\left[-1 - \frac{\alpha}{\gamma(\beta - \alpha)}\right]}(\beta - x)^{\left[\frac{1-\gamma}{\gamma} - \frac{\alpha}{\gamma(\beta - \alpha)}\right]}$$

If

$$\mu \equiv \left[-1 - \frac{\alpha}{\gamma(\beta - \alpha)}\right] > -1$$

$$\nu \equiv \left[\frac{1 - \gamma}{\gamma} - \frac{\alpha}{\gamma(\beta - \alpha)}\right] > -1$$

then $s(x)w(x)$ vanishes at $x = \alpha$ and $x = \beta$, and $w(x)$ is integrable on the interval $[\alpha,\beta]$, which, of course, we identify with the interval $[a,b]$. With the substitution

$$\frac{2x - \alpha - \beta}{\beta - \alpha} \to x$$

we obtain apart from multiplicative factors

$$s = (1 - x^2)$$

$$w = (1 - x)^{\nu}(1 + x)^{\mu} \qquad \nu,\mu > -1$$

$$a = -1, \qquad b = +1$$

We leave it to the reader as an exercise to verify that when $s(x)$ has a double root, the boundary condition 10.7 cannot be satisfied, since in this case the function $s(x)w(x)$ cannot vanish at more than one point (the root of $s(x)$).

Summarizing, we can say that apart from a trivial linear transformation of the argument, the orthogonal polynomials introduced in Sec. 10.2 can be reduced, up to multiplicative constants, to three types of polynomials, called the **classical polynomials.** These are given in the table below.

TABLE 2

Interval	Weight Function	$s(x)$	Name of the Polynomial
$(-\infty,+\infty)$	e^{-x^2}	1	Hermite, $H_n(x)$
$[0, +\infty)$	$x^{\nu}e^{-x}$ $(\nu > -1)$	x	Laguerre, $L_n^{\nu}(x)$
$[-1, +1]$	$(1 - x)^{\nu}(1 + x)^{\mu}$ $(\nu,\mu > -1)$	$(1 - x^2)$	Jacobi, $P_n^{(\nu,\mu)}(x)$

The polynomials listed in the following table are up to constant factors, particular cases of Jacobi polynomials. For historical reasons and because they play important roles in applications, they have their own names.

TABLE 3

Interval	Weight Function	$s(x)$	Name of the Polynomial
$[-1, +1]$	$(1 - x^2)^{\lambda - 1/2}$ $(\lambda > -\frac{1}{2})$	$(1 - x^2)$	Gegenbauer, $C_n^\lambda(x)$
$[-1, +1]$	1	$(1 - x^2)$	Legendre, $P_n(x)$
$[-1, +1]$	$(1 - x^2)^{-1/2}$	$(1 - x^2)$	Tchebichef of the first kind, $T_n(x)$
$[-1, +1]$	$(1 - x^2)^{1/2}$	$(1 - x^2)$	Tchebichef of the second kind, $U_n(x)$

Strictly speaking, the definition of each of the preceding polynomials contains also the specification of the constants in the corresponding Rodrigues formulae, or as one says, the standardization; this standardization will be specified later on. It is worthwhile to mention that it is not the requirement of orthonormality of the system of polynomials which in general is employed to fix the values of the constants K_n.

At this place we would like to caution the reader that he may find in the literature slightly different definitions of some of the polynomials listed above and, in particular, different standardizations. We shall keep the conventions used by Erdelyi *et al.*

10.4 The Recurrence Relations

We now show that any three consecutive orthogonal polynomials satisfy a functional relation

$$C_{(n+1)}(x) = (A_n x + B_n)C_{(n)}(x) - D_n C_{(n-1)}(x) \tag{10.9}$$

called a recurrence relation. A_n, B_n, and D_n are constants that depend on n only and which are determined by the class of polynomials considered.

The property of satisfying a recurrence relation is shared by all orthogonal polynomials (not only by the classical ones), and therefore the considerations of this subsection are more general than those of the rest of this section. The only property that will be used is the orthogonality relation

$$\int_a^b C_{(n)}(x) p_{(<n)} w(x)\, dx = 0 \tag{10.10}$$

Following Erdelyi *et al.*, the notation indicated below will be used.

$$k_n = \text{coefficient of } x^n \text{ in } C_{(n)}(x)$$

$$k'_n = \text{coefficient of } x^{n-1} \text{ in } C_{(n)}(x)$$

$$h_n = \int_a^b C_{(n)}^2 w\, dx$$

Let us consider the polynomial

$$C_{(n+1)}(x) - \left(\frac{k_{n+1}}{k_n}\right) x C_{(n)}(x) \tag{10.11}$$

which is clearly of degree $\leq n$ and therefore can be written as

$$C_{(n+1)}(x) - \left(\frac{k_{n+1}}{k_n}\right) x C_{(n)}(x) = \sum_{i=0}^{n} a_{(i)}^{(n)} C_{(i)}(x)$$

Multiplying both sides of the preceding equation by $wC_{(m)}$, taking m equal to $0, 1, 2, \cdots, n-2$ successively, and using the orthogonality relation 10.10, we get

$$a^{(n)}_{(m)} = 0 \qquad \text{for } m = 0, 1, \cdots, n-2$$

Thus

$$C_{(n+1)}(x) - \left(\frac{k_{n+1}}{k_n}\right) x C_{(n)}(x) = a^{(n)}_{(n)} \cdot C_{(n)}(x) + a^{(n)}_{(n-1)} \cdot C_{(n-1)}(x) \tag{10.12}$$

This is the recurrence formula we are looking for; there remains to find the constants $a^{(n)}_{(n)}$ and $a^{(n)}_{(n-1)}$.

Notice first that due to the orthogonality relation (Eq. 10.10)

$$h_n = \int_a^b C^2_{(n)} w\, dx \equiv k_n \int_a^b C_{(n)} x^n w\, dx \tag{10.13}$$

Now multiplying Eq. 10.12 by $[wC_{(n-1)}]$ and integrating, we get

$$h_{(n-1)} a^{(n)}_{(n-1)} = -\frac{k_{n+1}}{k_n} \int_a^b C_{(n)} C_{(n-1)} x w\, dx$$

$$= -\frac{k_{n+1}}{k_n} \cdot \frac{k_{n-1}}{k_n} \int_a^b C_{(n)} k_n x^n w\, dx$$

Therefore

$$a^{(n)}_{(n-1)} = -\frac{h_n}{h_{n-1}} \frac{k_{n+1} \cdot k_{n-1}}{k_n^2} \tag{10.14}$$

Comparing the coefficients of x^n in Eq. 10.12, we obtain

$$a^{(n)}_{(n)} = \frac{k'_{n+1}}{k_n} - \frac{k_{n+1} k'_n}{k_n^2} \tag{10.15}$$

Comparing Eq. 10.9 with Eq. 10.12, where the constants $a^{(n)}_{(n)}$ and $a^{(n)}_{(n-1)}$ are given by the above formulae, we find

$$A_n = \frac{k_{n+1}}{k_n}$$

$$B_n = \frac{k_{n+1}}{k_n}\left[\frac{k'_{n+1}}{k_{n+1}} - \frac{k'_n}{k_n}\right] \tag{10.16}$$

$$D_n = \frac{h_n}{h_{n-1}} \frac{k_{n+1} k_{n-1}}{k_n^2}$$

10.5 Differential Equations Satisfied by the Classical Polynomials

We keep our previous notation; $C_{(n)}(x)$ is a classical polynomial, which in the interval $[a,b]$ is orthogonal with weight $w(x)$ to any polynomial of degree $<n$. Since $dC_{(n)}/dx$ is a polynomial of degree $\leq (n-1)$, it follows according to Lemma 1 of Sec. 10.2 that the function

$$\frac{1}{w}\frac{d}{dx}\left[sw\frac{dC_{(n)}}{dx}\right]$$

is a polynomial of degree $\leq n$. Thus, we can write

$$\frac{d}{dx}\left[sw\,\frac{dC_{(n)}}{dx}\right] = -w\sum_{i=1}^{n}\lambda_{(n)}^{(i)}C_{(i)} \tag{10.17}$$

where the $\lambda^{(i)}$ are some numbers. Multiplying both sides of Eq. 10.17 by $C_{(m)}$ and integrating, we get

$$\int_a^b C_{(m)}\frac{d}{dx}\left[sw\,\frac{dC_{(n)}}{dx}\right]dx = -\lambda_{(n)}^{(m)}h_m \tag{10.18}$$

Integrating by parts, the LHS yields for $m < n$

$$\int_a^b C_{(m)}\frac{d}{dx}\left[sw\,\frac{dC_{(n)}}{dx}\right] = -\int_a^b sw\,\frac{dC_{(n)}}{dx}\,\frac{dC_{(m)}}{dx}\,dx$$

$$= \int_a^b wC_{(n)}\left\{\frac{1}{w}\frac{d}{dx}\left[sw\,\frac{dC_{(m)}}{dx}\right]\right\}dx$$

$$= 0$$

We have used the facts that sw vanishes at the ends of the integration interval (assumption (iii), Sec. 10.2) and that $C_{(n)}(x)$ is orthogonal to any polynomial of degree $<n$.

Thus, comparing with Eq. 10.18, we arrive at the result

$$\lambda_{(n)}^{(m)} = 0 \qquad \text{for } m < n$$

Putting for simplicity

$$\lambda_{(n)}^{(n)} \equiv \lambda_n$$

we can rewrite Eq. 10.17 in the form

$$\frac{d}{dx}\left[sw\,\frac{dC_{(n)}}{dx}\right] = -w\lambda_n C_{(n)} \tag{10.19}$$

This is the differential equation satisfied by a classical polynomial $C_{(n)}(x)$.

The constant λ_n can be easily found. Putting $m = n$ in Eq. 10.18, we obtain on the LHS

$$\int_a^b C_{(n)}\frac{d}{dx}\left[sw\,\frac{dC_{(n)}}{dx}\right]dx = \int_a^b C_{(n)}\left[\frac{d(sw)}{dx}\frac{dC_{(n)}}{dx} + sw\,\frac{d^2C_{(n)}}{dx^2}\right]dx$$

$$= \int_a^b wC_{(n)}\left[K_1C_{(1)}\frac{dC_{(n)}}{dx} + s\,\frac{d^2C_{(n)}}{dx^2}\right]dx$$

Because of the orthogonality property of the polynomials $C_{(n)}$ only the nth power of x in the nth degree polynomial in the square brackets contributes to the integral. Therefore, we have

$$\int_a^b C_{(n)}\frac{d}{dx}\left[sw\,\frac{dC_{(n)}}{dx}\right] = \left[nK_1\frac{dC_{(1)}}{dx} + \frac{1}{2}n(n-1)\frac{d^2s}{dx^2}\right]\int_a^b wC_{(n)}(k_nx^n)\,dx$$

$$= \left[nK_1\frac{dC_{(1)}}{dx} + \frac{1}{2}n(n-1)\frac{d^2s}{dx^2}\right]h_n$$

Comparing with Eq. 10.18, we get

$$\lambda_n = -n\left[K_1\frac{dC_{(1)}}{dx} + \frac{1}{2}(n-1)\frac{d^2s}{dx^2}\right]$$

10.6 The Classical Polynomials

In Sec. 10.2 we listed the classical polynomials without, however, specifying the constants K_n entering into the generalized Rodrigues formula. This specification, which amounts to a standardization of the polynomials, will be given below, together with a short (and incomplete) list of their properties, which follow from the previous discussion.

Before going further, we make a few general comments. Once the constants K_n have been fixed by some convention, the constants k_n and k'_n can, in principle, be found from the Rodrigues formula. The constants h_n, which determine the normalization of the polynomials, are given by

$$h_n = \frac{(-1)^n k_n n!}{K_n} \int_a^b s^n w \, dx$$

This follows immediately from the Rodrigues formula if one integrates n times by parts the integral

$$\begin{aligned} h_n &\equiv \int_a^b C_{(n)}^2 w \, dx \\ &= k_n \int_a^b C_{(n)} x^n w \, dx \\ &= \frac{k_n}{K_n} \int_a^b x^n \frac{d^n}{dx^n} (s^n w) \, dx \end{aligned}$$

We shall not give the calculations leading to the particular values of the constants k_n, k'_n, h_n, which are listed below, but will indicate only the results.

Hermite Polynomials $H_n(x)$

Standardization

$$K_n = (-1)^n$$

Constants

$$k_n = 2^n, \quad k'_n = 0, \quad h_n = \sqrt{\pi} 2^n n!$$

Rodrigues formula

$$H_n(x) = (-1)^n e^{x^2} \frac{d^n}{dx^n} (e^{-x^2})$$

Differential equation (Hermite's)

$$\frac{d^2}{dx^2} H_n(x) - 2x \frac{d}{dx} H_n(x) + 2n H_n(x) = 0$$

Recurrence formula

$$H_{n+1}(x) = 2x H_n(x) - 2n H_{n-1}(x)$$

Laguerre Polynomials $L_n^\nu(x)$

Standardization

$$K_n = n!$$

Constants

$$k_n = \frac{(-1)^n}{n!}, \quad k'_n = -\left(\frac{n+\nu}{n}\right)k_n, \quad h_n = \frac{\Gamma(n+\nu+1)}{n!}$$

Rodrigues formula

$$L_n^\nu(x) = \frac{1}{n!} x^{-\nu} e^x \frac{d^n}{dx^n}(e^{-x}x^{\nu+n})$$

Differential equation

$$x \frac{d^2}{dx^2} L_n^\nu(x) + (\nu + 1 - x)\frac{d}{dx} L_n^\nu(x) + nL_n^\nu(x) = 0$$

Recurrence formula

$$(n+1)L_{n+1}^\nu(x) = (2n + \nu + 1 - x)L_n^\nu(x) - (n+\nu)L_{n-1}^\nu(x)$$

Jacobi Polynomials $P_n^{(\nu,\mu)}(x)$

Standardization

$$K_n = (-2)^n n!$$

Constants*

$$k_n = 2^{-n}\binom{2n+\nu+\mu}{n}, \qquad k'_n = \frac{n(\nu-\mu)}{2n+\nu+\mu} k_n$$

$$h_n = \frac{2^{\nu+\mu+1}\Gamma(n+\nu+1)\Gamma(n+\mu+1)}{(2n+\nu+\mu+1)n!\Gamma(n+\nu+\mu+1)}$$

Rodrigues formula

$$P_n^{(\nu,\mu)}(x) = \frac{(-1)^n}{2^n n!}(1-x)^{-\nu}(1+x)^{-\mu}\frac{d^n}{dx^n}\left\{(1-x)^{\nu+n}(1+x)^{\mu+n}\right\}$$

Differential equation

$$(1-x^2)\frac{d^2}{dx^2} P_n^{(\nu,\mu)}(x) + [\mu - \nu - (\nu+\mu+2)x]\frac{d}{dx} P_n^{(\nu,\mu)}(x)$$

$$+ n(n+\nu+\mu+1)P_n^{(\nu,\mu)}(x) = 0$$

---

* The symbol $\binom{y}{x}$ should be read

$$\binom{y}{x} = \frac{\Gamma(y+1)}{\Gamma(x+1)\Gamma(y-x+1)}$$

which for integer y and x reduces to the well-known expression

$$\frac{y!}{x!(y-x)!}$$

Recurrence formula

$$2(n+1)(n+\nu+\mu+1)(2n+\nu+\mu)P_{n+1}^{(\nu,\mu)}(x)$$
$$=(2n+\nu+\mu+1)[(2n+\nu+\mu)(2n+\nu+\mu+2)x+\nu^2-\mu^2]P_n^{(\nu,\mu)}(x)$$
$$-2(n+\nu)(n+\mu)(2n+\nu+\mu+2)P_{n-1}^{(\nu,\mu)}(x)$$

Gegenbauer Polynomials $C_n^\lambda(x)$

Standardization

$$K_n=(-2)^n n!\,\frac{\Gamma(n+\lambda+\frac{1}{2})\Gamma(2\lambda)}{\Gamma(n+2\lambda)\Gamma(\lambda+\frac{1}{2})}$$

Constants

$$k_n=\frac{2^n}{n!}\frac{\Gamma(n+\lambda)}{\Gamma(\lambda)},\qquad k'_n=0$$

$$h_n=\frac{\sqrt{\pi}\,\Gamma(n+2\lambda)\Gamma(\lambda+\frac{1}{2})}{(n+\lambda)n!\Gamma(\lambda)\Gamma(2\lambda)}$$

Rodrigues formula

$$C_n^\lambda(x)=\frac{(-1)^n\Gamma(n+2\lambda)\Gamma(\lambda+\frac{1}{2})}{2^n n!\Gamma(n+\lambda+\frac{1}{2})\Gamma(2\lambda)}(1-x^2)^{-\lambda+1/2}\frac{d^n}{dx^n}[(1-x^2)^{n+\lambda-1/2}]$$

Differential equation

$$(1-x^2)\frac{d^2}{dx^2}C_n^\lambda(x)-(2\lambda+1)x\frac{d}{dx}C_n^\lambda(x)+n(n+2\lambda)C_n^\lambda(x)=0$$

Recurrence formula

$$(n+1)C_{n+1}^\lambda(x)=2(n+\lambda)xC_n^\lambda(x)-(n+2\lambda-1)C_{n-1}^\lambda(x)$$

Legendre Polynomials $P_n(x)$

Standardization

$$K_n=(-2)^n n!$$

Constants

$$k_n=\frac{2^n\Gamma(n+\frac{1}{2})}{n!\Gamma(\frac{1}{2})},\qquad k'_n=0$$

$$h_n=(n+\tfrac{1}{2})^{-1}$$

Rodrigues formula*

$$P_n(x)=\frac{(-1)^n}{2^n n!}\frac{d^n}{dx^n}[(1-x^2)^n]$$

---

* The generalized Rodrigues formula is a generalization of this particular formula, originally derived by Rodrigues.

Differential equation

$$(1 - x^2)\frac{d^2}{dx^2} P_n(x) - 2x \frac{d}{dx} P_n(x) + n(n+1)P_n(x) = 0$$

Recurrence formula

$$(n+1)P_{n+1}(x) = (2n+1)xP_n(x) - nP_{n-1}(x)$$

The first few Legendre polynomials are

$$P_0(x) = 1$$

$$P_1(x) = x$$

$$P_2(x) = \tfrac{1}{2}(3x^2 - 1)$$

$$P_3(x) = \tfrac{1}{2}(5x^3 - 3x)$$

$$P_4(x) = \tfrac{1}{8}(35x^4 - 30x^2 + 3)$$

The Rodrigues formula yields immediately

$$P_n(x) = (-1)^n P_n(-x)$$

Noticing that $P_0(1) = P_1(1) = 1$, from the recurrence formula we obtain by induction

$$P_n(1) = 1$$

Tchebichef Polynomials of the First Kind $T_n(x)$

Standardization

$$K_n = (-1)^n \frac{(2n)!}{2^n n!}$$

Constants

$$k_n = 2^{n-1}, \quad k'_n = 0, \quad h_n = \frac{\pi}{2}$$

Rodrigues formula

$$T_n(x) = \frac{(-1)^n 2^n n!}{(2n)!} (1 - x^2)^{\frac{1}{2}} \frac{d^n}{dx^n} [(1 - x^2)^{n-1/2}]$$

Differential equation

$$(1 - x^2) \frac{d^2}{dx^2} T_n(x) - x \frac{d}{dx} T_n(x) + n^2 T_n(x) = 0$$

Recurrence formula

$$T_{n+1}(x) = 2xT_n(x) - T_{n-1}(x)$$

Tchebichef Polynomials of the Second Kind $U_n(x)$

Standardization

$$K_n = \frac{(-1)^n (2n+1)!}{2^n (n+1)!}$$

Constants

$$k_n = 2^n, \quad k'_n = 0, \quad h_n = \frac{\pi}{2}$$

Rodrigues formula

$$U_n(x) = \frac{(-1)^n 2^n (n+1)!}{(2n+1)!} (1-x^2)^{-\frac{1}{2}} \frac{d^n}{dx^n} [(1-x^2)^{n+1/2}]$$

Differential equation

$$(1-x^2)\frac{d^2}{dx^2} U_n(x) - 3x \frac{d}{dx} U_n(x) + n(n+2)U_n(x) = 0$$

Recurrence formula

$$U_{n+1}(x) = 2xU_n(x) - U_{n-1}(x)$$

10.7 The Expansion of Functions in Series of Orthogonal Polynomials

According to the discussion of Sec. 8, the vectors

$$\frac{1}{\sqrt{h_i}} |C_{(i)}\rangle \in L^2_w(a,b) \qquad (i = 0, 1, 2, \cdots)$$

which are represented by the orthogonal polynomials

$$\frac{1}{\sqrt{h_i}} C_{(i)}(x) \qquad (i = 0, 1, 2, \cdots)$$

form an orthonormal basis in $L^2_w(a,b)$, provided the interval $[a,b]$ is finite. Indeed these vectors can be obtained by the orthogonalization of the vectors (9.1) which, at least in the case of a finite interval $[a,b]$, form a basis of $L^2_w(a,b)$; this follows directly from the Weierstrass theorem. Therefore the expansion

$$|f\rangle = \sum_{i=0}^{\infty} f^i \frac{1}{\sqrt{h_i}} |C_{(i)}\rangle$$

$$f^i = \frac{1}{\sqrt{h_i}} \langle C_{(i)}|f\rangle$$

is valid for any $|f\rangle \in L^2_w(a,b)$ when the interval $[a,b]$ is of finite length. In the language of analysis, one says that the partial sums

$$\sum_{i=0}^{n} f^i \frac{1}{\sqrt{h_i}} C_{(i)}(x)$$

converge in the mean to $f(x)$

$$\lim_{n\to\infty} \int_a^b \left| f(x) - \sum_{i=0}^{n} f^i \frac{1}{\sqrt{h_i}} C_{(i)}(x) \right|^2 w(x)\, dx = 0 \tag{10.20}$$

This result, of course, applies immediately to those classical polynomials that are defined on a finite interval and which can be reduced to within a constant factor, to the Jacobi polynomials and in particular to the Gegenbauer, the Legendre, or the Tchebichef polynomials.

In the proof of the Weierstrass theorem, the assumption about the finiteness of the interval $[a,b]$ was essential. Therefore, in the case of Hermite and Laguerre polynomials, one needs a separate argument to prove the validity of Eq. 10.20; this, in fact, can be done, but we shall omit the proof.

11 · TRIGONOMETRICAL SERIES

11.1 An Orthonormal Basis in $L_1^2(-\pi,\pi)$

Consider a function $f(\theta)$, which is continuous in the closed interval $[-\pi,\pi]$ and which satisfies

$$f(\pi) = f(-\pi) \tag{11.1}$$

Equation 11.1 is called a **periodicity condition.**

We define an auxiliary function $f(x,y)$ of two real variables x and y as

$$f(x,y) = rf(\theta)$$

where r and θ are interpreted as polar coordinates of the point with Cartesian coordinates x and y

$$x = r\cos\theta$$

$$y = r\sin\theta$$

$f(x,y)$ is continuous for

$$-1 \leq x,y \leq 1$$

Therefore, according to the generalized version of the Weierstrass theorem (see Sec. 9), $f(x,y)$ can be uniformly approximated in this domain by functions of the type

$$f_{(n)}(x,y) = \sum_{0\leq m_i,m_j\leq n} a_{(ij)}^{(n)} x^{m_i} y^{m_j}$$

In particular, on the circle of unit radius $r = 1$, we have

$$f_{(n)}(x,y) \equiv f_{(n)}(\theta) = \sum_{0\leq m_i,m_j\leq n} a_{(ij)}^{(n)} \cos^{m_i}\theta \sin^{m_j}\theta \tag{11.2}$$

It is convenient to replace the trigonometric functions by exponential functions of imaginary argument

$$\cos\theta = \frac{1}{2}(e^{i\theta} + e^{-i\theta})$$
$$\sin\theta = \frac{1}{2i}(e^{i\theta} - e^{-i\theta}) \tag{11.3}$$

Inserting Eqs. 11.3 into Eq. 11.2, and rearranging terms, we get

$$f_{(m)}(\theta) = \sum_{m=-n}^{n} a_{(m)}^{(n)} e^{im\theta} \tag{11.4}$$

where the constants $a_{(m)}^{(n)}$ are linear combinations of the constants $a_{(ij)}^{(n)}$.

Summarizing: For any $\varepsilon > 0$, there exists a function of the type 11.4, where n is determined by the magnitude of ε such that

$$|f(\theta) - f_{(n)}(\theta)| < \varepsilon \quad \text{for } \theta \in [-\pi,\pi]$$

provided $f(\theta)$ is continuous and satisfies Eq. 11.1.

An argument quite analogous to the one that followed the proof of the Weierstrass theorem* establishes that the continuous functions

$$e_{(m)}(\theta) = \frac{1}{\sqrt{2\pi}} e^{im\theta} \qquad (m = 0, \pm 1, \pm 2, \cdots) \tag{11.5}$$

represent vectors $|e_m\rangle$, which form a basis in the space $L_w^2(-\pi,\pi)$. Moreover, it is easy to see that these vectors form an orthonormal set with $w(\theta) = 1$

$$\langle e_m|e_n\rangle = \frac{1}{2\pi}\int_{-\pi}^{\pi} e^{-im\theta}e^{in\theta}d\theta = \delta_{nm}$$

Instead of the vectors $|e_m\rangle$, one can use as a basis in $L_1^2(-\pi,\pi)$ certain of their linear combinations

$$|e_0^+\rangle = |e_0\rangle$$

$$|e_m^+\rangle = \frac{1}{\sqrt{2}}[|e_m\rangle + |e_{-m}\rangle] \qquad (m = 1, 2, \cdots)$$

$$|e_m^-\rangle = \frac{1}{\iota\sqrt{2}}[|e_m\rangle - |e_{-m}\rangle] \qquad (m = 1, 2, \cdots)$$

which also have the orthonormality property and which are represented by the trigonometric functions

$$|e_0^+\rangle \rightarrow \frac{1}{\sqrt{2\pi}}$$

$$|e_m^+\rangle \rightarrow \frac{1}{\sqrt{\pi}} \cos m\theta \qquad (m = 1, 2, \cdots) \tag{11.6}$$

$$|e_m^-\rangle \rightarrow \frac{1}{\sqrt{\pi}} \sin m\theta \qquad (m = 1, 2, \cdots)$$

11.2 The Convergence Problem

According to the results obtained in the preceding subsection, any vector $|f\rangle \in L_1^2(-\pi,\pi)$ is a limit vector of a series of vectors

$$|f_n\rangle = a_{(0)}^{(n)+}\,|e_0^+\rangle + \sum_{m=1}^{n} [a_{(m)}^{(n)+}\,|e_m^+\rangle + a_m^{(n)-}\,|e_m^-\rangle] \tag{11.7}$$

---

* The only difference is the requirement $f(\pi) = f(-\pi)$. Notice, however, that any continuous function $f(\theta)$ defined in $[-\pi,\pi]$ can be considered as a limit of functions satisfying this requirement; for instance,

$$f_{(n)}(\theta) = \begin{cases} f(\pi) + n(\theta + \pi)[f(-\pi + 1/n) - f(\pi)] & -\pi \leq \theta \leq -\pi + 1/n \\ f(\theta) \qquad -\pi + 1/n < \theta \leq \pi \end{cases}$$

or equivalently, of vectors

$$|f_n\rangle = \sum_{m=-n}^{n} a_m^{(n)} \, |e_m\rangle$$

We have (compare Sec. 8)

$$\rho^2(|f\rangle,|f_n\rangle) = \left\{\langle f|f\rangle - \sum_{m=-n}^{n} |\langle e_m|f\rangle|^2\right\} + \sum_{m=-n}^{n} |a_{(m)}^{(n)} - \langle e_m|f\rangle|^2 \qquad (11.8)$$

According to Bessel's inequality 6.5, the first term is positive. Therefore, for a fixed n, the distance $\rho(|f\rangle,|f_n\rangle)$ between $|f\rangle$ and $|f_n\rangle$ is minimized when the $a_{(m)}^{(n)}$ are chosen as the Fourier coefficients

$$a_{(m)}^{(n)} = \langle e_m|f\rangle \qquad (m = 0, \pm 1, \pm 2, \cdots, \pm n)$$

In other words, for a fixed n

$$\sum_{m=-n}^{n} \langle e_m|f\rangle e_{(m)}(\theta)$$

is the best approximation in the mean of the function $f(\theta)$ by a linear combination of functions $e_{(m)}(\theta)$ $(m = 0, \pm 1, \cdots, \pm n)$.

Since the first term in Eq. 11.8 tends to zero as $n \to \infty$ (Parseval's equation), one has

$$\lim_{n\to\infty} a_{(m)}^{(n)} = \langle e_m|f\rangle \qquad \text{for any } m \qquad (11.9)$$

In the limit $n \to \infty$, we write

$$|f\rangle = \sum_{m=-\infty}^{\infty} \langle e_m|f\rangle|e_m\rangle \qquad (11.10)$$

as explained in Sec. 6, where the orthogonal expansions of vectors were discussed.

Here we examine under what conditions the vector equation 11.10 implies the functional equation for the corresponding functions. The problem is quite general, but we limit ourselves to trigonometric series. Thus, given a function $f(\theta)$ defined on the interval $[-\pi,\pi]$ and which represents a vector $|f\rangle \in L_1^2(-\pi,\pi)$, can one write:

$$f(\theta) = \sum_{-\infty}^{\infty} \langle e_m|f\rangle \frac{1}{\sqrt{2\pi}} e^{im\theta} \qquad (11.11)$$

or, in other words, does the sum on the RHS converge to $f(\theta)$?

Two facts have been established in the preceding discussion. First, the series on the RHS of Eq. 11.11 converges in the mean to $f(\theta)$

$$\lim_{n\to\infty} \int_{-\pi}^{\pi} \left| f(\theta) - \sum_{-n}^{n} \langle e_m|f\rangle \frac{1}{\sqrt{2\pi}} e^{im\theta}\right|^2 d\theta = 0 \qquad (11.12)$$

Secondly, as follows from the generalized version of the Weierstrass theorem, any continuous function $f(\theta)$ that satisfies the periodicity condition $f(-\pi) = f(\pi)$ can be uniformly approximated by linear combinations of functions $(1/\sqrt{2\pi})e^{im\theta}$

$$\left| f(\theta) - \sum_{m=-n}^{n} a_{(m)}^{(n)} \frac{1}{\sqrt{2\pi}} e^{im\theta}\right| \xrightarrow[n\to\infty]{} 0$$

However, in spite of Eq. 11.9, it is not necessarily true that the expansion 11.11 of $f(\theta)$ or, as one says, its **Fourier series** converges, since it may happen that

$$\lim_{n\to\infty}\left\{\sum_{m=-n}^{n} a_{(m)}^{(n)} \frac{1}{\sqrt{2\pi}} e^{im\theta}\right\} \neq \sum_{m=-\infty}^{\infty}\left\{\lim_{n\to\infty} a_{(m)}^{(n)}\right\} \frac{1}{\sqrt{2\pi}} e^{im\theta}$$

for a set of arguments θ of zero total measure; this would not contradict Eq. 11.12.

One has to add additional conditions on a function to ensure the convergence of its Fourier series. These conditions will be given later in their more general form. First we prove a theorem that is less general but whose proof is elementary.

Theorem 1. Let a function $f(\theta)$ and its derivative be continuous for $-\pi \leq \theta \leq \pi$, and let it satisfy the periodicity condition

$$f(\pi) = f(-\pi)$$

Then the Fourier series

$$\frac{1}{\sqrt{2\pi}} \sum_{-\infty}^{+\infty} f^m e^{im\theta} \tag{11.13}$$

with

$$f^m = \frac{1}{\sqrt{2\pi}} \int_{-\pi}^{\pi} f(\theta) e^{-im\theta}\, d\theta$$

converges uniformly to $f(\theta)$ in the interval $[-\pi,\pi]$.

Proof. $f(\theta)$ and $df/d\theta$ represent vectors $|f\rangle$ and $|df/d\theta\rangle$, which belong to $L_1^2(-\pi,\pi)$. Thus $|df/d\theta\rangle$ can be decomposed

$$\left|\frac{df}{d\theta}\right\rangle = \sum_{-\infty}^{+\infty} \left\langle e_m \middle| \frac{df}{d\theta}\right\rangle \left| e_m \right\rangle$$

with

$$\begin{aligned}
\left\langle e_m \middle| \frac{df}{d\theta}\right\rangle &= \frac{1}{\sqrt{2\pi}} \int_{-\pi}^{\pi} \frac{df}{d\theta} e^{-ir\theta}\, d\theta \\
&= \frac{1}{\sqrt{2\pi}} f(\theta) e^{-im\theta}\Big|_{-\pi}^{\pi} + \frac{im}{\sqrt{2\pi}} \int_{-\pi}^{\pi} f(\theta) e^{-im\theta}\, d\theta \\
&= \frac{im}{\sqrt{2\pi}} \int_{-\pi}^{\pi} f(\theta) e^{-im\theta}\, d\theta \\
&= imf^m
\end{aligned}$$

Parseval's equation for $|df/d\theta\rangle$ reads

$$\begin{aligned}
\left\langle \frac{df}{d\theta} \middle| \frac{df}{d\theta}\right\rangle &= \sum_{m=-\infty}^{\infty} \left|\left\langle e_m \middle| \frac{df}{d\theta}\right\rangle\right|^2 \\
&= \sum_{m=-\infty}^{\infty} \left| mf^m \right|^2 < \infty
\end{aligned} \tag{11.14}$$

According to the Cauchy-Schwarz inequality we have for any partial sum

$$\left|\sum_{k\le|m|\le l} f^m e^{im\theta}\right|^2 \le \sum_{k\le|m|\le l}\left|f^m\right|^2 = \sum_{k\le|m|\le l}\frac{1}{m^2}\left|mf^m\right|^2 \le \sqrt{\sum_{k\le|m|\le l}\left|mf^m\right|^2}\sqrt{\sum_{k\le|m|\le l}\frac{1}{m^2}} \tag{11.15}$$

Thus, on account of Eq. 11.14, the expression on the RHS of inequality 11.15 can be made arbitrarily small when k is large enough; moreover, it does not depend on θ. This proves, by virtue of Cauchy's criterion, that the Fourier series 11.13 converges uniformly.

The individual terms of the Fourier series are continuous functions; therefore, because of the uniform convergence, the series tends to a continuous function that must be identical to $f(\theta)$. In fact, due to the convergence in the mean to $f(\theta)$ of the Fourier series, the two functions could differ at most on a set of total measure zero. However, two continuous functions differing only on a set of arguments of measure zero must clearly be identical.

It has been noted previously that the theorem just proved is the simplest one and that the conditions for the convergence of Fourier series can be appreciably weakened. This can be seen by examining more carefully the proof given above. We did not make use, in fact, of the continuity of $df/d\theta$; we required that $|df/d\theta|^2$ be integrable in the Lebesgue sense (a rather weak condition) and noted that the rule of integration by parts could be used to find the relation between the Fourier coefficients of $f(\theta)$ and $df/d\theta$. (This condition may be shown to mean roughly that $f(\theta)$ does not make very violent oscillations.) Thus, the theorem is valid also in the case when $df/d\theta$, instead of being continuous, makes a finite number of finite jumps.*

Furthermore, the conditions of periodicity and of the continuity of $f(\theta)$ can be abandoned. Provided $df/d\theta$ satisfies the conditions just mentioned and $f(\theta)$ has only discontinuities of the first kind**

$$\lim_{n\to\infty}\left\{\frac{1}{\sqrt{2\pi}}\sum_{m=-n}^{n} f^m e^{im\theta}\right\} = \begin{cases}\frac{1}{2}[f(\theta+0)+f(\theta-0)], & -\pi<\theta<\pi\\ \frac{1}{2}[f(\pi)+f(-\pi)], & \theta=\pm\pi\end{cases}$$

the convergence is uniform in every closed interval where $f(\theta)$ is continuous. The proof of this result† consists in expressing $f(\theta)$ as a sum of two functions, one of which is continuous and the other discontinuous. It turns out that the discontinuous function can always be chosen in such a way that the proof of the theorem for it becomes very easy.

We shall not enter further into the discussion of different sets of conditions that ensure the convergence of Fourier series. We limit ourselves to give, without proof,†† a theorem which holds for a very general class of functions.

---

* This happens, for example, at a point where $f(\theta)$ has a cusp.

** $f(\theta)$ has a discontinuity of the first kind at $\theta=\theta_0$ if both limits $f(\theta_0\pm 0)=\lim_{\varepsilon\to 0} f(\theta_0\pm\varepsilon)$ exist, but $f(\theta_0+0)\neq f(\theta_0-0)$.

† See Courant and Hilbert, *Methods of Mathematical Physics*, Vol. I, Interscience Publishers, Inc., 1953, p. 71.

†† See E. C. Titchmarsh, *Theory of Functions*, Oxford University Press, New York, 1964.

Theorem 2. The Fourier series of a function $f(\theta)$ that is of bounded variation* for $-\pi \leq \theta \leq \pi$ converges to

$$\tfrac{1}{2}[f(\theta+0)+f(\theta-0)] \qquad \text{for } -\pi<\theta<\pi$$

$$\tfrac{1}{2}[f(\pi)+f(-\pi)] \qquad \text{for } \theta=\pm\pi$$

Moreover, in every closed interval where $f(\theta)$ is continuous, the convergence is uniform.

EXAMPLE 1

Consider the function

$$f(\theta)=\begin{cases}0 & 0<\theta\leq\pi\\ 1 & -\pi\leq\theta<0\end{cases}$$

We now find the Fourier expansion of this function. The Fourier coefficients of $f(\theta)$ are

$$\begin{aligned} f^m &= \frac{1}{\sqrt{2\pi}}\int_{-\pi}^{\pi} f(\theta)e^{-im\theta}\,d\theta \\ &= \frac{1}{\sqrt{2\pi}}\int_{-\pi}^{0} e^{-im\theta}\,d\theta \\ &= \left[\frac{i}{m\sqrt{2\pi}}\right][1-(-1)^m] \qquad \text{for } m\neq 0 \end{aligned}$$

$$f^0=\frac{1}{\sqrt{2\pi}}\int_{-\pi}^{0} d\theta=\sqrt{\frac{\pi}{2}}$$

Hence

$$\begin{aligned} f(\theta) &= \frac{1}{2}+\frac{i}{2\pi}\sum_{\substack{m=-\infty\\ m\neq 0}}^{+\infty}\frac{[1-(-1)^m]}{m}e^{im\theta} \\ &= \frac{1}{2}+\frac{i}{2\pi}\sum_{m\text{ odd}>0}\left[\frac{2}{m}e^{im\theta}-\frac{2}{m}e^{-im\theta}\right] \\ &= \frac{1}{2}-\frac{2}{\pi}\sum_{m\text{ odd}>0}\frac{\sin m\theta}{m} \end{aligned} \tag{11.16}$$

At $\theta=0$, which is a point of discontinuity of $f(\theta)$, one gets from Eq. 11.16, $f(0)=\frac{1}{2}$, which is equal to $\frac{1}{2}[f(+0)+f(-0)]$ in agreement with the preceding theorem.

The results of this section can be extended to include functions defined within arbitrary finite intervals. Thus, let $f(\theta)$ satisfy the condition of the preceding theorem in the interval $-l\leq\theta\leq l$. The transformation $(\pi/l)\theta\rightarrow\theta$ brings the interval $[-l,l]$ into $[-\pi,\pi]$, and therefore one has the Fourier expansion

$$\begin{aligned} f(\theta) &= \frac{1}{\sqrt{2l}}\sum_{-\infty}^{\infty} f^m e^{im(\pi/l)\theta} \\ f^m &= \frac{1}{\sqrt{2l}}\int_{-l}^{l} f(\theta)e^{-im(\pi/l)\theta}\,d\theta \end{aligned} \tag{11.17}$$

---

* A function $f(\theta)$ is said to be of **bounded variation** on an interval $[a,b,]$ if it can be written as the sum of two functions $f(\theta)=f_1(\theta)+f_2(\theta)$, where one of the functions is nondecreasing and bounded and the other is nonincreasing and bounded on $[a,b]$. It can be shown that such functions have the property that within $[a,b]$ both limits $f(\theta+0)$ and $f(\theta-0)$ exist.

The expansion 11.17 clearly is valid either for functions defined on $[-l,l]$ or for functions that are periodic with period $2l$: $f(\theta) = f(\theta + 2l)$.

If $f(\theta)$ is an even or an odd function of θ

$$f(\theta) = \pm f(-\theta) \qquad \theta \in [-l,l]$$

the Fourier series (Eq. 11.13) can be simplified. It is easy to verify that in these cases, the Fourier coefficients become

$f(\theta)$ even:

$$f^{(m)} = f^{(-m)} = \frac{2}{\sqrt{2l}} \int_0^l f(\theta) \cos m \frac{\pi}{l} \theta \, d\theta \tag{11.18}$$

$f(\theta)$ odd:

$$f^{(m)} = -f^{(-m)} = \frac{-2i}{\sqrt{2l}} \int_0^l f(\theta) \sin m \frac{\pi}{l} \theta \, d\theta \tag{11.19}$$

Combining the terms $\pm m$ in Eq. 11.13 and using Eqs. 11.18 and 11.19, we obtain

For $f(\theta)$ even

$$\begin{aligned} f(\theta) &= a^0 + \sum_{m=1}^{\infty} a^m \cos m \frac{\pi}{l} \theta \\ a^0 &= \frac{1}{l} \int_0^l f(\theta) \, d\theta \\ a^m &= \frac{2}{l} \int_0^l f(\theta) \cos m \frac{\pi}{l} \theta \, d\theta \qquad (m > 0) \end{aligned} \tag{11.20}$$

For $f(\theta)$ odd

$$\begin{aligned} f(\theta) &= \sum_{m=1}^{\infty} b^m \sin m \frac{\pi}{l} \theta \\ b^m &= \frac{2}{l} \int_0^l f(\theta) \sin m \frac{\pi}{l} \theta \, d\theta \end{aligned} \tag{11.21}$$

Equations 11.20 and 11.21 are called the **Fourier cosine** and **sine series**, respectively. They can be used for the expansion of functions defined on $[0,l]$.

EXAMPLE 2

Consider the function defined on $[0,\pi]$

$$f(\theta) = \theta^2 \qquad 0 \leq \theta \leq \pi$$

Since $f(\theta)$ is an even function of θ, it admits the expansion 11.20 and we have

$$a^{\circ} = \frac{1}{\pi} \int_0^{\pi} \theta^2 \, d\theta = \frac{\pi^2}{3}$$

$$a^m = \frac{2}{\pi} \int_0^{\pi} \theta^2 \cos m\theta \, d\theta = \frac{4}{m^2} (-1)^m$$

Hence

$$\theta^2 = \frac{\pi^2}{3} + 4 \sum_{m=1}^{\infty} \frac{(-1)^m}{m^2} \cos m\theta$$

In the particular case $\theta = 0$, we have

$$\frac{\pi^2}{12} = 1 - \frac{1}{4} + \frac{1}{9} - \frac{1}{16} + \cdots + (-1)^{m-1}\frac{1}{m^2} + \cdots$$

12 · THE FOURIER TRANSFORM

In this section we extend some of the preceding results to functions that are defined in the interval $(-\infty, +\infty)$. This includes, of course, functions that are not periodic.

Let us rewrite the Eqs. 11.17 of the preceding section as

$$f(\theta) = \frac{1}{\sqrt{2\pi}} \sum_{t=-\infty}^{+\infty} F(t)e^{it\theta}\,\delta t$$
$$F(t) = \frac{1}{\sqrt{2\pi}} \int_{-l}^{l} f(\theta)e^{-it\theta}\,d\theta \qquad (12.1)$$

where we have replaced the summation over m by a summation over $t = m(\pi/l)$ and where we have put $\pi/l = \delta t$. In the limit as $l \to \infty$, we might be tempted to write the above sum as an integral provided $|f(\theta)|$ is integrable in the interval $(-\infty, +\infty)$; we would then have

$$f(\theta) = \frac{1}{\sqrt{2\pi}} \int_{-\infty}^{\infty} F(t)e^{it\theta}\,dt \qquad (12.2)$$

$$F(t) = \frac{1}{\sqrt{2\pi}} \int_{-\infty}^{\infty} f(\theta)e^{-it\theta}\,d\theta \qquad (12.3)$$

$F(t)$ is called the **Fourier transform** of $f(\theta)$.*

The validity of the relations 12.2 and 12.3 depends on the correctness of the limiting transition that was performed in order to arrive at these relations. As in the case of Fourier series, some additional conditions have to be imposed on the function $f(\theta)$ to give to Eqs. 12.2 and 12.3 a precise meaning. There exists a wide variety of possible choices for such conditions,** each one having some merit and some disadvantage. We give below a very important theorem on the subject.

Theorem 1. Let $|f(\theta)|$ be integrable in the infinite interval $(-\infty, \infty)$. Then

$$\frac{1}{2}[f(\theta+0) + f(\theta-0)] = \frac{1}{2\pi} \lim_{\Lambda\to\infty} \int_{-\Lambda}^{\Lambda} dt \int_{-\infty}^{\infty} f(r)e^{it(\theta-r)}\,dr \qquad (12.4)$$

provided $f(\theta)$ is of bounded variation in an interval $[a,b]$ including θ. Moreover, if $f(\theta)$ is continuous in this interval, then the integral on the RHS of Eq. 12.4 converges uniformly to $f(\theta)$ in $[a,b]$.

---

* Clearly, if the relations 12.2 and 12.3 are correct, $f(-\theta)$ is a Fourier transform of $F(t)$.

** See, for example, E. C. Titchmarsh, *Introduction to the Theory of Fourier Integrals*, Oxford University Press, New York, 1937.

The relations 12.2 and 12.3 between $f(\theta)$ and its Fourier transform $F(t)$ are quite symmetrical in form. Theorem 1 does not preserve this symmetry; it imposes conditions on $f(\theta)$ that ensure a strong convergence, but the properties of $F(t)$ are left undetermined. In fact, Theorem 1 is an analog of the corresponding theorem for the Fourier series. However, while the function expanded in a Fourier series and the corresponding expansion coefficients are quite different mathematical objects, a function and its Fourier transform are objects of exactly the same type, and therefore their reciprocal relation is of much interest. This relation is established by the following theorem due to Plancherel.

Theorem 2. Let $|f(\theta)|^2$ be integrable in the interval $(-\infty,\infty)$. The integral

$$F(t,\Lambda) = \frac{1}{\sqrt{2\pi}} \int_{-\Lambda}^{\Lambda} f(\theta)e^{-i\theta t}\, d\theta$$

converges in the mean, as $\Lambda \to \infty$, to a function $F(t)$ whose square modulus $|F(t)|^2$ is also integrable in $(-\infty,\infty)$. Furthermore, the integral

$$f(\theta,\Lambda) = \frac{1}{\sqrt{2\pi}} \int_{-\Lambda}^{\Lambda} F(t)e^{i\theta t}\, dt$$

converges in the mean to $f(\theta)$ and

$$\int_{-\infty}^{\infty} |f(\theta)|^2\, d\theta = \int_{-\infty}^{\infty} |F(t)|^2\, dt \tag{12.5}$$

The functions $f(\theta)$ and $F(t)$ play perfectly symmetrical roles in the Plancherel theorem; the convergence ensured is, however, only a convergence in the mean. Notice that Eq. 12.5 is a direct generalization of Parseval's equation.

We omit the proofs of the two theorems of this section. Instead, we introduce in the next section the notion of so-called generalized functions, for which the theory of Fourier transforms becomes particularly simple.*

EXAMPLE

We calculate the Fourier transform of the function

$$f(x) = \frac{1}{1+x^2}$$

From Eq. 12.3 we have

$$F(t) = \frac{1}{\sqrt{2\pi}} \int_{-\infty}^{\infty} \frac{e^{-ixt}}{1+x^2}\, dx$$

This integral can be evaluated by using the calculus of residues and Jordan's lemma (Chap. I, Sec. 22). For $t > 0$, the contour of integration must be closed in the lower half of the complex plane, and for $t < 0$, the contour must be closed in the upper half of the complex plane.

---

* We follow the method of presentation of the subject as given in M. J. Lighthill, *Introduction to Fourier Analysis and Generalized Functions*, Cambridge University Press, New York, 1959.

Jordan's lemma ensures that the contributions from the infinite semicircles vanish in both cases. Thus

$$F(t)=\frac{-1}{\sqrt{2\pi}}\int_{\circlearrowright}\frac{e^{-izt}}{1+z^2}\,dz \qquad \text{for } t>0 \tag{12.6}$$

and

$$F(t)=\frac{1}{\sqrt{2\pi}}\int_{\circlearrowleft}\frac{e^{-izt}}{1+z^2}\,dz \qquad \text{for } t<0 \tag{12.7}$$

The integrand has simple poles at $z=\pm i$; the pole at $z=i$ contributes to Eq. 12.7, and the pole $z=-i$ contributes to Eq. 12.6.

Hence

$$F(t)=\frac{-2\pi i}{\sqrt{2\pi}}\operatorname{Res}\left\{\frac{e^{-izt}}{(z+i)(z-i)}\right\}_{z=-i}$$

$$=\sqrt{\frac{\pi}{2}}\,e^{-t} \qquad (t>0)$$

and

$$F(t)=\frac{2\pi i}{\sqrt{2\pi}}\operatorname{Res}\left\{\frac{e^{-izt}}{(z+i)(z-i)}\right\}_{z=i}$$

$$=\sqrt{\frac{\pi}{2}}\,e^{t} \qquad (t<0)$$

It is easy to verify that Eq. 12.2 holds, for

$$\frac{1}{\sqrt{2\pi}}\int_{-\infty}^{\infty}F(t)e^{ixt}\,dt=\frac{1}{\sqrt{2\pi}}\left\{\int_{-\infty}^{0}\sqrt{\frac{\pi}{2}}\,e^{t+ixt}\,dx+\int_{0}^{\infty}\sqrt{\frac{\pi}{2}}\,e^{-t+ixt}\,dt\right\}$$

$$=\frac{1}{1+x^2}=f(x)$$

Although the Plancherel theorem ensures only a convergence in the mean of the preceding integral to $f(x)$, in this particular case the function is so regular that the integral converges to $f(x)$ in the usual sense.

13 · AN INTRODUCTION TO THE THEORY OF GENERALIZED FUNCTIONS

13.1 Preliminaries

We start with an example in order to give to the reader some intuitive understanding of why the notion of a generalized function may be useful in applications.

The concept of a point charge is commonly used in electrostatics. This can be justified by the fact that at sufficiently large distances, as compared to the dimensions of a charged body, its electric field becomes independent of the detailed structure of the source. It is then practically given by the well-known Coulomb law, which involves only the total charge of the body. Thus, to a distant observer the situation looks as if the whole charge were concentrated at the center of mass of the body, at least if he is not making very precise measurements to detect its multipole moments. The notion of a point charge is an elegant and useful simplification which evolves quite naturally.

More generally, if one wants to get rid of unimportant structural details, it is very useful to introduce the idealized picture of material points bearing a charge, or a gravitational mass, or electric and magnetic multipole moments, etc. However, one faces a formal difficulty, which we illustrate by again referring to the simple electrostatic problem.

A charged body is characterized by its charge distribution, given by some density function $D(x,y,z)$. Integrating over the whole volume V of the source, one gets

$$\int_{V} D(x,y,z)dV = \text{total charge}$$

The point charge picture corresponds to letting $V \to 0$ while keeping the total charge constant. Thus, as $V \to 0$, $D(x,y,z)$ must become more and more peaked, and in the limit of a point charge, $D(x,y,z)$ will vanish everywhere except at one point, where it will be infinite. Obviously the concept of such a "function" goes beyond the framework of conventional analysis.

In general, the currents entering for instance, Maxwell's equations have the singular behavior just described when the material point idealization is used. Of course one can circumvent the difficulty in several ways, but the most elegant (and the most practical!) manner of treating the problem consists in properly generalizing the notion of a function.

Let us examine a simple mathematical example of a limiting process involving all the difficulties we encountered in trying to define the "charge distribution" of a point charge. Consider the functions

$$D_n(x) = \sqrt{\frac{n}{\pi}}\, e^{-nx^2} \qquad n = 1, 2, \cdots$$

As $n \to \infty$, $D_n(x)$ behaves very irregularly

$$\lim_{n\to\infty} D_n(x) = \begin{cases} \infty & x = 0 \\ 0 & x \neq 0 \end{cases}$$

However, according to Gauss' integral formula (Chapter I, Eq. 22.20)

$$\int_{-\infty}^{\infty} D_n(x)\, dx = 1$$

for all n, so that the functions $D_n(x)$ have, as $n \to \infty$, the properties required for the charge distribution in the limit $V \to 0$ and for a point charge located at the origin. For a smooth enough function $f(x)$, the integral

$$\int_{-\infty}^{\infty} D_n(x)f(x)\, dx$$

exists for any n and tends to a well-defined limit as $n \to \infty$

$$\int_{-\infty}^{\infty} D_n(x)f(x)\, dx \underset{n\to\infty}{\longrightarrow} f(0)$$

There is nothing strange in that since the operations of taking a limit and of integrating do not, in general, commute. This is precisely the origin of the trouble in our physical example. One can, however, overcome the difficulty by considering the sequence of "ordinary" functions $D_n(x)$ $(n = 1, 2, \cdots)$ as defining a "generalized function" $\delta(x)$ whose rule of integration is **by definition**

$$\int_{-\infty}^{\infty} \delta(x)f(x)\,dx = \lim_{n\to\infty} \int_{-\infty}^{\infty} D_n(x)f(x)\,dx$$

Notice that now we take advantage of the noncommutation of the limiting and integrating operations; we state that, by convention, the limiting transition is meaningful only **after** the integration has been carried out because it does not imply that the integral of the limit should be meaningful.

Anticipating slightly, we can say that many of the operations defined for ordinary functions can be extended to generalized functions. One simply requires that when such an operation is applied to a sequence of functions that defines a generalized function, it will always yield a sequence of functions that again defines a generalized function. For example, the sequence of derivatives $dD_n(x)/dx$ $(n = 1, 2, \cdots)$ defines the generalized function which we call the derivative $d\delta(x)/dx$ of $\delta(x)$.

The usefulness of generalized functions is that one can handle them in much the same way as ordinary functions. It is true that a generalized function is defined by certain rules which tell us how to manipulate a sequence of functions; this sequence is then treated as a single mathematical entity. This should not shock the reader who has already encountered many examples where mathematical objects have been defined by prescribing rules of composition on certain well-defined sets of simpler objects. Doesn't the introduction of complex numbers simply correspond to defining rules of composition for pairs of real numbers?

13.2 Definition of a Generalized Function

In the preceding subsection we merely sketched the main idea underlying the notion of a generalized function. We now give a more systematic presentation.

First of all, we must determine precisely the class of functions we use to define generalized functions. Following Lighthill, we call good a function satisfying the following condition: A function $g(x)$ that is differentiable everywhere any number of times is called a **good function** if it and its derivatives vanish as $|x| \to \infty$ faster than any power of $1/|x|$.

The notion of a fairly good function is also used: A function $f(x)$ that is differentiable everywhere any number of times is called a **fairly good function** if its modulus and that of its derivatives does not increase faster than some power of $|x|$ as $|x| \to \infty$.

An example of a fairly good function is provided by any polynomial, and the function

$$D_n(x) = \sqrt{\frac{n}{\pi}}\, e^{-nx^2}$$

is an example of a good function. The mth-order derivative of $D_n(x)$ is

$$\frac{d^m}{dx^m} D_n(x) = \frac{(-1)^m}{n^{m/2}} H_m(\sqrt{n}x) D_n(x) \tag{13.1}$$

where $H_m(\sqrt{n}x)$ is the Hermite polynomial of the mth degree (compare with the

Rodriguez formula for Hermite's polynomials). For any m, $(d^m/dx^m)D_n(x)$ is the product of a polynomial and of e^{-nx^2}, and therefore tends to zero faster than any power of $1/|x|$ as $|x| \to \infty$.

We can now give a rigorous definition of generalized functions or **distributions**, as one also calls them.

K One defines a generalized function $\chi(x)$ as a sequence of good functions $h_n(x)$ such that, for any good function $g(x)$, the limit

$$\lim_{n\to\infty} \int_{-\infty}^{\infty} h_n(x)g(x)\,dx \underset{\text{def}}{=} \int_{-\infty}^{\infty} \chi(x)g(x)\,dx$$

exists. Two generalized functions $\alpha(x)$ and $\beta(x)$ are considered to be equal if the corresponding sequences satisfy

$$\lim_{n\to\infty} \int_{-\infty}^{\infty} a_n(x)g(x)\,dx = \lim_{n\to\infty} \int_{-\infty}^{\infty} b_n(x)g(x)\,dx \tag{13.2}$$

for any good function $g(x)$. In other words, two sequences $a_n(x)$, $b_n(x)$ satisfying Eq. 13.2 define the same distribution.

It is not difficult to see that the definition K truly generalizes the notion of a function. By that we mean: For an ordinary function that satisfies not very restrictive integrability conditions, one can construct without much difficulty at least one sequence that fulfills the requirements of the definition. A trivial example is afforded by a good function; the sequence, all of whose members are simply equal to this good function, defines a generalized function. Given a fairly good function $f(x)$, the sequence

$$f_n(x) = f(x)e^{-x^2/n^2} \qquad (n = 1, 2, \cdots) \tag{13.3}$$

defines a distribution, since $f(x)$ behaves at infinity at most as some polynomial and therefore, because of the exponential factor e^{-x^2/n^2}, $f_n(x)$ is a good function.

We also have

$$\lim_{n\to\infty} \int_{-\infty}^{\infty} f_n(x)g(x)\,dx = \int_{-\infty}^{\infty} f(x)g(x)\,dx$$

More generally, it is sufficient for $|f(x)|/(1 + x^2)^N$ to be integrable in $(-\infty, +\infty)$ for some finite N to ensure the possibility of constructing a sequence of good functions $f_n(x)$ satisfying

$$\lim_{n\to\infty} \int_{-\infty}^{\infty} f_n(x)g(x)\,dx = \int_{-\infty}^{\infty} f(x)g(x)\,dx \tag{13.4}$$

for an arbitrary good function $g(x)$. The integral on the RHS of Eq. 13.4 is understood in the ordinary sense. This integral exists because the integrated function does not behave worse than $|f(x)|$ for finite values of its argument, while for $|x| \to \infty$, it vanishes faster than any power of $1/|x|$.* A sequence $f_n(x)$ can be constructed explicitly. For

* One has

$$f(x)g(x) = \frac{f(x)}{(1 + x^2)^N}\,[(1 + x^2)^N g(x)]$$

The first factor vanishes at infinity for some N because otherwise $|f(x)|/(1 + x^2)^N$ would not be integrable. The second factor is clearly a good function.

instance, Lighthill gives the example

$$f_n(x) = \int_{-\infty}^{\infty} f(t)S[n(t-x)]ne^{-t^2/n^2}dt \tag{13.5}$$

where $S(y)$ is a good function that satisfies

$$S(y) = \begin{cases} > 0 & \text{for } -1 < y < 1 \\ 0 & \text{otherwise} \end{cases}$$

and which is normalized to unity

$$\int_{-1}^{1} S(y)\, dy = 1$$

In particular we may take

$$S(y) = \text{const } e^{-1/(1-y^2)} \qquad \text{for } -1 < y < 1$$

Because of the assumed properties of $S(y)$, the integration in Eq. 13.5 extends in fact over the interval $(x - 1/n, x + 1/n)$, i.e., over a neighborhood of x only. $f_n(x)$ is a smooth function, differentiable any number of times; notice that the operation of integration with the smooth kernel that was chosen smudges the eventual discontinuities of $f(x)$. Furthermore, if the integrand in Eq. 13.5 does not vanish, then necessarily

$$|x| - 1 < |t| < |x| + 1$$

so that for arbitrary m

$$\left|\frac{d^m f_n(x)}{dx^m}\right| = \left|\int_{-\infty}^{\infty} f(t)(-n)^m \left\{\frac{d^m S(y)}{dy^m}\right\}_{y=n(t-x)} ne^{-t^2/n^2}\, dt\right|$$

$$\leq n^{m+1} \max \left|\frac{d^m S(y)}{dy^m}\right| e^{-(|x|-1)^2/n^2} [1 + (|x|+1)^2]^N \times \int_{-\infty}^{\infty} \frac{|f(t)|}{(1+t^2)^N}\, dt \xrightarrow[|x|\to\infty]{} 0$$

the convergence being faster than that of any power of $1/|x|$, owing to the factor $e^{-(|x|-1)^2/n^2}$. This establishes the "goodness" of $f_n(x)$. Similarly, one can show that Eq. 13.4 is satisfied.

The fact that a wide class of ordinary functions can be treated as distributions justifies using the same notation for ordinary and generalized functions. For the convenience of the reader, however, we denote in this section generalized functions by Greek letters. Of course it does not make much sense to replace smooth enough functions by sequences of good functions. This would be entirely unnecessary and would not add anything new. However, as we shall see later on, there exist ordinary functions for which certain operations (for example, differentiation) are meaningful only in the generalized sense.

It may happen that at least one of the equivalent sequences defining a generalized function converges uniformly to an ordinary function in some neighborhood of a point $x = x_0$. In this case, $x = x_0$ will be called a **regular point** of the generalized function and the limit of the corresponding sequence at $x = x_0$ will be called the **local value** of the generalized function at this point. For example, the sequence

$$D_n(x) = \sqrt{\frac{n}{\pi}}\, e^{-nx^2} \qquad (n = 1, 2, \cdots)$$

converges uniformly to zero in every interval that does not include the point $x = 0$, and therefore the distribution $\delta(x)$ satisfies $\delta(x) = 0$ locally for any $x \neq 0$.

In general, the equation

$$\chi(x_0) = c$$

will mean that the generalized function $\chi(x)$ is locally equal to c at $x = x_0$. Of course a distribution that has a local value everywhere reduces practically to some continuous function.

If two equivalent sequences converge uniformly in the neighborhood of $x = x_0$, they determine the same local value of the corresponding generalized function. In fact, suppose that in some interval $(x_0 - \varepsilon, x_0 + \varepsilon)$

$$\lim_{n\to\infty} f_n^{(1)}(x) = f^{(1)}(x)$$

$$\lim_{n\to\infty} f_n^{(2)}(x) = f^{(2)}(x)$$

$f^{(1)}(x)$ and $f^{(2)}(x)$ are continuous functions and one can choose ε such that their difference in $(x_0 - \varepsilon, x_0 + \varepsilon)$ has a given sign. Let $S(x_0,x) > 0$ be a good function which vanishes outside of $(x_0 - \varepsilon, x_0 + \varepsilon)$. Then

$$0 = \left|\lim_{n\to\infty}\int_{-\infty}^{\infty} [f_n^{(1)}(x) - f_n^{(2)}(x)]S(x_0,x)\,dx\right|$$

$$= \left|\int_{-\infty}^{\infty} [f^{(1)}(x) - f^{(2)}(x)]S(x_0,x)\,dx\right|$$

$$= \int_{-\infty}^{\infty} |f^{(1)}(x) - f^{(2)}(x)|\; S(x_0,x)\,dx$$

which implies that

$$f^{(1)}(x) = f^{(2)}(x) \qquad \text{for} \quad x \in (x_0 - \varepsilon, x_0 + \varepsilon)$$

13.3 Handling Generalized Functions

The operations defined for generalized functions are similar to the operations defined in conventional analysis. We define them one by one.

(i) *Addition of Generalized Functions*

Let the sequences $a_n(x)$ and $b_n(x)$ define the generalized functions $\alpha(x)$ and $\beta(x)$. The generalized function $\chi(x)$ defined by the sequence

$$h_n(x) = a_n(x) + b_n(x)$$

is called the sum of $\alpha(x)$ and $\beta(x)$

$$\chi(x) = \alpha(x) + \beta(x)$$

The definition is self-consistent, since the sum of two good functions is itself a good function and the integral

$$\int_{-\infty}^{\infty} \chi(x)g(x)\,dx = \lim_{n\to\infty}\int_{-\infty}^{\infty} h_n(x)g(x)\,dx$$

$$= \int_{-\infty}^{\infty} \alpha(x)g(x)\,dx + \int_{-\infty}^{\infty} \beta(x)g(x)\,dx$$

certainly exists for any good function $g(x)$; furthermore, if there exist several equivalent sequences that define $\alpha(x)$ or $\beta(x)$, or both, the integral does not depend on the particular choice of the sequences $a_n(x)$ and $b_n(x)$.

(ii) *The Multiplication of a Generalized Function by a Fairly Good Function*

Let the sequence $a_n(x)$ define the distribution $\alpha(x)$ and let $f(x)$ be a fairly good function. The generalized function $\chi(x)$ defined by the sequence

$$h_n(x) = f(x)a_n(x)$$

is called a product of $f(x)$ and $\alpha(x)$.

$$\chi(x) = f(x)\alpha(x)$$

A product of a good function and of a fairly good function is itself a good function; we leave to the reader the verification that the definition of the product is self-consistent.

In particular one may multiply generalized functions by numbers, since a number is a trivial example of a fairly good function.

Caution. The product of two generalized functions is **not** defined. This is the major difference between the formalism of conventional analysis and that of distribution theory. The difficulty is due to the fact that the convergence of

$$\int_{-\infty}^{\infty} a_n(x)g(x)\,dx \qquad \text{and} \qquad \int_{-\infty}^{\infty} b_n(x)g(x)\,dx$$

as $n \to \infty$, does not imply the convergence of

$$\int_{-\infty}^{\infty} [a_n(x)b_n(x)]g(x)\,dx$$

(iii) *Differentiation of a Generalized Function*

If the sequence $a_n(x)$ defines a distribution $\alpha(x)$, the sequence of derivatives

$$h_n(x) = \frac{d}{dx}\,a_n(x)$$

defines a generalized function $\chi(x)$, where

$$\chi(x) = \frac{d}{dx}\,\alpha(x)$$

The consistency of the definition follows from

$$\int_{-\infty}^{\infty} h_n(x)g(x)\,dx = \int_{-\infty}^{\infty} \frac{d}{dx}\,a_n(x)g(x)\,dx$$
$$= -\int_{-\infty}^{\infty} a_n(x)\frac{dg(x)}{dx}\,dx \tag{13.7}$$

and from the fact that the derivative of a good function is itself a good function. This ensures that the functions $h_n(x)$ are good functions and that the first integral in Eq. 13.7

exists and has the same value for all equivalent sequences $a_n(x)$ defining $\alpha(x)$. Equation 13.7 gives immediately

$$\int_{-\infty}^{\infty} \frac{d\alpha(x)}{dx} g(x)\, dx = -\int_{-\infty}^{\infty} \alpha(x) \frac{dg(x)}{dx}\, dx$$

for any good function $g(x)$.

We see that a generalized function always has a derivative. This is a very important fact. It follows in particular that an ordinary function, treated in the sense of the theory of distributions, always has a derivative.

(iv) *Linear Change of Argument*

If the distribution $\alpha(x)$ is defined by the sequence $a_n(x)$, the distribution $\alpha(ax + b)$ is defined by the sequence $a_n(ax + b)$.

We leave to the reader the verification of the consistency of this definition as well as the verification that

(a) $$\frac{d}{dx}[\alpha(x) + \beta(x)] = \frac{d}{dx}\alpha(x) + \frac{d}{dx}\beta(x)$$

where $\alpha(x)$ and $\beta(x)$ are generalized functions, that

(b) $$\frac{d}{dx}[f(x)\alpha(x)] = \frac{df(x)}{dx}\alpha(x) + f(x)\frac{d\alpha(x)}{dx}$$

where $f(x)$ is a fairly good function and $\alpha(x)$ a distribution, and that

(c) $$\frac{d}{dx}\alpha(ax + b) = a\left\{\frac{d\alpha(y)}{dy}\right\}_{y=ax+b}$$

where $\alpha(y)$ is a distribution.

These definitions allow one to work directly with distributions rather than with sequences of functions.

13.4 The Fourier Transform of a Generalized Function

We first show that the Fourier transform of a good function is itself a good function.

Let $g(x)$ be a good function. Its Fourier transform is given by

$$G(t) = \frac{1}{\sqrt{2\pi}} \int_{-\infty}^{\infty} g(x) e^{-itx}\, dx$$

A derivative of the mth order of $G(t)$ is

$$\frac{d^m G(t)}{dt^m} = \frac{1}{\sqrt{2\pi}} \int_{-\infty}^{\infty} (-ix)^m g(x) e^{-itx}\, dx$$

Integrating by parts n times, one gets

$$\left|\frac{d^m G(t)}{dt^m}\right| = \left|\frac{1}{\sqrt{2\pi}} \frac{1}{(-it)^n} \int_{-\infty}^{\infty} \frac{d^n}{dx^n}[g(x)(-ix)^m] e^{-itx}\, dx\right|$$

$$\leq \frac{1}{\sqrt{2\pi}} \frac{1}{|t|^n} \int_{-\infty}^{\infty} \left|\frac{d^n}{dx^n}[x^m g(x)]\right| dx$$

Since n is completely arbitrary, it can be chosen arbitrarily large. Thus, $|d^mG(t)/dt^m| \to 0$ faster than any power of $1/|t|$, which proves that $G(t)$ is a good function.

We can now prove the following lemma.

Lemma. Let $g(x)$ be a good function. Then

$$G(t) = \frac{1}{\sqrt{2\pi}} \int_{-\infty}^{\infty} g(x)e^{-itx}\, dx$$

implies

$$g(x) = \frac{1}{\sqrt{2\pi}} \int_{-\infty}^{\infty} G(t)e^{itx}\, dt$$

Proof. We introduce an auxiliary function

$$\tilde{g}(y) = \frac{1}{\sqrt{2\pi}} \int_{-\infty}^{\infty} G(t)e^{ity - \frac{1}{2}\varepsilon t^2}\, dt \qquad (\varepsilon > 0)$$

which is a good function because $G(t)$ and $e^{-\varepsilon t^2}$ are good functions and so is their product. Let us compare $g(y)$ and $\tilde{g}(y)$. Using Gauss' integral formula (Chapter I, Eq. 22.20), one gets

$$\begin{aligned}
|\tilde{g}(y) - g(y)| &= \left| \frac{1}{2\pi} \int_{-\infty}^{\infty} dx\, g(x) \int_{-\infty}^{\infty} e^{it(y-x) - \frac{1}{2}\varepsilon t^2}\, dt - \int_{-\infty}^{\infty} \frac{1}{\sqrt{2\pi\varepsilon}} e^{-(y-x)^2/2\varepsilon} g(y)\, dx \right| \\
&= \left| \int_{-\infty}^{\infty} \frac{1}{\sqrt{2\pi\varepsilon}} e^{-(y-x)^2/2\varepsilon} [g(x) - g(y)]\, dx \right| \\
&\leq \left\{ \max \left| \frac{dg}{dy} \right| \right\} \int_{-\infty}^{\infty} \frac{1}{\sqrt{2\pi\varepsilon}} e^{-(y-x)^2/2\varepsilon} |y - x|\, dx \\
&= \sqrt{\frac{2\varepsilon}{\pi}} \left\{ \max \left| \frac{dg}{dy} \right| \right\}
\end{aligned}$$

Thus, $\tilde{g}(y)$ tends uniformly to $g(y)$ as $\varepsilon \to 0$, and therefore

$$\begin{aligned}
g(x) &= \lim_{\varepsilon \to 0} \frac{1}{\sqrt{2\pi}} \int_{-\infty}^{\infty} G(t)e^{itx - \frac{1}{2}\varepsilon t^2}\, dt \\
&= \frac{1}{\sqrt{2\pi}} \int_{-\infty}^{\infty} \lim_{\varepsilon \to 0} \left\{ G(t)e^{itx - \frac{1}{2}\varepsilon t^2}\, dt \right\} \\
&= \frac{1}{\sqrt{2\pi}} \int_{-\infty}^{\infty} G(t)e^{itx}\, dt
\end{aligned}$$

We are now in the position to prove the essential result of this subsection.

Theorem. Let the sequence $f_n(x)$ $(n = 1, 2, \cdots)$ define a generalized function $\varphi(x)$. Then the sequence of Fourier transforms of the sequence $f_n(x)$

$$F_n(t) = \frac{1}{\sqrt{2\pi}} \int_{-\infty}^{\infty} f_n(x)e^{-itx}\, dx$$

defines a generalized function $\Phi(t)$, which is called the Fourier transform of $\varphi(x)$. Furthermore, the Fourier transform of $\Phi(t)$ is $\varphi(-x)$.

Proof. The "goodness" of $F_n(t)$ follows from the goodness of $f_n(x)$. We show that the limit of the integral

$$\int_{-\infty}^{\infty} F_n(t)g(t)\,dt$$

exists for any good function $g(t)$ and does not depend on the choice of the sequence defining $\varphi(x)$. Let

$$g(t) = \frac{1}{\sqrt{2\pi}} \int_{-\infty}^{\infty} G(x)e^{itx}\,dx$$

One has

$$\int_{-\infty}^{\infty} F_n(t)g(t)\,dt = \frac{1}{\sqrt{2\pi}} \int_{-\infty}^{\infty} \int_{-\infty}^{\infty} f_n(x)g(t)e^{-itx}\,dt\,dx \tag{13.8}$$

and also

$$\int_{-\infty}^{\infty} f_n(x)G(x)\,dx = \frac{1}{\sqrt{2\pi}} \int_{-\infty}^{\infty} \int_{-\infty}^{\infty} f_n(x)g(t)e^{-itx}\,dt\,dx \tag{13.9}$$

Since the integrals on the RHS of Eqs. 13.8 and 13.9 exist, we have

$$\int_{-\infty}^{\infty} F_n(t)g(t)\,dt = \int_{-\infty}^{\infty} f_n(x)G(x)\,dx$$

$G(x)$ is a good function, thus the integral on the RHS (and therefore also the integral on the LHS) has a limit that is independent of the choice of the sequence defining $\varphi(x)$. It follows that the sequences $F_n(t)$ $(n = 1, 2, \cdots)$ obtained by taking Fourier transforms of the members of equivalent sequences defining $\varphi(x)$ are also equivalent and define some generalized function $\Phi(t)$.

Moreover, the relationship between $\varphi(x)$ and $\Phi(t)$ is fully symmetrical and therefore, by applying the preceding lemma to every member of a sequence defining $\Phi(t)$, one immediately gets that the Fourier transform of $\Phi(t)$ is $\varphi(-x)$.

We write formally

$$\Phi(t) = \frac{1}{\sqrt{2\pi}} \int_{-\infty}^{\infty} \varphi(x)e^{-itx}\,ax \tag{13.10}$$

for the Fourier transform of $\varphi(x)$. The meaning of the integral is determined by the preceding theorem; inserting instead of the generalized function under the integral, a member of a sequence defining this distribution, one gets a member of a sequence defining its Fourier transform. According to the preceding theorem, the inversion formula is always valid

$$\varphi(x) = \frac{1}{\sqrt{2\pi}} \int_{-\infty}^{\infty} \Phi(t)e^{itx}\,dt \tag{13.11}$$

It is seen that the theory of Fourier transform becomes extremely simple for generalized functions; every generalized function has a Fourier transform which is again a generalized function, and the inversion formula holds without any restrictions.

13.5 The Dirac δ Function

We have already encountered in our examples the sequence

$$D_n(x) = \sqrt{\frac{n}{\pi}}\, e^{-nx^2} \qquad (n = 1, 2, \cdots)$$

It defines a distribution denoted by $\delta(x)$ and called the δ function. The δ function was first introduced by Dirac, on the basis of rather intuitive arguments, long before the theory of distributions was developed by L. Schwartz. The δ function has several remarkable properties, which we list below.

(i) For any good function $g(x)$, one has

$$\int_{-\infty}^{\infty} \delta(x) g(x)\, dx = g(0) \tag{13.12}$$

In fact

$$\begin{aligned} &\left| \int_{-\infty}^{\infty} \sqrt{\frac{n}{\pi}}\, e^{-nx^2} g(x)\, dx - g(0) \right| \\ &= \left| \int_{-\infty}^{\infty} \sqrt{\frac{n}{\pi}}\, e^{-nx^2} [g(x) - g(0)]\, dx \right| \\ &\le \max\left\{ \left| \frac{dg}{dx} \right| \right\} \int_{-\infty}^{\infty} \sqrt{\frac{n}{\pi}}\, |x|\, e^{-nx^2}\, dx \\ &= \frac{1}{\sqrt{n\pi}} \max\left\{ \left| \frac{dg}{dx} \right| \right\} \underset{n\to\infty}{\to} 0 \end{aligned}$$

which proves Eq. 13.12.

(ii) For an arbitrary fairly good function $f(x)$, one has

$$f(x)\, \delta(x) = f(0)\, \delta(x) \tag{13.13}$$

In fact

$$\int_{-\infty}^{\infty} [f(x)\, \delta(x)] g(x)\, dx = f(0) g(0) = \int_{-\infty}^{\infty} [f(0)\, \delta(x)] g(x)\, dx$$

(iii) Consider the following "step function"

$$H(x) = \begin{cases} 1 & x > 0 \\ 0 & x < 0 \end{cases}$$

According to the results of our discussion on the relation between ordinary and generalized functions (Sec. 13.2), we treat $H(x)$ as a distribution $\theta(x)$ with local values 1 and 0 for $x > 0$ and $x < 0$, respectively

$$\theta(x) = \begin{cases} 1 & x > 0 \\ 0 & x < 0 \end{cases}$$

For any good function $g(x)$, one has

$$\int_{-\infty}^{\infty} \frac{d\theta(x)}{dx} g(x)\, dx = -\int_{-\infty}^{\infty} \theta(x) \frac{dg(x)}{dx}\, dx$$

$$= -\int_{-\infty}^{\infty} H(x) \frac{dg(x)}{dx}\, dx$$

$$= -\int_{0}^{\infty} \frac{dg(x)}{dx}\, dx = g(0)$$

Thus

$$\frac{d\theta(x)}{dx} = \delta(x) \tag{13.14}$$

In contrast to the discontinuous function $H(x)$, which is not differentiable at $x = 0$, the generalized function $\theta(x)$ is differentiable everywhere, and this differentiation yields again a generalized function for which there is, however, no analog among ordinary functions.

(iv) The Fourier transform of the δ function is easily found by using Gauss' integral formula.

$$\sqrt{\frac{n}{\pi}}\, e^{-nx^2} = \frac{1}{2\pi} \int_{-\infty}^{\infty} e^{itx - t^2/4n}\, dt$$

Since the sequence $I_n(t) = e^{-t^2/4n}$ defines the generalized function $\xi(t) = 1$,* we can write

$$\delta(x) = \frac{1}{2\pi} \int_{-\infty}^{\infty} e^{itx}\, dt \tag{13.15}$$

The inversion formula yields

$$1 = \int_{-\infty}^{\infty} \delta(x) e^{-itx}\, dx$$

For $t = 0$, this becomes

$$\int_{-\infty}^{\infty} \delta(x)\, dx = 1$$

or more generally, taking account of Eq. 13.13

$$\int_{-\infty}^{\infty} f(x)\, \delta(x)\, dx = f(0) \tag{13.16}$$

where $f(x)$ is any fairly good function.

---

* For any good function $g(x)$, one has

$$\int_{-\infty}^{\infty} \xi(x) g(x)\, dx = \int_{-\infty}^{\infty} g(x)\, dx$$

Since it is only the behavior of the integrand around $x = 0$ that is relevant, Eq. 13.16 may be considered to be valid* for a much wider class of functions $f(x)$, provided these functions are smooth enough at $x = 0$. In fact, it can be shown, by generalizing the theory presented here, that Eq. 13.16 remains valid, provided $f(x)$ is continuous in the neighborhood of $x = 0$.

(v) We now show that

$$\delta(y(x)) = \sum_{i=1} \frac{\delta(x - x_i)}{\left|\frac{dy}{dx}\right|_{x=x_i}} \tag{13.17}$$

where x_i $(i = 1, \cdots, s)$ are the roots of the equation

$$y(x) = 0$$

Since $\delta(x) = 0$, except at $x = 0$, Eq. 13.16 can be written as

$$\int_{-\varepsilon}^{\varepsilon} \delta(x) f(x)\, dx = f(0)$$

One has

$$\int_{-\infty}^{\infty} \delta(y(x)) g(x)\, dx = \sum_{i=1}^{s} \int_{-\varepsilon}^{\varepsilon} \delta(y) g(x(y)) \frac{dy}{\left|\frac{dy}{dx}\right|}$$

$$= \sum_{i=1}^{s} \frac{g(x_i)}{\left|\frac{dy}{dx}\right|_{x=x_i}} \qquad (\varepsilon > 0)$$

This is precisely the result one would obtain by inserting the RHS of Eq. 13.17 under the integral, instead of $\delta(y(x))$.

14 · LINEAR OPERATORS IN INFINITE-DIMENSIONAL SPACES

14.1 Introduction

In the preceding chapter we introduced linear operators quite generally. However, after a brief discussion of their properties, we limited the scope of the discussion to operators defined in finite-dimensional vector spaces. We found that the generalized eigenvectors of any such operator span the space, and in particular we showed that the set of all linearly independent eigenvectors of any Hermitian operator forms a basis of the space.

A general examination of the conditions under which these results can be extended to infinite-dimensional spaces would lead us far beyond the framework of this book. Therefore, in this section, we limit our considerations to a particular class of operators for which the results of Chapter II can be extended without too much difficulty.

* Remember that, strictly speaking, Eq. 13.16 means that replacing $\delta(x)$ under the integral by a member of a sequence defining $\delta(x)$, and integrating, one gets an expression that tends to $f(0)$ as $n \to \infty$

Although from a purely mathematical point of view this will be a rather severe limitation, it turns out that a great many of the operators one encounters in physical applications either belong to, or can be reduced to, ones belonging to this class.

We begin with a general classification of operators in order to be able to define the class of operators we shall deal with. Then we prove several important results concerning the operators that belong to this class, and finally we discuss certain particular operators defined in a function space. This will allow us to apply the general results obtained to those operators that are of particular interest to a physicist. Throughout this section we consider only complete vector spaces.

14.2 Compact Sets

Before entering into a discussion of the properties of linear operators, we need to acquaint the reader with an important property of certain infinite sets.

First we prove the following theorem.

Theorem 1 (Bolzano–Weierstrass). From any infinite set of numbers one can select at least one infinite convergent sequence, provided there exists a common upper bound for the moduli of the numbers belonging to the set.

Proof. The numbers belonging to the set in question can be represented by points in the complex plane. According to the conditions of the theorem, all these points are located within some finite region of the complex plane and consequently can be enclosed within a square (let us denote it by A) of finite area D.

We divide A into four equal squares. Obviously, at least one of these four squares must enclose an infinite number of points belonging to the set; we denote this square by A_1. We now divide A_1 into four equal squares, from which we choose the square that encloses an infinite number of points belonging to the set; we denote this square by A_2. Repeating the argument n times, we obtain a sequence of squares $A, A_1, A_2, \cdots, A_n$ with areas $D, D/4, D/4^2, \cdots, D/4^n$, respectively, each square enclosing an infinite number of points belonging to the set. As $n \to \infty$, there remains a point, common to all squares A_j, which is evidently an accumulation point of the set.

Consider an infinite sequence of circles, centered at this accumulation point. with radii $r, r^2, r^3, \cdots, r^n, \cdots$, where $r < 1$ so that $r^n \to 0$ as $n \to \infty$. Since the radii are centered at an accumulation point of the set, there cannot exist a number N such that for $n > N$ the circle with radius r^n fails to enclose any point belonging to the set. On the contrary, every circle encloses, in fact, an infinite number of these points. Hence, we can construct an infinite sequence of points converging to the accumulation point by taking successively one element of the set out of the annular regions between neighboring circles, except when accidentally such an annular region does not contain any element of the set.

More intuitively, the Bolzano–Weierstrass theorem states that when an infinity of points are enclosed within a finite domain, they must accumulate somewhere within or on the boundary of this domain. This result can be extended without difficulty to abstract, finite-dimensional spaces.

Corollary. From an infinite set of vectors in an N-dimensional space S_N, one can select at least one infinite convergent sequence, provided the lengths of the vectors have a common upper bound.

Proof. We choose an orthonormal basis in S_N. Thus, each vector is represented by a set of N complex numbers. The square of the length of a vector $|a\rangle$ is

$$\langle a|a\rangle = |a^1|^2 + |a^2|^2 + \cdots + |a^N|^2$$

We consider an infinite set $T \subset S_N$ of vectors such that

$$\langle a|a\rangle < b \qquad |a\rangle \in T$$

where b is some real positive number independent of $|a\rangle$. Of course similar inequalities hold for individual components of any vector $|a\rangle \in T$.

$$|a^j|^2 < b \qquad j = 1, 2, 3, \cdots, N$$

In particular, the first component of vectors belonging to T satisfy the conditions of the Bolzano–Weierstrass theorem. Therefore, there exists an infinite sequence T_1 of vectors $|a_j\rangle$ $(j = 1, 2, \cdots)$ and a number c^1 such that as $n \to \infty$

$$|a^1_{(n)} - c^1| \to 0 \qquad |a_n\rangle \in T_1 \subset T$$

Analogously among the vectors belonging to T_1, we select an infinite sequence T_2 of vectors whose second component converges to a number c^2. Repeating the argument N times, we obtain an infinite sequence T_N of vectors whose components converge respectively to the numbers $c^1, c^2, \cdots, c^N$. These numbers, in turn, determine a vector in S_N, which is clearly a limit vector of an infinite sequence of vectors of T. This proves the corollary.

The boundedness of the lengths of an infinite set of vectors in an infinite-dimensional space no longer ensures the existence of a convergent sequence of vectors belonging to the set. For example, take an infinite sequence of orthonormal vectors $|e_j\rangle$.

$$\langle e_i|e_j\rangle = \delta_{ij} \qquad i, j = 1, 2, \cdots$$

The distance between $|e_i\rangle$ and $|e_j\rangle$

$$\rho(|e_i\rangle, |e_j\rangle) = \sqrt{((\langle e_i| - \langle e_j|)(|e_i\rangle - |e_j\rangle))}$$
$$= \begin{cases} \sqrt{2} & i \neq j \\ 0 & i = j \end{cases}$$

is finite for any $i \neq j$, and the infinite sequence $|e_j\rangle$ $(i = 1, 2, \cdots)$ has no limit vector.

An infinite set of vectors having the property that any of its infinite subsets has a limit vector is called a **compact** set. Thus, the above corollary can be reformulated by saying that every infinite, bounded set of vectors in a finite-dimensional space is compact.

14.3 The Norm of a Linear Operator. Bounded Operators

The concept of a linear operator may be regarded as a generalization of the concept of a number. The multiplication of a vector by a number a is equivalent to its multiplication by the operator aE, and the algebra of linear operators is the same as the algebra of numbers except for the commutative law of multiplication which, as we know, is not postulated for operators. We shall now introduce a notion that will apply to operators; it is a generalization of the notion of the absolute value of a number and it has similar formal properties.

Consider a linear vector space S in which a scalar product has been defined. S is then a metric space where the distance is defined as (see Sec. 6, Chapter II)

$$\rho(|1\rangle, |2\rangle) = \sqrt{(\langle 1| - \langle 2|)(|1\rangle - |2\rangle)}$$

We see from the definition that

$$\rho(|1\rangle, |2\rangle) = \rho(|2\rangle, |1\rangle) = \rho(|1\rangle - |2\rangle, 0) \tag{14.1}$$

which has an obvious intuitive interpretation; the LHS is the distance between "points" determined by the vectors $|1\rangle$ and $|2\rangle$, and the RHS is the length of the vector that joins these "points." In fact $\rho(|a\rangle, 0) = \sqrt{\langle a|a\rangle}$.

The distance of the "point" determined by the "radius vector" $|a\rangle$ from the "origin" is the length of the "radius vector." Let A denote a linear operator defined in S. We shall study the properties of those vectors that are obtained by multiplying the vectors of S by the operator A. First of all, we compare the lengths of $A\,|\rangle$ with the length of $|\rangle \in$ S. The upper limit of the ratio of these lengths is called the **norm** of A and is denoted by $\|A\|$

$$\|A\| \underset{\text{def}}{=} \sup \frac{\rho(A\,|\rangle, 0)}{\rho(|\rangle, 0)} \tag{14.2}$$

In Eq. 14.2 "sup" denotes the upper limit of $\rho(A\,|\rangle, 0)/\rho(|\rangle, 0)$ which may or may not reach $\|A\|$ for any $|\rangle$, although it may approach it arbitrarily closely. From the very definition of $\|A\|$, it is evident that for an arbitrary vector $|\rangle \in$ S, one has

$$\rho(A\,|\rangle, 0) \leq \|A\| \cdot \rho(|\rangle, 0) \tag{14.3}$$

Let us observe that $\|A\| = 0$ if and only if $A = 0$, for this means that $\rho(A\,|\rangle, 0) = 0$ and therefore $A\,|\rangle = 0$ for any $|\rangle \in$ S. But this is just the definition of the null operator.

Using the symmetry property of the distance ρ

$$\rho(|1\rangle, |2\rangle) = \rho(|2\rangle, |1\rangle) \tag{14.4}$$

we have

$$\begin{aligned} \rho[(A - B)\,|\rangle, 0] &= \rho(A\,|\rangle, B\,|\rangle) \\ &= \rho(B\,|\rangle, A\,|\rangle) \\ &= \rho[(B - A)\,|\rangle, 0] \end{aligned} \tag{14.5}$$

and therefore

$$\|A - B\| = \|B - A\| \tag{14.6}$$

Finally, if we take three operators A, B, and C, one has by virtue of the triangle inequality (see Sec. 6, Chapter II)

$$\rho(A\,|\rangle, C\,|\rangle) + \rho(C\,|\rangle, B\,|\rangle) \geq \rho(A\,|\rangle, B\,|\rangle) \tag{14.7}$$

Using Eq. 14.1, we can rewrite this inequality as

$$\rho[(A - C)\,|\rangle, 0] + \rho[(C - B)\,|\rangle, 0] \geq \rho[(A - B)\,|\rangle, 0] \tag{14.8}$$

Comparing 14.8 with the definition (Eq. 14.2) of the norm of an operator, we immediately find

$$\|A - C\| + \|C - B\| \geq \|A - B\| \tag{14.9}$$

This is the triangle inequality for the norm of a linear operator.

Comparing the results of the above discussion with the definition E (Sec. 6, Chapter II) of a metric space, we find that a set of linear operators forms a metric space if we consider the norm $\|A - B\|$ of $A - B$ as the "distance" between A and B. This is the origin of the similarity between the norm of an operator and the modulus of a complex number (which is in fact the distance between the point representing the number and the origin). When the action of an operator reduces to the multiplication of $|\rangle$ by a number a, say, by putting $A = aE$ into Eq. 14.2 we have

$$\|aE\| = \sup_{|\rangle \in S} \frac{\sqrt{\langle | \bar{a}a | \rangle}}{\sqrt{\langle | \rangle}} = |a| \tag{14.10}$$

The norm of aE is equal to the modulus of a. However, the analogy between the norm of an operator and the modulus of a number is not complete; for example, the reader may verify that

$$\|A \cdot B\| \leq \|A\| \cdot \|B\| \tag{14.11}$$

The norm of a linear operator defined in an arbitrary linear vector space is not necessarily finite. When $\|A\|$ is finite, then A is called a **bounded** operator. In this section, unless stated otherwise, we consider bounded operators only.

14.4 Sequences of Operators

We have shown that the norm $\|A - B\|$ of the operator $A - B$ has all the formal properties of an abstract distance between A and B, and therefore may be regarded as a measure of the difference between A and B. Hence, it is natural to consider infinite sequences of operators and to speak of the convergence of a given sequence to some operator if the distances between the members of the sequence and this operator (which are pure numbers) tend to zero.

More specifically, we say that the sequence of operators

$$A_1, \quad A_2, \quad \cdots, \quad A_n, \quad \cdots$$

converges to the operator A if, for any $\varepsilon > 0$, there exists a number N such that

$$\|A - A_n\| < \varepsilon \qquad \text{whenever } n > N$$

We now verify that effectively, as $n \to \infty$, the action of A_n on an arbitrary vector $|\rangle$ differs less and less from the action of A on that vector. Compare the two vectors $A|\rangle$ and $A_n|\rangle$. Using Eq. 14.1 and the inequality 14.3, one has

$$\rho(A|\rangle, A_n|\rangle) = \rho[(A - A_n)|\rangle, 0] \leq \|A - A_n\| \, \rho(|\rangle, 0)$$

Hence

$$\|A - A_n\| \to 0$$

implies

$$A_n|\rangle \to A|\rangle$$

14.5 Completely Continuous Linear Operators

Consider an infinite-dimensional linear vector space S_∞ and an orthonormal basis in S_∞:

$$\begin{aligned} &|e_j\rangle \qquad j = 1, 2, \cdots \\ &\langle e_i|e_j\rangle = \delta_{ij} \qquad i, j = 1, 2, \cdots \end{aligned} \tag{14.12}$$

The operators of the form (compare the notation of Sec. 8 in Chapter 2)

$$A_N = \sum_{m,n=M}^{M+N} |e_m\rangle A_n^m(N) \langle e_n| \tag{14.13}$$

are called **finite-dimensional**, for they operate in fact in the N-dimensional subspace spanned by the vectors $|e_M\rangle, |e_{M+1}\rangle, |e_{M+2}\rangle, \cdots, |e_{M+N}\rangle$. The array of numbers $A_n^m(N)$ forms the matrix that represents A_N in this subspace. This can be verified by multiplying Eq. 14.13 from the right by $|e_k\rangle$ ($M \le k \le M+N$) and using Eq. 14.12

$$A_N\,|e_k\rangle = \sum_{m=M}^{M+N} A_k^m(N)\,|e_m\rangle \qquad k = M, M+1, \cdots, M+N \tag{14.14}$$

This is the usual way of defining matrix elements. Notice that each operator defined in a finite-dimensional space can be put in the form given in Eq. 14.13.

The operator A_N^+ adjoint to A_N is given by

$$\begin{aligned} A_N^+ &= \left\{\sum_{m,n=M}^{M+N} |e_m\rangle\, A_n^m(N)\,\langle e_n|\right\}^+ \\ &= \sum_{m,n=M}^{M+N} \{|e_m\rangle[\langle e_n|\, A_n^m(N)]\}^+ \\ &= \sum_{m,n=M}^{M+N} |e_n\rangle\, \bar{A}_n^m(N)\,\langle e_m| \end{aligned} \tag{14.15}$$

We now show that every finite-dimensional operator is bounded. We first calculate the square of the length of the vector $A_N\,|\rangle$. Using the Cauchy-Schwarz inequality (in the particular form 7.3) twice, we get

$$\begin{aligned} \langle|\, A_N^+ A_N|\rangle &= \sum_{klm} \langle|e_l\rangle \bar{A}_l^k(N) A_m^k(N)\langle e_m|\rangle \\ &\le \left\{\sum_k \left|\sum_l \langle|e_l\rangle \bar{A}_l^k(N)\right|^2\right\}^{1/2} \left\{\sum_n \left|\sum_m A_m^n(N)\langle e_m|\rangle\right|^2\right\}^{1/2} \\ &\le \left\{\sum_k \left[\sum_p \left|\langle|e_p\rangle\right|^2\right] \cdot \left[\sum_l \left|A_l^k(N)\right|^2\right]\right\}^{1/2} \cdot \left\{\sum_n \left[\sum_m \left|A_m^n\right|^2\right] \cdot \left[\sum_q \left|\langle e_q|\rangle\right|^2\right]\right\}^{1/2} \end{aligned} \tag{14.16}$$

Notice, however, that

$$\sum_j |\langle|e_j\rangle|^2 = \langle|\rangle \tag{14.17}$$

Hence

$$\langle|A_N^+ A_N|\rangle \le \left(\sum_{k,l} |A_k^l(N)|^2\right)\langle|\rangle \tag{14.18}$$

or finally

$$\|A_N\| = \sup_{|\rangle \in S} \frac{\sqrt{\langle|\, A_N^+ A_N\,|\rangle}}{\sqrt{\langle|\rangle}} \le \sqrt{\sum_{k,l=1}^{N} |A_k^l\,(N)\,|^2} \tag{14.19}$$

Therefore, A_N is bounded.

We are now in a position to define a very important class of linear operators, which are limits of sequences of finite-dimensional operators:

L An operator A is said to be **completely continuous** if there exists a sequence of finite-dimensional operators that converges to A.

Obviously, every finite-dimensional operator is completely continuous. If a completely continuous operator A is not finite-dimensional, then by the definition L, there exists an infinite sequence $A_1, A_2, \cdots, A_n$ of finite-dimensional operators such that

$$\lim_{n \to \infty} \|A - A_n\| = 0 \tag{14.20}$$

One can easily show that a completely continuous operator is bounded. The triangle inequality 14.9 yields

$$\|A - A_n\| + \|A_n\| \geq \|(A - A_n) + A_n\| = \|A\| \tag{14.21}$$

It is therefore sufficient (because of Eq. 14.20) to prove that $\lim_{n \to \infty} \|A_n\|$ exists. To this end, notice that for m,n large enough, we have

$$\begin{aligned} \|A - A_n\| &< \varepsilon/2 \\ \|A - A_m\| &< \varepsilon/2 \end{aligned} \tag{14.22}$$

Therefore

$$\begin{aligned} \|A_m - A_n\| &\leq \|A_n - A\| + \|A - A_m\| \\ &= \|A - A_n\| + \|A - A_m\| < \varepsilon \end{aligned} \tag{14.23}$$

On the other hand, the triangle inequality 14.7 leads also to

$$\begin{aligned} \|A_m - A_n\| &\geq \|A_m\| - \|A_n\| \\ \|A_m - A_n\| &\geq \|A_n\| - \|A_m\| \end{aligned} \tag{14.24}$$

Combining inequalities 14.23 and 14.24, we obtain

$$\big| \|A_m\| - \|A_n\| \big| < \varepsilon \tag{14.25}$$

for m and n large enough. Since the norm of an operator is a pure number, this implies, by virtue of the Cauchy criterion, that the numerical sequence $\|A_1\|, \|A_2\|, \cdots, \|A_n\|, \cdots$ converges. Hence, $\|A\|$ is finite, and this proves the boundedness of every completely continuous operator.

To close this subsection, we prove a very important property of completely continuous operators, which, in fact, is frequently used as their defining property.

Theorem 2. Let A be a completely continuous linear operator. Then the set of all vectors $A|\rangle$, where $|\rangle$ satisfies

$$\rho(|\rangle, 0) < \text{const} \tag{14.26}$$

is compact.

In other words, from any infinite set of vectors $A|\rangle$ with $|\rangle$ satisfying 14.26, one can select a convergent sequence.

Proof. The proof is immediate if A is finite-dimensional. In this case $A|\rangle$ belongs to a finite-dimensional subspace, and since A is bounded, $\rho(A|\rangle, 0)$ is also bounded (see inequality 14.3). Hence, the theorem follows from the corollary we proved in Sec. 14.3.

Now take an arbitrary, completely continuous operator. It is a limit of a sequence

$$A_1, \quad A_2, \quad \cdots, \quad A_n, \quad \cdots \tag{14.27}$$

of finite dimensional operators. Let

$$|1\rangle, \quad |2\rangle, \quad \cdots, \quad |k\rangle, \quad \cdots$$

be an arbitrary infinite sequence of vectors satisfying 14.26. We select from the sequence $A_1\,|k\rangle$ $(k = 1, 2, \cdots)$ a convergent infinite sequence $A_1\,|k,1\rangle$ $(k = 1, 2, \cdots)$. Then we select from the sequence $A_2\,|k,1\rangle$ $(k = 1, 2, \cdots)$ a convergent infinite sequence $A_2\,|k,2\rangle$ $(k = 1, 2, \cdots)$. We repeat indefinitely the process, always selecting from the sequence $A_{l+1}\,|k,l\rangle$ $(k = 1, 2, \cdots)$ an infinite convergent sequence $A_{l+1}\,|k,l+1\rangle$ $(k = 1, 2, \cdots)$. Consider now the infinite sequence

$$|1, 1\rangle, \quad |2, 2\rangle, \quad \cdots, \quad |l-1, l-1\rangle, \quad |l,l\rangle, \quad |l+1, l+1\rangle, \quad \cdots \tag{14.28}$$

Multiplying each vector of the sequence by any one of the operators of the sequence 14.27 (A_n say), we get a convergent sequence, since all the vectors in the sequence 14.28 have been selected out of the sequence $|k,n\rangle$ $(k = 1, 2, \cdots)$ with the possible exception of $|1, 1\rangle, |2, 2\rangle, \cdots, |n-1, n-1\rangle$.

For n large enough, we have

$$\|A - A_n\| < \varepsilon$$

On the other hand, the convergence of the sequence $A_n\,|1, 1\rangle, A_n\,|2, 2\rangle, \cdots$ means that for p and q large enough, one must have

$$\rho(A_n\,|p,p\rangle, A_n\,|q,q\rangle) < \varepsilon$$

Hence, using the triangle inequality and inequality 14.3, we get

$$\begin{aligned} \rho(A|p,p\rangle, A|q,q\rangle) &= \rho[A(|p,p\rangle - |q,q\rangle), 0] \\ &\leq \rho[A_n(|p,p\rangle - |q,q\rangle), 0] \\ &\quad + \rho[(A - A_n)(|p,p\rangle - |q,q\rangle), 0] \\ &\leq \varepsilon + \|A - A_n\|(|p,p\rangle - |q,q\rangle, 0) \\ &= \varepsilon[1 + \rho(|p,p\rangle, |q,q)]\rangle \end{aligned}$$

But since the vectors $|p,p\rangle$ and $|q,q\rangle$ satisfy 14.26, $\rho(|p,p\rangle, |q,q\rangle) \leq \rho(|p,p\rangle, 0) + \rho(|q,q\rangle, 0) < 2(\text{const})$, it follows that $\rho(A\,|p,p\rangle, A\,|q,q\rangle)$ can be made arbitrarily small for p and q sufficiently large. Since in this section we consider only complete vector spaces, this proves that the sequence $A\,|1, 1\rangle, A\,|2, 2\rangle, \cdots, A\,|n,n\rangle, \cdots$ is convergent. The theorem is proved.

14.6 The Fundamental Theorem on Completely Continuous Hermitian Operators

Theorem 3. Let H be a Hermitian, completely continuous operator. Then:

(i) There exists at least one eigenvector of H with a nonzero eigenvalue.

(ii) For an arbitrary $\varepsilon > 0$, there can exist only a finite number of mutually orthogonal and normalized eigenvectors of H with eigenvalues that lie outside the interval $[-\varepsilon,\varepsilon]$. In particular, to a given nonzero eigenvalue of H, there can correspond only a finite number of eigenvectors.

(iii) The eigenvectors of H span the space.

Proof. We shall prove successively the three parts of Theorem 3.

(i) Since the operator H is completely continuous, it must also be bounded, as we have shown in the preceding subsection. Let us take an infinite sequence of vectors $|h_n\rangle = H|n\rangle$ ($n = 1, 2, \cdots$) with $|n\rangle$ normalized to unity.

$$\langle n|n\rangle = 1 \qquad n = 1, 2, \cdots \tag{14.29}$$

We impose the following condition on the vectors $|h_n\rangle$ ($n = 1, 2, \cdots$)

$$\lim_{n\to\infty} \rho(|h_n\rangle, 0) = \|H\| \tag{14.30}$$

By virtue of the theorem of the preceding subsection, $|n\rangle$ can be so chosen that the sequence $|h_n\rangle$ converges; we denote by $|h\rangle$ the corresponding limit vector

$$\lim_{n\to\infty} \rho(|h_n\rangle, |h\rangle) = 0 \tag{14.31}$$

From the triangle inequalities

$$\rho(|h_n\rangle, |h\rangle) + \rho(|h\rangle, 0) \geq \rho(|h_n\rangle, 0)$$

$$\rho(|h_n\rangle, |h\rangle) + \rho(|h_n\rangle, 0) \geq \rho(|h\rangle, 0)$$

we obtain

$$\rho(|h_n\rangle, |h\rangle) \geq |\rho(|h\rangle, 0) - \rho(|h_n\rangle, 0)| \tag{14.32}$$

and because of Eqs. 14.30 and 14.31

$$\rho(|h\rangle, 0) = \lim_{n\to\infty} \rho(|h_n\rangle, 0) = \|H\| \tag{14.33}$$

Equation 14.33 together with inequality 14.3 yields

$$\rho(H|h\rangle, 0) \leq \|H\|^2 \tag{14.34}$$

On the other hand, again using inequality 14.3

$$\begin{aligned}\rho(H|h_n\rangle, H|h\rangle) &= \rho[H(|h_n\rangle - |h\rangle), 0] \\ &\leq \|H\rho\|(|h_n\rangle - |h\rangle, 0) \\ &= \|H\|\rho(|h_n\rangle, |h\rangle)\end{aligned} \tag{14.35}$$

Therefore, from Eq. 14.31 it follows that for n large enough and for an arbitrary $\varepsilon > 0$

$$\rho(H\,|h_n\rangle, H\,|h\rangle) < \varepsilon \tag{14.36}$$

The triangle inequality

$$\rho(H|h_n\rangle, H|h\rangle) + \rho(H|h\rangle, 0) \geq \rho(H|h_n\rangle, 0)$$

together with inequality 14.36 gives

$$\rho(H|h\rangle, 0) \geq \rho(H|h_n\rangle, 0) - \varepsilon \tag{14.37}$$

But since H is Hermitian, we have

$$\rho(H|h_n\rangle, 0) = \sqrt{\langle h_n|\, H^2\, |h_n\rangle} = \sqrt{\langle n|\, H^4\, |n\rangle} \tag{14.38}$$

Using the Cauchy-Schwarz inequality, we obtain

$$\langle n|\, H^2\, |n\rangle \leq \sqrt{\langle n|n\rangle}\sqrt{\langle n|\, H^4\, |n\rangle} = \sqrt{\langle n|\, H^4\, |n\rangle} \tag{14.39}$$

where we have taken account of the normalization (Eq. 14.29) of the $|n\rangle$.

Collecting 14.37, 14.38, and 14.39, we obtain

$$\rho(H\,|h\rangle, 0) \geq \langle n|H^2\,|n\rangle - \varepsilon = \rho^2(|h_n\rangle, 0) - \varepsilon \tag{14.40}$$

Letting $\varepsilon \to 0$, $n \to \infty$, this leads to

$$\rho(H\,|h\rangle, 0) \geq \|H\|^2 \tag{14.41}$$

Comparing with inequality 14.34, we see that

$$\rho(H\,|h\rangle, 0) = \|H\|^2 \tag{14.42}$$

It is now easy to prove that $|h\rangle$ is an eigenvector of H^2.

Notice first that the Cauchy-Schwarz inequality

$$\langle h|\,H^2\,|h\rangle \leq \sqrt{\langle h|h\rangle}\,\sqrt{\langle h|\,H^4\,|h\rangle} \tag{14.43}$$

may be rewritten as

$$\rho^2(H\,|h\rangle, 0) \leq \rho(|h\rangle, 0) \cdot \rho(H^2\,|h\rangle, 0) \tag{14.44}$$

Using 14.3, 14.33, and 14.42, we have

$$\begin{aligned}\rho(|h\rangle, 0)\rho(H^2\,|h\rangle, 0) &\leq \rho(|h\rangle, 0) \cdot \|H\| \cdot \rho(H\,|h\rangle\,,0)\\ &= \|H\|^2\rho(H\,|h\rangle, 0) = \rho^2(H|h\rangle, 0)\end{aligned}$$

Hence, 14.44 is in fact not an inequality but an equation, which in turn implies that

$$\langle h|\,H^2\,|h\rangle = \sqrt{\langle h|h\rangle}\,\sqrt{\langle h|\,H^4\,|h\rangle}$$

or

$$(\langle h|\,H^2\,|h\rangle)^2 = \langle h|h\rangle\,\langle h|\,H^4\,|h\rangle \tag{14.45}$$

Now we know that the length of a vector is nonzero unless the vector is a null vector. Therefore, the equation

$$(\langle h|\,H^2 - \langle h|\,\lambda)(H^2\,|h\rangle - \lambda\,|h\rangle) = 0 \tag{14.46}$$

with real λ can have a solution with respect to λ if and only if

$$H^2\,|h\rangle - \lambda\,|h\rangle = 0 \tag{14.47}$$

However, Eq. 14.46 is equivalent to

$$\lambda^2\langle h|h\rangle - 2\lambda\,\langle h|\,H^2\,|h\rangle + \langle h|\,H^4\,|h\rangle = 0 \tag{14.48}$$

Equation 14.45 guarantees that a solution of Eq. 14.48 with respect to λ exists and, moreover, that it is unique. Since Eq. 14.42 can also be written as

$$\sqrt{\langle h|\,H^2\,|h\rangle} = \|H\|^2$$

we find, using Eq. 14.45, that

$$\lambda = \|H\|^2$$

Hence

$$H^2\,|h\rangle - \|H\|^2\,|h\rangle = 0$$

Therefore

$$(H - \|H\|)(H + \|H\|)\,|h\rangle = 0$$

One has either

$$(H + \|H\|)\,|h\rangle \neq 0 \qquad \text{or} \qquad (H + \|H\|)\,|h\rangle = 0$$

In the first case, $(H + \|H\|)\,|h\rangle$ is an eigenvector of H with eigenvalue $\|H\|$, and in the second case, $|h\rangle$ is an eigenvector of H with eigenvalue $-\|H\|$.

This proves that **if H is a completely continuous, Hermitian operator, then it always has an eigenvector with eigenvalue equal to $+\|H\|$ or $-\|H\|$.**

(ii) Suppose now that the theorem is not true and that there exists an infinite orthonormal set of eigenvectors $|h_\alpha\rangle$ of H

$$H\,|h_\alpha\rangle = h_\alpha\,|h_\alpha\rangle$$

such that

$$|h_\alpha| > \varepsilon > 0 \qquad \text{for any } \alpha$$

Because H is completely continuous, the infinite set of all vectors $H|h_\alpha\rangle$ must be compact, i.e., it must contain a convergent sequence. This is, however, impossible, since the distance between two vectors $H|h_\alpha\rangle$ and $H|h_{\alpha'}\rangle$ is always finite unless $|h_\alpha\rangle = |h_{\alpha'}\rangle$.

$$\begin{aligned}\rho(H\,|h_\alpha\rangle, H\,|h_{\alpha'}\rangle) &= \rho(h_\alpha\,|h_\alpha\rangle, h_{\alpha'}\,|h_{\alpha'}\rangle)\\ &= \sqrt{h_\alpha^2 + h_{\alpha'}^2} > \varepsilon\end{aligned}$$

We did not assume that all vectors $|h_\alpha\rangle$ are different and therefore a subspace spanned by the eigenvectors of H that correspond to a given nonzero eigenvalue is always finite-dimensional.

This proves part (ii) of the theorem; notice that we did not make use of the hermiticity of H and thus the result holds for any completely continuous linear operator.

(iii) It is evident that H cannot have an eigenvalue h' such that $|h'| > \|H\|$, for if $|h'\rangle$ were the corresponding eigenvector, one would have

$$\sqrt{\frac{\langle h'|\,H^2\,|h'\rangle}{\langle h'|h'\rangle}} = |h'| > \|H\|$$

in contradiction with the definition of $\|H\|$.

We proved in part (i) that there exists at least one eigenvector $|h_0\rangle$ with eigenvalue $h_0 = \pm\|H\|$. Consider now the operator

$$H_1 = H - \sum_j |h_0,j\rangle\, h_0\, \langle j,h_0| \tag{14.49}$$

The sum on j extends over the finite [see part (ii)] set of orthonormal eigenvectors corresponding to the eigenvalue h_0

$$\begin{aligned}H\,|h_0,j\rangle &= h_0\,|h_0,j\rangle\\ \langle j,h_0|h_0,k\rangle &= \delta_{jk}\end{aligned} \tag{14.50}$$

Since h_0 is real and the sum in Eq. 14.49 is finite, it is evident that H_1 is both Hermitian and completely continuous. It can easily be verified that

$$\begin{aligned}\rho^2(H_1|\rangle,0) &= \langle|\,H_1^2\,|\rangle\\ &= \langle|H^2|\rangle - h_0^2 \sum_j |\langle|h_0,|j\rangle|^2 \le \langle|\,H^2\,|\rangle\\ &= \rho^2(H|\rangle,0)\end{aligned}$$

Therefore

$$\|H_1\| < \|H\| \tag{14.51}$$

One has an alternative. Either

$$\|H_1\| = 0 \qquad \text{or} \qquad \|H_1\| > 0$$

When $\|H_1\| = 0$, then $H_1 = 0$ and

$$H = \sum_j |h_0,j\rangle\, h_0\, \langle j,h_0|$$

On the other hand, assuming that $\|H_1\| > 0$, and applying the results of (i) and (ii) to H_1, we find that H_1 has an eigenvector with eigenvalue $h_1 = \pm\|H_1\|$. Let us multiply both sides of the equation

$$H_1\, |h_1\rangle = h_1\, |h_1\rangle$$

by $\langle k,h_0|$. One obtains

$$\langle k,h_0|\, H_1\, |h_1\rangle = h_1 \langle k,h_0|h_1\rangle \tag{14.52}$$

However, using Eq. 14.49, one has

$$\begin{aligned}\langle k,h_0|\, H_1\, |h_1\rangle &= \langle k,h_0| \left(H - \sum_j |h_0,j\rangle\, h_0\, \langle j,h_0|\right)|h_1\rangle \\ &= \langle k,h_0| \left(E - \sum_j |h_0,j\rangle\langle j,h_0|\right) H\, |h_1\rangle = 0\end{aligned}$$

Comparing the preceding result with Eq. 14.52, we find for any k

$$\langle k,h_0|h_1\rangle = 0$$

But this implies that $|h_1\rangle$ is also an eigenvector of H with eigenvalue h_1 for

$$\begin{aligned}H\, |h_1\rangle &= \left(H - \sum_j |h_0,j\rangle\, h_0\, \langle j,h_0|\right)|h_1\rangle \\ &= H_1\, |h_1\rangle = h_1\, |h_1\rangle\end{aligned}$$

We now examine the operator

$$H_2 = H - \sum_j |h_0,j\rangle\, h_0\, \langle j,h_0| - \sum_j |h_1,j\rangle\, h_1\, \langle j,h_1|$$

and repeat the previous reasoning. It is clear that after a finite number (n, say) of steps, one finds that either

$$H - \sum_{k=1}^{n} \sum_j |h_k,j\rangle\, h_k\, \langle j,h_k| = 0$$

or the process continues indefinitely. In the latter case, one obtains an infinite sequence of operators H_n

$$H_n = H - \sum_{k=1}^{n} \sum_j |h_k,j\rangle\, h_k\, \langle j,h_k| \qquad n = 1, 2, \cdots$$

However, since (see inequality 14.51)

$$\|H\| > \|H_1\| > \|H_2\| > \cdots \|H_n\| > \cdots > 0$$

one must have $\lim_{n\to\infty} \|H_n\| = 0$. In any case, we can write

$$H = \sum_k \sum_j |h_k,j\rangle\, h_k\, \langle j,h_k| \tag{14.53}$$

When the summation over k is infinite, it is understood in the sense that the norm of the RHS tends to $\|H\|$. This allows us to write (see Sec. 14.4)

$$H\,|\rangle = \sum_k \sum_j |h_k,j\rangle\, h_k\, \langle j,h_k|\rangle \qquad \text{for any } |\rangle \in S_\infty \tag{14.54}$$

Thus, any vector $H|\rangle$ can be expanded in a (finite or infinite) series of eigenvectors of H.

Now take an arbitrary vector $|\rangle \in S_\infty$ and consider the vector

$$|'\rangle = |\rangle - \sum_k \sum_j |h_k,j\rangle\, \langle j,h_k|\rangle$$

According to Eq. 14.54, one has

$$H\,|'\rangle = H\,|\rangle - \sum_k \sum_j |h_k,j\rangle\, h_k\, \langle j,h_k|\rangle = 0 \tag{14.55}$$

Thus, $|'\rangle$ is an eigenvector of H with eigenvalue zero.* We rewrite Eq. 14.55 as

$$|\rangle = |'\rangle + \sum_k \sum_j |h_k,j\rangle\, \langle j,h_k|\rangle \tag{14.56}$$

which can now be interpreted as the decomposition of an arbitrary vector $|\rangle \in S_\infty$ into eigenvectors of H.

14.7 A Convenient Notation

We introduce in this section a notation commonly used by physicists and which has the merit of being simple. It has, however, the disadvantage of being introduced formally and therefore of lacking in rigor, at least in this presentation.

Let us assume that the scalar product is given by

$$\langle f|g\rangle = \int_a^b w(x)\bar{f}(x)g(x)\,dx \tag{14.57}$$

It has already been noted that this can be regarded as a generalization of the expression

$$\langle a|b\rangle = \sum_{j=1}^{N} \bar{a}^j b^j \tag{14.58}$$

for the scalar product in an N-dimensional space in which an orthonormal basis has been chosen. We also stated at the beginning of this chapter that the set of numbers $f(x)$ which makes up a function may be regarded as the components of an abstract vector $|f\rangle$ with respect to some basis, which, however, was left undefined. We now formally introduce such a basis; it will consist of vectors $|x\rangle$, where the "index" x that labels these vectors ($a \le x \le b$) is continuous. Hence, we write in analogy** to Chapter II, Eq. 18.2

$$f(x) \underset{\text{def}}{=} \langle x|f\rangle \tag{14.59}$$

* There may be an infinite number of such eigenvectors.

** The equality is considered in the sense that it holds for all x values except for sets of measure zero.

The continuity of x gives rise to difficulties in defining the normalization of $|x\rangle$. To maintain the analogy between Eq. 14.59 and Chapter II, Eq. 18.2, we assume that two distinct basis vectors $|x\rangle$ and $|x'\rangle$ are orthogonal.

$$\langle x'|x\rangle = 0 \qquad \text{for } x' \neq x$$

and we write instead of Eq. 18.1 in Chapter II

$$|f\rangle = \int_a^b dx\, w(x) f(x)\, |x\rangle \tag{14.60}$$

If we pursue the analogy further, we should have, multiplying from the left by $\langle x'|$

$$\langle x'|f\rangle = f(x') = \int_a^b dx\, w(x) f(x)\, \langle x'|x\rangle \tag{14.61}$$

Therefore, $w(x)\,\langle x'|x\rangle$ has the properties of the δ distribution and evidently one cannot have $\langle x|x\rangle = 1$. We put

$$\begin{aligned}\langle x|x'\rangle &= \frac{1}{\sqrt{w(x)w(x')}}\,\delta(x - x') = \frac{1}{w(x)}\,\delta(x - x') \\ &= \frac{1}{w(x')}\,\delta(x - x')\end{aligned} \tag{14.62}$$

Consider now the identity operator E in a finite-dimensional space S_N. If $|e_i\rangle$ $(i = 1, 2, \cdots, N)$ is an orthonormal basis in S_N, the operator E can be written as

$$E = \sum_{i=1}^{N} |e_i\rangle\, \langle e_i| \tag{14.63}$$

The validity of Eq. 14.63 follows from the fact that multiplying this relation from the right by an arbitrary vector $|a\rangle \in S_N$

$$|a\rangle = \sum_{j=1}^{N} a^j\, |e_j\rangle$$

one obtains, since $\langle e_i|e_j\rangle = \delta_{ij}$

$$\begin{aligned}E\,|a\rangle &= \sum_{i,j=1}^{N} a^j\, |e_i\rangle\, \langle e_i|e_j\rangle \\ &= \sum_{j=1}^{N} a^j\, |e_j\rangle = |a\rangle\end{aligned} \tag{14.64}$$

and similarly

$$\langle a|\, E = \langle a| \tag{14.65}$$

Equations 14.64 and 14.65 are the defining relations for E.

In a function space, the identity operator can be written in the form

$$E = \int_a^b dx\, |x\rangle w(x)\, \langle x| \tag{14.66}$$

which is a generalization of Eq. 14.63. Multiplying Eq. 14.66 from the left by $\langle x'|$ and from the right by $|x''\rangle$, we get

$$\delta(x' - x'') = \int_a^b dx\, \delta(x' - x)\, \delta(x - x'') \tag{14.67}$$

Actually we have rigorously shown that

$$f(x') = \int_a^b dx\, f(x)\, \delta(x - x')$$

only for a restricted class of functions $f(x)$ that are sufficiently smooth in the neighborhood of $x = x'$. We did not prove that Eq. 14.67 is meaningful. In this section we simply treat $\delta(x - x')$ as a formal generalization of the Kronecker delta, δ_{ij}, for Eqs. 14.61 and 14.67 are the continuous index generalizations of

$$a^i = \sum_j a^j\, \delta_{ij} \qquad \text{and} \qquad \delta_{ij} = \sum_k \delta_{ik}\, \delta_{kj}$$

This can be justified, but it necessitates a deeper and more abstract formulation of the theory of generalized functions than the one we presented in our brief introduction where we tried to reconcile simplicity with rigor at the cost of generality.

Notice that a vector $|x\rangle$ is itself not represented by a function as seen from Eq. 14.62. It belongs to the larger linear space of generalized functions; remember that multiplication by numbers and addition of generalized functions are well-defined operations that again yield generalized functions. The fact that one may have a set of vectors which form a basis for a function space without belonging themselves to the space is typical for the cases when the vectors are labeled by a continuous index. For example, the Fourier transformation

$$f(x) = \frac{1}{\sqrt{2\pi}} \int_{-\infty}^{+\infty} F(t) e^{itx}\, dt$$

may be regarded as representing a decomposition of a vector $|f\rangle$ into vectors represented by functions $(1/\sqrt{2\pi})e^{itx}$.

When $|f\rangle \in L_1^2(-\infty, +\infty)$, $F(t)$ is also a well-defined square integrable function (Plancherel's theorem), but $(1/\sqrt{2\pi})e^{itx}$ does not represent a vector of $L_1^2(-\infty, +\infty)$, since its modulus is not square-integrable in the infinite interval $(-\infty, +\infty)$.

Let vectors $|e_m\rangle$ $(m = 1, 2, \cdots)$ form an orthonormal basis in S_∞ and let these vectors be represented by the functions $e_{(m)}(x)$

$$e_{(m)}(x) = \langle x | e_m \rangle \qquad m = 1, 2, \cdots$$

In analogy to Eq. 14.63, we write the identity operator E as

$$E = \sum_{m=1}^{\infty} |e_m\rangle \langle e_m|$$

Multiplying from the left by $\langle x'|$, from the right by $|x\rangle$, and using Eq. 14.62, we find

$$[w(x)w(x')]^{-\frac{1}{2}}\, \delta(x' - x) = \sum_{m=1}^{\infty} e_{(m)}(x')\bar{e}_{(m)}(x) \tag{14.68}$$

This relation is often useful.

14.8 Integral and Differential Operators

The action of an operator in a function space is represented by operations applied to functions that may involve, in particular, integrations and differentiations.

We call an operator an **integral operator** if, using the notation of the preceding subsection, it can be written formally as a double integral

$$K = \int_a^b \int_a^b dx''\, dx'\, |x''\rangle w(x'') K(x'',x') w(x')\, \langle x'| \tag{14.69}$$

This is a direct generalization of the expression

$$\sum_{i,j=1}^{N} |e_i\rangle\, K^i_j\, \langle e_j|$$

for a linear operator in an N-dimensional space; $K(x'',x')$ plays the role of a matrix element. From Eq. 14.69 we find that the operator equation $|g\rangle = K|f\rangle$ has as its analytical counterpart (see Eqs. 14.59 and 14.61)

$$g(x) = \int_a^b dx'\, w(x')K(x,x')f(x') \tag{14.70}$$

This has the form of an integral representation for the function $g(x)$ and justifies calling $K(x,x')$ the kernel of the operator K. The hermiticity of K is expressed by the relation satisfied by the kernel

$$\overline{K}(x',x) = K(x,x') \tag{14.71}$$

which is a direct analog of the identity 18.14 of the preceding chapter, which defines a Hermitian matrix.

We now examine the conditions that $K(x,x')$ must satisfy in order for K to be completely continuous. Let $|e_m\rangle$ be vectors of an orthonormal basis in $L^2_w(a,b)$. Suppose that $K(x,x')$ can be expanded in a double series of functions $\langle x|e_m\rangle$

$$K(x,x') = \sum_{m,n} K^m_n\, \langle x|e_m\rangle\, \overline{\langle x'|e_n\rangle} \tag{14.72}$$

which converges in the mean

$$\lim_{N\to\infty} \int_a^b \int_a^b dx\, dx'\, w(x)w(x')|K(x,x') - K_N(x,x')|^2 = 0 \tag{14.73}$$

where

$$K_N(x,x') = \sum_{m,n=1}^{N} K^m_n\, \langle x|e_m\rangle\, \overline{\langle x'|e_n\rangle} \tag{14.74}$$

The function $K_N(x,x')$ defines an integral operator

$$K_N = \int_a^b \int_a^b dx''\, dx'\, |x''\rangle\, w(x'')K_N(x'',x')w(x')\, \langle x'| \tag{14.75}$$

Inserting Eq. 14.74 into Eq. 14.75, we find

$$\begin{aligned} K_N &= \int_a^b \int_a^b dx''\, dx'\, w(x'')w(x')\, |x''\rangle \sum_{m,n=1}^{N} K^m_n\, \langle x''|e_m\rangle\, \langle e_n|x'\rangle\, \langle x'| \\ &= \int_a^b dx''\, |x''\rangle\, w(x'')\, \langle x''| \sum_{m,n=1}^{N} |e_m\rangle K^m_n \langle e_n| \int_a^b dx'\, |x'\rangle w(x')\, \langle x'| \\ &= \sum_{m,n=1}^{N} |e_m\rangle K^m_n \langle e_n| \end{aligned} \tag{14.76}$$

In the last step we used the fact that

$$\int_a^b dx'\, |x'\rangle\, w(x')\, \langle x'| = E$$

Hence, K_N is a finite-dimensional operator.

We now calculate the norm of the operator

$$K - K_N = \int_a^b \int_a^b dx''\, dx'\, |x''\rangle w(x'')[K(x'',x') - K_N(x'',x')]w(x')\,\langle x'|$$

One has

$$\begin{aligned} h(x) &\equiv \langle x|\,(K - K_N)\,|f\rangle \\ &= \int_a^b dx'\, w(x')[K(x,x') - K_N(x,x')]f(x') \end{aligned} \tag{14.77}$$

Applying the Cauchy-Schwarz inequality to the integral on the RHS (compare Eq. 3.4), we get

$$|h(x)|^2 \le \int_a^b |K(x,x') - K_N(x,x')|^2 w(x')\, dx' \cdot \int_a^b |f(x'')|^2 w(x'')\, dx''$$

Integrating both sides of the preceding inequality with respect to x, one obtains

$$\begin{aligned} \int_a^b |h(x)|^2 w(x)\, dx \le \int_a^b \int_a^b & dx\, dx'\, w(x)w(x') \\ & \times |K(x,x') - K_N(x,x')|^2 \\ & \times \int_a^b |f(x'')|^2 w(x'')\, dx'' \end{aligned} \tag{14.78}$$

This, together with Eq. 14.73, leads to

$$\begin{aligned} 0 &\le \frac{\int_a^b |h(x)|^2 w(x)\, dx}{\int_a^b |f(x)|^2 w(x)\, dx} \\ &\le \int_a^b \int_a^b dx\, dx'\, w(x)w(x')|K(x,x') - K_N(x,x')|^2 \underset{N\to\infty}{\to} 0 \end{aligned} \tag{14.79}$$

However

$$\begin{aligned} \frac{\int_a^b dx\, w(x)\,|h(x)|^2}{\int_a^b dx\, w(x)\,|f(x)|^2} &= \frac{\int_a^b dx\,\langle f|\,(K^+ - K_N^+)\,|x\rangle\, w(x)\,\langle x|\,(K - K_N)\,|f\rangle}{\int_a^b dx\,\langle f|x\rangle\, w(x)\,\langle x|f\rangle} \\ &= \frac{\langle f|(K^+ - K_N^+)(K - K_N)\,|f\rangle}{\langle f|f\rangle} \end{aligned} \tag{14.80}$$

In the last step we again used Eq. 14.66. The vector $|f\rangle$ was completely arbitrary and a comparison of Eq. 14.80 and inequality 14.79 shows that

$$\|K - K_N\| \underset{N\to\infty}{\to} 0$$

Thus, K is a completely continuous operator, provided its kernel $K(x,x')$ can be expanded in a series (Eq. 14.72). In the case $w(x) = 1$, one can, for example, choose for the functions $\langle x|e_m\rangle$, the trigonometric functions

$$\langle x|e_m\rangle = \frac{1}{\sqrt{b-a}}\, e^{im\pi\left[\frac{2x-a-b}{b-a}\right]} \qquad m = 0, \pm 1, \cdots$$

which are orthonormal on the interval $[a,b]$ with weight unity. Since only the convergence in the mean of the sum in Eq. 14.72 is required, the conditions to be satisfied by $K(x,x')$ are very weak. For this type of convergence of a Fourier series of a function of one variable, it is sufficient for the function to have a square integrable modulus. Similarly, in order to ensure the convergence in the mean of the series in Eq. 14.72, one must require that the double integral

$$\int_a^b \int_a^b |K(x,x')|^2\, w(x)\, w(x')\, dx\, dx'$$

exists.

To summarize, we restate the general theorem of the preceding subsection for the particular case of integral operators:

Theorem 4 (Hilbert). (i) An integral eigenvalue equation*

$$\int_a^b K(x,x') \cdot f(x')w(x') \cdot dx' = k \cdot f(x) \tag{14.81}$$

has at least one nontrivial solution, provided

$$K(x,x') = \bar{K}(x',x)$$

and

$$\int_a^b \int_a^b |K(x,x')|^2 w(x)w(x')\, dx\, dx' < \infty$$

The above two conditions ensure that the integral operator is completely continuous and Hermitian.

(ii) Outside any finite interval $[-\varepsilon,\varepsilon]$ there can be only a finite number of eigenvalues k and the number of orthonormal eigenfunctions $f_m(x)$ of Eq. 14.81 corresponding to a given eigenvalue k_m is finite.

(iii) Any function that can be represented as

$$g(x) = \int_a^b K(x,x')h(x')w(x')\, dx'$$

can be expanded in a Fourier series

$$g(x) = \sum_{m,n} \langle n,k_m|g\rangle f_m^{(n)}(x) \tag{14.82}$$

(this is the counterpart of the vector equation 14.54). The convergence in the mean of this series is evident. In fact, it can be proved that the expansion 14.82 converges uniformly provided

$$\int_a^b |K(x,x')|^2 w(x')\, dx' \qquad < \text{const}$$

where the constant is independent of x.

---

* An integral equation is one in which the unknown function appears under the integral sign.

Before closing this section, we make a few brief comments on the so-called differential operators, which will be discussed in more detail in the next chapter.

An operator L is called a **differential** operator if

$$\langle x|\, L\, |f\rangle = a_0(x)f(x) + a_{(1)}(x)\frac{df(x)}{dx} + \cdots + a_{(n)}(x)\frac{d^nf(x)}{dx^n}$$

We also write

$$\langle x|\, L\, |f\rangle = L_x \cdot f(x)$$

with

$$L_x = a_0(x) + a_{(1)}(x)\frac{d}{dx} + \cdots + a_{(n)}(x)\frac{d^n}{dx^n}$$

The subscript x indicates that the differentiation involved in the definition of the operator refers to the particular variable x of the function $f(x)$, which could also depend on other parameters.

If we require that $L\,|f\rangle$ belongs to a well-defined function space, we must impose certain restrictions on $|f\rangle$ in order to ensure the differentiability of $f(x)$. It is clear, for example, that if $f(x)$ is discontinuous, then $\langle x|L|f\rangle$ is a generalized function rather than a function.

Differential operators are, in general, not bounded except when very restrictive conditions are imposed on the space of functions in which the operators operate (but then the space is no longer complete). This is easily understandable, since a derivative of a function may be very large (in fact, infinite) for an integrable function. Hence, there is no reason for

$$\frac{\rho(L\,|f\rangle, 0)}{\rho(|f\rangle, 0)} = \sqrt{\frac{\displaystyle\int_a^b dx\, w(x)|L_x f(x)|^2}{\displaystyle\int_a^b dx\, w(x)|f(x)|^2}}$$

to have an upper bound. Therefore the theory of this section does not apply directly to differential operators. However, a differential operator L may have an inverse operator L^{-1}, which is not only bounded but even completely continuous. L^{-1} is then in fact an integral operator, as will be seen in Chapter IV. Since the eigenvalue equation

$$L\,|f\rangle = \lambda\,|f\rangle$$

can be rewritten as

$$L^{-1}\,|f\rangle = \frac{1}{\lambda}\,|f\rangle$$

when L^{-1} exists, many properties of the eigenvectors of L (in particular, their completeness) may be deduced from a consideration of the properties of L^{-1}. These problems will be discussed in detail in the next chapter.

CHAPTER IV

DIFFERENTIAL EQUATIONS

Part I Ordinary Differential Equations

1 · INTRODUCTION

An ordinary differential equation of the Nth order is a relation of the form

$$F\left(x,u,\frac{du}{dx},\cdots,\frac{d^N u}{dx^N}\right)=0 \tag{1.1}$$

Equation 1.1 is called a **linear** differential equation if F is a linear function of its arguments

$$u,\quad \frac{du}{dx},\cdots,\frac{d^N u}{dx^N}$$

The most general form of such an equation is

$$q(x)+r(x)u+s(x)\frac{du}{dx}+\cdots+t(x)\frac{d^N u}{dx^N}=0 \qquad (t(x)\neq 0) \tag{1.2}$$

where $u(x), q(x), \cdots, t(x)$ are functions of the variable x. If the term $q(x)$, which does not multiply the function $u(x)$, or one of its derivatives, is zero, the equation is said to be **homogeneous**; otherwise, it is said to be **inhomogeneous** and $q(x)$ is called the **inhomogeneous term.**

In the next few sections it will be assumed that the variable x is real. This assumption will greatly simplify the theory of Green's functions, which we shall develop later, and particularly certain of its algebraic aspects. On the other hand, starting with Sec. 13, we shall examine an important class of differential equations in the complex domain. Then we shall adopt a completely different point of view; the emphasis will be on the properties of analytic functions, and the algebraic aspects of the problem will be forgotten.

It should be understood that when one seeks a solution of a differential equation, one must specify the class of functions to which the solution should belong. For example, one may seek a solution which is N-times differentiable, or infinitely differentiable, or differentiable only in the sense of generalized functions. If the class of admissible function is too restricted, the equation may have no solutions at all belonging to this class. Conversely, if the class is too large, the equation may very well have solutions which, however, may be of no interest from the physical point of view.

EXAMPLE

Consider the 1st order differential equation

$$\frac{du}{dx} = |x|$$

Among the class of differentiable functions, the particular function

$$u(x) = \begin{cases} \frac{1}{2}x^2 & x \geq 0 \\ -\frac{1}{2}x^2 & x \leq 0 \end{cases}$$

is clearly a solution of the equation. However, as can be immediately seen from the equation itself, the second derivative of $u(x)$ has a discontinuity at $x = 0$, and its third derivative exists only in the sense of generalized functions. Hence, for example, a solution of the equation which is three times differentiable in the ordinary sense, does not exist at all.

Given an Nth order differential equation, we shall, in this chapter and unless stated otherwise, look for solutions that are N-times differentiable. More generally, we will as a rule, assume that all the functions that are preceded by a differentiation symbol, are sufficiently differentiable so that all differentiations involved have meaning in the ordinary sense. At times, however, it will be useful to take advantage of the formalism of generalized functions; in the cases where the differentiation symbol d/dx is to be understood in the generalized sense, the reader will be cautioned.

The solution of Eq. 1.1 (if it exists) is not unique. Additional information must be given about the function $u(x)$ and its first $(N - 1)$ derivatives,* and this information constitutes the so-called **boundary conditions**. The boundary conditions may be given in many different forms, and for each form the existence and uniqueness of a solution must be proved.

There is one type of boundary condition, however, for which a unique solution of Eq. 1.1 is ensured under rather weak conditions. The existence theorem is so general that it is worthwhile to merely state it. We must first introduce a definition.

Let $f(y_1, y_2, \cdots, y_N)$ be a function of N arguments $y_1, y_2, \cdots, y_N$ and suppose that the argument y_i, say, varies within an interval

$$c_i - \eta \leq y_i \leq c_i + \eta \tag{1.3}$$

where η is a positive number and c_i is some constant.

Let $y_{(i)}^{(1)}$ and $y_{(i)}^{(2)}$ be two arbitrary points of the interval (1.3). Then, if there exists a positive number k such that

$$|f(y_1, y_2, \cdots, y_i^{(2)}, y_{i+1}, \cdots, y_n) - f(y_1, y_2, \cdots, y_i^{(1)}, y_{i+1}, \cdots, y_n)| \leq k|y_i^{(2)} - y_i^{(1)}| \tag{1.4}$$

the function f is said to obey a **Lipschitz condition** with respect to the argument y_i in the interval (1.3).

---

* Equation 1.1 can be solved for $d^N u/dx^N$ in the neighborhood of $x = x_0$ under very general conditions, namely, if the partial derivative of F with respect to $d^N u/dx^N$ does not vanish at

$$x = x_0, \quad u = u(x_0), \quad \cdots, \quad \left.\frac{d^N u}{dx^N}\right|_{x=x_0}$$

Once u and its first $(N - 1)$ derivatives are known at some point, the Nth and higher derivatives of $u(x)$ can be found by taking successively higher-order derivatives of the equation itself.

Let us now write Eq. 1.1 in the form (see footnote on p. 258)

$$\frac{d^N u}{dx^N} = H\left(x,u,\frac{du}{dx},\cdots,\frac{d^{N-1}u}{dx^{N-1}}\right) \tag{1.5}$$

and suppose that the boundary conditions consist in prescribing the values of the function $u(x)$ and of its first $(N-1)$ derivatives at some point x_0 of the interval $[a,b]$.

$$u(x_0) = c_0$$
$$\left.\frac{du}{dx}\right|_{x=x_0} = c_1, \cdots, \qquad \frac{d^{N-1}u}{dx^{N-1}} = c_{N-1} \tag{1.6}$$

$c_0, c_1, \cdots, c_{N-1}$ being given constants. Then we have the following theorem due to Cauchy and Lipschitz, which applies to equations of the type of Eq. 1.5 even when H is not a linear function of its arguments.

Theorem. Consider the differential equation (1.5) together with the boundary conditions (1.6). If the function H in Eq. 1.5 is continuous and if there exists a positive number η such that, whenever $a \leq x \leq b$, H obeys Lipschitz conditions with respect to its arguments $u, du/dx, \cdots, d^{N-1}u/dx^{N-1}$, when these arguments vary within the intervals

$$c_0 - \eta \leq u \leq c_0 + \eta$$
$$c_1 - \eta \leq \frac{du}{dx} \leq c_1 + \eta \tag{1.7}$$
$$\cdots\cdots\cdots\cdots$$
$$c_{N-1} - \eta \leq \frac{d^{N-1}u}{dx^{N-1}} \leq c_{N-1} + \eta$$

a solution of Eq. 1.5 satisfying 1.6, exists at all points of the interval $[a,b]$ and is unique.*

The boundary conditions (1.6) are one type of possible boundary conditions. Other, more general types of boundary conditions are equally important in applications. To give an example, consider the solution of the general linear second-order differential equation

$$a(x)\frac{d^2u}{dx^2} + b(x)\frac{du}{dx} + c(x)u = f(x) \tag{1.8}$$

in the interval $[a,b]$ and let there be given boundary conditions of the form

$$B_1(u) \equiv \alpha_1 u(a) + \beta_1 \left.\frac{du}{dx}\right|_{x=a} + \gamma_1 u(b) + \delta_1 \left.\frac{du}{dx}\right|_{x=b} = \sigma_1 \tag{1.9}$$

$$B_2(u) \equiv \alpha_2 u(a) + \beta_2 \left.\frac{du}{dx}\right|_{x=a} + \gamma_2 u(b) + \delta_2 \left.\frac{du}{dx}\right|_{x=b} = \sigma_2 \tag{1.10}$$

where $\alpha_i, \beta_i, \gamma_i, \delta_i$ and σ_i are constants that can be specified in any way as long as Eqs. 1.9 and 1.10 are linearly independent. If $\sigma_1 = \sigma_2 = 0$, the boundary conditions are said to be **homogeneous**. If either σ_1 or σ_2, or both, differs from zero, the boundary

* The Lipschitz conditions are needed to ensure the uniqueness of the solution.

conditions are said to be **inhomogeneous**. We shall study in great detail equations of the type of Eq. 1.8. The conditions (1.9) and (1.10) are the most general linear boundary conditions that can be associated with a second-order linear differential equation, and in Secs. 8 and 10 we shall give the necessary and sufficient conditions for the existence and uniqueness of a solution of Eq. 1.8 that obeys the conditions 1.9 and 1.10.

2 · SECOND-ORDER DIFFERENTIAL EQUATIONS; PRELIMINARIES

We shall now and in the rest of Part I of this chapter consider ordinary differential equations of order not higher than 2.

We can dispose immediately of the first-order case by a direct integration. Let the equation be

$$a(x)\frac{du}{dx} + b(x)u(x) = f(x) \qquad (a(x) \neq 0) \tag{2.1}$$

Putting

$$\frac{b(x)}{a(x)} \equiv \frac{1}{p(x)}\frac{dp(x)}{dx} \tag{2.2}$$

the equation becomes

$$\frac{d}{dx}(pu) = \frac{p(x)f(x)}{a(x)} \tag{2.3}$$

Equations 2.2 and 2.3 can be immediately integrated to give

$$u(x) = \frac{1}{p(x)}\left[\int_{x_0}^{x} dx' \, \frac{p(x')f(x')}{a(x')} + u(x_0)\right] \tag{2.4}$$

$$p(x) = \exp\left(\int_{x_0}^{x} dx \, \frac{b(x')}{a(x')}\right) \tag{2.5}$$

where x_0 is an arbitrary initial point and where the constant in Eq. 2.4 is determined by the single boundary condition imposed on $u(x)$.

Consider now the second-order linear, inhomogeneous differential equation

$$a(x)\frac{d^2u}{dx^2} + b(x)\frac{du}{dx} + c(x)u = q(x) \qquad a(x) \neq 0 \tag{2.6}$$

and the associated homogeneous equation

$$a(x)\frac{d^2u}{dx^2} + b(x)\frac{du}{dx} + c(x)u = 0 \qquad (a(x) \neq 0) \tag{2.7}$$

Let $u_1(x)$ and $u_2(x)$ be two solutions of Eq. 2.7. Because of the linearity of this equation, $c_1u_1 + c_2u_2$, where c_1 and c_2 are arbitrary constants, is also a solution of Eq. 2.7. We shall show that if u_1 and u_2 are two linearly independent solutions of Eq. 2.7, then any solution u of Eq. 2.7 can be expressed as a linear combination of u_1 and u_2; i.e.,

$$u(x) = c_1u_1(x) + c_2u_2(x) \tag{2.8}$$

The constants c_1 and c_2 reflect the arbitrariness in the solution of a differential equation when boundary conditions have not been specified. Now u_1 and u_2 are linearly independent if the relation

$$\alpha u_1(x) + \beta u_2(x) = 0 \tag{2.9}$$

implies

$$\alpha = \beta = 0$$

Differentiating Eq. 2.9, there follows

$$\alpha \frac{du_1(x)}{dx} + \beta \frac{du_2(x)}{dx} = 0 \tag{2.10}$$

The Eqs. 2.9 and 2.10 imply that for u_1, u_2 to be linearly independent, it is sufficient that

$$W(u_1,u_2) = \begin{vmatrix} u_1 & u_2 \\ \dfrac{du_1}{dx} & \dfrac{du_2}{dx} \end{vmatrix} = u_1 \frac{du_2}{dx} - u_2 \frac{du_1}{dx} \neq 0 \tag{2.11}$$

$W(u_1,u_2)$ is called the **Wronskian** of the solutions u_1 and u_2.

It is easy to show that if $W(u_1,u_2) = 0$, then u_1 and u_2 are necessarily linearly dependent, for $W(u_1,u_2) = 0$ means

$$u_1 \frac{du_2}{dx} - u_2 \frac{du_1}{dx} = 0$$

which can be easily integrated to give

$$u_2 = \text{const} \times u_1$$

Hence, u_1, u_2 are linearly dependent.

On the other hand, if u, u_1, u_2 are any three solutions of Eq. 2.7, i.e.,

$$\begin{aligned} a \frac{d^2u}{dx^2} + b \frac{du}{dx} + cu &= 0 \\ a \frac{d^2u_1}{dx^2} + b \frac{du_1}{dx} + cu_1 &= 0 \\ a \frac{d^2u_2}{dx^2} + b \frac{du_2}{dx} + cu_2 &= 0 \end{aligned} \tag{2.12}$$

then Eqs. 2.12 have a nontrivial solution for a,b,c only if

$$\begin{vmatrix} u & u_1 & u_2 \\ \dfrac{du}{dx} & \dfrac{du_1}{dx} & \dfrac{du_2}{dx} \\ \dfrac{d^2u}{dx^2} & \dfrac{d^2u_1}{dx^2} & \dfrac{d^2u_2}{dx^2} \end{vmatrix} = 0 \tag{2.13}$$

But Eq. 2.13 implies that u, u_1, and u_2 are linearly dependent. Hence, the most general solution of Eq. 2.7 can always be written in the form of Eq. 2.8.

As two linearly independent solutions of Eq. 2.7, u_1 and u_2 are called a **fundamental set of solutions** of the equation. Hence, we have shown

Theorem 1. The most general solution of a homogeneous linear differential equation of the second order is of the form

$$u(x) = c_1 u_1(x) + c_2 u_2(x)$$

where c_1 and c_2 are arbitrary complex constants, and u_1,u_2 are a fundamental set of solutions of Eq. 2.7, i.e., solutions satisfying the condition

$$W(u_1,u_2) \neq 0$$

Consider now the inhomogeneous equation 2.6. Suppose that u_p is any particular solution of Eq. 2.6. Then the substitution

$$u = u_p + y$$

transforms Eq. 2.6 into the homogeneous equation

$$a\frac{d^2y}{dx^2} + b\frac{dy}{dx} + cy = 0$$

whose most general solution is a linear combination of a fundamental set of its solutions. Hence, we have

Theorem 2. The most general solution of the general second-order inhomogeneous differential equation (Eq. 2.6) is

$$u(x) = c_1 u_1(x) + c_2 u_2(x) + u_p(x)$$

where u_p is **any** particular solution of Eq. 2.6, c_1,c_2 are arbitrary complex constants, and u_1,u_2 are a fundamental set of solutions of the associated homogeneous equation.

If one is able to find any solution u_1 of Eq. 2.7, then a second linearly independent solution can always be found by using a method called the **method of variation of constants.** Let the second solution be written as

$$u_2(x) = u_1(x)h(x) \tag{2.14}$$

where $h(x)$ is to be determined. Inserting Eq. 2.14 into Eq. 2.7 and remembering that u_1 satisfies the same equation, we find the following equation for $h(x)$

$$\frac{d^2h(x)}{dx^2} + \left(\frac{2(du_1/dx) + pu_1}{u_1}\right)\frac{dh(x)}{dx} = 0 \tag{2.15}$$

where

$$p(x) \equiv \frac{b(x)}{a(x)} \qquad (a(x) \neq 0)$$

Equation 2.15 can be written as

$$\frac{d}{dx}\ln\frac{dh}{dx} = -p(x) - 2\frac{d}{dx}\ln u_1(x)$$

which can be integrated to give

$$\frac{dh}{dx} = \frac{\text{const}}{u_1^2(x)}\exp\left(\int_{x_0}^{x} p(x')\,dx'\right) \tag{2.16}$$

where x_0 is an arbitrary initial point. A further integration of Eq. 2.16 yields $h(x)$, and inserting this value into Eq. 2.14, we obtain the second solution

$$u_2(x) = \text{const}\ u_1(x) \int_{x_0}^{x} \frac{dx''}{u_1^2(x'')} \exp\left(-\int_{x_0}^{x''} p(x')\,dx'\right) \tag{2.17}$$

Using Eq. 2.16, the Wronskian of the two solutions u_1 and u_2 is

$$W(u_1,u_2) = u_1^2 \frac{dh}{dx}$$

Hence

$$W(u_1,u_2) = \text{const}\ \exp\left(-\int_{x_0}^{x} p(x')\,dx'\right) \tag{2.18}$$

Thus the Wronskian never vanishes, and u_1 and u_2 are a fundamental set of solutions of the equation. Equation 2.17 is known as **Liouville's formula.** Since any solution of Eq. 2.7 is a linear combination of u_1 and u_2, the Wronskian will always have the form of Eq. 2.18 for any fundamental set of solutions of Eq. 2.7.

If we have a fundamental set of solutions of the homogeneous equation, a particular solution of the inhomogeneous equation can also be obtained, again using the method of variation of constants.

Putting $u_p = u_1 v$ into Eq. 2.6, where u_1 is a solution of the homogeneous equation, we find $[f(x) = q(x)/a(x)]$

$$\frac{d^2v}{dx^2} + \left[p(x) + \frac{2}{u_1}\frac{du_1}{dx}\right]\frac{dv}{dx} = \frac{f(x)}{u_1(x)} \tag{2.19}$$

We note that

$$p(x) + \frac{2}{u_1}\frac{du_1}{dx} \equiv \frac{1}{R(x)}\frac{dR}{dx} \tag{2.20}$$

where

$$R(x) \equiv \frac{u_1^2}{W(u_1,u_2)} = \left[\frac{d}{dx}\left(\frac{u_2}{u_1}\right)\right]^{-1} \tag{2.21}$$

Thus, using Eqs. 2.20 and 2.21, one can solve Eq. 2.19 for dv/dx

$$\begin{aligned}\frac{dv}{dx} &= \frac{1}{R(x)} \int_{x_0}^{x} dx' \frac{R(x')f(x')}{u_1(x')} = \frac{d}{dx}\left(\frac{u_2}{u_1}\right) \int_{x_0}^{x} dx' \frac{u_1(x')f(x')}{W[u_1(x'),u_2(x')]} \\ &= \frac{d}{dx}\left[\frac{u_2}{u_1} \int_{x_0}^{x} dx' \frac{u_1(x')f(x')}{W[u_1(x'),u_2(x')]}\right] - \frac{u_2(x)f(x)}{W[u_1(x),u_2(x)]}\end{aligned} \tag{2.22}$$

Integrating Eq. 2.22 and multiplying by u_1, we find the particular solution

$$u_p(x) = u_2(x) \int_{x_0}^{x} dx' \frac{u_1(x')f(x')}{W[u_1(x'),u_2(x')]} - u_1(x) \int_{x_0}^{x} dx' \frac{u_2(x')f(x')}{W[u_1(x'),u_2(x')]} \tag{2.23}$$

We see, therefore, that a knowledge of one solution of a linear homogeneous differential equation of the second order is sufficient to find the most general solution of the inhomogeneous equation. The fundamental difficulty resides in finding a solution of the homogeneous equation; this will be discussed in Secs. 14 through 20.

The method just described, whereby the complete solution of a differential equation is engendered by one solution only of the homogeneous equation, cannot be easily generalized either to equations of higher order or to partial differential equations. For this reason we shall consider linear second-order equations from a different point of view, and this simple example will serve to introduce a method of solution known as the **method of Green's functions**. This method can be easily generalized to problems where the previously described method fails.

Green's functions play a very important role in mathematics and in theoretical physics. In particular, with their help, we shall be able to reduce the eigenvalue problem associated with a differential operator to the more tractable eigenvalue problem for an integral operator.

In the next few sections we shall be concerned with **homogeneous** boundary conditions only. It will be seen in Sec. 11 that the solution of a problem with inhomogeneous boundary conditions can be obtained by a simple extension of the results derived for the case of homogeneous boundary conditions. The reader, however, may find in the following section a preliminary illustration of the reason why the problem with inhomogeneous boundary conditions is not inherently different from the problem with homogeneous boundary conditions.

3 · THE TRANSITION FROM LINEAR ALGEBRAIC SYSTEMS TO LINEAR DIFFERENTIAL EQUATIONS—DIFFERENCE EQUATIONS

Suppose that the argument x of a function $u(x)$ is increased by a positive amount h. One defines the **first difference** of u, which one denotes by Δu as

$$\Delta u \equiv u(x+h) - u(x) \tag{3.1}$$

Here, h is a finite quantity, not necessarily infinitesimally small. Similarly, the **second difference** of u, denoted by $\Delta^2 u$, is given by

$$\begin{aligned} \Delta^2 u \equiv \Delta(\Delta u) &= u(x+2h) - u(x+h) - u(x+h) + u(x) \\ &= u(x+2h) - 2u(x+h) + u(x) \end{aligned} \tag{3.2}$$

Higher differences can be obtained in an analogous manner. An equation that involves differences of various orders is called **a difference equation**. An example of such an equation is

$$\Delta^2 u + 2\Delta u = x$$

It will not be our task here to discuss difference equations and their solutions. The reader is referred to any of a number of books on the subject. Rather, we shall use them to set up a link between systems of algebraic equations and differential equations. The link is best delineated by supposing that the argument x is specified only at a discrete set of equally spaced points

$$x_k = kh \qquad (k = 0, 1, 2, \cdots, n) \tag{3.3}$$

It is then convenient to introduce the notation

$$u(x_k) = u(kh) \equiv u_k \qquad (k = 0, 1, \cdots, n) \tag{3.4}$$

In terms of u_k, Eqs. 3.1 and 3.2 become respectively

$$\Delta u_k = u_{k+1} - u_k \tag{3.5}$$

and

$$\Delta^2 u_k = u_{k+2} - 2u_{k+1} + u_k \tag{3.6}$$

Consider now a linear differential equation of the second order with constant coefficients (this limitation is not at all essential but simply saves writing). Take

$$\frac{d^2u}{dx^2} + \frac{du}{dx} + u = f(x) \qquad (0 \le x \le 1) \tag{3.7}$$

We can approximate Eq. 3.7 by a linear difference equation, in which the smaller the h, the better the approximation. Thus, using Eqs. 3.5 and 3.6, we obtain

$$u_{k+2} - (2-h)u_{k+1} + (1 - h + h^2)u_k = h^2 f(x) \qquad (k = 0, 1, \cdots, n-2) \tag{3.8}$$

Notice, that the equations with $k = n - 1$ and $k = n$ cannot be written down because they involve the quantities u_{n+1} and u_{n+2}, which have not been specified in the interval we are considering. This is why we write Eq. 3.8 only for $k = 0, 1, \cdots, n-2$. Setting $\alpha \equiv 2 - h$ and $\beta \equiv 1 - h + h^2$, we have

$$\begin{aligned}
\beta u_0 - \alpha u_1 + u_2 &= h^2 f(0) \\
\beta u_1 - \alpha u_2 + u_3 &= h^2 f(h) \\
\beta u_2 - \alpha u_3 + u_4 &= h^2 f(2h) \\
\cdots\cdots\cdots\cdots & \\
\beta u_{n-2} - \alpha u_{n-1} + u_n &= h^2 f(nh - 2h)
\end{aligned} \tag{3.9}$$

Equations 3.9 represent a set of $n - 1$ equations with $n + 1$ unknowns. Therefore, the solution of Eqs. 3.9 is not unique. It does become unique, however, if one specifies **two** boundary conditions. As examples, the boundary conditions

$$u(0) = u(1) = 0$$

are equivalent to

$$u_0 = u_n = 0 \qquad (nh = 1)$$

and the boundary conditions

$$u(0) = 0 \qquad \left.\frac{du}{dx}\right|_{x=0} = 1$$

are equivalent to

$$u = 0 \qquad \frac{u_1 - u_0}{h} = 1 \qquad (nh = 1)$$

In all cases, two boundary conditions will fix the values of two of the $(n + 1)$ coefficients u_j, and only then can the system of Eqs. 3.9 have a well-determined (unique) solution.

To be specific, we now suppose that u_0 and u_n are given by the boundary conditions. Let us write Eq. 3.9 in matrix form

$$\mathbf{M}\mathbf{u} = \mathbf{g} \tag{3.10}$$

where

$$\mathbf{M} = \begin{pmatrix} -\alpha & 1 & 0 & 0 & \cdots & 0 & 0 & 0 \\ \beta & -\alpha & 1 & 0 & \cdots & 0 & 0 & 0 \\ 0 & \beta & -\alpha & 1 & \cdots & 0 & 0 & 0 \\ \cdot & \cdot & \cdot & \cdot & \cdot & \cdot & \cdot & \cdot \\ 0 & 0 & & & & 0 & \beta & -\alpha \end{pmatrix} \tag{3.11}$$

$$\mathbf{u} = \begin{pmatrix} u_1 \\ u_2 \\ \vdots \\ u_{n-1} \end{pmatrix} \qquad \text{and} \qquad \mathbf{g} = \begin{pmatrix} h^2 f(0) - \beta u_0 \\ h^2 f(h) \\ \vdots \\ h^2 f(nh - 2h) - u_n \end{pmatrix}$$

It can be shown under fairly general conditions that as n becomes larger and larger, the solution $u(x)$ of the differential equation becomes better and better approximated by the set of numbers $u_1, u_2, \cdots, u_{n-1}$, which constitute the solution of the system of algebraic equations.

Let us observe that the problem of solving the system of algebraic equations 3.9 is equivalent to finding the matrix $\mathbf{M}^{-1}$ inverse to $\mathbf{M}$; this inverse matrix exists if and only if $\det \mathbf{M} \neq 0$. The matrix $\mathbf{M}$ is determined not only by Eqs. 3.9, but also by the boundary conditions imposed on the solution. However, possible inhomogeneities in the boundary conditions will not enter the matrix $\mathbf{M}$ but rather the column vector $\mathbf{g}$. Therefore, if we have succeeded in solving the problem with **homogeneous** boundary conditions, i.e., finding the matrix $\mathbf{M}^{-1}$, the problem with inhomogeneous conditions will present no further difficulties. Assuredly, the solution $\mathbf{M}^{-1}\mathbf{g}$ will be different, since $\mathbf{g}$ will be different, but the matrix $\mathbf{M}^{-1}$ will be the same. This intuitive example illustrates the general fact that the problem of solving a differential equation with inhomogeneous boundary conditions can be reduced to solving an inhomogeneous equation with homogeneous boundary conditions, but in which the inhomogeneous term of the equation has been modified.

4 · GENERALIZED GREEN'S IDENTITY

In Sec. 14 of the preceding chapter we defined a linear differential operator L as an operator in a function space whose action on vectors $|u\rangle$ of this space is represented by a differential form $L_x u(x)$, where

$$L_x = a_0 + a_1 \frac{d}{dx} + a_2 \frac{d^2}{dx^2} + \cdots + a_N \frac{d^N}{dx^N} \tag{4.1}$$

The expression 4.1 for L_x has a purely formal character. It indicates the operations that will be involved in obtaining the differential form $L_x u$, but is itself only symbolic.

The symbolic expression L_x represents the differential operator L defined in an abstract function space in the sense that $L_x u$ represents the vector $L|u\rangle$. However, the differential form $L_x u$ is meaningful for **any** sufficiently differentiable function $u(x)$,* whereas the vectors $|u\rangle$ for which $L|u\rangle$ is meaningful, must belong to a specific function space, called the **domain** of the operator L. As we shall see, the functions that represent the vectors of the domain of L must not only be sufficiently differentiable, but must also satisfy certain additional conditions (namely, boundary conditions) which in effect restrict considerably the class of admissible functions.

Since every linear differential equation involves some differential form of the type $L_x u$, it is of interest to consider certain formal properties of the kind of symbolic expression such as that given by Eq. 4.1, without at all considering possible algebraic aspects of the problem. To stress this point, we shall call every expression of the type of 4.1 a **formal differential operator.****

We consider in this chapter the cases where L_x is a formal differential operator of the first or second order (the order of L_x is by definition the order of the highest derivative in $L_x u$), since these are the most important cases for physical applications. Many of the results obtained, however, can be easily extended to operators of higher order, and our restriction is mainly for reasons of economy of space.

---

* Actually, it will be required that $u(x)$ is sufficiently differentiable so that the form $L_x u$ does not contain derivatives of δ functions, although the δ function itself may be present.

** In the literature, where algebraic problems are not considered, a formal differential operator is simply called a differential operator.

Suppose that there exists a formal differential operator L_x^+ with the property that the quantity

$$w[\bar{v}L_x u - u\overline{(L_x^+ v)}] \tag{4.2}$$

is proportional to a perfect differential for any sufficiently differentiable functions $u = u(x)$ and $v = v(x)$; $w = w(x)$ is some positive definite function over an interval $[a,b]$. More precisely, the following relation should hold.

$$w[\bar{v}L_x u - u\overline{(L_x^+ v)}] = \frac{d}{dx}\{Q[u,\bar{v}]\} \tag{4.3}$$

for some function $Q[u,\bar{v}]$ which depends bilinearly on $u,\bar{v}$, and their derivatives, du/dx, $d\bar{v}/dx$. In general

$$Q[u,\bar{v}] = Au\bar{v} + Bu\frac{d\bar{v}}{dx} + C\frac{du}{dx}\bar{v} + D\frac{du}{dx}\frac{d\bar{v}}{dx} \tag{4.4}$$

where A,B,C, and D are some functions of x. Equation 4.3 is called the **Lagrange identity**. One calls L_x^+ the **formal adjoint** of L_x with respect to a weight w (the reason for this naming will become apparent).

Integrating Eq. 4.3 over the interval $[a,b]$, we get

$$\int_a^b dx\, w[\bar{v}L_x u] - \int_a^b dx\, w[u\overline{(L_x^+ v)}] = Q[u,\bar{v}]\,|_{x=b} - Q[u,\bar{v}]\,|_{x=a} \tag{4.5}$$

Equation 4.5 is known as the **generalized Green's identity** and the expression on the RHS is called a **boundary** or **surface** term.

We shall later give examples of how one can obtain in practice the formal adjoint of an operator. A glance at Eq. 4.5, however, gives us the clue immediately. We can see that after a sufficient number of partial integrations, a formal differential operator L_x, which originally operated on the function u to the right of it, will be transformed to another operator, its formal adjoint L_x^+, which will operate on the other function $\bar{v}$ that was originally to the left of L_x. Hence, the method needed to obtain the formal adjoint of an operator is the method of partial integrations. The surface term is then just the integrated term that results from the partial integration. Equation 4.5 shows that in addition to partial integrations, a complex conjugation may be involved in the definition of L_x^+.

In the case where $L_x = L_x^+$, the formal differential operator is said to be **self-adjoint**.

EXAMPLE 1

Consider the operator

$$L_x = \frac{d}{dx}$$

By a partial integration we obtain

$$\int_a^b dx\, \bar{v}\left[\frac{d}{dx}u\right] - \int_a^b u\overline{\left[-\frac{dv}{dx}\right]} = [u\bar{v}]\Big|_a^b \tag{4.6}$$

Comparing with Eq. 4.5, we see that the formal adjoint of $L_x = d/dx$, with respect to a weight $w = 1$, is

$$L_x^+ = -\frac{d}{dx}$$

EXAMPLE 2

The operator $L_x = i(d/dx)$ is similar to the one considered in the preceding example except that now L_x has an imaginary coefficient and the process of finding the formal adjoint of L_x also involves a complex conjugation. Similarly to Eq. 4.6, one has

$$\int_a^b dx\, \bar{v}\left[i\frac{d}{dx}u\right] - \int_a^b dx\, u\,\overline{\left[i\frac{d}{dx}v\right]} = i[u\bar{v}]_a^b \tag{4.7}$$

Hence, for $w = 1$, $L_x^+ = i(d/dx) = L_x$, and the operator is self-adjoint.*

5 · GREEN'S IDENTITY AND ADJOINT BOUNDARY CONDITIONS

Let us return to the generalized Green's identity (Eq. 4.5) and suppose that the function $u(x)$ satisfies **homogeneous** boundary conditions

$$\begin{aligned} B_1(u) &\equiv \alpha_1 u(a) + \beta_1 \frac{du}{dx}\bigg|_{x=a} + \gamma_1 u(b) + \delta_1 \frac{du}{dx}\bigg|_{x=b} = 0 \\ B_2(u) &\equiv \alpha_2 u(a) + \beta_2 \frac{du}{dx}\bigg|_{x=a} + \gamma_2 u(b) + \delta_2 \frac{du}{dx}\bigg|_{x=b} = 0 \end{aligned} \tag{5.1}$$

Such functions $u(x)$ define a linear vector space, since any linear combination of functions satisfying Eqs. 5.1 is a function that also satisfies Eqs. 5.1.

We shall inquire about the conditions that are to be imposed on the function v in order to make the surface term in Eq. 4.5 vanish. It is easy to see that this can be achieved if one prescribes homogeneous boundary conditions on v that are of the same general form as Eqs. 5.1. In fact, from the equation

$$Q[u,\bar{v}]\,|_{x=a} - Q[u,\bar{v}]\,|_{x=b} = 0 \tag{5.2}$$

one can eliminate two of the four quantities $u(a)$, $u(b)$, $\frac{du}{dx}\Big|_{x=a}$, $\frac{du}{dx}\Big|_{x=b}$, and then the coefficients in front of the two remaining quantities will have the form

$$\alpha\bar{v}(a) + \beta\frac{d\bar{v}}{dx}\bigg|_{x=a} + \gamma\bar{v}(b) + \delta\frac{d\bar{v}}{dx}\bigg|_{x=b}$$

where $\alpha,\beta,\gamma,\delta$ are constants. Setting these two coefficients equal to zero and taking complex conjugates of the equations, we obtain two **homogeneous** boundary conditions for the function v. These conditions imposed on v are said to be **adjoint** to the conditions 5.1.

EXAMPLE 1

Consider again Example 1 of Sec. 4, assuming that the boundary condition is

$$u(b) = \tfrac{1}{4}u(a)$$

(there is only one such condition, since L_x is of the first order). The surface term is

$$u(b)\bar{v}(b) - u(a)\bar{v}(a)$$

* Notice that the factor i cannot simply be cancelled out in Eq. 4.7, since it is an integral part of the operator and is therefore needed to find its adjoint.

and vanishes if

$$\frac{\bar{v}(b)}{\bar{v}(a)} = \frac{u(a)}{u(b)} = 4$$

Thus, the adjoint boundary condition is

$$v(b) = 4v(a) \tag{5.3}$$

EXAMPLE 2

Consider now Example 2 of Sec. 4 with the boundary condition

$$u(a) = u(b) \tag{5.4}$$

The surface term is, apart from a factor i, the same as in the preceding example and vanishes when

$$v(a) = v(b) \tag{5.5}$$

Notice that the adjoint boundary condition is the same as the original one, which was not the case in the preceding example.

We have already mentioned that the homogeneous boundary conditions imposed on functions $u(x)$ define a certain function space; let us denote it by U. Similarly, let us denote by V the function space defined by the adjoint boundary conditions satisfied by functions $v(x)$. Suppose that there is given a formal differential operator L_x. In the space U, and **only** in that space a **differential operator** L (without a subscript x) will be defined by the requirement

$$\langle x| L |u\rangle = L_x u(x) \qquad \text{for } |u\rangle \in \mathrm{U}$$

In other words, L is represented by the formal differential operator L_x, but **only** when L_x acts on the functions that represent vectors of U.

Similarly, in the space V, we define the operator L^+ as

$$\langle x| L^+ |v\rangle = L_x^+ v(x) \qquad \text{for } |v\rangle \in V$$

and again L^+ is represented by L_x^+ when and only when L_x^+ acts on functions representing vectors of V. Since the adjoint boundary conditions have been defined so as to eliminate the surface term in Eq. 4.5, we obtain the following identity, known as **Green's identity**

$$\int_a^b dx\, w\bar{v}(L_x u) - \int_a^b dx\, wu\overline{(L_x^+ v)} = 0 \tag{5.6}$$

Equation 5.6 is valid when $u(x)$ satisfies the homogeneous boundary conditions 5.1 and $v(x)$ satisfies the homogeneous boundary conditions adjoint to those given by Eqs. 5.1.

With the definition of the scalar product given by Eq. 5.2, Chapter III, Eq. 5.6 can be given an obvious algebraic meaning. In fact, using the formalism of Sec. 14.7, Chapter III (Eqs. 14.59 and 14.66), we have

$$\begin{aligned}\langle v| L |u\rangle - \overline{\langle u| L^+ |v\rangle} &= \int_a^b dx\, w(x)\langle v|x\rangle\langle x| L |u\rangle - \int_a^b dx\, w(x)\overline{\langle u|x\rangle}\,\overline{\langle x| L^+ |v\rangle} \\ &= \int_a^b dx\, w(x)\bar{v}[L_x u] - \int_a^b dx\, w(x)u\overline{[L_x^+ v]}\end{aligned} \tag{5.7}$$

Hence, we see that Green's identity (Eq. 5.6) is equivalent to

$$\langle v|\, L\, |u\rangle = \overline{\langle u|\, L^+\, |v\rangle} \tag{5.8}$$

which is the defining relation for L^+ (see Eq. 9.7 Chapter II). This relation, which from a purely algebraic point of view looks trivial, since it merely defines L^+, is not at all trivial from the analytic point of view. In fact, in order to arrive at it, it was necessary to restrict the class of admissible functions.

It is important to realize that had we chosen different homogeneous boundary conditions (i.e., different constants in Eqs. 5.1), the domain of L and hence of L^+ would have been different. Thus, the form of L_x does not determine uniquely the differential operator L; the boundary conditions prescribed are part and parcel of the definition of the operator L itself.

As usual, we call L a **Hermitian** operator if

$$L = L^+ \tag{5.9}$$

The definition (5.9) implies that the formal differential operator L_x is self-adjoint

$$L_x = L_x^+ \tag{5.10}$$

and that the domains of L and of L^+ are identical

$$U = V \tag{5.11}$$

This latter equation means that the boundary conditions satisfied by the functions $u(x)$ and $v(x)$ are exactly the same.

EXAMPLE 3

Example 2 of Secs. 4 and 5 shows that the formal differential operator $L_x = i(d/dx)$, together with the boundary condition $u(a) = u(b)$, defines a Hermitian differential operator $L = L^+$, since

$$L_x^+ = i\frac{d}{dx} = L_x$$

and the adjoint boundary conditions are again $v(a) = v(b)$.

On the other hand, the reader will verify that choosing for the boundary condition the condition $u(a) = 2u(b)$ leads to the adjoint boundary condition $v(a) = \frac{1}{2}v(b)$. Thus, in spite of the fact that $L_x^+ = L_x$, $L^+ \neq L$, since the domains of L and L^+ are different.

In some cases it may happen that the surface term vanishes identically and independently of the given boundary conditions. In those cases, certain regularity conditions on $u(x)$, $v(x)$, and their derivatives at the end points of the interval considered are the appropriate conditions that should be imposed, and they play the role of boundary conditions.

6 · SECOND-ORDER SELF-ADJOINT OPERATORS

The most important differential operators which occur in physical problems are of the second order. We shall therefore devote special attention to these operators.

Consider the general second-order formal differential operator

$$L_x \equiv a(x)\frac{d^2}{dx^2} + b(x)\frac{d}{dx} + c(x) \qquad (\text{Re } a(x) > 0) \tag{6.1}$$

where $a(x)$, $b(x)$, and $c(x)$ are, in general, complex functions of the real variable x. Then the rule for differentiating a product of functions yields the relations

$$\bar{v}a(x)\frac{d^2u}{dx^2} - u\frac{d^2(a\bar{v})}{dx^2} = \frac{d}{dx}\left[a\bar{v}\frac{du}{dx} - u\frac{d}{dx}(a\bar{v})\right]$$
$$\bar{v}b(x)\frac{du}{dx} + u\frac{d(b\bar{v})}{dx} = \frac{d}{dx}[bu\bar{v}] \tag{6.2}$$

Hence with $w(x) = 1$

$$L_x^+ v = \frac{d^2}{dx^2}(\bar{a}v) - \frac{d}{dx}(\bar{b}v) + \bar{c}v$$

and therefore L_x^+ is given by

$$L_x^+ = \bar{a}\frac{d^2}{dx^2} + \left(2\frac{d\bar{a}}{dx} - \bar{b}\right)\frac{d}{dx} + \left(\frac{d^2\bar{a}}{dx^2} - \frac{d\bar{b}}{dx} + \bar{c}\right) \tag{6.3}$$

and Eq. 4.5 takes the form

$$\int_a^b dx[\bar{v}(L_x u) - u(L_x^+ v)] = \left[a\bar{v}\frac{du}{dx} - u\frac{d}{dx}(a\bar{v}) + b\bar{v}u\right]\Bigg|_{x=a}^{x=b} \tag{6.4}$$

Comparing Eq. 6.3 with Eq. 6.1, we see that L_x is self-adjoint with respect to a weight unity if $a(x)$, $b(x)$, and $c(x)$ are real, and if $da/dx = b$, in which case Eq. 6.4 becomes

$$\int_a^b dx[\bar{v}(L_x u) - u\overline{(L_x^+ v)}] = \left[a\left(\bar{v}\frac{du}{dx} - u\frac{d\bar{v}}{dx}\right)\right]\Bigg|_a^b \tag{6.5}$$

and Eq. 6.1 can be written

$$L_x u = \frac{d}{dx}\left[a\frac{du}{dx}\right] + cu \tag{6.6}$$

We shall assume henceforth that $a(x)$, $b(x)$, and $c(x)$ are real functions and that $a(x)$ is positive definite throughout $[a,b]$.

The role of second-order self-adjoint operators is particularly important, since one can prove that all second-order formal differential operators with real coefficients are self-adjoint, provided the weight function $w(x)$ has been properly chosen.

We first show that when $p(x)$ and $w(x)$ are taken to be real, the operator L_x defined by

$$L_x u = \frac{1}{w(x)}\frac{d}{dx}\left[p\frac{du}{dx}\right] + cu, \qquad w(x) > 0 \tag{6.7}$$

is self-adjoint with respect to a weight $w(x)$. The proof is simple. We note that calculating the adjoint of L_x with respect to a weight $w(x)$ is the same as calculating the adjoint of wL_x with respect to a weight unity. Using the method described above, we find that

$$wL_x^+ v = \frac{d}{dx}\left[p\frac{dv}{dx}\right] + wcv \tag{6.8}$$

i.e.,

$$L_x^+ v = \frac{1}{w}\frac{d}{dx}\left[p\frac{dv}{dx}\right] + cv \tag{6.9}$$

Therefore, the operator L_x defined by Eq. 6.7 is self-adjoint and we have

$$\int_a^b dx\, w(x)[\bar{v}(L_x u) - u(\overline{L_x^+ v})] = \left[p(x)\left(\bar{v}\frac{du}{dx} - u\frac{dv}{dx}\right)\right]\Bigg|_a^b \tag{6.10}$$

We now show that for a general second-order operator (6.1) with real coefficients $L_x u$ can be written in the form of Eq. 6.7. In fact, in order for the definitions 6.1 and 6.7 to be equivalent, we must have

$$\frac{p}{w} = a, \qquad b = \frac{1}{w}\frac{dp}{dx} \tag{6.11}$$

which means

$$\frac{1}{p}\frac{dp}{dx} = \frac{b}{a} \qquad \text{and} \qquad p(x) = \exp\left(\int_{x_0}^x dx' \frac{b(x')}{a(x')}\right) \tag{6.12}$$

Therefore, with the choice

$$w(x) = \frac{p(x)}{a(x)} = \frac{1}{a(x)} \exp\left(\int_{x_0}^x dx' \frac{b(x')}{a(x')}\right) \tag{6.13}$$

the two definitions (6.1 and 6.7) are equivalent. Since we have shown that L_x defined by Eq. 6.7 is self-adjoint with respect to the weight $w(x)$, we have proved the following.

Theorem. Every linear, formal differential operator of the second order with real coefficients is self-adjoint, provided the weight function $w(x)$ is chosen as

$$w(x) = \frac{1}{a(x)} \exp\left(\int_{x_0}^x dx' \frac{b(x')}{a(x')}\right)$$

and $a(x) > 0$.

When the conditions of the preceding theorem are satisfied, and in particular when the weight function is chosen as in Eq. 6.13, the surface term in the generalized Green's identity has the form (see Eqs. 6.10 and 6.11)

$$Q[u,\bar{v}]\Bigg|_a^b = \left[w(x)a(x)\left(\bar{v}\frac{du}{dx} - u\frac{d\bar{v}}{dx}\right)\right]\Bigg|_a^b \tag{6.14}$$

and, as stated in the theorem

$$L_x = a\frac{d^2}{dx^2} + b\frac{d}{dx} + c$$

is self-adjoint. Moreover, with the following boundary conditions on $u(x)$, L_x defines a Hermitian differential operator L:

(i) $u(a) = u(b) = 0$ (Dirichlet conditions)

(ii) $\left.\frac{du}{dx}\right|_{x=a} = \left.\frac{du}{dx}\right|_{x=b} = 0$ (Neumann conditions)

(iii) $\alpha u(a) - \left.\frac{du}{dx}\right|_{x=a} = \beta u(b) - \left.\frac{du}{dx}\right|_{x=b} = 0$ (α,β real) (general unmixed conditions)

(iv) $u(a) = u(b)$

$\left.\frac{du}{dx}\right|_{x=a} = \left.\frac{du}{dx}\right|_{x=b}$ (periodic conditions, $p(a) = p(b)$ assumed)

The reader can easily verify that, independently of the form of the functions $a(x)$ and $w(x)$ in Eq. 6.14, the foregoing boundary conditions lead to identical adjoint boundary conditions on $v(x)$.

7 · GREEN'S FUNCTIONS

In this section we seek solutions of ordinary second-order differential equations, using a method known as the **method of Green's function.**

Let our problem be to find the solution of the inhomogeneous equation

$$L_x u(x) = f(x) \tag{7.1}$$

with **homogeneous** boundary conditions imposed on $u(x)$ of the type given in Eqs. 5.1.

We consider simultaneously the problem of solving the equation

$$L_x^+ v(x) = h(x) \tag{7.2}$$

where L_x^+ is the adjoint of L_x and $v(x)$ satisfies boundary conditions adjoint to those imposed on $u(x)$. The functions $f(x)$ and $h(x)$ are arbitrary.

Because we are limiting our considerations to functions $u(x)$ and $v(x)$, which satisfy homogeneous boundary conditions (adjoint with respect to one another), our problem is equivalent to the algebraic problem of finding the solutions of the operator equations

$$L\,|u\rangle = |f\rangle \tag{7.3}$$

and

$$L^+\,|v\rangle = |h\rangle \tag{7.4}$$

where L (respectively L^+) is defined by L_x (respectively L_x^+) together with the boundary conditions on $u(x)$ (respectively $v(x)$), as explained in Sec. 5.

Provided there exists an operator G satisfying

$$LG = E \tag{7.5}$$

where E is the identity operator, the solution of Eq. 7.3 is

$$|u\rangle = G|f\rangle$$

If the operator G does exist, it is called **Green's operator.**

Let us tentatively write G as an integral operator of the type of Eq. 14.69 of Chapter III

$$G = \int_a^b \int_a^b dx'\, dx''\, |x'\rangle w(x') G(x', x'') w(x'') \langle x''| \tag{7.6}$$

Using this form for G, we get from Eq. 7.5 (see 14.61 and 14.62, Chapter III)

$$\begin{aligned} \langle x|\,LG\,|y\rangle &= L_x\langle x|\,G\,|y\rangle = L_x G(x,y) \\ &= \langle x|\,E\,|y\rangle = \langle x|y\rangle = \frac{\delta(x-y)}{w(x)} \end{aligned} \tag{7.7}$$

Therefore, the kernel $G(x,y)$ of the operator G satisfies the differential equation

$$L_x G(x,y) = \frac{\delta(x-y)}{w(x)} \tag{7.8}$$

which is to be understood in the sense of the theory of generalized functions.

Notice that the first step of Eq. 7.7 is really meaningful only when the vector $G|y\rangle$ on which L acts belongs to the domain of L; therefore, $G(x,y)$, as a function of x, **must satisfy the same homogeneous boundary conditions as those imposed on** $u(x)$.

Anticipating slightly, it will be seen that even though Eq. 7.8 is an equation in which the differentiations are meant in the generalized sense, nevertheless $G(x,y)$ is a function in the elementary sense; $G(x,y)$ is called the **Green's function** associated with the differential equation (Eq. 7.1). It satisfies the same equation as $u(x)$ except that the inhomogeneous term $f(x)$ has been replaced by a δ function divided by a weight $w(x)$.

Similarly, one can repeat the previous reasoning for Eq. 7.2; denoting by $g(x,y)$ the kernel of the integral operator that is a right inverse of L^+, we obtain

$$L_x^+ g(x,y) = \frac{\delta(x-y)}{w(x)} \tag{7.9}$$

$g(x,y)$ is known as the **adjoint Green's function** associated with the differential equation (Eq. 7.1), and as it was the case for $G(x,y)$, $g(x,y)$ as a function of x **must satisfy the same homogeneous conditions as** $v(x)$.

Whenever a solution of Eq. 7.8 exists, the integral operator G is well defined and is a (right) inverse of the differential operator L. Similarly, when a solution of Eq. 7.9 exists, an integral operator inverse to L^+ exists.

It is now easy to obtain the solutions of Eqs. 7.1 and 7.2 in terms of the Green's functions. Using Eqs. 7.1 and 7.9, Green's identity (Eq. 5.6) with $v \equiv g(x,y)$ leads to the solution

$$u(y) = \int_a^b dx\, w(x)\bar{g}(x,y)f(x) \tag{7.10}$$

Similarly, using Eqs. 7.2 and 7.8, Eq. 5.6 leads to the solution of the adjoint equation (Eq. 7.2)

$$v(y) = \int_a^b dx\, w(x)\bar{G}(x,y)h(x) \tag{7.11}$$

Let us summarize the preceding discussion. If the solutions of Eqs. 7.8 and 7.9 exist, the solutions of Eqs. 7.1 and 7.2 with homogeneous boundary conditions are given respectively by Eqs. 7.10 and 7.11. The Green's function $G(x,y)$ [respectively $g(x,y)$] satisfies Eq. 7.8 (respectively 7.9) and obeys the same homogeneous boundary conditions as the associated function $u(x)$ [respectively $v(x)$].

We now consider the properties of the Green's functions and the conditions for the existence of unique solutions of Eqs. 7.1 and 7.2.

8 · PROPERTIES OF GREEN'S FUNCTIONS

Let us assume that $G(x,y)$ and $g(x,y)$ exist and are unique, and let us inquire about their properties. The problem of the existence and uniqueness of Green's functions (for second-order differential operators) will be discussed in the next section. It will become apparent that either $G(x,y)$ and $g(x,y)$ both exist and are unique, or they do not exist at all.

(i) Putting $u = G(y',y)$ and $v = g(y',x)$ into Green's identity (Eq. 5.6), changing the integration variable there from x to y', and using Eqs. 7.8 and 7.9, we get

$$\int_a^b dy'\, g(y',x)\delta(y'-y) = \int_a^b dy'\, \bar{G}(y',y)\delta(y'-x)$$

i.e.,

$$g(y,x) = \bar{G}(x,y) \tag{8.1}$$

We see that $\bar{G}(x,y)$ must satisfy the adjoint boundary conditions with respect to its second argument y. Another consequence of Eq. 8.1 is that the uniqueness of $G(x,y)$ implies the uniqueness of $g(x,y)$, and vice versa.

Equation 8.1 shows that the same Green's function can be used to solve both the original differential equation (Eq. 7.1) and the adjoint equation (Eq. 7.2). Using Eq. 8.1, we can rewrite the solutions of Eqs. 7.10 and 7.11 as

$$u(y) = \int_a^b dx\, w(x)G(y,x)f(x) \tag{8.2}$$

and

$$v(y) = \int_a^b dx\, w(x)g(y,x)h(x) \tag{8.3}$$

(ii) Until now we only assumed the existence of $G(x,y)$ and $g(x,y)$. Let us now suppose that $G(x,y)$ [and therefore $g(x,y)$] is unique. Then, if L is Hermitian

$$G(x,y) = g(x,y) \tag{8.4}$$

since in this case Eq. 7.1 is identical to Eq. 7.2 and the boundary conditions associated with Eq. 7.1 are identical to those associated with Eq. 7.2.

Equations 8.1 and 8.4 show that for a Hermitian differential operator L, the Green's function obeys the relation

$$G(x,y) = \bar{G}(y,x) \tag{8.5}$$

Moreover, from Eq. 8.5 it follows that if the coefficients in L_x are real, then $G(x,y)$, being itself real, is symmetric

$$G(x,y) = G(y,x) \tag{8.6}$$

(iii) The following discussion will apply to Green's functions for second-order differential equations; the corresponding discussion for a first-order equation would be trivial and, in fact, of little practical interest.

Since

$$L_x G(x,y) = \frac{\delta(x-y)}{w(x)} \tag{8.7}$$

when $G(x,y)$ exists, it follows that $G(x,y)$ satisfies the homogeneous equation

$$L_x G(x,y) = 0 \tag{8.8}$$

at all points of the interval $a \leq x \leq b$ except at the point $x = y$. What happens at that point? We have

$$L_x G(x,y) = a(x)\frac{\partial^2 G}{\partial x^2} + b(x)\frac{\partial G}{\partial x} + c(x)G = \frac{\delta(x-y)}{w(x)} \tag{8.9}$$

and we assume that $a(x) > 0$.

Equation 8.9 has meaning only in the sense of generalized functions. For the time being, we shall proceed, however, as if $G(x,y)$ were itself an ordinary function, and we shall determine the conditions that it must then satisfy. In the next section, it will be shown how an ordinary function that satisfies these conditions can be explicitly constructed.

Let

$$p(x) \equiv \exp\left(\int_{x_0}^{x} dx' \frac{b(x')}{(ax')}\right)$$

Then

$$\frac{1}{p(x)}\frac{dp}{dx} = \frac{b(x)}{a(x)}$$

and Eq. 8.9 can be cast into the form

$$\frac{a(x)}{p(x)}\frac{\partial}{\partial x}\left[p\frac{\partial G}{\partial x}\right] + c(x)G = \frac{\delta(x-y)}{w(x)} \tag{8.10}$$

Dividing both sides of this equation by $a(x)/p(x)$ and using the property 13.13, Chapter III of the δ function, we have

$$\frac{\partial}{\partial x}\left[p\frac{\partial G}{\partial x}\right] = p(y)\frac{\delta(x-y)}{w(y)a(y)} - \frac{p(x)c(x)}{a(x)}G(x,y) \tag{8.11}$$

Integrating Eq. 8.11 and using the fact that the δ function is the derivative in the generalized sense of the step function $H(x)$ (see Eq. 13.14, Chapter III), we find

$$p(x)\frac{\partial G}{\partial x} = \frac{p(y)}{w(y)a(y)}H(x-y) - \int_a^x dx' \frac{p(x')c(x')}{a(x')}G(x',y) + \text{const.} \tag{8.12}$$

Notice that with the supposition that $G(x,y)$ is an ordinary function, Eq. 8.12 is already meaningful in the ordinary sense and not only in the sense of distributions.

The RHS of Eq. 8.12 is a continuous function of x except for the first term (because of the presence of the step function $H(x-y)$), which makes $\partial G(x,y)/\partial x$ discontinuous at the point $x = y$. This discontinuity can be easily calculated directly from Eq. 8.12, using the defining property of $H(x-y)$

$$H(x) = \begin{cases} 1 & x > 0 \\ 0 & x < 0 \end{cases}$$

One has

$$\lim_{\varepsilon\to+0}\left\{p(y+\varepsilon)\frac{\partial G(x,y)}{\partial x}\bigg|_{x=y+\varepsilon} - p(y-\varepsilon)\frac{\partial G(x,y)}{\partial x}\bigg|_{x=y-\varepsilon}\right\}$$

$$= \lim_{\varepsilon\to+0}\frac{p(y)}{w(y)a(y)}[H(+\varepsilon) - H(-\varepsilon)] = \frac{p(y)}{w(y)a(y)}$$

Hence, since $p(x)$ is continuous (as can be seen from its definition)

$$\lim_{\varepsilon \to +0} \left\{ \frac{\partial G(x,y)}{\partial x}\bigg|_{x=y+\varepsilon} - \frac{\partial G(x,y)}{\partial x}\bigg|_{x=y-\varepsilon} \right\} = \frac{1}{w(y)a(y)} \tag{8.13}$$

Hence, $\partial G(x,y)/\partial x$ suffers a discontinuity of magnitude $1/w(y)a(y)$ at $x = y$.

A further integration of Eq. 8.12 shows that $G(x,y)$ itself, unlike its first derivative, is continuous at $x = y$.

9 · CONSTRUCTION AND UNIQUENESS OF GREEN'S FUNCTIONS

We are now in the position to explain how a Green's function for a second-order differential equation can be constructed. We have noted that except at the point $x = y$, the Green's function satisfies the homogeneous differential equation

$$L_x G(x,y) = 0 \tag{9.1}$$

in the entire interval $[a,b]$.

Suppose that u_1 and u_2 are a fundamental set of solutions of Eq. 9.1. We can express the most general solution of Eq. 9.1, valid in the intervals to the left and to the right of the point $x = y$, as

$$\begin{aligned} u_<(c_1,c_2,x) &= c_1 u_1(x) + c_2 u_2(x) \qquad \text{for } a \le x < y \\ u_>(d_1,d_2,x) &= d_1 u_1(x) + d_2 u_2(x) \qquad \text{for } y < x \le b \end{aligned} \tag{9.2}$$

where c_1,c_2,d_1,d_2 are still arbitrary constants. Thus

$$G(x,y) = \begin{cases} u_<(c_1,c_2,x) & \text{for } a \le x < y \\ u_>(d_1,d_2,x) & \text{for } y < x \le b \end{cases} \tag{9.3}$$

The four constants in Eq. 9.3, which may, of course, depend upon the parameter y, must now be determined.

As already explained, $G(x,y)$ as a function of x must obey the same homogeneous boundary conditions as those associated with the differential equation (Eq. 7.1) whose solution we are seeking

$$\begin{aligned} B_1(G) &= 0 \\ B_2(G) &= 0 \end{aligned} \tag{9.4}$$

Furthermore, according to the results of the preceding section, $G(x,y)$ must be continuous at $x = y$, whereas its derivative at that point should have a jump of magnitude $1/a(y)w(y)$. These two conditions together with the two boundary conditions (Eqs. 9.4) imposed on $G(x,y)$ determine the four constants c_1,c_2,d_1,d_2, and hence the Green's function (Eq. 9.3).

There is an exceptional case, however, when the foregoing construction must fail. This is when the homogeneous equation $L_x u = 0$ has a nontrivial solution satisfying both boundary conditions. On the other hand, if the equation $L_x u = 0$ has no nontrivial solutions satisfying both boundary conditions, the construction of the Green's function is always possible.

To show this, we consider first the conditions that $G(x,y)$ must satisfy at the point $x = y$. Using Eqs. 9.3 and 8.13, these conditions imply

$$c_1u_1(y) + c_2u_2(y) = d_1u_1(y) + d_2u_2(y)$$
$$c_1\frac{du_1}{dx}\bigg|_{x=y} + c_2\frac{du_2}{dx}\bigg|_{x=y} = d_1\frac{du_1}{dx}\bigg|_{x=y} + d_2\frac{du_2}{dx}\bigg|_{x=y} - \frac{1}{a(y)w(y)} \tag{9.5}$$

That is

$$(c_1 - d_1)u_1(y) + (c_2 - d_2)u_2(y) = 0$$
$$(c_1 - d_1)\frac{du_1(y)}{dy} + (c_2 - d_2)\frac{du_2(y)}{dy} = \frac{-1}{a(y)w(y)} \tag{9.6}$$

These algebraic equations can always be solved with respect to $(c_1 - d_1)$ and $(c_2 - d_2)$ because the determinant of this system of equations is just the Wronskian $W(u_1,u_2)$ of u_1,u_2, which cannot vanish for any y because u_1 and u_2 are a fundamental set of solutions of $L_xu = 0$.

Let

$$c_1 - d_1 \equiv c$$
$$c_2 - d_2 \equiv d$$

where c and d are now known functions of y. Eliminating c_1 and c_2 from $G(x,y)$, we get

$$G(x,y) = G_s(x,y) + d_1u_1(x) + d_2u_2(x) \tag{9.7}$$

where

$$G_s(x,y) = \begin{cases} c(y)u_1(x) + d(y)u_2(x) & \text{for } a \le x < y \\ 0 & \text{for } y < x \le b \end{cases} \tag{9.8}$$

Imposing on $G(x,y)$ the homogeneous conditions (9.4), we find, owing to the linearity of these conditions

$$B_1(G) = B_1(G_s) + d_1B_1(u_1) + d_2B_1(u_2) = 0$$
$$B_2(G) = B_2(G_s) + d_1B_2(u_1) + d_2B_2(u_2) = 0 \tag{9.9}$$

These equations have a solution if and only if

$$\begin{vmatrix} B_1(u_1) & B_1(u_2) \\ B_2(u_1) & B_2(u_2) \end{vmatrix} \neq 0 \tag{9.10}$$

The necessary and sufficient condition for 9.10 to be satisfied is that the relations

$$\alpha B_1(u_1) + \beta B_1(u_2) = 0$$
$$\alpha B_2(u_1) + \beta B_2(u_2) = 0 \tag{9.11}$$

do not hold for any constants α and β. But, again using the linearity of the boundary conditions, we can rewrite Eq. 9.11 as

$$B_1(\alpha u_1 + \beta u_2) = 0$$
$$B_2(\alpha u_1 + \beta u_2) = 0 \tag{9.12}$$

These equations would imply the existence of a nontrivial solution $\alpha u_1 + \beta u_2$ of the equation $L_xu = 0$, satisfying both boundary conditions.

The condition that the equation has no nontrivial solutions not only ensures the possiblity of constructing a Green's function by the method we have described, but it also guarantees the uniqueness of the Green's function. For if there existed two Green's functions obeying the two boundary conditions, their difference $\tilde{G}$ would satisfy the homogeneous equation $L_x\tilde{G} = 0$ and the same boundary conditions.

We state these results as a theorem.

Theorem. Consider the 2^d order linear equation

$$L_x u = f(x) \tag{9.13}$$

with homogeneous boundary conditions

$$\begin{aligned} B_1(u) &= 0 \\ B_2(u) &= 0 \end{aligned} \tag{9.14}$$

Provided the homogeneous equation $L_x u = 0$ has no nontrivial solutions satisfying the boundary conditions (9.14), the Green's function associated with Eq. 9.13 exists and is unique. The solution of Eq. 9.13 given by

$$u(x) = \int_a^b dy\, w(y)G(x,y)f(y) \tag{9.15}$$

is unique.*

EXAMPLE 1

$$\begin{aligned} &\frac{d^2u}{dx^2} = f(x) \\ &u(0) = u(a) = 0 \qquad 0 \leq x \leq a \end{aligned} \tag{9.16}$$

The Green's function obeys the equation

$$\frac{d^2G(x,y)}{dx^2} = \delta(x-y) \qquad \text{(we take } w(x) \equiv 1\text{)}$$

A fundamental set of solutions of the homogeneous equation $d^2u/dx^2 = 0$ is $u_1 = 1$ and $u_2 = x$, and therefore

$$\begin{aligned} u_< &= c_1 + c_2 x \qquad 0 \leq x < y \\ u_> &= d_1 + d_2 x \qquad y < x \leq a \end{aligned}$$

The boundary conditions give $c_1 = 0$ and $d_1 = -ad_2$. Hence

$$u_< = c_2 x \qquad \text{and} \qquad u_< = d_2(x-a) \tag{9.17}$$

The continuity of $G(x,y)$ at $x = y$ yields the relation

$$c_2 y = d_2(y-a) \tag{9.18}$$

and the discontinuity of dG/dx leads to

$$d_2 - c_2 = 1 \tag{9.19}$$

* The uniqueness of $u(x)$ is proved in the same manner as the uniqueness of $G(x,y)$.

Equations 9.17, 9.18, and 9.19 determine $G(x,y)$

$$G(x,y)=\begin{cases}\dfrac{(y-a)x}{a} & x\le y\\[2ex] \dfrac{y(x-a)}{a}\,. & x\ge y\end{cases}$$

The solution of Eq. 9.16 is therefore

$$u(x)=\int_0^a G(x,y)f(y)\,dy=\frac{(x-a)}{a}\int_0^x dy\,yf(y)+\frac{x}{a}\int_x^a dy\,(y-a)f(y)$$

By putting, for example, $f(y)=1$ into this equation, it can be easily verified after an integration that this is indeed a solution of Eq. 9.16.

EXAMPLE 2

We have seen that the operator L_x in Eq. 6.7 was self-adjoint with respect to a norm with weight $w(x)$ given by Eq. 6.13. Equivalently, wL_x is self-adjoint with respect to a norm with weight unity. We solve the general equation

$$\frac{d}{dx}\left[p\,\frac{du}{dx}\right]+wcu=f(x)\qquad a\le x\le b \tag{9.20}$$

with $p(x)\neq 0$ and under the homogeneous boundary conditions

$$u(a)=\alpha\left.\frac{du}{dx}\right|_{x=a} \tag{9.21}$$

$$u(b)=\beta\left.\frac{du}{dx}\right|_{x=b} \tag{9.22}$$

Let u_1 and u_2 be a fundamental set of solutions of the homogeneous equation $wL_xu=0$, and let

$$\begin{aligned}u_<(c_1,c_2,x)&=c_1u_1+c_2u_2\\ u_>(d_1,d_2,x)&=d_1u_1+d_2u_2\end{aligned} \tag{9.23}$$

be the solutions of that equation valid in the left $(a\le x<y)$ and right $(y<x\le b)$ intervals, respectively. Then

$$G(x,y)=\begin{cases}u_<(c_1,c_2,x) & a\le x<y\\ u_>(d_1,d_2,x) & y<x\le b\end{cases} \tag{9.24}$$

The boundary conditions (Eqs. 9.21 and 9.22) give

$$\begin{aligned}c_1u_1(a)+c_2u_2(a)&=\alpha\left[c_1\left.\frac{du_1}{dx}\right|_{x=a}+c_2\left.\frac{du_2}{dx}\right|_{x=a}\right]\\ d_1u_1(b)+d_2u_2(b)&=\beta\left[d_1\left.\frac{du_1}{dx}\right|_{x=b}+d_2\left.\frac{du_2}{dx}\right|_{x=b}\right]\end{aligned} \tag{9.25}$$

From Eqs. 9.21 and 9.22 we see that the functions

$$U_<(x)=\left[u_1(x)+\frac{u_1(a)-\alpha\left.\dfrac{du_1}{dx}\right|_{x=a}}{\alpha\left.\dfrac{du_2}{dx}\right|_{x=a}-u_2(a)}\,u_2(x)\right]$$

and

$$U_>(x) = \left[u_1(x) + \frac{u_1(b) - \beta \left.\dfrac{du_1}{dx}\right|_{x=b}}{\beta \left.\dfrac{du_2}{dx}\right|_{x=b} - u_2(b)} u_2(x) \right]$$

are solutions of the homogeneous equation $L_x G = 0$ and satisfy respectively the left- and right-hand boundary conditions. The conditions that G be continuous at $x = y$ and that dG/dx has a discontinuity of magnitude $1/p(y)$ at $x = y$ yields the two relations

$$c_1 U_<(y) = d_1 U_>(y)$$
$$d_1 \left.\frac{dU_>(x)}{dx}\right|_{x=y} - c_1 \left.\frac{dU_<}{dx}\right|_{x=y} = \frac{1}{p(y)} \tag{9.26}$$

Whence

$$c_1 = \frac{U_>(y)}{p(y)\left[U_<(y)\left.\dfrac{dU_>}{dx}\right|_{x=y} - U_>(y)\left.\dfrac{dU_<}{dx}\right|_{x=y}\right]}$$
$$d_1 = \frac{U_<(y)}{p(y)\left[U_<(y)\left.\dfrac{dU_>}{dx}\right|_{x=y} - U_>(y)\left.\dfrac{dU_<}{dx}\right|_{x=y}\right]}$$

and so we have

$$G(x,y) = \begin{cases} \dfrac{U_<(x)U_>(y)}{p(y)\left[U_<(y)\left.\dfrac{dU_>}{dx}\right|_{x=y} - U_>(y)\left.\dfrac{dU_<}{dx}\right|_{x=y}\right]} & a \le x < y \\[2ex] \dfrac{U_<(y)U_>(x)}{p(y)\left[U_<(y)\left.\dfrac{dU_>}{dx}\right|_{x=y} - U_<(y)\left.\dfrac{dU_<}{dx}\right|_{x=y}\right]} & y \le x < b \end{cases} \tag{9.27}$$

The factor in brackets in the denominator of Eq. 9.27 is simply the Wronskian of the two solutions $U_<$ and $U_<$

$$W = U_<(y)\left.\frac{dU_>}{dx}\right|_{x=y} - \left.\frac{dU_<}{dx}\right|_{x=y} U_>(y)$$

Since $U_<$ and $U_>$ obey the homogeneous equation it follows from Eq. 9.20, with $f(x) = 0$, that

$$\frac{d}{dx}\left\{p(x)\left[U_<(x)\frac{dU_>}{dx} - \frac{dU_<}{dx}U_>(x)\right]\right\} = 0$$

so that

$$p(x)\left[U_<(x)\frac{dU_>}{dx} - \frac{dU_<}{dx}U_>(x)\right] = p(y)\left[U_<(y)\left.\frac{dU_>}{dx}\right|_{x=y} - \left.\frac{dU_<}{dx}\right|_{x=y} U_>(y)\right] = \text{const} \tag{9.28}$$

If $U_<$ and $U_>$ are linearly independent solutions of $L_x u = 0$, then $W(U_<, U_>) \neq 0$, and the constant in Eq. 9.28 is nonzero. Conversely, if the constant is zero, then $U_<$ and $U_>$ are linearly dependent solutions of $L_x u = 0$; in that case, there exists a nonzero constant η such that

$$U_<(x) = \eta U_>(x) \tag{9.29}$$

But since $U_>(b) - \beta\, du_>/dx|_{x=b} = 0$, it follows from Eq. 9.29 that

$$U_<(b) - \beta \left.\frac{dU_>}{dx}\right|_{x=b} = \eta\left[U_>(b) - \beta \left.\frac{dU_>}{dx}\right|_{x=b}\right] = 0 \tag{9.30}$$

so that

$$U_<(b) = \beta \left.\frac{dU_>}{dx}\right|_{x=b}$$

and from Eq. 9.21

$$U_<(a) = \alpha \left.\frac{dU_<}{dx}\right|_{x=a}$$

Hence, $U_<$ is a solution of the homogeneous equation $L_x u = 0$, and satisfies the associated boundary conditions.

The reader will also note that because of Eq. 9.28, $G(x,y)$ is indeed a symmetric function of x and y.

EXAMPLE 3

The differential equation

$$x^2 \frac{d^2u}{dx^2} + x \frac{du}{dx} - (1 - \nu^2 x^2)u = 0 \tag{9.31}$$

with $u(0) = u(1) = 0$, can be cast into the form

$$L_x u \equiv \frac{d}{dx}\left[x \frac{du}{dx}\right] - \frac{1}{x} u = -\nu^2 x u \tag{9.32}$$

Consider the right side of Eq. 9.32 as an inhomogeneous term. Then, with the results of Example 2, we can easily obtain a Green's function for $L_x = x(d^2/dx^2) + (d/dx) - (1/x)$. A fundamental set of solutions of $L_x u = 0$ is x and $1/x$. Hence

$$u_< = x \qquad \text{and} \qquad u_> = \frac{1}{x} - x$$

are solutions of $L_x u = 0$, which obey respectively the left and right boundary conditions. From Eq. 9.27, since $p(x) = x$ and since

$$u_< \frac{du_>}{dx} - \frac{du_<}{dx} u_> = -\frac{2}{x}$$

we have

$$G(x,y) = \begin{cases} \dfrac{x}{2y}(y^2 - 1) & 0 \le x \le y \\[2ex] \dfrac{y}{2x}(x^2 - 1) & y \le x \le 1 \end{cases}$$

and therefore the solution of Eq. 9.31 can be expressed as

$$u(x) = -\nu^2 \int_0^1 dy\, G(x,y) y u(y)$$

With the aid of the Green's function we have therefore transformed a differential equation into an integral equation. In some cases an integral equation is more easily solved than the corresponding differential equation.

We now show that the condition for the nonexistence of a trivial solution of the equation

$$L_x u = 0 \tag{9.33}$$

obeying the boundary conditions, is not only a sufficient condition for the existence of the Green's function $G(x,y)$, but it is also necessary.

Suppose that $G(x,y)$ exists. Then from Green's identity it follows that the adjoint homogeneous equation cannot have a nontrivial solution satisfying the adjoint boundary conditions. For if $v(x)$ were a solution of

$$L_x^+ v = 0 \tag{9.34}$$

one would have, setting $u = G(x,y)$ in Green's identity

$$\int_a^b dx\, w(x)\bar{v}(x)[L_x G(x,y)] = \int_a^b dx\, w(x)G(x,y)\overline{[L_x^+ v(x)]} = 0 \tag{9.35}$$

Using Eq. 7.8, Eq. 9.35 would yield

$$\int_a^b dx\, v(x)\delta(x - y) = v(y) = 0 \tag{9.36}$$

and this would contradict the existence of a nontrivial solution of Eq. 9.34.

Since Eq. 9.34 has no nontrivial solutions, there exists a unique adjoint Green's function $g(x,y)$. Using Eq. 8.1, we see that $G(x,y)$ must also be unique. This, however, implies that Eq. 9.33 has no nontrivial solutions, since such a solution could always be added to $G(x,y)$ and hence $G(x,y)$ could not be unique.

The condition that Eq. 9.33 has no nontrivial solutions is a necessary condition for the inhomogeneous equation (Eq. 7.1) to have a unique solution satisfying Eqs. 9.4. However, even in the case when Eq. 9.33 does have a nontrivial solution, it is still possible under certain conditions to generalize the considerations of this section.

Notice that the existence of a nontrivial solution of Eq. 9.33 satisfying Eqs. 9.4 implies the existence of a nontrivial solution of Eq. 9.34 satisfying the adjoint boundary conditions. Otherwise, $g(x,y)$ would exist and, again using Green's identity, we could show that this would contradict the existence of a nontrivial solution of Eq. 9.33. Let $|v_i\rangle$ $(i \le 2)$ be the vectors representing the nontrivial solution of Eq. 9.34. Multiplying the vector equation

$$L|u\rangle = |f\rangle$$

from the left by $\langle v_i|$, we have

$$\langle v_i|\, L\, |u\rangle = \langle v_i|f\rangle$$

Hence

$$\langle v_i|f\rangle = \overline{\langle u|\, L^+\, |v_i\rangle} = 0 \qquad (i \le 2) \tag{9.37}$$

is a necessary condition for the existence of a solution of the equation $L_x u = f(x)$ under homogeneous boundary conditions. Similarly, we find that

$$\langle u_i|h\rangle = 0 \qquad (i \le 2) \tag{9.38}$$

is a necessary condition for the existence of a solution of the adjoint problem

$$L_x^+ v = h$$

with adjoint boundary conditions. In Eq. 9.38, $|u_i\rangle$ denotes a vector that represents a nontrivial solution of Eqs. 9.33 and 9.4.

If condition 9.35 is satisfied, the generalization of the theorem of this section can be achieved by properly generalizing the notion of the Green's function. The interested reader can find a relevant discussion in the next section.

10 · GENERALIZED GREEN'S FUNCTION

In previous sections we discussed a method for constructing the Green's function for a differential equation which, as we have seen, succeeds if the associated adjoint homogeneous equation with the prescribed homogeneous boundary conditions has no nontrivial solutions. We also stated that if the adjoint homogeneous equation did possess nontrivial solutions, a solution of the equation nevertheless existed, but only if the orthogonality conditions (Eq. 9.37) were satisfied. However, we mentioned that even in this case, the method of Green's functions had to be generalized. This section deals with such a generalization.

Let v_i $(i \leq 2)$ be a set of orthonormalized nontrivial solutions of the second-order adjoint homogeneous equation $L_x^+ v = 0$, and let u_i $(i \leq 2)$ be a set of orthonormalized* nontrivial solutions of the second-order equation $L_x u = 0$. The so-called **generalized** Green's function and adjoint Green's function are defined by the equations

$$L_x G'(x,y) = \frac{\delta(x-y)}{w(x)} - \sum_i \bar{v}_i(x) v_i(y) \tag{10.1}$$

$$L_x g'(x,y) = \frac{\delta(x-y)}{w(x)} - \sum_i \bar{u}_i(x) u_i(y) \tag{10.2}$$

which replace Eqs. 7.8 and 7.9 when $u_i, v_i \neq 0$.

The solutions of Eqs. 10.1 and 10.2 are, however, not yet unique because one can still add to $G'(x,y)$ any linear combination of the u_i and to $g'(x,y)$ any linear combination of the v_i. Uniqueness is achieved by imposing the additional orthogonality conditions

$$\langle u_i|G'\rangle = 0 \qquad (i \leq 2) \tag{10.3}$$

$$\langle v_i|g'\rangle = 0 \qquad (i \leq 2) \tag{10.4}$$

The reader can verify that all the definitions are now self-consistent, that the solutions of Eqs. 7.1 and 7.2 are still given by Eqs. 8.2 and 8.3, respectively, with the Green's functions replaced by the generalized ones, and that all the usual properties of the ordinary Green's functions are also properties of the generalized ones.

EXAMPLE

Consider the differential equation

$$\frac{d^2u}{dx^2} = f(x) \qquad -a \leq x \leq a \tag{10.5}$$

with the boundary conditions

$$u(a) = u(-a)$$
$$\left.\frac{du}{dx}\right|_{x=a} = \left.\frac{du}{dx}\right|_{x=-a} \tag{10.6}$$

---

* The sets can be orthonormalized, since the solutions are supposed to be linearly independent.

The homogeneous equation $d^2u/dx^2 = 0$ has a nontrivial solution, $u_1 = \text{const}$, which obeys the two boundary conditions 10.6. The constant is determined by the normalization condition

$$\int_{-a}^{a} u_1^2 \, dx = 1$$

Thus

$$u_1 = \frac{1}{\sqrt{2a}} \tag{10.7}$$

The generalized Green's function satisfies ($w = 1$)

$$\frac{d^2G'(x,y)}{dx^2} = \delta(x-y) - \frac{1}{2a} \tag{10.8}$$

We proceed as before and find the most general solution of Eq. 10.8 without the δ function, valid for $x \leq y$ and for $x \geq y$. The boundary conditions, the continuity of $G'(x,y)$ at $x = y$ and the discontinuity of $\partial G'/\partial x$ at that point determine in this case only three of the four arbitrary constants

$$G'(x,y) = \begin{cases} \dfrac{-x^2}{4a} + \left(\dfrac{y-a}{2a}\right)x + \alpha & x \leq y \\[2ex] \dfrac{-x^2}{4a} + \left(\dfrac{y+a}{2a}\right)x + (\alpha - y) & x \geq y \end{cases} \tag{10.9}$$

The constant α is determined by the additional condition 10.3. Finally

$$G'(x,y) = \frac{-1}{4a}(x-y)^2 + \frac{1}{2}|x-y| - \frac{a}{6}$$

11 · SECOND-ORDER EQUATIONS WITH INHOMOGENEOUS BOUNDARY CONDITIONS

Up to now we have considered inhomogeneous differential equations, but the boundary conditions were assumed to be homogeneous. Consider now the case where the boundary conditions have the inhomogeneous form

$$\begin{aligned} B_1(u) &= \sigma_1 \\ B_2(u) &= \sigma_2 \end{aligned} \tag{11.1}$$

The solution of this problem can be carried out in two steps. First we find as before the Green's function associated with the equation

$$L_x u = f \tag{11.2}$$

under the homogeneous boundary conditions

$$\begin{aligned} B_1(u) &= 0 \\ B_2(u) &= 0 \end{aligned} \tag{11.3}$$

obtained from Eqs. 10.1 by setting $\sigma_1 = \sigma_2 = 0$.

Having found this Green's function, we return to the **generalized** Green's identity, which is a pure analytic identity and which holds for **any** boundary conditions.

Putting $v(x) = g(x,y) = \bar{G}(y,x)$ into the generalized Green's identity (Eq. 4.5), we find

$$u(y) = \int_a^b dx\, w(x)G(y,x)f(x) + Q[u(x),G(y,x)]\Big|_{x=a}^{x=b} \tag{11.4}$$

In particular, choosing $w(x)$ according to Eq. 6.13, the surface term $Q\,[u,G]$ is given by Eq. 6.14

$$u(y) = \int_a^b dx\, w(x)G(y,x)f(x) + \left\{w(x)a(x)\left[G(y,x)\frac{du}{dx} - u\frac{\partial G(x,y)}{\partial x}\right]\right\}\Big|_{x=a}^{x=b} \tag{11.5}$$

The surface term contains the values of $u(x)$ and du/dx at $x = a$ and at $x = b$. The reader may ask how two boundary conditions can determine these four quantities. Naturally, they do not. However, with their help, one can eliminate from the surface term in Eq. 11.5 two of these quantities, and the coefficients of the other two quantities will automatically be zero because $G(y,x) = \bar{g}(x,y)$ and the boundary conditions adjoint to the conditions 11.3, which $g(x,y)$ satisfies, were just defined so as to make these coefficients vanish.

The net result is that the surface term will contain only $G(y,x)$ and $\partial G(y,x)/\partial x$ evaluated at the end points $x = a$ and $x = b$ and the constants appearing in the boundary conditions (Eq. 11.1), including σ_1 and σ_2.

Setting $f(x) = 0$ in Eq. 11.2, we also obtain the solution of the homogeneous equation $L_x u = 0$ with inhomogeneous boundary conditions.

EXAMPLE

Let us consider once more Example 1, p. 279, but now with the inhomogeneous conditions

$$u(0) = \sigma_1; \qquad u(a) = \sigma_2 \tag{11.6}$$

The homogeneous boundary conditions to be imposed on $G(x,y)$, corresponding to Eq. 11.6, are

$$G(x,0) = 0; \qquad G(x,a) = 0 \tag{11.7}$$

We have

$$\frac{\partial G(x,y)}{\partial y}\Big|_{y=0} = \frac{x-a}{a}; \qquad \frac{\partial G(x,y)}{\partial y}\Big|_{y=a} = \frac{x}{a} \tag{11.8}$$

Hence, inserting Eqs. 11.6, 11.7, and 11.8 into Eq. 11.5 yields the solution

$$u(x) = \int_0^a dy\, G(x,y)f(y) + \left[(\sigma_2 - \sigma_1)\frac{x}{a} + \sigma_1\right]$$

12 · THE STURM-LIOUVILLE PROBLEM

Let L be a Hermitian differential operator

$$L = L^+$$

The investigation of the eigenvalue equation

$$L\,|u_\lambda\rangle = \lambda\,|u_\lambda\rangle \tag{12.1}$$

forms the content of the so-called Sturm-Liouville problem. The analytical counterpart of Eq. 12.1 is a differential eigenvalue equation

$$L_x u_\lambda(x) = \lambda u_\lambda(x) \tag{12.2}$$

where the functions $u_\lambda(x)$ are subjected to certain ("self-adjoint") homogeneous boundary conditions. These boundary conditions together with differentiability requirements define a certain function space, the domain of L, which we shall denote by U. As before, we limit our considerations to the case when L_x is a second-order operator.

The theorem of Sec. 9 asserted the existence and uniqueness of the solution of

$$L_x u(x) = f(x) \tag{12.3}$$

with homogeneous boundary conditions imposed on $u(x)$. For any $f(x)$, this solution is

$$u(x) = \int_a^b dy\, w(y)G(x,y)f(y) \tag{12.4}$$

and the existence of $G(x,y)$ is ensured by the condition that Eq. 12.2 has no nontrivial solutions with eigenvalue $\lambda = 0$.

We shall assume in what follows that the last condition is fulfilled. Then Green's integral operator G not only exists, but is a unique right inverse of L (otherwise one might have solutions of Eq. 12.3 other than that given by Eq. 12.4). Hence, G is the operator inverse to L (see Sec. 9, Chapter II)

$$LG = GL = E \tag{12.5}$$

and the eigenvalue equation (Eq. 12.1) is equivalent to

$$G\,|u_\lambda\rangle = \frac{1}{\lambda}\,|u_\lambda\rangle \tag{12.6}$$

(Remember that we assumed $\lambda \neq 0$.) By a direct construction it was shown that the kernel $G(x,y)$ of G is a continuous function of both x and y. Therefore, the integral

$$\int_a^b\int_a^b dx\, dy\, w(x)w(y)|G(x,y)|^2$$

exists, and G is a completely continuous integral operator. Furthermore, G is Hermitian; this follows directly from the hermiticity of L, for Eq. 12.5 yields

$$LG^+ = G^+L = E$$

and since the operator inverse to L is unique, we have

$$G = G^+$$

We are now in a position to discuss the Sturm-Liouville problem.

First we note that since L is Hermitian, all its eigenvalues are real and its eigenvectors corresponding to different eigenvalues are orthogonal (cf. theorem of Sec. 11.1, Chapter II)

$$\langle u_\lambda | u_{\lambda'}\rangle = 0 \qquad \text{for } \lambda \neq \lambda' \tag{12.7}$$

In the language of analysis, Eq. 12.7 reads

$$\int_a^b dx\, w(x)u_\lambda(x)\bar{u}_{\lambda'}(x) = 0 \qquad (\lambda \neq \lambda') \tag{12.8}$$

Because of the equivalence between the eigenvalue equations 12.1 and 12.6, the theorem of Sec. 14.6 of the preceding chapter, which is fully applicable to G, leads immediately to the following results:

(i) There exists at least one eigenvector of L with a finite eigenvalue.

One has in fact an even stronger result. Notice that G cannot have an eigenvalue zero, for if $|u_0\rangle \neq 0$ were the corresponding eigenvector, one would have

$$LG\,|u_0\rangle = 0$$

This is in contradiction to Eq. 12.5, which requires that

$$LG\,|u_0\rangle = E\,|u_0\rangle = |u_0\rangle \neq 0$$

The fact that G has no vanishing eigenvalue implies, according to the results of Sec. 14.6 of Chapter III, that G has an infinite (but, of course, enumerable) number of eigenvalues that have zero as an accumulation point. Hence:

(i) L has an infinity of eigenvalues and their absolute values are not bounded.

(ii) For an arbitrary $\eta > 0$, there can exist only a finite number of mutually orthogonal and normalized eigenvalues that lie within the interval $[-\eta,\eta]$; to a given eigenvalue corresponds a finite number of eigenvectors.

(iii) The eigenvectors of L span the space $L_w^2(a,b)$. Any vector $|f\rangle \in L_w^2(a,b)$ can be expanded in a series (the index k distinguishes between different eigenvectors corresponding to the same eigenvalue)

$$|f\rangle = \sum_{\lambda,k} \langle k,u_\lambda|f\rangle\,|u_\lambda,k\rangle \tag{12.9}$$

In terms of functions, Eq. 12.9 reads

$$f(x) = \sum_{\lambda,k} \langle k,u_\lambda|f\rangle u_\lambda^{(k)}(x) \tag{12.10}$$

where $f(x)$ is a function which represents a vector belonging to the function space $L_w^2(a,b)$.

We shall not discuss in detail the problem of the convergence of the expansion 12.10. The convergence in the mean is evident. To have the expansion 12.10 converge uniformly, it is necessary to impose certain restrictive conditions on $f(x)$. It can be proved that these conditions are essentially the same as those that ensure the uniform convergence of a trigonometric series (see Sec. 11.2, Chapter III). Notice that these conditions do not require that $f(x)$ be twice differentiable in the elementary sense; for example, it is sufficient for $f(x)$ to be continuous and to have a derivative with discontinuities of the first kind (in this case the second derivative of $f(x)$ will be a generalized function).

13 · EIGENFUNCTION EXPANSION OF GREEN'S FUNCTIONS

As in the preceding section, we consider a Hermitian differential operator L that has only nonzero eigenvalues. Take the equation

$$(L_x - l)u(x) = f(x) \qquad a \leq x \leq b \tag{13.1}$$

where again we suppose that L_x is of second order, and l is a constant.

It may be that the solution of the homogeneous equation

$$(L_x - l)u = 0 \tag{13.2}$$

cannot be easily obtained but that the eigenvalues λ_n $(n = 0, 1, \cdots)$ and eigenvectors $|\lambda_n,k\rangle$ of the operator L alone are known; i.e., we know the set of eigenfunctions $u_n^{(k)}(x)$ which satisfy the equations

$$L_x u_n^{(k)}(x) = \lambda_n u_n^{(k)}(x) \qquad (n = 0, 1, \cdots) \tag{13.3}$$

together with the ("self-adjoint") homogeneous boundary conditions associated with Eq. 13.1. In that case, we can find a solution of Eq. 13.1.

The Green's function relative to Eq. 13.1 satisfies

$$(L_x - l)G_l(x,y) = \frac{\delta(x-y)}{w(x)} \tag{13.4}$$

The index l on $G_l(x,y)$ indicates that $G_l(x,y)$ is the Green's function for a given value of l.

Let us expand $G(x,y)$ in a series of eigenfunctions $u_n^{(k)}(x)$.

$$G(x,y) = \sum_{n=0,k}^{\infty} a_n^{(k)}(y)u_n^{(k)}(x) \tag{13.5}$$

Using Green's identity one has

$$\begin{aligned}\lambda_n a_n^{(k)}(y) &\equiv \lambda_n \int_a^b dx\, w(x)\bar{u}_n^{(k)}(x)G_l(x,y) \\ &= \int_a^b dx\, G_l(x,y)[\overline{(L_x - l)u_n^{(k)}(x)}] + la_n^{(k)}(y) \\ &= \bar{u}_n^{(k)}(y) + la_n^{(k)}(y)\end{aligned} \tag{13.6}$$

Therefore

$$a_n^{(k)}(y) = \frac{\bar{u}_n^{(k)}(y)}{\lambda_n - l} \tag{13.7}$$

Hence, the Green's function is

$$G_l(x,y) = \sum_{n=0,k}^{\infty} \frac{\bar{u}_n^{(k)}(y)u_n^{(k)}(x)}{\lambda_n - l} \tag{13.8}$$

The expansion (13.8) is valid, provided $\lambda_n \neq l$ for any n. However $\lambda_n = l$ means that there is a nontrivial solution of the homogeneous equation associated with Eq. 13.1, and we know that under this circumstance a Green's function in the ordinary sense does not exist.

The form of the Green's function obtained here is very different from the forms that we have been used to up to now. In contradistinction to all previous derivations, we have obtained a Green's function in the form of an infinite series. This is the price that one has to pay for not knowing the solutions of the complete homogeneous equation (Eq. 13.2). Of course the two procedures are the same, and one can exploit this equivalence to obtain useful information.

Suppose that we analytically continue Eq. 13.8 to complex values of l. Then $G_l(x,y)$ may be regarded as an analytic function of l with simple poles at $l = \lambda_n$ $(n = 0, 1, \cdots)$. The residues at these poles are

$$-\sum_k \bar{u}_n^{(k)}(y)u_n^{(k)}(x)$$

Therefore

$$\frac{1}{2\pi i}\int_C G_l(x,y)\,dl = -\sum_{n=0,k}^{\infty} \bar{u}_n^{(k)}(y)u_n^{(k)}(x) \tag{13.9}$$

where the path C is a circle extending to infinity so that it includes all the eigenvalues λ_n.

But, by Eq. 14.68, Chapter III, the sum on the right side of Eq. 13.9 is simply $\delta(x-y)$. Hence, we have the result

$$\frac{1}{2\pi i}\int_C G_l(x,y)\,dl = -\delta(x-y) \tag{13.10}$$

We illustrate these properties by an example.

EXAMPLE

$$\frac{d^2u}{dx^2} + lu(x) = f(x)$$
$$u(0) = u(a) = 0 \tag{13.11}$$

We first find the Green's function for Eq. 13.11 by the methods of Sec. 9. A fundamental set of solutions of the homogeneous equation

$$\frac{d^2u}{dx^2} + lu = 0$$

is $\sin\sqrt{l}x$ and $\cos\sqrt{l}x$. By a now familiar calculation, we obtain the Green's function

$$G_l(x,y) = \frac{1}{\sqrt{l}\sin\sqrt{l}a}\{\sin\sqrt{l}x\sin\sqrt{l}(a-y)H(y-x)$$
$$+ \sin\sqrt{l}(a-x)\sin\sqrt{l}yH(x-y)\} \tag{13.12}$$

It is easy to check that the only singularities of G_l come from those values of l for which $\sin\sqrt{l}a = 0$, i.e., for the values $l = \lambda_n$ where

$$\lambda_n = \frac{n^2\pi^2}{a^2} \qquad (n = 0, 1, 2, \cdots) \tag{13.13}$$

The residues of G_l at these (simple) poles are

$$-\frac{2}{a}\sin\frac{n\pi}{a}x\sin\frac{n\pi}{a}y[H(y-x)+H(x-y)] = -\frac{2}{a}\sin\frac{n\pi}{a}x\sin\frac{n\pi}{a}y$$

Comparing with Eq. 13.8 shows that

$$u_n(x) = \sqrt{\frac{2}{a}}\sin\frac{n\pi}{a}x \tag{13.14}$$

These are indeed the orthonormal eigenfunctions of the equation

$$\frac{d^2u}{dx^2} + \lambda_n u_n = 0$$

with the boundary conditions $u_n(a) = u_n(0) = 0$. Finally, the formula 13.10 is translated here into the relation

$$\frac{2}{a} \sum_{n=0}^{\infty} \sin \frac{n\pi}{a} x \sin \frac{n\pi}{a} y = \delta(x - y) \tag{13.15}$$

which is just a particular case of the general result (Eq. 14.68) of Chapter III. The LHS of Eq. 13.15 does not converge for $x = y$, and this relation must be interpreted in the sense that after multiplying it by any continuous function, $g(y)$ say, the integration over y must be carried out **before** the summation of the series is effected. Doing this, we find

$$\frac{2}{a} \sum_{n=1}^{\infty} \sin \frac{n\pi}{a} x \int_0^a g(y) \sin \frac{n\pi}{\alpha} y \, dy = g(x)$$

which is just the Fourier sine series for $g(x)$ over the interval $[0, a]$ (see Eq. 11.21, Chapter III).

14 · SERIES SOLUTIONS OF LINEAR DIFFERENTIAL EQUATIONS OF THE SECOND ORDER THAT DEPEND ON A COMPLEX VARIABLE

14.1 Introduction

We know from elementary calculus that the solutions of differential equations with constant coefficients can be expressed in terms of elementary functions. This is usually no longer true when the coefficients are functions of the independent variable. In this case, the solutions lead to transcendental functions, which can be expressed either in terms of infinite series or in terms of definite integrals.

In the next few sections we shall consider second-order homogeneous differential equations in which the independent variable is complex. The interest here will be on the analytic properties of the solutions of these equations.

The reader should have no difficulty in convincing himself that all results of Sec. 2 that were derived for differential equations which depended on a real variable, also hold when the variable is complex.

14.2 Classification of Singularities

Consider the general homogeneous differential equation of the second order, written in such a way that the coefficient of d^2u/dz^2 is unity.

$$\frac{d^2u}{dz^2} + p(z)\frac{du}{dz} + q(z)u = 0 \tag{14.1}$$

We limit our considerations to the cases where the functions $p(z)$ and $q(z)$ are analytic in a certain region R, except perhaps at an enumerable number of points of R where these functions may have isolated singularities.

If $p(z)$ and $q(z)$ are analytic at a point z_0 of R, z_0 is called an **ordinary point** of the differential equation. We shall show in the next section that in a neighborhood of an ordinary point, Eq. 14.1 has two linearly independent analytic solutions.

If z_0 is an isolated singularity of either $p(z)$ or $q(z)$, or both, z_0 is called a **singular point** of the equation. In general, at least one of the solutions of Eq. 14.1 will then have a singularity at $z = z_0$.

It turns out that the behavior of the solutions of Eq. 14.1 in the environment of a singular point z_0 depends in a crucial way upon the nature of the singularities of $p(z)$ or $q(z)$, or both, at that point. This is why the following distinction is made.

If $p(z)$ has a pole of order no greater than 1, and if $q(z)$ has a pole of order no greater than 2 at $z = z_0$, the point z_0 is called a **regular singular point** of the equation.

In all other singular cases, z_0 is called an **irregular singular point** of the equation.

It will be shown in Sec. 16 that when z_0 is a regular singular point, the two fundamental solutions of Eq. 14.1 in an environment of z_0 have either the form

$$\begin{aligned} u_1(z) &= (z - z_0)^{r_1} \sum_{n=0}^{\infty} c_n^{(1)}(z - z_0)^n \\ u_2(z) &= (z - z_0)^{r_2} \sum_{n=0}^{\infty} c_n^{(2)}(z - z_0)^n \qquad (r_1 \neq r_2) \end{aligned} \tag{14.2}$$

or the form

$$\begin{aligned} u_1(z) &= (z - z_0)^{r_1} \sum_{n=0}^{\infty} c_n^{(1)}(z - z_0)^n \\ u_2(z) &= (z - z_0)^{r_2} \sum_{n=0}^{\infty} c_n^{(2)}(z - z_0)^n + \text{const}\, u_1(z)\ln(z - z_0) \end{aligned} \tag{14.3}$$

It can be shown (but we shall skip the proof) that if z_0 is an irregular singular point, the two solutions of Eq. 14.1 will have again one of the two forms 14.2 or 14.3, but at least one of the sums will extend from $n = -\infty$ to $n = +\infty$. In that case, the theory we develop in the following sections will fail; however, this point cannot be pursued here.

The next section will be concerned with the solutions of Eq. 14.1 that are valid in a neighborhood of an ordinary point, and the subsequent sections will deal with solutions that are valid in a neighborhood of a regular singular point.

14.3 Existence and Uniqueness of the Solution of a Differential Equation in the Neighborhood of an Ordinary Point

Let z_0 be an ordinary point of R; we seek a solution of Eq. 14.1, valid in the neighborhood of this point, and which satisfies the boundary conditions

$$u(z_0) = a \qquad \left.\frac{du}{dz}\right|_{z=z_0} = b \tag{14.4}$$

It is usual to transform Eq. 14.1 into a more tractable form by making the substitution

$$u(z) = f(z)\exp\left(-\frac{1}{2}\int_{z_0}^{z} p(\xi)d\xi\right) \tag{14.5}$$

Putting Eq. 14.5 into Eq. 14.1, it is easy to see that $f(z)$ satisfies the differential equation

$$\frac{d^2 f}{dz^2} + k(z)f(z) = 0 \tag{14.6}$$

where

$$k(z) \equiv q(z) - \frac{1}{4} p^2(z) - \frac{1}{2} \frac{dp}{dz} \tag{14.7}$$

We now define a sequence of functions analytic in R

$$f_0(z), \quad f_1(z), \quad f_2(z), \quad \cdots$$

which will be taken to satisfy the recurrence relation

$$\frac{d^2 f_n(z)}{dz^2} + k(z) f_{n-1}(z) = 0 \qquad (n = 1, 2, \cdots) \tag{14.8}$$

and show that the functions $f_n(z)$ can be used to construct a solution of the differential equation 14.1 ($f_n(z_0) = f_n'(z_0) = 0$, $n > 0$, is assumed).

By integrating Eq. 14.8 twice with respect to z, one first obtains*

$$\frac{df_n(z)}{dz} = -\int_{z_0}^{z} k(z') f_{n-1}(z')\, dz' \tag{14.9}$$

and then

$$f_n(z) = -\int_{z_0}^{z} dz'' \int_{z_0}^{z''} dz'\, k(z') f_{n-1}(z')\, dz' \tag{14.10}$$

Integrating Eq. 14.10 by parts, we get

$$\begin{aligned} f_n(z) &= -\left\{ z'' \int_{z_0}^{z''} k(z') f_{n-1}(z')\, dz' \right\} \Bigg|_{z''=z_0}^{z''=z} \\ &\quad + \int_{z_0}^{z} z''\, k(z'') f_{n-1}(z'')\, dz'' \\ &= \int_{z_0}^{z} (z' - z) k(z') f_{n-1}(z')\, dz' \end{aligned} \tag{14.11}$$

We now make the conjecture that

$$|f_m(z)| \leq \max|f_0|\, [\max|k|]^m \frac{|z - z_0|^{2m}}{m!} \tag{14.12}$$

for any $z \in$ R. Obviously, this relation holds for $m = 0$. We shall prove the validity of 14.12 by induction. Suppose that it holds for $m = n - 1$. Integrating Eq. 14.11 along a straight line joining z_0 to z, with parametric equation

$$z' = z_0 + (z - z_0)t \qquad 0 \leq t \leq 1$$

---

* We recall that (see Eqs. 16.4 and 16.1, Chapter I)

$$\frac{d}{dz} \int_{z_0}^{z} f(z')\, dz' = f(z)$$

one has, using 14.12 with $m = n - 1$

$$\begin{aligned}
|f_n(z)| &= |z - z_0|^2 \left| \int_0^1 k[z'(t)](t-1) f_{n-1}[z'(t)]\, dt \right| \\
&\le |z - z_0|^2 \int_0^1 |k[z'(t)]|(1-t)|f_{n-1}[z'(t)]|\, dt \\
&\le \max|f_0|\, [\max|k|]^n \frac{|z - z_0|^{2n}}{(n-1)!} \int_0^1 (1-t)t^{2n-2} dt \\
&\le \max|f_0|\, [\max|k|]^n \frac{|z - z_0|^{2n}}{n!}
\end{aligned}$$

which proves 14.12.

It is now evident that the series

$$f(z) \equiv \sum_{n=0}^{\infty} f_n(z) \tag{14.13}$$

converges uniformly when $z \in \mathrm{R}$. This follows from Weierstrass' criterion, since every term in the series (14.13) is dominated by every term in the series

$$\max|f_0| \sum_{n=0}^{\infty} \frac{[\max|k|\, |z - z_0|^2]^n}{n!}$$

which converges uniformly to $\max|f_0| \exp(\max|k|\, |z - z_0|^2)$. There remains to show that if we choose

$$f_0(z) = a + b(z - z_0) \tag{14.14}$$

then the series in 14.13 will converge to $u(z)$, i.e., to the solution of the differential equation with the prescribed boundary conditions.

Since the series in 14.13 converges uniformly, we can differentiate it term by term. Then, using Eq. 14.8 and taking account of Eq. 14.14

$$\frac{d^2}{dz^2}\left\{\sum_{n=0}^{\infty} f_n(z)\right\} = \frac{d^2}{dz^2}\left\{\sum_{n=1}^{\infty} f_n(z)\right\} = \sum_{n=1}^{\infty} \frac{d^2}{dz^2} f_n(z) = -k(z) \sum_{n=1}^{\infty} f_{n-1}(z)$$

Hence

$$\frac{d^2 f(z)}{dz^2} + k(z) f(z) = 0$$

Obviously, $f(z)$ will satisfy the initial conditions, for from Eqs. 14.9 and 14.10

$$f_n(z_0) = \left.\frac{df_n(z)}{dz}\right|_{z=z_0} = 0 \qquad (n = 1, 2, \cdots)$$

whereas from Eq. 14.14

$$f_0(z_0) = a \qquad \left.\frac{df_0}{dz}\right|_{z=z_0} = b$$

Therefore, $f(z)$ obeys the differential equation (14.6) with the boundary conditions 14.4.

Let $\tilde{f}(z)$ be another such solution and put

$$w(z) \equiv f(z) - \tilde{f}(z)$$

Then $w(z)$ also satisfies the equation

$$\frac{d^2w}{dz^2} + k(z)w(z) = 0 \tag{14.15}$$

but with the boundary conditions

$$w(z_0) = \left.\frac{dw(z)}{dz}\right|_{z=z_0} = 0 \tag{14.16}$$

From Eqs. 14.15 and 14.16 it follows that $d^2w/dz^2 = 0$ and also, by successively differentiating Eq. 14.15, that all higher-order derivatives of $w(z)$ at $z = z_0$ vanish. Therefore, by Taylor's theorem

$$w(z) \equiv 0$$

and hence

$$f(z) = \tilde{f}(z)$$

Thus

$$f(z) = \sum_{n=0}^{\infty} f_n(z)$$

is the **unique** analytic solution of Eq. 14.6 which satisfies the prescribed boundary conditions.

Suppose now that all points of a region $R' \subset R$ are ordinary points of the equation. By the method just described, we can find the solutions of Eq. 14.1 in the neighborhood of an arbitrary point $z_0 \in R'$. This solution can be analytically continued along any path within R' by the method described in Sec. 26, Chapter I. It is evident that this analytic continuation yields a solution of the equation, since for any $z \in R'$ the LHS of the differential equation is an analytic function, and therefore if it vanishes in the neighborhood of the point z_0, it vanishes everywhere in R'. However, if R' is multiply-connected (for example, if R' coincides with R except at a set of isolated points that are singular points of $p(z)$ or $q(z)$, or both), the result of the analytic continuation from z_0 to a given point of R' will depend on the path along which this continuation has been carried out. In other words, an analytic continuation of the solution generally defines a multivalued function with branch points located at the singular points of the equation.

Since the solution of a differential equation in a neighborhood of an ordinary point is analytic, it can be expanded in a power series about that point

$$u(z) = \sum_{n=0}^{\infty} c_n(z - z_0)^n \tag{14.17}$$

The coefficients c_n can be determined from a recurrence relation obtained by imposing the condition that 14.17 satisfies the differential equation.

EXAMPLE

Consider again Hermite's differential equation (see Sec. 10.6, Chapter III)

$$\frac{d^2u}{dz^2} - 2z\frac{du}{dz} + 2\lambda u = 0 \tag{14.18}$$

The only singular point of this equation is the point at infinity. The solution of Eq. 14.18 about $z = 0$ can be written as a power series

$$u(z) = \sum_{n=0}^{\infty} c_n z^n \tag{14.19}$$

Inserting Eq. 14.19 into Eq. 14.18, we obtain

$$\sum_{n=0}^{\infty} [(n+2)(n+1)c_{n+2} - 2nc_n + 2\lambda c_n]\, z^n = 0 \tag{14.20}$$

Since Eq. 14.19 should satisfy Eq. 14.18 for all values of z, the coefficient of z^n in Eq. 14.20 must be set equal to zero, giving the recurrence relation

$$c_{n+2} = \frac{2(n-\lambda)}{(n+2)(n+1)}\, c_n \tag{14.21}$$

Starting from the coefficient c_0, we can generate from Eq. 14.21 all even coefficients $c_2, c_4, \cdots$, which can be expressed in terms of c_0.

On the other hand, starting from the coefficient c_1, we generate from Eq. 14.21 all odd coefficients $c_3, c_5, \cdots$, which are expressible in terms of c_1.

Thus, we have two solutions in the form of power series: One solution contains even powers of z only and the other solution contains odd powers of z only; i.e.,

$$\begin{aligned} u_1(z) &= \sum_{n=0} c_{2n} z^{2n} \\ u_2(z) &= \sum_{n=0}^{\infty} c_{2n+1} z^{2n+1} \end{aligned} \tag{14.22}$$

It can be seen from Eq. 14.21 that if λ is an integer, all coefficients beginning with $c_{\lambda+2}$ vanish. Thus, if λ is an even integer, $u_1(z)$ is a polynomial, and if λ is an odd integer, $u_2(z)$ is a polynomial. These are the Hermite polynomials considered in Sec. 10.6, Chapter III.

14.4 Solution of a Differential Equation in a Neighborhood of a Regular Singular Point

We now consider second-order linear differential equations that have at least one regular singular point, and we seek solutions to these equations in a neighborhood of one such point. Consider the equation

$$L_z u = \frac{d^2u}{dz^2} + p(z)\frac{du}{dz} + q(z)u = 0 \tag{14.23}$$

We shall assume that all points of a region D are ordinary points of Eq. 14.23 except $z = z_0$, which is a regular singular point of Eq. 14.23. This means that the functions

$$A(z) \equiv (z - z_0)p(z) \qquad \text{and} \qquad B(z) \equiv (z - z_0)^2 q(z)$$

are analytic everywhere in D including the point z_0, and can therefore be expanded in a Taylor series about z_0

$$A(z) = \sum_{n=0}^{\infty} a_n (z - z_0)^n \qquad B(z) = \sum_{n=0}^{\infty} b_n (z - z_0)^n \tag{14.24}$$

where

$$a_n = \frac{1}{2\pi i}\int_\Gamma \frac{A(z')}{(z' - z_0)^{n+1}}\, dz \qquad b_n = \frac{1}{2\pi i}\int_\Gamma \frac{B(z')}{(z' - z_0)^{n+1}}\, dz \tag{14.25}$$

and Γ is a circle of radius R surrounding z_0.

We now try to find a solution of Eq. 14.23 in the form of a series

$$u(z) = (z - z_0)^r \sum_{n=0}^{\infty} c_n (z - z_0)^n \tag{14.26}$$

where r and c_n ($n = 0, 1, \cdots$) are constants that are to be determined. Inserting 14.26 into Eq. 14.23, using Eqs. 14.24, and multiplying through by $(z - z_0)^{2-r}$, one finds

$$\sum_{n=0}^{\infty} (n + r)(n + r - 1)(z - z_0)^n c_n + \left\{\sum_{n=0}^{\infty} a_n (z - z_0)^n\right\}\left\{\sum_{n=0}^{\infty} c_n (n + r)(z - z_0)^n\right\} + \left\{\sum_{n=0}^{\infty} b_n (z - z_0)^n\right\}\left\{\sum_{n=0}^{\infty} c_n (z - z_0)^n\right\} = 0$$

By applying the rule for multiplying two power series,* one obtains

$$\sum_{n=0}^{\infty}\left\{(n + r)(n + r - 1)c_n + \sum_{m=0}^{n} [(m + r)a_{n-m} + b_{n-m}]c_m\right\}(z - z_0)^n = 0$$

In order that $L_z u(z) = 0$, the coefficients c_n and r should obey the relation

$$(n + r)(n + r - 1)c_n + \sum_{m=0}^{n} [(m + r)a_{n-m} + b_{n-m}]c_m = 0$$

or separating the complete term in c_n

$$[(n + r)(n + r - 1) + (n + r)a_0 + b_0]c_n + \sum_{m=0}^{n-1} [(m + r)a_{n-m} + b_{n-m}]c_m = 0 \tag{14.27}$$

Let us define the functions

$$\lambda_0(r) \equiv r(r - 1) + ra_0 + b_0$$
$$\lambda_i(r) \equiv \begin{cases} ra_i + b_i, & i > 0 \\ 0, & i < 0 \end{cases} \tag{14.28}$$

Then Eq. 14.27 can be rewritten as

$$\lambda_0(r + n)c_n = -\sum_{m=0}^{n-1} \lambda_{n-m}(r + m)c_m \tag{14.29}$$

These equations serve to determine successively the coefficients c_n (except c_0, which remains arbitrary). The first equation obtained from Eq. 14.29 by setting $n = 0$ is particularly important

$$\lambda_0(r)c_0 = 0$$

and since we can always choose $c_0 \neq 0$ (because r in Eq. 14.26 is an arbitrary parameter and c_0 can always be defined as the first nonvanishing coefficient in Eq. 14.26)

$$\lambda_0(r) \equiv r^2 + (a_0 - 1)r + b_0 = 0 \tag{14.30}$$

---

* If $f = \sum_{n=0}^{\infty} a_n z^n$ and $g = \sum_{n=0}^{\infty} b_n z^n$, then $f.g = \sum_{n=0}^{\infty} c_n z^n$, where $c_n = \sum_{m=0}^{n} a_{n-m} b_m$.

Equation 14.30 is of fundamental importance in determining the solutions to the differential equation, and is called the **indicial equation** associated with this equation. Let r_1 and r_2 be the roots of Eq. 14.30; r_1 and r_2 may or may not be distinct. In either case, we shall label these roots in such a way that

$$\operatorname{Re} r_1 \geq \operatorname{Re} r_2 \tag{14.31}$$

which, solving Eq. 14.30, implies that

$$\operatorname{Re} r_1 > \tfrac{1}{2}\operatorname{Re}(1 - a_0) \tag{14.32}$$

We now show

(i) There always exists one solution of Eq. 14.30, given by Eq. 14.26, with $r = r_1$.

(ii) With $r = r_2$, Eq. 14.26 yields a second linearly independent solution of Eq. 14.23, **provided the roots r_1 and r_2 do not differ by an integer.**

In order to prove (i) we note first that $\lambda_0(r_1 + n)$ for $n > 0$ can never vanish. That this is so follows from the fact that λ_0 vanishes only for values of its argument equal to r_1 or r_2. Therefore, for $n > 0$, λ_0 can vanish only if $r_1 + n = r_2$. But because of inequality 14.31, this condition cannot be realized. Therefore, in Eq. 14.29 with $r = r_1$, we can divide through by $\lambda_0(r_1 + n)$ and obtain

$$c_n = -\sum_{m=0}^{n-1} \frac{\lambda_{n-m}(r_1 + m)}{\lambda_0(r_1 + n)} c_m \qquad (n = 1, 2, \cdots) \tag{14.33}$$

Or, writing out explicitly a few of these relations

$$c_1 = \frac{-\lambda_1(r_1)}{\lambda_0(r_1 + 1)} c_0$$

$$c_2 = \frac{-\lambda_2(r_1)}{\lambda_0(r_1 + 2)} c_0 - \frac{\lambda_1(r_1 + 1)}{\lambda_0(r_1 + 2)} c_1$$

$$c_3 = \frac{-\lambda_3(r_1)}{\lambda_0(r_1 + 3)} c_0 - \frac{\lambda_2(r_1 + 1)}{\lambda_0(r_1 + 3)} c_1 - \frac{\lambda_1(r_1 + 2)}{\lambda_0(r_1 + 3)} c_2$$

$$\cdots\cdots\cdots\cdots\cdots\cdots\cdots\cdots\cdots$$

These relations show that it is possible to determine successively all the coefficients c_n and express them in terms of c_0 only. It remains to be proved that the series 14.26 converges.

Now $A(z)$ and $B(z)$ are bounded, since they are analytic in D and therefore there exist two numbers Λ_1 and Λ_2 such that

$$|A(z)| < \Lambda_1 \qquad \text{and} \qquad |B(z)| < \Lambda_2 \qquad \text{for } z \in \mathrm{D} \tag{14.34}$$

Equations 14.25 imply that

$$|a_n| \leq \frac{\Lambda_1}{R^n} \qquad \text{and} \qquad |b_n| \leq \frac{\Lambda_2}{R^n} \tag{14.35}$$

Hence

$$\begin{aligned} |\lambda_{n-m}(r_1 + m)| &= |(r_1 + m)a_{n-m} + b_{n-m}| \\ &\leq \frac{(|r_1| + m)\Lambda_1 + \Lambda_2}{R^{n-m}} \end{aligned} \tag{14.36}$$

Also

$$\lambda_0(r_1 + n) = n^2 + n(2r_1 - 1 + a_0)$$

and because of Eq. 14.32, we have that

$$|\lambda_0(r_1 + n)| \geq n^2 \tag{14.37}$$

Putting 14.36 and 14.37 into Eq. 14.33 yields

$$\begin{aligned} |c_n| &\leq \sum_{m=0}^{n-1} \frac{(|r_1| + m)\Lambda_1 + \Lambda_2}{n^2 R^{n-m}} |c_m| \\ &\leq \frac{K}{n} \sum_{m=0}^{n-1} \frac{|c_m|}{R^{n-m}} \qquad (n = 1, 2, \cdots) \end{aligned} \tag{14.38}$$

where we have set

$$\begin{aligned} \frac{(|r_1| + m)\Lambda_1 + \Lambda_2}{n} &\leq |r_1| + \left(\frac{m}{n}\right)\Lambda_1 + \Lambda_2 \\ &\leq (|r_1| + \Lambda_1 + \Lambda_2) < K \end{aligned}$$

The second inequality follows, since in the summation (Eq. 14.38) we always have $m < n$. We may assume, without any loss of generality, that $K > 1$. Then, from inequality 14.38, one has

$$|c_1| \leq \frac{K}{R} |c_0|$$

$$|c_2| \leq \frac{K}{2} |c_0| \left\{\frac{1}{R^2} + \frac{K}{R^2}\right\} \leq \frac{K^2}{R^2} |c_0|$$

$$|c_3| \leq \frac{K}{3} \frac{|c_0|}{R^3} \{1 + K + K^2\} \leq \frac{K^3}{R^3} |c_0|$$

....

$$|c_n| \leq \left(\frac{K}{R}\right)^n |c_0|$$

Thus the series 14.26 converges, since it is dominated by the geometric series

$$\sum_{n=0}^{\infty} \left(\frac{K}{R}\right)^n (z - z_0)^n$$

whose radius of convergence is greater than or equal to (R/K).

To prove (ii) we note that if r_1 and r_2 do **not** differ by an integer, then $\lambda_0(r_2 + n)$ can never vanish, and one can successively determine, as before, all the coefficients c_n in terms of the single coefficient c_0. We therefore have in this case a second solution of Eq. 14.23 in the form of Eq. 14.26, setting $r = r_2$ both in the exponent of Eq. 14.26 and in the recurrence relations 14.29 that determine the new set of coefficients c_n. It is easy to verify that when $r_1 \neq r_2$, the Wronskian of the two solutions does not vanish and that therefore these solutions are linearly independent.

Suppose now that the two roots r_1 and r_2 of the indicial equation differ by an integer, $r_1 - r_2 = N$. In that case, $\lambda_0(r_2 + n)$ will vanish when $n = N$, for then

$$\lambda_0(r_2 + N) = \lambda_0(r_1) = 0$$

Therefore, all the coefficients c_n with $n \geq N$ can no longer be determined from the recurrence relation 14.29. This means that a second solution of Eq. 14.23 cannot be found in this manner.

In order to find a second linearly independent solution, we make use of Eq. 2.16, extended to complex variables, which can be rewritten as

$$\frac{d}{dz}\left(\frac{u_2}{u_1}\right) = \frac{c}{u_1^2}\exp\left(-\int_{z_0}^{z} p(z')\,dz'\right) \tag{14.39}$$

Using Eq. 14.24 and the solution 14.26 for u_1, Eq. 14.39 leads to

$$\begin{aligned}\frac{d}{dz}\left(\frac{u_2}{u_1}\right) &= \frac{c}{\left\{(z-z_0)^{r_1}\sum_{n=0}^{\infty} c_n(z-z_0)^n\right\}^2}\exp\left(-a_0\ln(z-z_0) - \sum_{n=1}^{\infty}\frac{a_n}{n}(z-z_0)^n\right)\\ &= c(z-z_0)^{-a_0-2r_1}F(z)\end{aligned} \tag{14.40}$$

where

$$F(z) \equiv \frac{\exp\left(-\sum_{n=1}^{\infty}\frac{a_n}{n}(z-z_0)^n\right)}{\left\{\sum_{n=0}^{\infty} c_n(z-z_0)^n\right\}^2} \tag{14.41}$$

is an analytic function of z and can therefore be expanded in a Taylor series about z_0

$$F(z) = \sum_{n=0}^{\infty} d_n(z-z_0)^n \tag{14.42}$$

On the other hand, the indicial equation 14.30 gives

$$r_1 + r_2 = 1 - a_0$$

and hence

$$(2r_1 + a_0) = 1 + N \tag{14.43}$$

Putting 14.42 and 14.43 into Eq. 14.40, we find

$$\begin{aligned}\frac{d}{dz}\left(\frac{u_2}{u_1}\right) = \text{const}\bigg\{&\frac{d_0}{(z-z_0)^{N+1}} + \frac{d_1}{(z-z_0)^N} + \cdots + \frac{d_N}{(z-z_0)}\\ &+ d_{N+1} + d_{N+2}(z-z_0) + \cdots\bigg\}\end{aligned} \tag{14.44}$$

Integrating Eq. 14.44 gives

$$\left(\frac{u_2}{u_1}\right) = \text{const}\ \ln(z-z_0) + (z-z_0)^{-N}\sum_{n=0}^{\infty} d_n(z-z_0)^n \tag{14.45}$$

The arbitrary integration constant was absorbed into the other coefficients. Finally

$$u_2(z) = \text{const}\ u_1(z)\ln(z-z_0) + (z-z_0)^{r_2}\sum_{n=0}^{\infty} f_n\,(z-z_0)^n \tag{14.46}$$

The coefficients that appear in Eq. 14.46 can be determined as before from a recurrence relation obtained by putting Eq. 14.46 into Eq. 14.23.

Let us note that in the environment of a singular point, at least one of the two independent solutions of the equation is a multivalued function, since either one of the roots of the indicial equation is not an integer or (when both roots are integers) the logarithm in the second solution (Eq. 14.46) introduces a branch cut. This is precisely the conclusion we reached at the end of the preceding section when we discussed the analytic continuation of a solution valid in a neighborhood of an ordinary point of the equation.

Observe, however, that when the indicial equation has an integer root, the equation always has one single-valued solution; moreover, if this root is positive, the solution is analytic.

15 · SOLUTION OF DIFFERENTIAL EQUATIONS USING THE METHOD OF INTEGRAL REPRESENTATIONS

15.1 General Theory

In the previous sections we showed how one could obtain series solutions of differential equations that were valid in some environment of either an ordinary point or a regular singular point z_0. Such solutions can always be found and are meaningful within their circle of convergence. The radius of convergence of the series, however, may be very small, and then one has to continue analytically the solution if one wants to enlarge its domain of validity.

There is another method for solving differential equations, which leads to a solution in which the analytic continuation is, so to speak, already built into it. The method consists in finding a solution to the differential equation in the form of an integral representation.

Suppose that we have found a series solution of the differential equation that is valid in some region R and which satisfies the boundary conditions imposed at an ordinary point in R. Suppose also that one can find a kernel $K(z,t)$ a function $v(t)$, and limits a and b such that

$$u(z) = \int_a^b K(z,t)v(t)\,dt \tag{15.1}$$

is a solution of the differential equation that satisfies the same boundary conditions but which is analytic in some region D that includes R. There will certainly be some neighborhood of the point for which the boundary conditions have been specified where the series solution of the differential equation and 15.1 will coincide. This follows because the solution of a differential equation with prescribed boundary conditions is unique. On the other hand, it may be, as is very often the case, that the region D in which the integral representation 15.1 is analytic and convergent is larger than the region R in which the series solution is defined. Then, according to Sec. 26 of Chapter I, since the two solutions coincide in $D \cap R$, Eq. 15.1 represents the analytic continuation of the series solution in the larger region, consisting of the union $R + D$ of the regions R and D.

EXAMPLE 1

The series

$$\sum_{n=0}^{\infty} z^n \tag{15.2}$$

converges to $1/(1-z)$ but only for $|z|<1$. On the other hand, the integral

$$\int_0^\infty e^{zt}e^{-t}\,dt \tag{15.3}$$

is an analytic function of z in the larger region

$$\operatorname{Re} z < 1 \tag{15.4}$$

where it converges. But for $|z|<1$, upon expanding the exponential under the integral 15.3 we have,

$$\int_0^\infty e^{zt}e^{-t}\,dt = \sum_{n=0}^\infty \frac{z^n}{n!}\int_0^\infty t^n e^{-t}\,dt$$

$$= \sum_{n=0}^\infty z^n = \frac{1}{1-z}$$

where we have used Eq. 32.4 of Chapter I. Hence, the integral representation, which reduces to the series 15.2 for $|z|<1$, represents the analytic continuation of the power series in the half-plane $\operatorname{Re} z<1$.

Suppose that we seek a solution of the differential equation

$$L_z u(z) = 0 \tag{15.5}$$

where L_z is a formal differential operator with respect to z. The method we use to find a solution of the differential equation in the form of an integral representation will succeed if we can find an operator M_t that is a formal differential operator with respect to the integration variable t and which has the property that

$$L_z K(z,t) = M_t K(z,t) \tag{15.6}$$

EXAMPLE 2

Take $K(z,t)=z^2t$ and $L_z = z(d/dz)$. Then the operator $M_t = 2t(d/dt)$ has the property 15.6.

If such an operator M_t can be found, then from Eq. 15.1 we have

$$L_z u(z) = \int_a^b [L_z K(z,t)]v(t)\,dt = \int_a^b [M_t K(z,t)]v(t)\,dt \tag{15.7}$$

On the other hand, if M_t has an adjoint M_t^+, then it satisfies the Lagrange identity

$$v(t)[M_t K(z,t)] - K(z,t)[M_t^+ v(t)] = \frac{\partial}{\partial t}[Q(K,v)] \tag{15.8}$$

where $Q(K,v)$ is a bilinear function of K,v and their derivatives. Putting Eq. 15.8 in Eq. 15.7, we have

$$L_z u(z) = \int_a^b K(z,t)M_t{}^+ v(t)\,dt + [Q(K,v)]\Bigg|_{t=a}^{t=b} \tag{15.9}$$

Hence, by choosing the end points a,b of the path of integration such that

$$[Q(K,v)]\Bigg|_{t=a}^{t=b} = 0 \tag{15.10}$$

and taking for $v(t)$ the solution of the adjoint equation

$$M_t^+ v(t) = 0 \tag{15.11}$$

we have, on account of Eq. 15.9, $L_z u(z) = 0$, so that 15.1 is a solution of the differential equation. The hope, of course, is that Eq. 15.11 will be easier to solve than Eq. 15.5.

15.2 Kernels of Integral Representations

In later sections we shall show how in practice one obtains solutions of differential equations using the foregoing method. We should note, however, that the chances of finding an operator M_t that satisfies Eq. 15.6 and which at the same time allows for an easy solution of Eq. 15.11 can be greatly enhanced by choosing judiciously the kernel $K(z,t)$. In this connection we list here a few of the more useful kernels that often appear in the solutions of the differential equations of physics.

(i) A linear differential equation with coefficients that are linear functions of the variable z can always be solved by using the kernel

$$K(z,t) = e^{zt} \tag{15.12}$$

called a **Laplace kernel.**

(ii) A differential equation in which the coefficient of $d^l u/dx_l$ is a polynomial of degree l can be solved with the kernel

$$K(z,t) = K(z - t) = (z - t)^u \tag{15.13}$$

where u is a complex number. The kernel 15.13 is called an **Euler kernel.**

(iii) An equation of the type

$$z^n H_1\left(z\frac{d}{dz}\right)u + H_2\left(z\frac{d}{dz}\right)u = 0$$

where H_1 and H_2 are functions of the product $z(d/dz)$ can be solved with kernels that depend only on the factor z^t

$$K(z,t) = K(z^t) \tag{15.14}$$

The kernel 15.14 is then called a **Mellin kernel.**

(iv) In connection with Bessel's equation, which will be discussed in Sec. 20, the kernel

$$K(z,t) = \left(\frac{z}{2}\right) e^{(t - z^2/4t)} \tag{15.15}$$

is very often useful.

16 · FUCHSIAN EQUATIONS WITH THREE REGULAR SINGULAR POINTS

A **Fuchsian equation** is an equation that has only regular singular points. In what follows, we consider Fuchsian equations that have at most three singular points throughout the entire complex plane, including the point at infinity. This type of equation is very general, and we shall see that by specializing its parameters, we can recover some of the most important equations of mathematical physics. The equation can be written quite generally as

$$\frac{d^2u}{dz^2}+\left(\frac{1-\alpha-\alpha'}{z-z_1}+\frac{1-\beta-\beta'}{z-z_2}+\frac{1-\gamma-\gamma'}{z-z_3}\right)\frac{du}{dz}$$
$$+\left[\frac{(z_1-z_2)(z_1-z_3)\alpha\alpha'}{z-z_1}+\frac{(z_2-z_1)(z_2-z_3)\beta\beta'}{z-z_2}\right.$$
$$\left.+\frac{(z_3-z_1)(z_3-z_2)\gamma\gamma'}{z-z_3}\right]\frac{u}{(z-z_1)(z-z_2)(z-z_3)}=0 \tag{16.1}$$

where*

$$\alpha+\alpha'+\beta+\beta'+\gamma+\gamma'=1$$

Equation 16.1 is known as the **equation of Riemann**. The indicial equation of Eq. 16.1 relative to the singularity at z_1 is

$$r(r-1)+(1-\alpha-\alpha')r+\alpha\alpha'=0 \tag{16.2}$$

and the two roots of Eq. 16.2 are

$$r_1=\alpha \qquad r_2=\alpha'$$

Similarly, β,β' and γ,γ' are respectively the roots of the indicial equation relative to the singular points z_2 and z_3. A solution of Eq. 16.1 is conventionally represented by the symbol

$$P\begin{Bmatrix} z_1 & z_2 & z_3 & \\ \alpha & \beta & \gamma & z \\ \alpha' & \beta' & \gamma' & \end{Bmatrix} \tag{16.3}$$

which is called a **Riemann *P* symbol** where the six quantities α,α', β,β', γ,γ' are called the exponents of the equation. The first row contains the singular points of the equation, and under the singular points are placed the roots of the indicial equation relative to the singularity. In the last column is placed the independent variable z. Of course any constant multiplied by 16.3 is also a solution of Eq. 16.1.

The nine parameters of 16.3, which characterize the Riemann equation, can be reduced to three by making the following transformations on the equation

$$v(z)=(z-z_1)^r(z-z_2)^s(z-z_3)^t u(z) \tag{16.4}$$

with $r+s+t=0$, and

$$z'=\frac{Az+B}{Cz+D} \tag{16.5}$$

From the transformation 16.4, one obtains the relation

$$(z-z_1)^r(z-z_2)^s(z-z_3)^t P\begin{Bmatrix} z_1 & z_2 & z_3 & \\ \alpha & \beta & \gamma & z \\ \alpha' & \beta' & \gamma' & \end{Bmatrix}$$
$$=P\begin{Bmatrix} z_1 & z_2 & z_3 & \\ \alpha+r & \beta+s & \gamma+t & z \\ \alpha'+r & \beta'+s & \gamma'+t & \end{Bmatrix} \qquad r+s+t=0 \tag{16.6}$$

* This condition ensures that $z=\infty$ is a regular point. See P. M. Morse and H. Feshbach *Methods of Theoretical Physics*, Vol. I, McGraw-Hill Book Co., New York, 1953, p. 538.

From the homographic transformation 16.5, one obtains

$$P\begin{Bmatrix} z_1 & z_2 & z_3 & \\ \alpha & \beta & \gamma & z \\ \alpha' & \beta' & \gamma' & \end{Bmatrix} = P\begin{Bmatrix} z_1' & z_2' & z_3' & \\ \alpha & \beta & \gamma & z' \\ \alpha' & \beta' & \gamma' & \end{Bmatrix} \tag{16.7}$$

where

$$z_1' = \frac{Az_1 + B}{Cz_1 + D}; \quad z_2' = \frac{Az_2 + B}{Cz_2 + D}; \quad z_3' = \frac{Az_3 + B}{Cz_3 + D} \tag{16.8}$$

The results 16.6 and 16.7 can be obtained in a straightforward manner by substituting 16.4 and 16.5 into Eq. 16.1, but the calculation is space consuming and not very instructive. The reader may verify them easily.

We would like to emphasize the importance of the identities, 16.6 and 16.7; they are of much use in relating the solutions of the Riemann equation valid about different points and in transforming solutions of different Riemann equations into one another. We shall use these identities to reduce the nine-parameter Riemann equation to a three-parameter equation.

Let the points z'_1, z'_2, z'_3 be respectively $0, \infty$ and 1. Then from Eq. 16.8 one can express the ratios A/D, B/D, and C/D in terms of z_1, z_2, z_3.

$$\frac{A}{D} = \frac{(z_3 - z_2)}{z_2(z_1 - z_3)}; \quad \frac{B}{D} = \frac{z_1(z_2 - z_3)}{z_2(z_1 - z_3)}; \quad \frac{C}{D} = -\frac{1}{z_2} \tag{16.9}$$

Substituting these values into 16.5, one has

$$z' = \frac{(z_3 - z_2)(z - z_1)}{(z_3 - z_1)(z - z_2)} \tag{16.10}$$

Now, letting $r = -\alpha$, $s = (\alpha + \gamma)$, $t = -\gamma$ in Eq. 16.6 leads to

$$P\begin{Bmatrix} z_1 & z_2 & z_3 & \\ \alpha & \beta & \gamma & z \\ \alpha' & \beta' & \gamma' & \end{Bmatrix} = \left(\frac{z - z_1}{z - z_2}\right)^{\alpha}\left(\frac{z - z_3}{z - z_2}\right)^{\gamma} P\begin{Bmatrix} 0 & \infty & 1 & \\ 0 & \alpha + \beta + \gamma & 0 & z' \\ \alpha' - \alpha & \alpha + \gamma + \beta' & \gamma' - \gamma & \end{Bmatrix} \tag{16.11}$$

with z' given by Eq. 16.10. Introducing the notation

$$a \equiv \alpha + \beta + \gamma \qquad b \equiv \alpha + \gamma + \beta' \qquad c \equiv 1 + \alpha - \alpha'$$

and using the fact that the sum of the exponents of the P symbol of the right side of Eq. 16.11 is equal to 1, we have

$$P\begin{Bmatrix} z_1 & z_2 & z_3 & \\ \alpha & \beta & \gamma & z \\ \alpha' & \beta' & \gamma' & \end{Bmatrix} = \left(\frac{z - z_1}{z - z_2}\right)^{\alpha}\left(\frac{z - z_3}{z - z_2}\right)^{\gamma} P\begin{Bmatrix} 0 & \infty & 1 & \\ 0 & a & 0 & z' \\ 1 - c & b & c - a - b & \end{Bmatrix} \tag{16.12}$$

Equation 16.12 shows that the solution of the general nine-parameter Riemann equation can be expressed in terms of the solution of a three-parameter equation with regular singular points at 0, 1, and ∞. The equation corresponding to the Riemann P symbol

$$P\begin{Bmatrix} 0 & \infty & 1 & \\ 0 & a & 0 & z \\ 1 - c & b & c - a - b & \end{Bmatrix} \tag{16.13}$$

can be written down by inspection (compare 16.1, 16.3, and 16.13); it is

$$z(1-z)\frac{d^2u}{dz^2} + [c-(a+b+1)z]\frac{du}{dz} - abu = 0 \tag{16.14}$$

Equation 16.14 is the very important **hypergeometric equation**; its solution, which is analytic at the origin, is denoted by the symbol $F(a,b;c;z)$.

Setting $F(a,b;c;z')$ in Eq. 16.12 in place of the P symbol, we obtain a solution of the general Riemann equation:

$$P\begin{Bmatrix} z_1 & z_2 & z_3 & \\ \alpha & \beta & \gamma & z \\ \alpha' & \beta' & \gamma' & \end{Bmatrix}$$

$$= \left(\frac{z-z_1}{z-z_2}\right)^{\alpha}\left(\frac{z-z_3}{z-z_2}\right)^{\gamma} F\left(\alpha+\beta+\gamma, \alpha+\gamma+\beta'; 1+\alpha-\alpha'; \frac{(z-z_1)(z_3-z_2)}{(z-z_2)(z_3-z_1)}\right) \tag{16.15}$$

17 · THE HYPERGEOMETRIC FUNCTION

17.1 Solutions of the Hypergeometric Equation

We pointed out in the preceding section that the hypergeometric equation

$$z(1-z)\frac{d^2u}{dz^2} + [c-(a+b+1)z]\frac{du}{dz} - abu = 0 \tag{17.1}$$

plays a particularly important role in applications. We first consider the series solution of this equation, valid near the regular singular point at the origin. From 16.13 we see that the roots of the indicial equation relative to the origin are 0 and $1 - c$.

Thus, there exists a solution that is analytic in the neighborhood of the origin and which can be normalized to unity. We call this solution the **hypergeometric function** and denote it by $F(a,b;c;z)$. The singularities of $F(a,b;c;z)$ are located at the singular points of the equation, i.e., at $z = 1$ and $z = \infty$. These are, in general, branch points, and in those cases we supplement the definition of $F(a,b;c;z)$ by taking the cut from $z = 1$ to $z = \infty$ along the positive real axis. Since $F(a,b;c;z)$ is analytic in the neighborhood of the origin, it may be represented there by a power series

$$F(a,b;c;z) = \sum_{n=0}^{\infty} c_n z^n \qquad (c_0 = 1) \tag{17.2}$$

Putting 17.2 into Eq. 17.1, one obtains the recurrence relation

$$c_n = \frac{(a+n-1)(b+n-1)}{n(c+n-1)} c_{n-1}$$

These coefficients can be determined successively if $c \neq 0, -1, -2, \cdots$, and we have

$$F(a,b;c;z) = 1 + \sum_{n=1}^{\infty} \frac{a(a+1)\cdots(a+n-1)b(b+1)\cdots(b+n-1)}{n!c(c+1)\cdots(c+n-1)} z^n$$

$$= \frac{\Gamma(c)}{\Gamma(a)\Gamma(b)} \sum_{n=0}^{\infty} \frac{\Gamma(a+n)\Gamma(b+n)}{\Gamma(c+n)\Gamma(n+1)} z^n \tag{17.3}$$

The series 17.3 is called the **hypergeometric series**. The name "hypergeometric" comes from the fact that the expansion of $F(1,b;b;z)$ about $z = 0$ is simply the geometric series.

Since $z = 1$ is the singularity nearest to the origin, the series expansion 17.3 will converge for $|z| < 1$. One can immediately note the important symmetry property

$$F(a,b;c;z) = F(b,a;c;z) \tag{17.4}$$

For $|z| < 1$, this property can evidently be deduced by inspection from the hypergeometric series, and by analytic continuation it must be valid everywhere where the function is well defined.

If $1 - c$ is **not** an integer, a second solution of Eq. 17.1 is of the form

$$z^{1-c}g_1(z) \tag{17.5}$$

where $g_1(z)$ is a power series in z. Putting 17.5 into Eq. 17.1 yields an equation for $g_1(z)$

$$z(z-1)\frac{d^2g_1}{dz^2} + [(a+b-2c+3)z + c - 2]\frac{dg_1}{dz} + (a-c+1)(b-c+1)g_1 = 0 \tag{17.6}$$

Comparing Eqs. 17.6 and 17.1 shows that $g_1(z)$ is in fact the hypergeometric function $F(b - c + 1, a - c + 1; 2 - c;z)$. Therefore, when $1 - c$ is not an integer, the general solution of Eq. 17.1 is

$$u = \alpha F(a,b;c;z) + \beta z^{1-c}F(b - c + 1, a - c + 1; 2 - c;z) \tag{17.7}$$

Similarly, 16.13 shows that if $(c - a - b)$ and $(a - b)$ are not integers, then

$$g_2(1-z) \qquad \text{and} \qquad (1-z)^{c-a-b}g_3(1-z) \tag{17.8}$$

where g_2 and g_3 are power series in $(1 - z)$, are solutions of Eq. 17.4 about $z = 1$; and

$$z^{-a}g_4\left(\frac{1}{z}\right) \qquad \text{and} \qquad z^{-b}g_5\left(\frac{1}{z}\right) \tag{17.9}$$

where g_4 and g_5 are power series in $1/z$, are solutions of Eq. 17.1 about the point at infinity.

We now show how the solutions 17.8 and 17.9 can be expressed in terms of $F(a,b;c;z)$. We do this by considering certain transformations on the Riemann P symbol.

First we note that the order of the columns in the P symbol is completely arbitrary, for this symbol represents only the association of a given singular point of the differential equation with the roots of the indicial equation relative to that singularity. Hence, the columns of the P symbol can be permuted at will. However, the RHS of 16.15 is not at all invariant under such a permutation. Therefore the $3! = 6$ permutations of the columns of the P symbol leads to six different solutions of Eq. 17.1. Furthermore, the Riemann equation (Eq. 16.1) is invariant under the transformations

$$\alpha \leftrightarrow \alpha'; \quad \beta \leftrightarrow \beta'; \quad \gamma \leftrightarrow \gamma' \tag{17.10}$$

However, because of 17.4, the RHS of 16.15 is invariant under the transformation $\beta \leftrightarrow \beta'$. Therefore only two of the preceding transformations are independent. This leads to four new solutions of Eq. 17.1. In total, therefore, we have $6 \times 4 = 24$ solutions of Eq. 17.1. These are known as **Kummer's solutions**.

Now consider Eq. 16.15. Since the P symbol represents an arbitrary solution of a differential equation, and since such a solution is determined only up to an arbitrary multiplicative constant, one can obviously multiply the right side of 16.5 by the constant $(-1)^\alpha z_2^{\alpha+\gamma}$ without affecting the result. Then, interchanging the first and last columns leads to the new solution

$$(-1)^\alpha z_2^{\alpha+\gamma}\left(\frac{z_3-z}{z_2-z}\right)^\gamma\left(\frac{z_1-z}{z_2-z}\right)^\alpha \times$$

$$F\left(\alpha+\beta+\gamma, \alpha+\gamma+\beta'; 1+\gamma-\gamma', \frac{(z-z_3)(z_3-z_2)}{(z-z_2)(z_1-z_3)}\right) \tag{17.11}$$

Choosing $z_1 = 0$, $z_3 = 1$, and letting $z_2 \to \infty$, 17.11 becomes

$$(1-z)^\gamma z^\alpha F(\alpha+\beta+\gamma, \alpha+\gamma+\beta'; 1+\gamma-\gamma'; 1-z) \tag{17.12}$$

The additional transformation $\gamma \leftrightarrow \gamma'$ gives rise to yet another solution of Eq. 17.1.

$$(1-z)^{\gamma'} z^\alpha F(\alpha+\beta+\gamma', \alpha+\gamma'+\beta'; 1+\gamma'-\gamma; 1-z) \tag{17.13}$$

By setting $\alpha = 0$, $\alpha' = 1-c$, $\beta = a$, $\beta' = b$, $\gamma = 0$, $\gamma' = c-a-b$, the Riemann P symbol in 16.5 will represent solutions of the hypergeometric equation; hence, making these substitutions in 17.12 and 17.13, we find that

$$F(a,b; a+b+1-c; 1-z) \tag{17.14}$$

and

$$(1-z)^{c-a-b}F(c-b, c-a, 1+c-a-b; 1-z) \tag{17.15}$$

are new solutions of the hypergeometric equation. These are just the solutions 17.8. Proceeding in an analogous manner, one can show that

$$z^{-a}F\left(a,a-c+1; a-b+1; \frac{1}{z}\right) \tag{17.16}$$

and

$$z^{-b}F\left(b,b-c+1; b-a+1; \frac{1}{z}\right) \tag{17.17}$$

are solutions corresponding to 17.9.

In the special cases when the various exponents of Eq. 17.1 differ by an integer, a second linearly independent solution can be readily found by the methods of Sec. 2. We shall not study these cases here.

17.2 Integral Representations for the Hypergeometric Function

We have already stressed the interest of finding solutions of differential equations in the form of integral representations. In this section we find an integral representation for the hypergeometric function which, according to Sec. 15.2, should be taken with an Euler kernel. Therefore, let

$$u(z) = \int_C (z-t)^\lambda v(t)\,dt \tag{17.18}$$

where we must determine λ, the function $v(t)$, and the contour C. We rewrite Eq. 17.1 in the symbolic form

$$L_z u(z) = 0$$

where

$$L_z \equiv z(1-z)\frac{d^2}{dz^2} + [c-(a+b+1)z]\frac{d}{dz} - ab \tag{17.19}$$

Applying L_z to Eq. 17.18, there results

$$L_z u(z) = \int_C \Big\{\lambda(\lambda-1)z(1-z) + \lambda[c-(a+b+1)z](z-t) - ab(z-t)^2\Big\}(z-t)^{\lambda-2}v(t)\,dt \tag{17.20}$$

We now choose λ in such a way that the coefficient of z^2 in the bracket of Eq. 17.20 vanishes. This yields the equation

$$\lambda(\lambda-1) + \lambda(a+b+1) + ab = 0 \tag{17.21}$$

From the two solutions of Eq. 17.21 we take $\lambda = -a$ corresponding to a particular integral representation 17.18. (We could, of course, have chosen the other root of Eq. 17.21, which would have led us to a different representation.) Then, after rearranging terms so that z appears in the combination $(z-t)$, i.e., in the same form as the kernel, Eq. 17.20 can be written as

$$\begin{aligned} L_z u &= -\int_C \{a(a+1)(t^2-t)(z-t)^{-a-2} \\ &\quad - a(z-t)^{-a-1}[(b-a-1)t+(a-c+1)]\}v(t)\,dt \\ &= -\int_C dt\, v(t)\left\{\left[(t^2-t)\frac{d^2}{dt^2} - [(b-a-1)t+(a-c+1)]\frac{d}{dt}\right](z-t)^{-a}\right\} \end{aligned} \tag{17.22}$$

The term in the bracket is just in the form of an operator M_t with respect to t, which acts on the kernel $(z-t)^{-a}$.

$$M_t = (t^2-t)\frac{d^2}{dt^2} - [(b-a-1)t+(a-c+1)]\frac{d}{dt} \tag{17.23}$$

According to Sec. 15.1, $v(t)$ should be a solution of the adjoint equation

$$M_t^+ v(t) \equiv \frac{d^2}{dt^2}[(t^2-t)v(t)] + \frac{d}{dt}[(b-a-1)t+(a-c+1)]v(t) = 0 \tag{17.23a}$$

This equation can be easily solved. Integrating once, one has

$$\frac{d}{dt}[(t^2-t)v(t)] = -[(b-a-1)t+(a-c+1)]v(t)$$

Putting

$$w(t) \equiv (t^2-t)v(t) \tag{17.24}$$

we have

$$\frac{d}{dt}w(t) = -\frac{[(b-a-1)t+(a-c+1)]}{t^2-t}w(t)$$

which can be immediately integrated; inserting its solution in 17.24, we find that

$$v(t) = kt^{a-c}(t-1)^{c-b-1} \tag{17.25}$$

where k is an integration constant.

The bilinear function $Q[K,v]$ can be readily calculated from the Lagrange identity, using Eq. 17.18 and the expressions for M_t and M_t^+ (Eqs. 17.23 and 17.23a)

$$Q[K,v] = akt^{a-c+1}(t-1)^{c-b}(z-t)^{-a-1} \tag{17.26}$$

$Q[K,v]$ has the property that it vanishes at $t = 1$ and at $t = \infty$ whenever Re $c >$ Re $b > 0$.

We must now choose the contour C. The only requirement is that $Q[K,v]$ returns to its initial value at the end of the contour. There are many different possibilities for choosing a contour, and to each choice for C will correspond a different solution of the equation.

Let us first consider the contour that extends from $t = 1$ to $t = \infty$ along the positive real axis. Then we find the solution of the hypergeometric equation

$$u(z) = k\int_1^\infty (t-z)^{-a}t^{a-c}(t-1)^{c-b-1}\,dt \tag{17.27}$$

Of course it is assumed that a,b, and c are such that this integral exists. Expanding the factor $(t-z)^{-a}$ in the integrand, we have for $|z| < 1$

$$(t-z)^{-a} = t^{-a}\sum_{n=0}^\infty \frac{\Gamma(a+n)}{\Gamma(a)\Gamma(n+1)}\left(\frac{z}{t}\right)^n \tag{17.28}$$

Inserting 17.28 into 17.27 and using Eqs. 32.6 and 32.8 of Chapter I, we find

$$\begin{aligned} u(z) &= k\sum_{n=0}^\infty \frac{\Gamma(a+n)}{\Gamma(a)\Gamma(n+1)}z^n\int_1^\infty t^{-c-n}(t-1)^{c-b-1}\,dt \\ &= \left\{k\frac{\Gamma(b)\Gamma(c-b)}{\Gamma(c)}\right\}\left\{\frac{\Gamma(c)}{\Gamma(a)\Gamma(b)}\sum_{n=0}^\infty\frac{\Gamma(a+n)\Gamma(b+n)}{\Gamma(c+n)\Gamma(n+1)}z^n\right\} \end{aligned} \tag{17.29}$$

Comparing with Eq. 17.3 and choosing

$$k = \frac{\Gamma(c)}{\Gamma(b)\Gamma(c-b)}$$

we obtain

$$\begin{aligned} u(z) &\equiv F(a,b;c;z) \\ &= \frac{\Gamma(c)}{\Gamma(b)\Gamma(c-b)}\int_1^\infty (t-z)^{-a}t^{a-c}(t-1)^{c-b-1}\,dt \end{aligned} \tag{17.30}$$

By making the substitution $t \to 1/t$, Eq. 17.30 can be transformed to give the so-called **Euler formula**

$$F(a,b;c;z) = \frac{\Gamma(c)}{\Gamma(b)\Gamma(c-b)}\int_0^1 (1-tz)^{-a}t^{b-1}(1-t)^{c-b-1}\,dt \tag{17.31}$$

This representation is valid for all values of b and c satisfying

$$\text{Re } c > \text{Re } b > 0 \tag{17.32}$$

and was derived under the condition that $|z| < 1$. However, by analytic continuation, 17.31 represents a single-valued function in the entire cut plane

$$|\arg(1 - z)| < \pi \tag{17.33}$$

The last condition ensures that the factor $(1 - tz)^{-a}$ in the integrand is well defined. It corresponds to choosing the branch cut that joins the two singular points of the hypergeometric equation at $z = 1$ and $z = \infty$ along the positive real axis.

We can also choose the contour C as a closed contour that encircles the singular points of the function $K(z,t)v(t)$. The choice of a closed contour will guarantee the vanishing of the contribution from $Q[K,v]$. Except for very particular values of a,b,c, $K(z,t)v(t)$ will have branch points at $t = 0$ and $t = 1$. Suppose for simplicity that z is not a real number between 0 and 1. Then, choosing the cut from $t = 0$ to $t = 1$ and fixing the value of the integrand at some point in the environment of the cut, we make $K(z,t)v(t)$ single-valued in some domain around the cut that does not contain the point z (for z itself may be a branch point of the integrand).

We now choose C as shown in Fig. 44. This complicated contour is unavoidable. It crosses the cut twice so that each branch point (whose orders are different) is encircled twice, but in opposite directions. Therefore, at the end of the cycle, we return to the original Riemann sheet and the contour is indeed closed. The point z must, of course, be outside the contour. Thus, we have a solution

$$u'(z) = k \int_C (t - z)^{-a} t^{a-c} (t - 1)^{c-b-1}\, dt \tag{17.34}$$

The contour can be shrunk so that it passes just above and just below the cut. The value that the integrand takes on the upper and lower lips of the cut differ by some constant which can be absorbed into a multiplicative constant. Thus, provided

$$a - c > -1 \qquad \text{and} \qquad c - b > 0 \tag{17.35}$$

we can reduce the representation 17.34 to an ordinary integral

$$\begin{aligned} u'(z) &= \text{const} \int_0^1 (t - z)^{-a} t^{a-c} (1 - t)^{c-b-1}\, dt \\ &= \text{const}\; z^{-a} \int_0^1 \left(1 - \frac{t}{z}\right)^{-a} t^{a-c} (1 - t)^{c-b-1}\, dt \end{aligned} \tag{17.36}$$

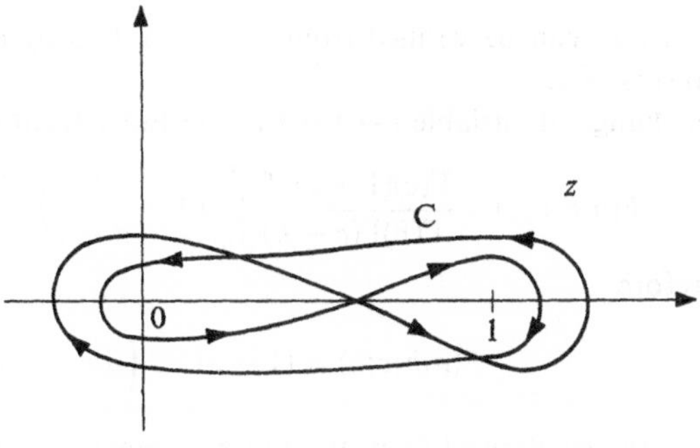

Fig. 44.

Comparing with the Euler formula (Eq. 17.31), we find

$$u'(z) = \text{const } z^{-a}F\left(a,a-c+1;a-b+1;\frac{1}{z}\right) \tag{17.37}$$

Apart from a constant this is just the solution 17.16 of the hypergeometric equation.

Because of the possibility of deforming the contour, a contour integral representation is usually valid for a much wider range of values of the parameters. This makes the use of such representations preferable in certain calculations (for example, when we seek asymptotic formulae for $F(a,b;c;z)$ for very large values of a,b, or c, using the method of steepest descent).

The reasoning that led from Eq. 17.34 to Eq. 17.37 can be inverted, and when applied to the Euler formula, it leads to

$$F(a,b;c;z) = \frac{-\Gamma(c)e^{-i\pi c}}{4\Gamma(b)\Gamma(c-b)\sin\pi b\sin\pi(c-b)} \times$$

$$\int_C t^{b-1}(1-t)^{c-b-1}(1-tz)^{-a}\,dt \tag{17.38}$$

where C again is the contour of Fig. 44. This representation is valid, provided

$$b,c-b \neq 1, 2, 3, \cdots$$

17.3 Some Further Relations Between Hypergeometric Functions

From the series solution 17.3 one can obtain directly many useful recurrence relations. We shall be satisfied here in listing only a few, and we refer the reader to Erdely *et al.* for a complete list of formulae. By differentiation of 17.3, one finds

$$\frac{d^n}{dz^n}F(a,b;c;z) = \frac{\Gamma(a+n)\Gamma(b+n)\Gamma(c)}{\Gamma(a)\Gamma(b)\Gamma(c+n)}F(a+n,b+n;c+n;z)$$

The six functions $F(a \pm 1, b;c;z)$, $F(a,b \pm 1; c;z)$, $F(a,b;c \pm 1; z)$ are called hypergeometric functions **contiguous** to $F(a,b;c;z)$. There exist 15 relations between these contiguous functions. We illustrate these relations with two examples.

$$[c-2a-(b-a)z]F(a,b;c;z)$$
$$+\,a(1-z)F(a+1,b;c;z)-(c-a)F(a-1,b;c;z)=0$$

$$(c-a-1)F(a,b;c;z)+aF(a+1,b;c;z)-(c-1)F(a,b;c-1;z)=0$$

These relations can be verified from the series 17.3 by comparing the coefficients of equal powers of z.

The change of variable $t \to 1-t$ in the Euler formula (Eq. 17.31) leads to

$$F(a,b;c;z) = \frac{\Gamma(c)(1-z)^{-a}}{\Gamma(b)\Gamma(c-b)}\int_0^1\left(1-\frac{tz}{z-1}\right)^{-a}t^{c-b-1}(1-t)^{b-1}\,dt$$

and therefore

$$F(a,b;c;z) = (1-z)^{-a}F\left(a,c-b;c;\frac{z}{z-1}\right) \tag{17.39}$$

This equality was derived from the Euler formula, which holds for Re $c >$ Re $b > 0$. However, it is valid for a much wider range of values of the parameters. It can be shown that the expansion of $F(a,b;c;z)$ in the hypergeometric series is uniformly convergent as a function of its parameters and of the variable z, provided $|z| < 1$ and $c \neq 0, -1, -2, \cdots$. Therefore, since a uniformly convergent series of analytic functions can be differentiated term by term, $F(a,b;c;z)$ is (for $|z| < 1$) an analytic function of a,b, and c for all values of these parameters (except when $c = 0, -1, -2, \cdots$).

In the region where

$$|z| < |z - 1| < 1 \tag{17.40}$$

both sides of Eq. 17.39 can be expanded in hypergeometric series, and since both sides are analytic functions of the parameters a,b, and c, this relation holds by analytic continuation for any a,b,c ($c \neq 0, -1, -2, \cdots$) and for z satisfying the inequality 17.40. We now continue Eq. 17.39, with fixed values of the parameters in the entire cut z plane.

The foregoing reasoning serves as a typical example of how the validity of a relation derived under restricted values of the parameters can be broadened by analytical continuation with respect to those parameters.

From the representation 17.31, with the aid of Eq. 32.7 of Chapter I, we find

$$\begin{aligned} F(a,b;c;1) &= \frac{\Gamma(c)}{\Gamma(b)\Gamma(c-b)} \int_0^1 t^{b-1}(1-t)^{c-b-a-1}\,dt \\ &= \frac{\Gamma(c)\Gamma(c-a-b)}{\Gamma(c-a)\Gamma(c-b)} \end{aligned} \tag{17.41}$$

The value of $F(a,b;c;1)$ given in Eq. 17.41 was derived from the Euler formula and Eq. 32.7, Chapter I; therefore, it holds for $\operatorname{Re} c > \operatorname{Re} b > 0$ and $\operatorname{Re}(c - b - a) > 0$. This formula cannot be extended by analytic continuation to less restrictive values of the parameters because the point $z = 1$ is just at the limit of convergence of the hypergeometric series. By a more elaborate calculation, it can be shown that this relation is in fact valid for

$$\operatorname{Re} c > \operatorname{Re}(a + b), \qquad c \neq 0, -1, -2, \cdots$$

We have shown that

$$F(a,b;a + b - c + 1; 1 - z)$$

and

$$(1 - z)^{c-a-b}F(c - b, c - a; c - a - b + 1; 1 - z)$$

are two linearly independent solutions of the hypergeometric equation about $z = 1$ when $c - a - b$ is not an integer. Therefore the function $F(a,b;c;z)$ must be expressible in terms of these solutions

$$\begin{aligned} F(a,b;c;z) &= \alpha F(a,b;a + b - c + 1; 1 - z) \\ &\quad + \beta(1 - z)^{c-a-b}F(c - b, c - a; c - a - b + 1; 1 - z) \end{aligned} \tag{17.42}$$

To determine the constants α and β, we set $z = 0$ and $z = 1$ successively in Eq. 17.42 and use Eqs. 17.41 and 17.3. We find

$$\begin{aligned} F(a,b;c,z) &= \frac{\Gamma(c)\Gamma(c-a-b)}{\Gamma(c-a)\Gamma(c-b)} F(a,b;a + b - c + 1, 1 - z) \\ &\quad + \frac{\Gamma(c)\Gamma(a+b-c)}{\Gamma(a)\Gamma(b)} (1 - z)^{c-a-b}F(c - b, c - a; c - a - b + 1, 1 - z) \end{aligned} \tag{17.43}$$

Equation 17.43 enables us to express the solution of the hypergeometric equation about the singular point at $z = 0$ in term of the solutions about $z = 1$. Similar relations exist which connect the hypergeometric series about other pairs of singular points of the equation.

18 · FUNCTIONS RELATED TO THE HYPERGEOMETRIC FUNCTION

By specializing some of the arguments of $F(a,b;c;z)$, one can relate this function to some elementary functions. For example

$$F(-a,b;b;\,-z) = (1 + z)^a$$

$$F\left(\frac{1}{2}, \frac{1}{2}; \frac{3}{2};\, z^2\right) = \frac{1}{z} \sin^{-1} z$$

$$F(1, 1; 2; -z) = \frac{1}{z} \log(1 + z)$$

These relations can be most easily verified from the series 17.3. Similarly, the solutions of a number of important equations of mathematical physics can be expressed in terms of hypergeometric functions.

18.1 The Jacobi Functions

The Jacobi functions are solutions of the equation

$$(1 - z^2)\frac{d^2u}{dz^2} + [\beta - \alpha - (\alpha + \beta + 2)z]\frac{du}{dz} + \lambda(\lambda + \alpha + \beta + 1)u = 0 \qquad (18.1)$$

which has already been encountered in Sec. 10, Chapter III, in connection with the Jacobi polynomials. The substitution $x = (1 - z)/2$ leads to an equation of the hypergeometric type, a particular solution of which is $F(-\lambda, \lambda + \alpha + \beta + 1, \alpha + 1; (1 - z)/2)$. With the conventional normalization, one defines **the Jacobi function of the first kind**

$$P_\lambda^{(\alpha,\beta)}(z) = \frac{\Gamma(\lambda + \alpha + 1)}{\Gamma(\lambda + 1)\Gamma(\alpha + 1)} F\left(-\lambda, \lambda + \alpha + \beta + 1, \alpha + 1; \frac{1 - z}{2}\right) \qquad (18.2)$$

When λ is a non-negative integer, $\lambda = n$ say, $P_n^{(\alpha,\beta)}(z)$ is a polynomial of degree n, as can be directly seen from the hypergeometric series

$$P_n^{(\alpha,\beta)}(z) = \frac{\Gamma(n + \alpha + 1)}{\Gamma(n + 1)\Gamma(n + \alpha + \beta + 1)} \sum_{m=0}^{\infty} \lim_{\lambda\to n}\left[\frac{\Gamma(m - \lambda)}{\Gamma(-\lambda)}\right] \times$$

$$\frac{\Gamma(n + \alpha + \beta + m + 1)}{\Gamma(\alpha + m + 1)\Gamma(m + 1)}\left(\frac{1 - z}{2}\right)^m$$

$$= \frac{\Gamma(n + \alpha + 1)}{\Gamma(n + 1)\Gamma(n + \alpha + \beta + 1)} \sum_{m=0}^{n} \frac{\Gamma(n + \alpha + \beta + m + 1)}{\Gamma(\alpha + m + 1)}\left(\frac{z - 1}{2}\right)^m$$

In the last step we used the relation

$$\lim_{\lambda\to n} \frac{\Gamma(m - \lambda)}{\Gamma(-\lambda)} = \begin{cases} (-1)^m\Gamma(m + 1) & m \le n \\ 0 & m > 0 \end{cases}$$

which follows directly from Eq. 32.5, Chapter I. It can be easily shown that $P_n^{(\alpha,\beta)}(z)$ is the Jacobi polynomial introduced on different grounds in Sec. 10 of Chapter III. With integer $\lambda = n$, the Jacobi polynomial satisfies Eq. 18.1, and therefore it can differ from $P_n^{\alpha,\beta}(z)$ as defined in Sec. 10.6, Chapter III, by at most a multiplicative constant, since one cannot have two linearly independent solutions of Eq. 18.1 that are both analytic in the neighborhood of the singular point of the equation at $z = 1$. That this multiplicative constant is indeed unity can be verified by comparing the coefficient of z^n in Eq. 18.2

$$k_n = \frac{1}{2^n}\frac{\Gamma(2n + \alpha + \beta + 1)}{\Gamma(n + 1)\Gamma(n + \alpha + \beta + 1)}$$

with the corresponding constant as given in Sec. 10.6 of Chapter III.

One of Kummer's 24 solutions of the hypergeometric equation is of the form (see Eq. 16.15)

$$(-x)^{a-c}(1 - x)^{c-a-b}F\left(1 - a, c - a; b - a + 1; \frac{1}{x}\right)$$

and it can be shown to be linearly independent of $F(a,b;c;x)$. Setting $a = -\lambda$, $b = \lambda + \alpha + \beta + 1$, $c = \alpha + 1$ and $x = (1 - z)/2$, we see that

$$\frac{1}{(z - 1)^{\lambda+\alpha+1}(z + 1)^{\beta}} F\left(\lambda + 1, \lambda + \alpha + 1; 2\lambda + \alpha + \beta + 2; \frac{2}{1 - z}\right)$$

is a second linearly independent solution of Eq. 18.1. The function

$$Q_\lambda^{(\alpha,\beta)}(z) = \frac{2^{\lambda+\alpha+\beta}\Gamma(\lambda + \alpha + 1)\Gamma(\lambda + \beta + 1)}{\Gamma(2\lambda + \alpha + \beta + 2)(z - 1)^{\lambda+\alpha+1}(z + 1)^{\beta}} \times$$
$$F\left(\lambda + 1, \lambda + \alpha + 1; 2\lambda + \alpha + \beta + 2; \frac{2}{1 - z}\right) \tag{18.3}$$

is known as the **Jacobi function of the second kind**. Note that it is not a polynomial.

18.2 The Gegenbauer Function

The solution of the Gegenbauer equation* (Chapter III, Sec. 10.6)

$$(1 - z^2)\frac{d^2u}{dz^2} - (2\mu + 1)z\frac{du}{dz} + \lambda(\lambda + 2\mu)u = 0 \tag{18.4}$$

which is proportional to $F(-\lambda, \lambda + 2\mu, \mu + \frac{1}{2}, (1 - z)/2)$, is called the Gegenbauer function $C_\lambda^\mu(z)$. The normalization is so chosen that

$$C_\lambda^\mu(z) = \frac{\Gamma(2\mu + \lambda)}{\Gamma(\lambda + 1)\Gamma(2\mu)} F\left(-\lambda, \lambda + 2\mu; \mu + \frac{1}{2}; \frac{1 - z}{2}\right) \tag{18.5}$$

$C_\lambda^\mu(z)$ reduces to the Gegenbauer polynomial for nonnegative integer λ and $\mu > 1/2$.

Comparing Eqs. 18.5 and 18.2, we find

$$C_\lambda^{\alpha+1/2}(z) = \frac{\Gamma(2\alpha + \lambda + 1)\Gamma(\alpha + 1)}{\Gamma(2\alpha + 1)\Gamma(\lambda + \alpha + 1)} P_\lambda^{(\alpha,\alpha)}(z) \tag{18.6}$$

A second linearly independent solution of Eq. 18.4 can be chosen as a function proportional to the Jacobi function of the second kind $Q_\lambda^{(\alpha,\alpha)}(z)$, with $\alpha = \mu - 1/2$.

---

* It is a particular case of the Jacobi equation, with $\alpha = \beta = \mu - \frac{1}{2}$.

18.3 The Legendre Functions

The Legendre equation

$$(1-z^2)\frac{d^2u}{dz^2} - 2z\frac{du}{dz} + \lambda(\lambda+1)u = 0 \tag{18.7}$$

has also been encountered in Sec. 10, Chapter III. It is a particular case of Jacobi's equation (Eq. 18.1). Its solution

$$P_\lambda(z) \equiv P_\lambda^{(0,0)}(z) = F\left(-\lambda, \lambda+1; 1; \frac{1-z}{2}\right) \tag{18.8}$$

is called the **Legendre function of the first kind**. With non-negative integer λ, $\lambda = n$ say, $P_n(z)$ is the Legendre polynomial of order n.

In analogy to Eq. 18.8, we define the **Legendre function of the second kind** by the relation

$$Q_\lambda(z) \equiv Q_\lambda^{(0,0)}(z) = \frac{2^\lambda[\Gamma(\lambda+1)]^2}{\Gamma(2\lambda+2)(z-1)^{\lambda+1}} F\left(\lambda+1, \lambda+1; 2\lambda+2; \frac{2}{1-z}\right)$$

It is a second solution of the Legendre equation, linearly independent of $P_\lambda(z)$.

19 · THE CONFLUENT HYPERGEOMETRIC FUNCTION

Consider the particular Riemann equation

$$\frac{d^2u}{dz^2} + \left(\frac{c}{z} + \frac{1-a-b}{z-z_2} + \frac{1-c-a+b}{z-z_3}\right)\frac{du}{dz} + \frac{abz_2(z_2-z_3)}{z(z-z_2)^2(z-z_3)}u = 0 \tag{19.1}$$

This equation has regular singular points at $z = 0$ and $z = z_2, z_3$. Let us set

$$z_2 = b = 2\alpha \qquad \text{and} \qquad z_3 = \alpha$$

in Eq. 19.1 and then let $\alpha \to \infty$ while keeping a and c fixed. The resulting equation is

$$z\frac{d^2u}{dz^2} + (c-z)\frac{du}{dz} - au = 0 \tag{19.2}$$

In the limiting process, two of the singular points, z_2 and z_3, have "coalesced" at infinity. An equation obtained from a Riemann equation by the coalescence of two singularities is called a **confluent Riemann equation**. In particular, Eq. 19.2 may be regarded as the limiting case of a hypergeometric equation in which the singularity at $z = 1$ has been "pushed" out to infinity while the singularities at $z = 0$ and $z = \infty$ remain. For that reason, Eq. 19.2 is called the **confluent hypergeometric equation**.

It is important to note that the point $z = \infty$ is no longer a regular singular point but an irregular singular point. This can be verified by making the substitution $z \to 1/\xi$ in Eq. 19.2 and seeing that $\xi = 0$ is an irregular singular point of the resulting equation.

Just as in the case of the hypergeometric equation, many important equations of mathematical physics are merely special cases of the confluent hypergeometric equation.

The roots of the indicial equation corresponding to the singular point of Eq. 19.2 at the origin are 0 and $1 - c$. Hence, there exists a solution that is analytic in the neighborhood of the origin and which can be normalized to unity. We call this solution the **confluent hypergeometric function** and denote it by $\Phi(a,c;z)$. Since the next singular point of the equation is at $z = \infty$, $\Phi(a,c;z)$ is an entire function.

We shall solve the confluent hypergeometric equation using the method of integral representations. Instead of using the Euler kernel $(z - t)^\mu$ we use the Laplace kernel e^{zt} and seek a solution in the form

$$u(z) = \int_C e^{zt} v(t)\, dt \tag{19.3}$$

Again we write Eq. 19.2 as

$$L_z u = 0$$

with

$$L_z = z\frac{d^2}{dz^2} + (c - z)\frac{d}{dz} - a \tag{19.4}$$

Operating on both sides of Eq. 19.3 with L_z, we find

$$\begin{aligned} L_z u &= \int_C dt\, v(t) e^{zt}[zt^2 + (c - z)t - a] \\ &= \int_C dt\, v(t)\left[(t^2 - t)\frac{\partial}{\partial t} + ct - a\right]e^{zt} \end{aligned}$$

$v(t)$ will be given by the solution of the adjoint equation

$$\frac{d}{dt}[(t^2 - t)v(t)] = (a - ct)v(t) \tag{19.5}$$

which can be immediately integrated

$$v(t) = kt^{a-1}(1 - t)^{c-a-1} \tag{19.6}$$

where k is an integration constant. Therefore

$$u(z) = k\int_C e^{zt}t^{a-1}(1 - t)^{c-a-1}\, dt \tag{19.7}$$

The function $Q[K,v]$, which appears in the surface term, is easily found to be

$$Q[K,v] = -[e^{zt}t^a(1 - t)^{c-a}]$$

and the path C in the representation 19.7 must be chosen in such a way that $Q[K,v]$ assumes equal values at its end points.

Let us first choose for C the segment of the real axis lying between $t = 0$ and $t = 1$. Then (provided $\operatorname{Re} c > \operatorname{Re} a > 0$)

$$u(z) = k\int_0^1 e^{zt}t^{a-1}(1 - t)^{c-a-1}\, dt \tag{19.8}$$

Expanding e^{zt} in powers of zt and using Eq. 32.6, Chapter I, we find

$$u(z) = k \sum_{n=0}^{\infty} \frac{z^n}{n!} \int_0^1 dt\, t^{a+n-1}(1-t)^{c-a-1}$$

$$= \left\{ k\,\frac{\Gamma(c-a)\Gamma(a)}{\Gamma(c)} \right\} \left\{ \frac{\Gamma(c)}{\Gamma(a)} \sum_{n=0}^{\infty} \frac{\Gamma(a+n)}{\Gamma(c+n)\Gamma(n+1)} z^n \right\}$$

Hence, setting

$$k = \frac{\Gamma(c)}{\Gamma(c-a)\Gamma(a)} \tag{19.9}$$

we find that $u(z)$ is an analytic function of z normalized at $z = 0$ to unity. This is just the way the confluent hypergeometric function has been defined. Therefore

$$\Phi(a,c;z) = \frac{\Gamma(c)}{\Gamma(a)} \sum_{n=0}^{\infty} \frac{\Gamma(a+n)}{\Gamma(c+n)\Gamma(n+1)} z^n \tag{19.10}$$

This series could, of course, have been obtained directly from the confluent hypergeometric equation; it is called the **confluent hypergeometric series** and is valid whenever c is not zero or a negative integer.

Comparing Eq. 19.9 with the hypergeometric series (Eq. 17.3), we see that

$$\Phi(a,c;z) = \lim_{b\to\infty} F\left(a,b;c,\frac{z}{b}\right) \tag{19.11}$$

Equation 19.8 can also be obtained by using the Euler formula

$$F\left(a,b;c;\frac{z}{b}\right) = F\left(b,a;c;\frac{z}{b}\right)$$

$$= \frac{\Gamma(c)}{\Gamma(c-a)\Gamma(a)} \int_0^1 dt \left(1 - \frac{tz}{b}\right)^{-b} t^{a-1}(1-t)^{c-a-1}$$

Since

$$\lim_{b\to\infty} \left(1 - \frac{tz}{b}\right)^{-b} = e^{zt}$$

we find Eq. 19.8 with k given by Eq. 19.9:

$$\Phi(a,c;z) = \frac{\Gamma(c)}{\Gamma(c-a)\Gamma(a)} \int_0^1 dt\, e^{zt} t^{a-1}(1-t)^{c-a-1} \qquad \operatorname{Re} c > \operatorname{Re} a > 0 \tag{19.12}$$

The relation 19.11 could have been arrived at directly by making the replacement $z \to z/b$ in the hypergeometric equation and then letting $b \to \infty$. The process would have transformed this equation into the confluent hypergeometric equation. However, it would still be necessary to prove that the same process, when applied to the hypergeometric function, yields a solution of the confluent hypergeometric equation. The indirect method we have used serves to justify this limiting process.

If c is not an integer, then

$$z^{1-c}F(b-c+1, a-c+1; 2-c;z)$$

is a second linearly independent solution of the hypergeometric equation, and a second linearly independent solution of the confluent hypergeometric equation is

$$z^{1-c} \lim_{b\to\infty} F\left(b-c+1, a-c+1; 2-c; \frac{z}{b}\right)$$

$$= z^{1-c} \lim_{b\to\infty} F\left(a-c+1, b-c+1; 2-c; \frac{z}{b}\right)$$

$$= z^{1-c}\,\Phi(a-c+1, 2-c; z) \tag{19.13}$$

Another solution of Eq. 19.2 appears in the literature. It is obtained by integrating Eq. 19.7 along the negative real axis from $t = -\infty$ to $t = 0$

$$u(z) = \text{const} \int_{-\infty}^{0} dt\, e^{zt} t^{a-1} (1-t)^{c-a-1} \qquad \operatorname{Re} a > 0 \tag{19.14}$$

This integral converges for $\operatorname{Re} z > 0$.

With the conventional normalization, we obtain the solution denoted by $\Psi(a,c;z)$

$$\Psi(a;c,z) = \frac{1}{\Gamma(a)} \int_{0}^{\infty} e^{-zt} t^{a-1} (1+t)^{c-a-1}\, dt \qquad \operatorname{Re} a > 0, \operatorname{Re} z > 0 \tag{19.15}$$

Since $\Phi(a;c,z)$ and $z^{1-c}\Phi(a-c+1, 2-c,z)$ form a fundamental set of solutions of Eq. 19.2 when c is not an integer, $\Psi(a;c,z)$ must be expressible as a linear combination of these solutions; i.e.,

$$\Psi(a,c;z) = \alpha\Phi(a,c;z) + \beta z^{1-c}\Phi(a-c+1, 2-c; z) \tag{19.16}$$

where α and β are constants, which we shall determine by examining the behavior of both sides of this equation for $z \to 0$ and for $z \to \infty$. Write Eq. 19.12 as

$$\Phi(a,c;z) = \frac{\Gamma(c)}{\Gamma(a)\Gamma(c-a)} \left[\int_{-\infty}^{1} e^{zt} t^{a-1} (1-t)^{c-a-1}\, dt - \int_{-\infty}^{0} e^{zt} t^{a-1} (1-t)^{c-a-1}\, dt \right]$$

We make the substitutions $t \to 1 - (t/z)$ and $t \to -t/z$ in the first and second integrals, respectively

$$\Phi(a,c;z) = \frac{\Gamma(c)}{\Gamma(a)\Gamma(c-a)} \left[z^{a-c} e^{z} \int_{0}^{\infty} e^{-t} t^{c-a-1} \left(1 - \frac{t}{z}\right)^{a-1} dt + (-z)^{-a} \int_{0}^{\infty} e^{-t} t^{a-1} \left(1 + \frac{t}{z}\right)^{c-a-1} dt \right] \tag{19.17}$$

As $z \to \infty$ along the positive real axis, we can neglect the second integral compared to the first integral, and we have the asymptotic behavior

$$\Phi(a,c;z) \sim \frac{\Gamma(c)}{\Gamma(a)\Gamma(c-a)} z^{a-c} e^{z} \int_{0}^{\infty} e^{-t} t^{c-a-1}\, dt$$

$$= \left(\frac{\Gamma(c)}{\Gamma(a)}\right) z^{a-c} e^{z} \qquad (\operatorname{Re} c > \operatorname{Re} a > 0) \tag{19.18}$$

since the last integral is simply $\Gamma(c-a)$.

In Eq. 19.18 we kept only the first term of a series obtained by expanding the factors $(1 \pm t/z)^{\text{const}}$ in the integrands in powers of t/z. This expansion diverges for $t > z$, but the contribution to the integral from this region of integration becomes negligible as $z \to \infty$. Consequently, Eq. 19.18 is the leading term of an asymptotic expansion of $\Phi(a,c;z)$. (For the definition of an asymptotic series, see the end of Sec. 31, Chapter I.)

A similar argument applied to $\Psi(a,c;z)$ yields (after making the substitution $zt \to t$ in Eq. 19.15)

$$\Psi(a,c;z) = \frac{z^{-a}}{\Gamma(a)} \int_0^\infty e^{-t} t^{a-1} \left(1 + \frac{t}{z}\right)^{c-a-1} dt$$

$$\sim z^{-a} \qquad (\text{Re}\, a > 0) \tag{19.19}$$

when $z \to \infty$ along the positive real axis. Putting 19.18 and 19.19 into Eq. 19.16, we see that the strong exponential behavior of the RHS is incompatible with the asymptotic behavior of the LHS unless α and β are chosen so as to cancel the exponentially increasing terms. This leads to the condition

$$\alpha = -\frac{\Gamma(a)\Gamma(2-c)}{\Gamma(c)\Gamma(a-c+1)} \beta \tag{19.20}$$

One can determine the constant α when $\text{Re}\, c < 1$ by setting $z = 0$ in Eq. 19.16. Since

$$\Phi(a,c;0) = 1$$

we have (using Eq. 19.15)

$$\alpha = \Psi(a,c;0) = \frac{1}{\Gamma(a)} \int_0^\infty t^{a-1}(1+t)^{c-a-1}\, dt$$

Upon making the change of integration variable $1 + t \to t$ and comparing with Eq. 32.8, Chapter I, we find

$$\alpha = \frac{1}{\Gamma(a)} B(1-c,a) = \frac{\Gamma(1-c)}{(\Gamma a - c + 1)} \tag{19.21}$$

Finally, with the help of Eq. 32.5, Chapter I, we arrive at

$$\Psi(a,c;z) = \frac{\Gamma(1-c)}{\Gamma(a-c+1)} \Phi(a,c;z)$$

$$+ \frac{\Gamma(c-1)}{\Gamma(a)} z^{1-c} \Phi(a-c+1, 2-c; z) \tag{19.22}$$

Strictly speaking, we have proved the validity of the preceding relation when $1 > \text{Re}\, c > \text{Re}\, a > 0$. However, it can be extended to a much wider range of variation of the parameters. It holds, when $1 - c$ is not an integer, in the cut z plane.

$$|\arg z| < \pi$$

A glance at Eq. 19.17 shows that when $z \to \infty$ along the **negative** real axis, the second term in brackets dominates, and one obtains the following asymptotic expansion for $\Phi(a,c;z)$.

$$\Phi(a,c;z) \sim \frac{\Gamma(c)}{\Gamma(c-a)} |z|^{-a} \tag{19.23}$$

This asymptotic behavior for $\Phi(a,c;z)$ is very different from the behavior 19.18 of this function where z went to infinity along the positive real axis. By taking $z \to \infty$ along an arbitrary line not coincident with the real axis, one would obtain still another asymptotic behavior for $\Phi(a,c;z)$. The dependence of the behavior of asymptotic series on the way the point at infinity is approached is called **Stokes' phenomenon.**

20 · FUNCTIONS RELATED TO THE CONFLUENT HYPERGEOMETRIC FUNCTION

We have mentioned that many of the often encountered equations of mathematical physics are special cases of the confluent hypergeometric equation. We list in this section the more important cases.

20.1 Parabolic Cylinder Functions; Hermite and Laguerre Polynomials

The parabolic cylinder functions are solutions of the Weber-Hermite differential equation

$$\frac{d^2u}{dz^2} + \left(v + \frac{1}{2} - \frac{1}{4}z^2\right)u = 0 \tag{20.1}$$

With the substitution

$$u(z) = e^{-\frac{1}{4}z^2}u'(z)$$

Eq. 20.1 is transformed into a confluent hypergeometric equation for $u'(z)$ with arguments

$$a = -\frac{v}{2}; \quad c = \frac{1}{2}$$

With the conventional normalization, the parabolic cylinder functions are defined by

$$D_v(z) = 2^{v/2}e^{-z^2/4}\Psi\left(-\frac{v}{2}, \frac{1}{2}; \frac{z^2}{2}\right) \tag{20.2}$$

If v is a non-negative integer n, the function $\Psi(-n/2, 1/2; z^2/2)$ cannot be defined by the integral 19.15, since this representation holds only for $\operatorname{Re} a > 0$, but it can be defined in terms of the functions Φ through the relation 19.22. Using the series solution for Φ, one sees that when n is a non-negative integer, $\Psi(-n/2, 1/2; z^2/2)$ is a polynomial of degree n. Therefore, the functions

$$H_n\left(\frac{z}{\sqrt{2}}\right) \equiv 2^{n/2}e^{z^2/4}D_n(z) = 2^n\Psi\left(\frac{-n}{2}, \frac{1}{2}; \frac{z^2}{2}\right)$$

are polynomials. They are the **Hermite polynomials** that have already been discussed in Sec. 10.6, Chapter III.

The functions $\Phi(-n,\mu+1;z)$, where n is a non-negative integer, are also polynomials of degree n and the functions

$$L_n^\mu(z) = \frac{\Gamma(n+\mu+1)}{\Gamma(n+1)\Gamma(\mu+1)}\Phi(-n,\mu+1;z)$$

are the **Laguerre polynomials.**

20.2 The Error Function

The **error function** plays an important role in statistics and is also related to the confluent hypergeometric function

$$\operatorname{Erf}(z) \equiv \int_0^z e^{-t^2}dt = z\Phi(\tfrac{1}{2},\tfrac{3}{2};-z^2)$$

This relation can be verified directly from the representation 19.12.

20.3 Bessel Functions

The **Bessel functions**, which are solutions of the equation

$$\frac{d^2u}{dz^2} + \frac{1}{z}\frac{du}{dz} + \left(1 - \frac{\nu^2}{z^2}\right)u = 0 \tag{20.3}$$

are among the most important functions that occur in physics. Therefore, we shall consider their properties in somewhat greater detail.

Putting $u = z^\nu e^{-iz}u'(z)$ in Eq. 20.3, we find

$$z\frac{d^2u'}{dz^2} + [(2\nu + 1) - 2iz]\frac{du'}{dz} - i(2\nu + 1)u' = 0 \tag{20.4}$$

A solution of Eq. 20.4 is the confluent hypergeometric function

$$\Phi(\nu + \tfrac{1}{2}, 2\nu + 1; 2iz)$$

and the function

$$J_\nu(z) = \frac{1}{\Gamma(\nu + 1)}\left(\frac{z}{2}\right)^\nu e^{-iz}\Phi\left(\nu + \frac{1}{2}, 2\nu + 1; 2iz\right) \tag{20.5}$$

which is a solution of Eq. 20.3, is called the **Bessel function of the first kind** and **of order** ν. With the aid of Eq. 19.10, one finds, after multiplying through by

$$e^{-iz} = 1 - iz + \frac{(iz)^2}{2!} + \cdots$$

that

$$J_\nu(z) = \frac{1}{\Gamma(\nu + 1)}\left(\frac{z}{2}\right)^\nu\left[1 - \frac{1}{(\nu + 1)}\left(\frac{z}{2}\right)^2 + \frac{1}{2!(\nu + 1)(\nu + 2)}\left(\frac{z}{2}\right)^4 + \cdots\right]$$

$$= \left(\frac{z}{2}\right)^\nu \sum_{m=0}^{\infty} \frac{(-1)^m}{m!\Gamma(\nu + m + 1)}\left(\frac{z}{2}\right)^{2m} \tag{20.6}$$

In our study of the confluent hypergeometric equation, we found that if $1 - c$ was not an integer, then

$$\Phi(a,c;z) \qquad \text{and} \qquad z^{1-c}\Phi(a - c + 1, 2 - c;z)$$

was a fundamental set of solutions of the equation.

Therefore, the Bessel function 20.5 and any function proportional to the function

$$\left(\frac{z}{2}\right)^\nu e^{-iz}(z)^{-2\nu}\Phi\left(\frac{1}{2} - \nu, 1 - 2\nu; 2iz\right) \tag{20.7}$$

form a fundamental set of solutions of Eq. 20.3, provided 2ν is not an integer. Since the function 20.7 is proportional to $J_{-\nu}(z)$, one can clearly take $J_\nu(z)$ and $J_{-\nu}(z)$ as the fundamental set of solutions of Eq. 20.3 when 2ν is not an integer.

When 2ν is an integer n, $J_n(z)$ and $J_{-n}(z)$ are no longer linearly independent. To see this, we note that

$$\Gamma(-n+m+1) = (m-n)! = \infty \qquad \text{for } m < n$$

and therefore the first $n-1$ terms in $J_{-n}(z)$ vanish (see Eq. 20.6). Hence

$$J_{-n}(z) = \left(\frac{z}{2}\right)^{-n} \sum_{m=n}^{\infty} \frac{(-1)^m}{m!\Gamma(-n+m+1)} \left(\frac{z}{2}\right)^{2m}$$

or, putting $m = l + n$,

$$J_{-n}(z) = (-1)^n \left(\frac{z}{2}\right)^{n} \sum_{l=0}^{\infty} \frac{(-1)^l}{l!\Gamma(n+l+1)} \left(\frac{z}{2}\right)^{2l} = (-1)^n J_n(z) \tag{20.8}$$

Hence, $J_n(z)$ and $J_{-n}(z)$ are linearly dependent. To find a second linearly independent solution when n is an integer, we first define a function $Y_\nu(z)$ for noninteger ν as

$$Y_\nu(z) = \frac{[J_\nu(z)\cos(\nu\pi) - J_{-\nu}(z)]}{\sin(\nu\pi)} \tag{20.9}$$

$Y_\nu(z)$ is called a **Bessel function of the second kind**, or **Neumann's function**. Obviously, when ν is not an integer, $J_\nu(z)$ and $Y_\nu(z)$ form a fundamental set of solutions of Bessel's equation, since $Y_\nu(z)$ is merely a linear combination of the fundamental set $J_\nu(z)$ and $J_{-\nu}(z)$. As ν tends to an integer n, both the numerator and the denominator in the RHS of Eq. 20.9 tend to zero on account of Eq. 20.8, but the ratio tends to a well-defined limit, which can be calculated from l'Hopital's rule

$$\begin{aligned} Y_n(z) &= \lim_{\nu\to n} Y_\nu(z) \\ &= \frac{1}{\pi} \lim_{\nu\to n} \left\{ \frac{\partial J_\nu(z)}{\partial \nu} - (-1)^n \frac{\partial J_{-\nu}(z)}{\partial \nu} \right\} \end{aligned}$$

From Eq. 20.6, one has

$$\frac{\partial J_\nu(z)}{\partial \nu} = J_\nu(z) \log\left(\frac{1}{2} z\right) - \sum_{m=0}^{\infty} (-1)^m \left(\frac{z}{2}\right)^{\nu+2m} \frac{\psi(\nu+m+1)}{m!\Gamma(\nu+m+1)}$$

where the function $\psi(z) \equiv (d/dz) \log \Gamma(z)$ has been defined in Sec. 32, Chapter I.

Similarly

$$\frac{\partial J_{-\nu}(z)}{\partial \nu} = -J_{-\nu}(z) \log\left(\frac{1}{2} z\right) + \sum_{m=0}^{\infty} (-1)^m \left(\frac{z}{2}\right)^{2m-\nu} \frac{\psi(-\nu+m+1)}{m!\Gamma(-\nu+m+1)}$$

Hence

$$\begin{aligned} Y_n(z) = \frac{2}{\pi} J_n(z) \log\left(\frac{1}{2} z\right) &- \frac{1}{\pi} \sum_{m=0}^{\infty} (-1)^m \left(\frac{z}{2}\right)^{2m+n} \frac{\psi(n+m+1)}{m!\Gamma(n+m+1)} \\ &- \frac{1}{\pi} \sum_{m=0}^{\infty} (-1)^{m+n} \left(\frac{z}{2}\right)^{2m-n} \frac{\psi(+m-n+1)}{m!\Gamma(m-n+1)} \end{aligned} \tag{20.10}$$

It is often useful, instead of working with the fundamental set $J_\nu(z)$ and $Y_\nu(z)$, to work with the linear combinations

$$\begin{aligned} H_\nu^{(1)}(z) &\equiv J_\nu(z) + iY_\nu(z) \\ H_\nu^{(2)}(z) &\equiv J_\nu(z) - iY_\nu(z) \end{aligned} \tag{20.11}$$

$H_\nu^{(1)}(z)$ and $H_\nu^{(2)}(z)$ are called **Bessel functions of the third kind** or **Hankel functions.** Because the Hankel functions have a different asymptotic behavior for large values of z than that of the functions $J_\nu(z)$ and $Y_\nu(z)$, they are the more appropriate functions to use for the solution of certain physical problems.

Bessel Functions of Imaginary Argument

The substitution $z \to iz$ transforms Eq. 20.3 to the equation

$$\frac{d^2u}{dz^2} + \frac{1}{z}\frac{du}{dz} - \left(1 + \frac{\nu^2}{z^2}\right)u = 0 \tag{20.12}$$

The Bessel functions $J_\nu(iz)$ and $J_{-\nu}(iz)$ are a fundamental set of solutions of Eq. 20.12 when 2ν is not an integer. However, the functions $I_\nu(z)$ and $I_{-\nu}(z)$ where

$$\begin{aligned} I_\nu(z) &\equiv e^{-i(\pi/2)\nu} J_\nu(iz) \\ &= \sum_{m=0}^{\infty} \left(\frac{z}{2}\right)^{2m+\nu} \frac{1}{m!\Gamma(m+\nu+1)} \end{aligned} \tag{20.13}$$

are more often used. They are known as **the modified Bessel functions of the first kind.** The function

$$K_\nu(z) = \frac{\pi}{2\sin(\nu\pi)}[I_{-\nu}(z) - I_\nu(z)] \tag{20.14}$$

which is also a solution of Eq. 20.12 when 2ν is not an integer, is known as the **modified Bessel function of the third kind.**

When 2ν is an integer n, $I_n(z)$ and $I_{-n}(z)$ no longer form a fundamental set of solutions of Eq. 20.12, since from Eq. 20.8

$$I_n(z) = I_{-n}(z) \tag{20.15}$$

In that case a fundamental set of solutions of Eq. 20.12 is provided by the functions $I_n(z)$ and $K_n(z)$, where

$$\begin{aligned} K_n(z) &= \lim_{\nu \to n} K_\nu(z) \\ &= \frac{(-1)^n}{2} \lim_{\nu \to n} \left[\frac{\partial I_{-\nu}(z)}{\partial \nu} - \frac{\partial I_\nu(z)}{\partial \nu}\right] \end{aligned} \tag{20.16}$$

From Eq. 20.14 and the above definition (Eq. 20.16), one easily obtains

$$\begin{aligned} K_n(z) &= (-1)^{n+1} I_n(z) \log\left(\frac{1}{2}z\right) \\ &+ \frac{1}{2}(-1)^n \sum_{m=0}^{\infty} \left(\frac{z}{2}\right)^{2m+n} \frac{\psi(m+n+1)}{m!(m+n)!} \\ &+ \frac{1}{2}(-1)^n \sum_{m=0}^{\infty} \left(\frac{z}{2}\right)^{2m-n} \frac{\psi(m-n+1)}{m!(m-n)!} \end{aligned} \tag{20.17}$$

Recurrence Relations

From the series expansion 20.6 for $J_\nu(z)$ one obtains easily the following recurrence relations.

$$\frac{d}{dz}\left\{\frac{J_\nu(z)}{z^\nu}\right\} = \frac{-J_{\nu+1}(z)}{z^\nu} \tag{20.18}$$

and

$$\frac{d}{dz}\{z^\nu J_\nu(z)\} = z^\nu J_{\nu-1}(z) \tag{20.19}$$

Eliminating $dJ_n(z)/dz$ from the preceding relations, we find

$$\frac{2\nu}{z}J_\nu(z) = J_{\nu-1}(z) + J_{\nu+1}(z) \tag{20.20}$$

The same relations are satisfied by the functions $Y_\nu(z)$, $H_\nu^{(1)}(z)$, and $H_\nu^{(2)}(z)$. Analogous relations can be derived for Bessel functions of imaginary argument.

Integral Representations for Bessel Functions

We mentioned (Sec. 15.2) that the kernel

$$K(z,t) = \left(\frac{z}{2}\right)^\nu e^{(t-z^2/4t)} \tag{20.21}$$

leads to useful integral representations for the Bessel functions. We note the identity

$$L_z K(z,t) = \left(\frac{d}{dt} - \frac{\nu+1}{t}\right)K(z,t) \tag{20.22}$$

where L_z is the Bessel operator (see Eq. 20.3)

$$L_z = \frac{d^2}{dz^2} + \frac{1}{z}\frac{d}{dz} + \left(1 - \frac{\nu^2}{z^2}\right) \tag{20.23}$$

Thus, according to the discussion of Sec. 15, an integral representation for the Bessel function is given by

$$J_\nu(z) = \frac{1}{2\pi i}\left(\frac{z}{2}\right)^\nu \int_C e^{(t-z^2/4t)}\, v(t)\, dt$$

where $v(t)$ is a solution of the equation

$$\left(\frac{d}{dt} + \frac{\nu+1}{t}\right)v(t) = 0$$

i.e.

$$v(t) = t^{-\nu-1}$$

and C is any closed path such that after its completion

$$Q[K,v] \equiv t^{-v-1}e^{(t-z^2/4t)}$$

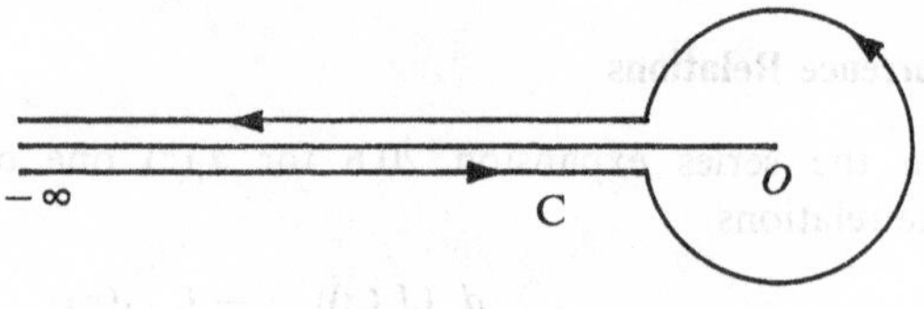

Fig. 45. The contour C for the representation in Eq. 20.24.

returns to its initial value. Hence, we have the representation

$$J_\nu(z) = \frac{1}{2\pi i}\left(\frac{z}{2}\right)^\nu \int_C t^{-\nu-1} e^{(t - z^2/4t)}\, dt \tag{20.24}$$

where C is the contour of Fig. 45 and we choose the branch of the integrand such that $|\arg t| < \pi$.

Setting $t = zu/2$ in Eq. 20.24 gives

$$J_\nu(z) = \frac{1}{2\pi i}\int_C u^{-\nu-1} e^{z[u-(1/u)]/2}\, du \tag{20.25}$$

which is an analytic function of z when Re $(zu) < 0$ as $u \to -\infty$ along the path of integration; i.e., Eq. 20.25 defines an analytic function of z for $|\arg z| < \pi/2$.

When ν is an integer, $\nu = n$, the contour C of Fig. 45 can be closed around the origin. It is then seen from Eq. 20.25 that $J_n(z)$ can be interpreted as the nth coefficient in a Laurent expansion of $e^{z[u-(1/u)]/2}$

$$e^{z[u-(1/u)]/2} = \sum_{n=-\infty}^{\infty} J_n(z) u^n \tag{20.26}$$

The exponential function of the LHS is called a **generating function** for the Bessel functions of integer order, since one can obtain from it all these functions.

By taking a circle of radius unity in Fig. 45 and making the transformation $u = e^w$, Eq. 20.25 becomes

$$J_\nu(z) = \frac{1}{2\pi i}\int_{C'} dw\, e^{z \sin hw - \nu w} \tag{20.27}$$

and the contour C is transformed into the contour C′ of Fig. 46.

We set $w = t \pm i\pi$ along the sides L_1 and L_2 and $w = \pm i\theta$ along L_3 and L_4. Thus, we obtain yet another integral representation for the Bessel function

$$J_\nu(z) = \frac{1}{\pi}\int_0^\pi \cos(\nu\theta - z\sin\theta)\, d\theta - \frac{\sin \nu\pi}{\pi}\int_0^\infty e^{-\nu t - z \sinh t}\, dt \tag{20.28}$$

which holds for $|\arg z| < \pi/2$.

From Eq. 20.28 and the definition (Eq. 20.9) of the Bessel function of the second kind, we have

$$\begin{aligned} Y_\nu(z) = {} & \frac{\cot \nu\pi}{\pi}\int_0^\pi \cos(\nu\theta - z\sin\theta)\, d\theta \\ & - \frac{\csc \nu\pi}{\pi}\int_0^\pi \cos(\nu\theta + z\sin\theta)\, d\theta \\ & - \frac{\cos \nu\pi}{\pi}\int_0^\infty e^{-\nu t - z \sin ht}\, dt - \frac{1}{\pi}\int_0^\infty e^{\nu t - z \sinh t}\, dt \end{aligned} \tag{20.29}$$

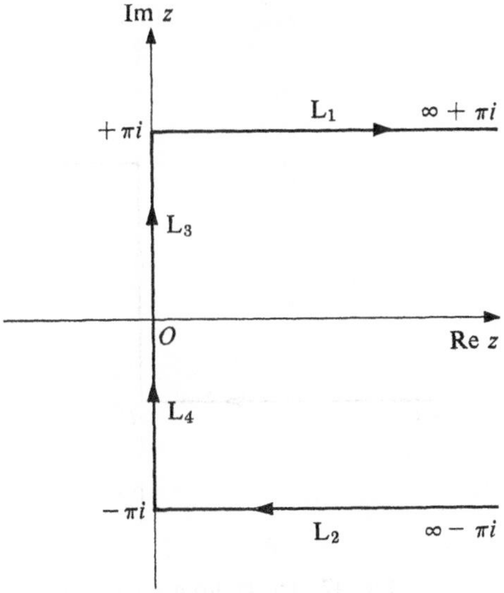

Fig. 46. The contour C′ for the representation in Eq. 20.27.

Making the substitution $\theta \to \pi - \theta$ in the second integral on the RHS of Eq. 20.29 and combining Eqs. 20.28 and 20.29, we find

$$\begin{aligned} J_\nu(z) + iY_\nu(z) = &\frac{1}{\pi}\int_0^\pi e^{i(z\sin\theta - \nu\theta)}\,d\theta \\ &+ \frac{1}{i\pi}\int_0^\infty e^{\nu t - z\sinh t}\,dt \\ &+ \frac{e^{-i\pi\nu}}{i\pi}\int_0^\infty e^{-\nu t - z\sinh t}\,dt \end{aligned} \tag{20.30}$$

Now consider the integral

$$I = \frac{1}{i\pi}\int_{-\infty}^{\infty + \pi i} e^{z\sinh w - \nu w}\,dw \tag{20.31}$$

for $|\arg z| < \pi/2$. Evaluating the integral 20.31 along the contour of Fig. 47 and setting $w = -t, i\theta, t + i\pi$ along the three parts of the contour, we obtain

$$I = \frac{1}{i\pi}\int_0^\infty e^{\nu t - z\sinh t}\,dt + \frac{1}{\pi}\int_0^\pi e^{i(z\sin\theta - \nu\theta)}\,d\theta + \frac{e^{-i\pi\nu}}{i\pi}\int_0^\infty e^{-\nu t - z\sinh t}\,dt \tag{20.32}$$

Comparing Eq. 20.32 with Eq. 20.30 and the definition (Eq. 20.11) of the Bessel function of the third kind, we deduce

$$H_\nu^{(1)}(z) = \frac{1}{i\pi}\int_{-\infty}^{\infty + i\pi} e^{z\sinh w - \nu w}\,dw \qquad |\arg z| < \frac{\pi}{2} \tag{20.33}$$

Similarly, changing i into $(-i)$ in the preceding equations leads to the integral representation for $H_\nu^{(2)}(z)$

$$H_\nu^{(2)}(z) = -\frac{1}{i\pi}\int_{-\infty}^{\infty - i\pi} e^{z\sinh w - \nu w}\,dw \qquad |\arg z <|\frac{\pi}{2} \tag{20.34}$$

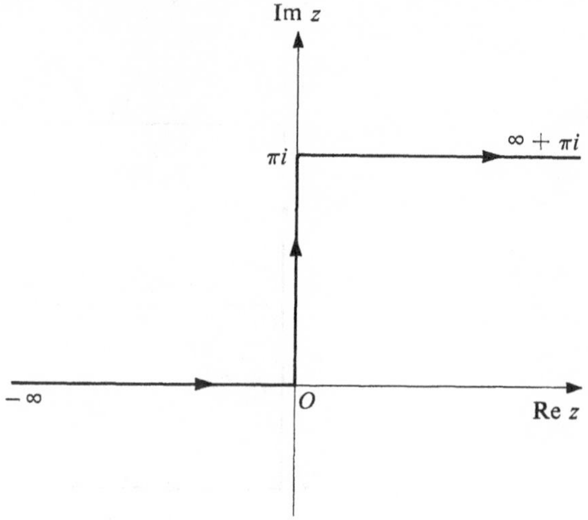

Fig. 47. The contour for the integral (Eq. 20.31).

Asymptotic Behavior of the Bessel Functions

Starting from the integral representations 20.27, 20.33, and 20.34, and using the method of steepest descent described in Sec. 31 of Chapter I we derive approximate expressions for the Bessel functions when either the argument z or the order ν is very large. To simplify the calculations, we take both $z\ (=x)$ and ν real. One can then distinguish three separate cases according as $\nu/x > 1$, $\nu/x < 1$, or $\nu/x \approx 1$. Here we discuss only the first two cases; we shall see that although the kernels in the three integral representations (Eqs. 20.27, 20.33, and 20.34) are the same, the integration limits are such that for the case $\nu/x < 1$, it is more convenient to study the particular function $J_\nu(x)$, while for the case $\nu/x > 0$, the functions $H_\nu^{(1)}(x)$ and $H_\nu^{(2)}(x)$ are the more appropriate functions to consider.

Bessel Functions of Large Order

Since $\nu > x$, we can set

$$\nu = x \cosh w_0 \qquad \text{with } w_0 > 0 \tag{20.35}$$

and Eq. 20.27 can be written as

$$J_\nu(x) = \frac{1}{2\pi i}\int_{\infty - \pi i}^{\infty + \pi i} e^{xf(w)}\, dw \tag{20.36}$$

where

$$f(w) \equiv [\sinh w - (\cosh w_0)w] \tag{20.37}$$

The saddle points of $f(w)$ are located at

$$w = \pm w_0 + 2\pi i n \qquad (n = 0, \pm 1, \pm 2, \cdots)$$

We consider the saddle point at $w = w_0$. Since $\operatorname{Im} f(w_0) = 0$, the path of constant $\operatorname{Im} f(w)$ is given by the equation

$$\operatorname{Im} f(w) = \operatorname{Im}[\sinh w - (\cosh w_0)w] = 0 \tag{20.38}$$

Putting $w = u + iv$ in Eq. 20.38 gives either $v = 0$ (which, however, leads to a divergent integral) or

$$\cosh u = \frac{v \cosh w_0}{\sin v} \tag{20.39}$$

Equation 20.39 defines two curves in the w-plane, which are symmetrical with respect to the v axis, as shown in Fig. 48. As in Eq. 31.18 of Chapter I, we define the real number

$$\tau^2 \equiv [\sinh w_0 - (\cosh w_0)w_0] - [\sinh w - (\cosh w_0)w] \tag{20.40}$$

The path of integration in the integral 20.36 can be deformed into C_0, and since the quantity on the RHS of Eq. 20.40 is indeed positive along C_0, this path is a path of steepest descent for $J_\nu(x)$.

In terms of the parameter τ, Eq. 20.36 becomes

$$J_\nu(\nu \operatorname{sech} w_0) = \frac{e^{\nu(\tanh w_0 - w_0)}}{2\pi i} \int_{-\infty}^{\infty} e^{-x\tau^2} \frac{dw(\tau)}{d\tau} d\tau \tag{20.41}$$

Expanding the RHS of Eq. 20.40 in powers of w about the saddle point w_0, we have

$$\tau^2 = \frac{-\sinh w_0}{2}(w - w_0)^2 - \frac{\cosh w_0}{6}(w - w_0)^3 - \frac{\sinh w_0}{24}(w - w_0)^4 + \cdots \tag{20.42}$$

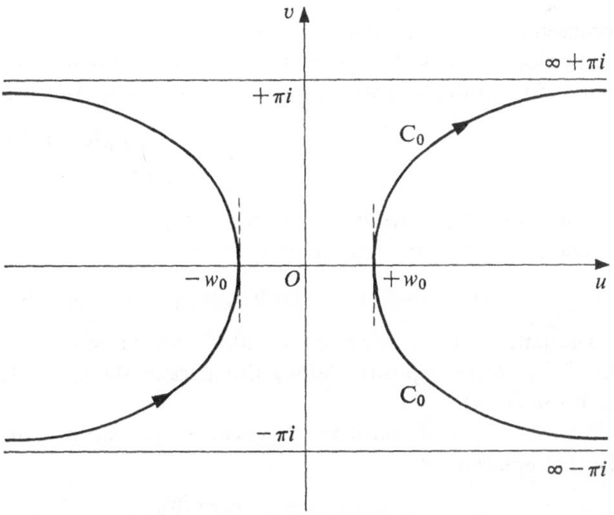

Fig. 48. The heavy lines are the curves defined by Eq. 20.39. The curve C_0 with $u > 0$ is the path of steepest descent.

Inverting Eq. 20.42, we find

$$w = w_0 + \sum_{m=1}^{\infty} c_m \tau^m \tag{20.43}$$

where

$$c_1 = \left(\frac{-2}{\sinh w_0}\right)^{1/2}; \quad c_2 = \frac{\coth w_0}{3 \sinh w_0} \tag{20.44}$$

$$c_3 = c_1^3 \frac{1}{24}\left[\frac{5}{3}\coth^2 w_0 - 1\right] \qquad \text{etc.}$$

The phase of c_1 can be determined by noticing that it is the value of the angle that the tangent to the curve $w(\tau)$ makes with the real axis in the limit as $\tau \to 0$. From Fig. 48 one sees that one must have

$$c_1 = i\left(\frac{2}{\sinh w_0}\right)^{1/2}$$

Comparing Eqs. 20.41 and 20.43 with Eqs. 31.1, 31.23, and 31.31 of Chapter I, we find

$$J_\nu(\nu \operatorname{sech} w_0) \sim \frac{e^{\nu(\tanh w_0 - w_0)}}{(2\pi\nu \tanh w_0)^{\frac{1}{2}}}\left[1 + \frac{1}{8\nu \tanh w_0}\left(1 - \frac{5}{3}\coth^2 w_0\right) + \cdots\right] \tag{20.45}$$

Bessel Functions of Large Argument

We now consider the case $x/\nu > 1$ when x is very large. Setting

$$x = \nu \sec w_0 \tag{20.46}$$

in Eq. 20.33,

$$H_\nu^{(1)}(x) = \frac{1}{\pi i}\int_{-\infty}^{\infty} e^{x \sinh w - \nu w} dw \tag{20.47}$$

we proceed as before and find that there are two saddle points at $w = \pm i w_0$. Focusing our attention on the saddle point at $w = +iw_0$ and putting $w = u + iv$, we obtain the equation of the path along which Im[$\sinh w - w \cos w_0$] is constant

$$\cosh u = \frac{\sin w_0 + (v - w_0)\cos w_0}{\sin v} \tag{20.48}$$

This equation defines two curves drawn in Fig. 49.

Again we define the real quantity

$$\tau^2 \equiv [\sinh w_0 - (\cosh w_0)w_0] - [\sinh w - (\cos w_0)w] \tag{20.49}$$

The path of integration in Eq. 20.47 can be deformed into C'_0, and since the RHS of Eq. 20.49 is positive along this part of the curve, C'_0 is a path of steepest descent for $H_\nu^1(x)$.

All the results obtained for $J_\nu(x)$ can be carried over to this case, provided we make the replacement

$$w_0 \to i w_0$$

and note that the angle that the tangent of C'_0 at w_0 makes with the real axis is now $\pi/4$ (see Fig. 49).

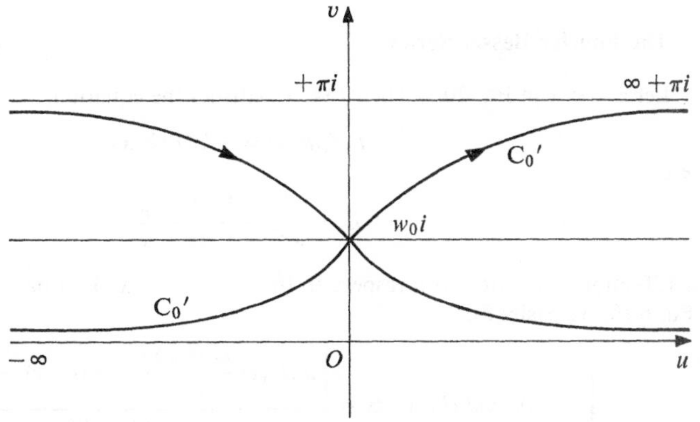

Fig. 49. C'_0 is the path of steepest descent for $H_\nu^{(1)}(x)$.

Hence, the coefficient c_1 is given by

$$c_1 = \left(\frac{2i}{\sin w_0}\right)^{1/2} = \left(\frac{2}{\sin w_0}\right)^{1/2} e^{i\pi/4}$$

and we obtain

$$H_\nu^{(1)}(x) \sim \sqrt{\frac{2}{\pi x \sin w_0}}\, e^{i[x \sin w_0 - \nu w_0 - \pi/4]} \times \left[1 + \frac{1}{8ix \sin w_0}\left(1 + \frac{5}{3}\cot^2 w_0\right) + \cdots\right] \tag{20.50}$$

When $x \gg \nu$, we have $w_0 \approx \pi/2 - \nu/x$, and so

$$H_\nu^{(1)}(x) \sim \sqrt{\frac{2}{\pi x}}\, e^{i[x - \nu(\pi/2) - (\pi/4)]}\left[1 + \frac{1-4\nu^2}{8ix} + \cdots\right] \tag{20.51}$$

By considering the other saddle point at $w = -iw_0$, we find a path of steepest descent, which is the mirror image of C'_0 with respect to the u axis. The end points of this path are appropriate for studying the asymptotic behavior of $H_\nu^{(2)}(x)$ for large x. We obtain for this function an asymptotic expression similar to that for $H_\nu^{(1)}(x)$, with the exception that i is replaced by $(-i)$

$$H_\nu^{(2)}(x) \sim \sqrt{\frac{2}{\pi x}}\, e^{-i[x - \nu(\pi/2) - (\pi/4)]}\left[1 - \frac{1-4\nu^2}{8ix} + \cdots\right] \tag{20.52}$$

The asymptotic behavior for large x of all the other Bessel functions can be obtained from Eqs. 20.51 and 20.52. For example, using the definitions (20.11), one finds

$$J_\nu(x) \sim \sqrt{\frac{2}{\pi x}}\left\{\cos\left(x - \nu\frac{\pi}{2} - \frac{\pi}{4}\right) + \frac{1-4\nu^2}{8x}\sin\left(x - \frac{\nu\pi}{2} - \frac{\pi}{4}\right) + \cdots\right\} \tag{20.53}$$

$$Y_\nu(x) \sim \sqrt{\frac{2}{\pi x}}\left\{\sin\left(x - \nu\frac{\pi}{2} - \frac{\pi}{4}\right) - \frac{1-4\nu^2}{8x}\cos\left(x - \frac{\nu\pi}{2} - \frac{\pi}{4}\right) + \cdots\right\} \tag{20.54}$$

The Fourier-Bessel Series

Let $z = k_m x$ in Eq. 20.3. Then $J_\nu(k_m x)$ satisfies the equation

$$L_x J_\nu(k_m x) = -k_m^2 J_\nu(k_m x) \tag{20.55}$$

where

$$L_x = \frac{d^2}{dx^2} + \frac{1}{x}\frac{d}{dx} - \frac{\nu^2}{x^2} \tag{20.56}$$

is a self-adjoint operator with respect to the weight $w = x$. Putting $u = J_\nu(k_m x)$, $v = J_\nu(k_{m'} x)$ in Eq. 6.10, we easily find

$$\int_0^1 x J_\nu(k_m x) J_\nu(k_{m'} x)\, dx = \frac{\left[J_\nu(k_m x)\dfrac{dJ_\nu(k_{m'}x)}{dx} - J_\nu(k_{m'}x)\dfrac{dJ_\nu(k_m x)}{dx}\right]_{x=1}}{k_m^2 - k_{m'}^2} \tag{20.57}$$

Let k_m and $k_{m'}$ be two zeros of J_ν; i.e.,

$$J_\nu(k_m) = J_\nu(k_{m'}) = 0 \tag{20.58}$$

From Eq. 20.57 we immediately obtain the orthogonality property of the Bessel functions on [0.1]

$$\int_0^1 dx\, x J_\nu(k_m x) J_\nu(k_{m'} x) = 0 \qquad \text{for } k_m \neq k_m \tag{20.59}$$

For $k_m = k_{m'}$, we obtain the normalization integral by using l'Hospital's rule, the recurrence relation 20.18 and Eq. 20.58. Thus

$$\int_0^1 dx\, x J_\nu(k_m x) J_\nu(k_{m'} x) = \tfrac{1}{2}[J_{\nu+1}(k_m)]^2 \delta_{k_m k_{m'}} \tag{20.60}$$

Let the boundary conditions associated with L_x be

$$\begin{aligned} &\text{(i)} && u|_{x=1} = 0 \\ &\text{(ii)} && u|_{x=0} \quad \text{and} \quad \left.\frac{du}{dx}\right|_{x=0} \quad \text{finite} \end{aligned} \tag{20.61}$$

Then the adjoint boundary conditions, which make the surface term in Eq. 6.10 vanish, are identical with Eq. 20.61, and L_x defines a Hermitian differential operator L. According to the results of Sec. 12, an infinite number of eigenvectors of this operator exist and span its domain. These eigenvectors are represented by the Bessel functions $J_\nu(k_m x)$, since these are the only functions analytic at the origin which satisfy the differential equation 20.55 and which vanish at $x = 1$.

Thus, a function $u(x)$ representing a vector of the domain of L can be expanded in the so-called Fourier-Bessel series

$$u(x) = \sum_m \left\{ \frac{2\int_0^1 dx'\, x' J_\nu(k_m x') u(x')}{[J_{\nu+1}(k_m)]^2} J_\nu(k_m x) \right. \tag{20.62}$$

We shall not enter into a discussion of the convergence of this series but state only that the conditions for the convergence of a Fourier-Bessel series are weaker than one would expect from our purely algebraic argument. We merely quote a result. When $\int_0^1 \sqrt{x}u(x)\, dx$ exists and $\nu \geq -\frac{1}{2}$, the series in 20.62 converges uniformly in $(a,b) \in [0,1]$, provided $u(x)$ is continuous in $[a,b]$.

Part II Introduction to Partial Differential Equations

21 · PRELIMINARIES

Let $u(x_i) = u(x_1,x_2,\cdots,x_N)$ be a function of the N variables $x_1,x_2,\cdots,x_N$. In analogy to the definition of an ordinary differential equation, a partial differential equation is defined as an expression of the form

$$F\left(x_i,u,\frac{\partial^{l+m+\cdots+n}}{\partial x_1^l\,\partial x_2^m\cdots\partial x_N^n}\right)=0 \tag{21.1}$$

Equation 21.1 is said to be of the Mth **order** if the highest partial derivative appearing in that equation is of the Mth order.

The most general **linear** partial differential equation involving $u(x_i)$ is of the form

$$q + ru + \sum_{0<j+k+\cdots+l\le M} s_{jk\cdots l}\frac{\partial^{j+k+\cdots+l}}{\partial x_1^j\,\partial x_2^k\cdots\partial x_N^l}\,u(x_i)=0 \tag{21.2}$$

where the coefficients q,r, and $s_{j,k,\cdots l}$ are, in general, functions of $x_1,x_2,\cdots,x_N$. An example of a partial differential equation, which we have already encountered, is Laplace's equation

$$\sum_{i=1}^{N}\frac{\partial^2 u}{\partial x_i^2}=0$$

which is a linear equation of the second order with constant coefficients.

We shall assume henceforth that all the variables x_i ($i = 1, 2, \cdots, N$) are real.

A set of N variables $x_1,x_2,\cdots,x_N$ may be considered as the components of a vector in a real N-dimensional space, and thus they determine a point in this space. It will often be convenient to regard a function $f(x_i)$ of N independent variables x_i ($i = 1, 2, \cdots, N$) as a function of a point in an N-dimensional space.

22 · THE CAUCHY-KOVALEVSKA THEOREM

As in the case of ordinary differential equations, the solution of a partial differential equation (if it exists) is uniquely specified only if one prescribes certain boundary conditions on the function which represents the solution, as well as on some of its derivatives. However, in the case of partial differential equations, the specification of the boundary conditions is a much more delicate matter, and it is very important that they be properly stated; otherwise, a solution may not exist at all, or if it exists, it may not be unique. In this connection there exists a theorem first found by Cauchy and then proved in its general form by S. Kovalevska.

Before formulating this theorem, a few remarks are in order. First we shall assume that Eq. 21.1 has been written in the form*

$$\frac{\partial^k u}{\partial x_1^k}=K\left(x_i,u,\frac{\partial^{l+m+\cdots+n}}{\partial x_1^l\,\partial x_2^m\cdots\partial x_N^n}\,u(x_i)\right) \tag{22.1}$$

with $l<k$ and $l+m+\cdots+n\le k$. K is not necessarily a linear function of its arguments.

* In general, this would necessitate a transformation of the independent variables in the original equation. See also the footnote on p. 258.

Next, we consider the boundary conditions that fix the values of $u, \partial u/\partial x_1, \cdots, \partial^{k-1}u/\partial x_1^{k-1}$ for a particular value of the variable x_1, at $x_1 = a_1$

$$\left.\frac{\partial^j u}{\partial x_1^j}\right|_{x_1=a_1} = L_j(x_2,x_3,\cdots,x_N) \qquad (j = 0, 1, \cdots, k-1) \tag{22.2}$$

Finally we shall say that a function of several variables $F(y_1,y_2,\cdots,y_L)$ is **analytic** at a point $y_i = y_{i0}$ $(i = 1, 2, \cdots, L)$ if it can be expanded in a power series

$$F(y_1,y_2,\cdots,y_L) = \sum_{r,s,\cdots t\geq 0} F_{r,s\cdots,t}(y_1 - y_{10})^r \cdots (y_L - y_{L0})^t \tag{22.3}$$

which is convergent for all $|y_i - y_{i0}|$ small enough. In the case of one variable, this definition coincides with the usual definition of analyticity at a point, since (as explained in Chapter I) a function that is single-valued and differentiable in a neighborhood of a point in the complex plane can be expanded in a power series about that point; conversely, a function that is expandable in a power series about a point represents a function that is analytic in a neighborhood of that point. It is clear that if one speaks about the analyticity of a function of a real variable, one is expressing the fact that the function is analytic in some environment of a given segment of the real axis. We can now state the Cauchy-Kovalevska theorem*

Theorem. The solution of Eq. 22.1 satisfying the boundary conditions (22.2) in some neighborhood of a point $x_i = a_i$ $(i = 1, 2, \cdots, N)$ exists and is unique and analytic in a neighborhood of that point, provided the function K is an analytic function of its arguments at

$$x_i = a_i \; (i = 1, 2, \cdots, N), \; u = u(a_i), \left.\frac{\partial^{l+m+\cdots+n}}{\partial x_1^l \cdots \partial x_N^n}\right|_{x_i=a_i}$$

and the functions L_j are analytic functions of their arguments at $x_i = a_i$ $(i = 1, 2, 3, \cdots, N)$.

23 · CLASSIFICATION OF SECOND-ORDER QUASILINEAR EQUATIONS

The boundary value problem stated in the conditions of the Cauchy-Kovalevska theorem, and which consists of fixing the values of the function $u(x_i)$ and of its partial derivatives with respect to x_1 (of order less than that of the equation) on a hyperplane $x_1 = \text{const}$, is a rather special boundary value problem. A more general problem consists of examining the possibility for the existence of a solution of the differential equation when, instead of specifying boundary values on a hyperplane $x_1 = \text{const}$, one considers an arbitrary hypersurface $S(x_1,x_2,\cdots,x_N) = 0$ on which one prescribes the values of the unknown function $u(x_i)$ and of its partial derivatives (again of order less than that of the equation) along a direction normal to the hypersurface**; such boundary conditions are called **Cauchy conditions**. It turns out that the existence of a solution with prescribed Cauchy boundary conditions is closely related to what one calls the **type** of the partial differential equation.

* This theorem actually holds for a system of partial differential equations, but for simplicity we consider a single equation only.

** Notice that the derivative in the directions tangent to the hypersurface can be evaluated, once the values the function takes along the hypersurface are known.

In what follows we shall limit ourselves to giving a classification of the second order, quasilinear equations, which play a particularly important role in physics; by quasilinear we mean linear with respect to the highest partial derivatives, in our case those of second order. Hence, we consider equations of the general form

$$\sum_{m,n=1}^{N} a_{mn}(x_i)\frac{\partial^2 u}{\partial x_m\,\partial x_n} + F\left(x_i, u, \frac{\partial u}{\partial x_j}\right) = 0 \tag{23.1}$$

Notice first that since

$$\frac{\partial^2 u}{\partial x_m\,\partial x_n} = \frac{\partial^2 u}{\partial x_n\,\partial x_m}$$

one has

$$\sum_{m,n=1}^{N} a_{mn}(x_i)\frac{\partial^2 u}{\partial x_m\,\partial x_n} = \frac{1}{2}\sum_{m,n=1}^{N} a_{mn}(x_i)\frac{\partial^2 u}{\partial x_m\,\partial x_n} + \frac{1}{2}\sum_{m,n=1}^{N} a_{nm}(x_i)\frac{\partial^2 u}{\partial x_n\,\partial x_m}$$

$$= \sum_{m,n=1}^{N} \tfrac{1}{2}[a_{mn}(x_i) + a_{nm}(x_i)]\frac{\partial^2 u}{\partial x_m\,\partial x_n}$$

In the last sum, the coefficient of $\partial^2 u/\partial x_m\,\partial x_n$ is symmetric with respect to the indices m and n, and therefore in Eq. 23.1 we can assume without any loss of generality that

$$a_{mn}(x_i) = a_{nm}(x_i)$$

For any set x_i ($i = 1, 2, \cdots, N$) the array of real numbers $a_{mn}(x_i)$ may be considered to form a symmetric matrix. But we proved in Chapter II, Sec. 24.2 that every real, symmetric matrix can be diagonalized by a suitable orthogonal transformation. In other words, there exists a set of N^2 real numbers $O_{kl}(k,l = 1, 2, \cdots, N)$ with the property

$$\sum_{j=1}^{N} O_{jm}O_{ji} = \delta_{mi} \tag{23.2}$$

and such that, for any given set of x_i, $x_i = x_{i0}$, say

$$\sum_{m,n=1}^{N} O_{km}a_{mn}(x_{i0})O_{ln} = a_k(x_{i0})\delta_{kl} \tag{23.3}$$

We now perform the linear transformation of the independent variables

$$y_j = \sum_{m=1}^{N} O_{jm}x_m \tag{23.4}$$

which, by using Eq. 23.2, can be immediately inverted

$$x_i = \sum_{j=1}^{N} O_{ji}y_j \tag{23.5}$$

Hence

$$\sum_{m,n=1}^{N} a_{mn}(x_i)\frac{\partial^2 u}{\partial x_m\,\partial x_n} = \sum_{k,l=1}^{N} a'_{kl}(y_j)\frac{\partial^2 u}{\partial y_k\,\partial y_l}$$

where

$$a'_{kl}(y_j) = \sum_{m,n=1}^{N} O_{km}a_{mn}[x(y_j)]O_{ln} \tag{23.6}$$

At $y_i = y_j(x_{i0}) = y_{i0}$, Eq. 23.6 yields

$$a'_{kl}(y_{j0}) = a_k(x_{i0})\delta_{kl}$$

and therefore

$$\left\{\sum_{k,l=1}^{N} a'_{kl}(y_j)\frac{\partial^2 u}{\partial y_k\,\partial y_l}\right\}_{y_j=y_{j0}} = \sum_k a_k(x_{i0})\frac{\partial^2 u}{\partial^2 y_k}\bigg|_{y_0=y_{j0}}$$

We see that mixed second-order derivatives disappear from the equation

$$\sum_{k,l=1}^{N} a'_{kl}(y_j)\frac{\partial^2 u}{\partial y_k\,\partial y_l} + F'\left(y_j, u, \frac{\partial u}{\partial y_m}\right) = 0 \tag{23.7}$$

obtained from Eq. 23.1 by the transformation 23.5, when $y_j = y_{j0}$. This fact is used for the classification of the equations we have considered:

(i) Equation 23.1 is of the **elliptic type** at the point $x_i = x_{i0} (i = 1, 2, \cdots, N)$ if all the coefficients $a_k(x_{i0})$ $(k = 1, 2, \cdots, N)$ are nonzero and have the same sign.

(ii) Equation 23.1 is of the **ultrahyperbolic type** at the point $x_i = x_{i0} (i = 1, 2, \cdots, N)$ if all the coefficients $a_k(x_{i0})$ $(k = 1, 2, \cdots, N)$ are nonzero but do not have the same sign. In particular, Eq. 23.7 is of the **hyperbolic type** at $x_i = x_{i0}$ $(i = 1, 2, \cdots, N)$ if only one coefficient among the $a_k(x_{i0})$ $(k = 1, 2, \cdots, N)$ has a sign different from all others.

(iii) Equation 23.1 is of the **parabolic type** at the point $x_i = x_{i0}$ $(i = 1, 2, \cdots, N)$ if at least one of the coefficients $a_k(x_{i0})$ is zero.

If a partial differential equation is of a given type at every point of some point set, it is said to be of this type throughout the set. For instance, in the case where the a_{mn} are constants, the type of the equation is the same at all points where the equation is meaningful.

EXAMPLE

The following equations, with $u = u(x,y)$ are respectively of elliptic, hyperbolic, and parabolic type for any x and y.

Laplace's equation

$$\frac{\partial^2 u}{\partial x^2} + \frac{\partial^2 u}{dy^2} = 0$$

Wave equation; y stands for the time coordinate:

$$\frac{\partial^2 u}{\partial x^2} - \frac{\partial^2 u}{\partial y^2} = 0$$

Diffusion equation; y stands for the time coordinate

$$\frac{\partial^2 u}{\partial x^2} - \frac{\partial u}{\partial y} = 0$$

24 · CHARACTERISTICS

Usually the relation that exists between the partial derivatives of a function $u(x_i)$ involved in a partial differential equation, together with Cauchy boundary conditions prescribed on a hypersurface $S(x_1, x_2, \cdots, x_N) = 0$, allows us to find on the hypersurface all the derivatives of $u(x_i)$ of order higher than those given by the boundary conditions. When it is possible to reconstruct uniquely the whole sequence of partial derivatives of $u(x_i)$ at a given point of the boundary hypersurface and furthermore, when an analytic solution of the differential equation in the neighborhood of this point exists, then we can obtain the coefficients of the power series expansion of $u(x_i)$ about this point by using the well-known relation between these coefficients and partial derivatives of the function. This amounts to constructing a unique analytic solution of the differential equation in a neighborhood of the point in question. There may exist, however, hypersurfaces with the property that everywhere on them Cauchy's boundary conditions, together with the differential equation itself, are not sufficient to yield uniquely the higher-order derivatives; such a hypersurface is called a **characteristic hypersurface.** It is clear that on a characteristic hypersurface, or even on a hypersurface that is somewhere tangent to it, the Cauchy boundary conditions are not the "proper" ones.

To be more specific, we shall examine in detail a quasilinear equation of second order in the simplest case where there are only two independent variables x and y. Such an equation has the general form

$$A(x,y)\frac{\partial^2 u}{\partial x^2} + 2B(x,y)\frac{\partial^2 u}{\partial x\,\partial y} + C(x,y)\frac{\partial^2 u}{\partial y^2} = D\left(x,y,\frac{\partial u}{\partial x},\frac{\partial u}{\partial y}\right) \tag{24.1}$$

Suppose that the Cauchy conditions are given along a regular curve (we shall call it the **boundary curve**) in the x,y plane; i.e., we assume that along the boundary curve, u and the derivatives $\partial u/\partial n$ in the direction of the normal to the curve are known.

It is convenient to define the boundary curve by the parametric equations

$$\begin{aligned} x &= X(t) \\ y &= Y(t) \end{aligned} \tag{24.2}$$

with $|t|$ measuring the distance along the curve from an arbitrary point on it (t itself may be positive or negative). The derivative $\partial u/\partial t$ in a direction tangent to the boundary curve can be easily calculated as

$$\frac{\partial u}{\partial t} = \frac{d}{dt}u[X(t),Y(t)]$$

Once we know $\partial u/\partial t$ and $\partial u/\partial n$, we can without difficulty find

$$\left.\frac{\partial u}{\partial x}\right|_{\substack{x=X(t)\\y=Y(t)}} \quad\text{and}\quad \left.\frac{\partial u}{\partial y}\right|_{\substack{x=X(t)\\y=Y(t)}}$$

for*

$$\begin{aligned} \frac{\partial u}{\partial n} &= \left.\frac{\partial u}{\partial x}\right|_{\substack{x=X(t)\\y=Y(t)}}\frac{dY}{dt} - \left.\frac{\partial u}{\partial y}\right|_{\substack{x=X(t)\\y=Y(t)}}\frac{dX}{dt} \\ \frac{\partial u}{\partial t} &= \left.\frac{\partial u}{\partial x}\right|_{\substack{x=X(t)\\y=Y(t)}}\frac{dX}{dt} + \left.\frac{\partial u}{\partial y}\right|_{\substack{x=X(t)\\y=Y(t)}}\frac{dY}{dt} \end{aligned} \tag{24.3}$$

* Notice that the vector with components dX/dt, dY/dt is tangent to the boundary curve and perpendicular to the vector with components dY/dt, $-dX/dt$. Furthermore, both vectors are of unit length. The derivatives $\partial u/\partial l$ in the direction of a unit vector $\vec{l}$ is

$$\frac{\partial u}{\partial l} = \vec{l}\cdot\vec{\nabla}u$$

The determinant of the system of algebraic equations 24.3 is

$$\left(\frac{dX}{dt}\right)^2 + \left(\frac{dY}{dt}\right)^2 = 1 \tag{24.4}$$

and therefore never vanishes, so that the system of equations has a unique solution.

Let us now look for an analytic solution of the partial differential equation (24.1) in the neighborhood of an arbitrary point x_0,y_0 lying on the boundary curve

$$u(x,y) = \sum_{m,n=0}^{\infty} u_{mn}(x - x_0)^m(y - y_0)^n \tag{24.5}$$

and let us inquire about the possibility of determining uniquely the coefficients u_{mn} of the expansion 24.5 using the differential equation 24.1 and the boundary data. In other words, we ask whether Cauchy's boundary conditions prescribed along an arbitrary curve are sufficient to find an analytic solution of Eq. 24.1 in the neighborhood of an arbitrary point on the boundary curve.

It is well known that

$$u_{mn} = \frac{1}{m!n!} \frac{\partial^{m+n}u}{\partial x^m \, \partial y^n}\bigg|_{\substack{x=x_0\\y=y_0}}$$

and therefore we are able to find the coefficients u_{mn} if we are able to find the higher-order derivatives of $u(x,y)$ at $x = x_0, y = y_0$ by using our input data, which are the values of u, $\partial u/\partial x$, $\partial u/\partial y$ along the boundary curve, and the differential equation 24.1.

Consider first the second-order derivatives. For any point x,y on the boundary curve, we have

$$\begin{aligned}\frac{d}{dt}\left(\frac{\partial u}{\partial x}\right) &= \frac{\partial^2 u}{\partial x^2}\frac{dX}{dt} + \frac{\partial^2 u}{\partial x\,\partial y}\frac{dY}{dt}\\ \frac{d}{dt}\left(\frac{\partial u}{\partial y}\right) &= \frac{\partial^2 u}{\partial x\,\partial y}\frac{dX}{dt} + \frac{\partial^2 u}{\partial y^2}\frac{dY}{dt}\end{aligned} \tag{24.6}$$

Equations 24.6 and Eq. 24.1 constitute a system of three linear algebraic equations with respect to

$$\frac{\partial^2 u}{\partial x^2}, \quad \frac{\partial^2 u}{\partial x\,\partial y}, \quad \frac{\partial^2 u}{\partial y^2}$$

There exists a unique solution to these equations if and only if the determinant

$$\begin{aligned}\Delta = \begin{vmatrix} A & 2B & C \\ \dfrac{dX}{dt} & \dfrac{dY}{dt} & 0 \\ 0 & \dfrac{dX}{dt} & \dfrac{dY}{dt}\end{vmatrix} &= A[X(t),Y(t)]\left(\frac{dY}{dt}\right)^2 \\ &\quad - 2B[X(t),Y(t)]\frac{dX}{dt}\frac{dY}{dt} + C[X(t),Y(t)]\left(\frac{dY}{dt}\right)^2\end{aligned} \tag{24.7}$$

is not zero.

It is easy to verify that the possibility of evaluating the partial derivatives of still higher-order will hinge again upon the condition $\Delta \neq 0$. For example, differentiating Eq. 24.1 with respect to x, we get

$$A\frac{\partial^3 u}{\partial x^3} + 2B\frac{\partial^3 u}{\partial x^2\,\partial y} + C\frac{\partial^3 u}{\partial x\,\partial y^2} = D'\left(x,y,u,\frac{\partial u}{\partial x},\frac{\partial u}{\partial y},\frac{\partial^2 u}{\partial x^2},\frac{\partial^2 u}{\partial x\,\partial y},\frac{\partial^2 u}{\partial y^2}\right)$$

The RHS is known on the boundary (when $\Delta \neq 0$), and the preceding equation together with the equations

$$\frac{d}{dt}\left(\frac{\partial^2 u}{\partial x^2}\right) = \frac{\partial^3 u}{\partial x^3}\frac{dX}{dt} + \frac{\partial^3 u}{\partial x^2\,\partial y}\frac{dX}{dt}$$

$$\frac{d}{dt}\left(\frac{\partial^2 u}{\partial x\,\partial y}\right) = \frac{\partial^3 u}{\partial x^2\,\partial y}\frac{dX}{dt} + \frac{\partial^3 u}{\partial x\,\partial y^2}\frac{dY}{dt}$$

which hold on the boundary curve, again require $\Delta \neq 0$ in order that a solution with respect to third-order derivatives will exist.

On the other hand, when $\Delta = 0$ at a point $x = x_0, y = y_0$, the higher-order derivatives of u, and thus the coefficients of the power series expansion 24.5, cannot be calculated. Hence, if $\Delta = 0$ along a curve, this is a characteristic curve.

One has $\Delta = 0$ along a curve determined by the (ordinary) differential equation (see Eq. 24.7)

$$A(x,y)\left(\frac{dy}{dx}\right)^2 - 2B(x,y)\frac{dy}{dx} + C(x,y) = 0 \tag{24.8}$$

which in fact is equivalent to the two equations

$$\frac{dy}{dx} = \frac{B + \sqrt{B^2 - AC}}{A} \tag{24.9}$$

and

$$\frac{dy}{dx} = \frac{B - \sqrt{B^2 - AC}}{A} \tag{24.10}$$

One can show that Eq. 24.1 is

(a) elliptic if $B^2 - AC < 0$.
(b) hyperbolic if $B^2 - AC > 0$.
(c) parabolic if $B^2 - AC = 0$.

In order to avoid calculations that are straightforward but cumbersome, we verify the foregoing statements when A,B, and C are nonzero constants. The transformations to be performed in the general case are essentially the same. Equations 24.9 and 24.10 lead to the following equations for the characteristics

$$\alpha(x,y) \equiv -\left(\frac{B + \sqrt{B^2 - AC}}{A}\right)x + y = \text{const}$$

$$\beta(x,y) \equiv -\left(\frac{B - \sqrt{B^2 - AC}}{A}\right)x + y = \text{const}$$

When $B^2 - AC < 0$, these are in fact not the equations of curves in the x,y plane. In this case, we make the transformation of independent variables

$$v = \frac{\alpha(x,y) + \beta(x,y)}{2}$$

$$w = \frac{\alpha(x,y) - \beta(x,y)}{2i}$$

One easily finds

$$\frac{\partial^2 u}{\partial x^2} = \frac{\partial^2 u}{\partial v^2}\frac{B^2}{A^2} + 2\frac{\partial^2 u}{\partial v\,\partial w}\frac{B\sqrt{AC - B^2}}{A^2} + \frac{\partial^2 u}{\partial w^2}\left(\frac{AC - B^2}{A^2}\right)$$

$$\frac{\partial^2 u}{\partial x\,\partial y} = -\frac{\partial^2 u}{\partial v^2}\frac{B}{A} = \frac{\partial^2 u}{\partial v\,\partial w}\frac{\sqrt{AC - B^2}}{A}$$

$$\frac{\partial^2 u}{\partial y^2} = \frac{\partial^2 u}{\partial v^2}$$

Inserting into Eq. 24.1, we get

$$\left(\frac{AC - B^2}{A}\right)\left(\frac{\partial^2 u}{\partial v^2} + \frac{\partial^2 u}{\partial w^2}\right) = D''\left(v,w,u,\frac{\partial u}{\partial v},\frac{\partial u}{\partial w}\right)$$

The coefficients of $\partial^2 u/\partial v^2$ and $\partial^2 u/\partial w^2$ are nonzero and have the same sign; the equation is elliptic.

When $B^2 - AC > 0$, we transform the variables

$$v = \frac{\alpha(x,y) + \beta(x,y)}{2}$$

$$w = \frac{\alpha(x,y) - \beta(x,y)}{2}$$

The reader will verify without difficulty that Eq. 24.1 becomes

$$\left(\frac{AC - B^2}{A}\right)\left(\frac{\partial^2 u}{\partial v^2} - \frac{\partial^2 u}{\partial w^2}\right) = D''\left(v,w,u,\frac{\partial u}{\partial v},\frac{\partial u}{\partial w}\right)$$

The coefficients of $\partial^2 u/\partial v^2$ and $\partial^2 u/\partial w^2$ are nonzero and of opposite sign; the equation is hyperbolic.

Finally, when $B^2 - AC = 0$, we put

$$v = \alpha(x,y) = \beta(x,y)$$

$$w = x$$

which leads to a parabolic equation

$$A\frac{\partial^2 u}{\partial w^2} = D''\left(v,w,u,\frac{\partial u}{\partial v},\frac{\partial u}{\partial w}\right)$$

25 · BOUNDARY CONDITIONS AND TYPES OF EQUATIONS

The Cauchy conditions are not the only boundary conditions that are of importance. On the contrary, with each physical problem there is associated a specific type of boundary conditions. Furthermore, different physical phenomena are described by equations of different types. Thus, electrostatic problems are governed by elliptic equations; the problem of wave propagation leads to hyperbolic equations; and the study of transport phenomena is associated with parabolic equations. If our mathematical description of physical processes is correct, then the requirement that the properly stated physical problem have a unique solution must be expected to have as its mathematical counterpart the existence of a close interrelation between types of differential equations and boundary conditions.

A detailed discussion of this fundamental problem would lead us far beyond the scope of this book. We shall therefore limit ourselves to a consideration of particular examples which will illustrate some characteristic situations.

The Cauchy conditions define on a hypersurface the values of a function and of its directional derivatives along the normal to the hypersurface. Two other very important types of boundary conditions, which are weaker than the Cauchy conditions, need to be defined.

The **Dirichlet condition** consists in prescribing only the values of a function on a hypersurface.

For the **Neumann condition**, only the values of the derivatives of a function along the normal to a hypersurface are specified.

25.1 One-dimensional Wave Equation

Consider the wave equation in one space-dimension

$$\frac{\partial^2 u}{\partial x^2} - \frac{1}{c^2}\frac{\partial^2 u}{\partial t^2} = 0 \tag{25.1}$$

where x is the space coordinate and t is the time. Equation 25.1 is of the hyperbolic type.

The equations of the characteristics are

$$\alpha = x + ct = \text{const}$$

$$\beta = x - ct = \text{const}$$

Choosing α and β as the independent variables, we rewrite Eq. 25.1 as

$$\frac{\partial^2 u}{\partial \alpha\, \partial \beta} = 0$$

The most general solution to the preceding equation is evidently

$$u = g(\alpha) + h(\beta) \equiv g(x + ct) + h(x - ct) \tag{25.2}$$

where g and h are arbitrary differentiable functions.

Let the Cauchy boundary conditions be prescribed for $t = 0$

$$u\big|_{t=0} = a(x) \qquad x_1 \le x \le x_2$$
$$\frac{\partial u}{\partial t}\bigg|_{t=0} = b(x)$$

Since t is the time, it is customary to call these conditions **initial conditions.**

From Eq. 25.2 we have

$$g(x) + h(x) = a(x) \tag{25.3}$$

and

$$c\frac{dg(x)}{dx} - c\frac{dh(x)}{dx} = b(x) \tag{25.4}$$

Hence

$$g(x) - h(x) = \frac{1}{c}\int_{x_1}^{x} b(x')\,dx' + \text{const} \tag{25.5}$$

Combining Eqs. 25.3 and 25.5, we find

$$g(x) = \frac{1}{2}\,a(x) + \frac{1}{2c}\int_{x_1}^{x} b(x')\,dx' + \text{const}$$

$$h(x) = \frac{1}{2}\,a(x) - \frac{1}{2c}\int_{x_1}^{x} b(x')\,dx' + \text{const}$$

Therefore the solution is

$$u(x,t) = \frac{1}{2}\left\{a(x+ct) + a(x-ct) + \frac{1}{c}\int_{x-ct}^{x+ct} b(x')\,dx'\right\} \tag{25.6}$$

Since the functions $a(x)$ and $b(x)$ are defined in the interval $[x_1,x_2]$ only, Eq. 25.6 is meaningful when

$$x_1 \le x \pm ct \le x_2 \tag{25.7}$$

which determines a rectangle in the x,t plane (Fig. 50). When $x_1 \to -\infty$ and $x_2 \to +\infty$, i.e., when the spatial domain is infinite, the rectangle covers the entire x,t plane and the solution is determined everywhere for any time.

If the initial conditions are specified along a finite segment of the x axis, one needs additional information to determine the solution everywhere. This information may consist in giving boundary values for all times; for example,

$$\begin{aligned} u(x_1,t) &= c(t) \\ u(x_2,t) &= d(t) \end{aligned} \tag{25.8}$$

To simplify the discussion, we consider the case where $x_1 = 0$ and $x_2 \to \infty$. Then it is sufficient to give only one boundary condition at $x = 0$; for example

$$u(0,t) = 0 \tag{25.9}$$

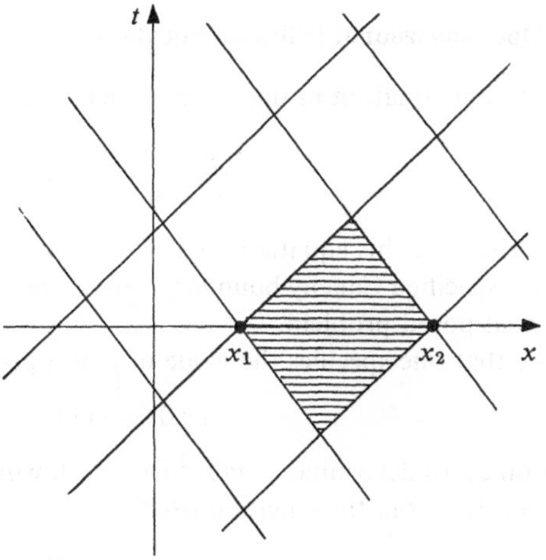

Fig. 50. The oblique lines are the characteristics $x \pm ct$ = constant. The Cauchy conditions on $[x_1,x_2]$ give a unique solution within the shaded rectangle delimited by the characteristics that pass through x_1 and x_2.

We replace our problem by one where the initial conditions are prescribed for all x: $-\infty \leq x \leq \infty$.

$$u\Big|_{t=0} = A(x)$$
$$\frac{\partial u}{\partial t}\Big|_{t=0} = B(x) \tag{25.10}$$

where

$$A(x) = a(x) \qquad \text{for } x > 0$$
$$B(x) = b(x) \qquad \text{for } x > 0 \tag{25.11}$$

The values of $A(x)$ and $B(x)$ for $x < 0$ are, for the moment, unknown.

The solution is given by Eq. 25.6 as

$$u(x,t) = \frac{1}{2}\left\{A(x+ct) + A(x-ct) + \frac{1}{c}\int_{x-ct}^{x+ct} B(x')\,dx'\right\} \tag{25.12}$$

To satisfy the boundary condition, we must have

$$A(ct) + A(-ct) + \frac{1}{c}\int_{-ct}^{ct} B(x')\,dx' = 0$$

Since the integral is an odd function of t, A must also be an odd function of its argument

$$A(x) = -A(-x) \tag{25.13}$$

and the same must be true for $B(x)$

$$B(x) = -B(-x) \tag{25.14}$$

The two requirements 25.13 and 25.14 together with Eq. 25.11 determine A and B for all arguments, and the solution given by Eq. 25.12 is valid without any restrictions on x and t.

25.2 The One-dimensional Diffusion Equation

The diffusion equation in one-space dimension is

$$\frac{\partial^2 u}{\partial x^2} - \frac{1}{a^2}\frac{\partial u}{\partial t} = 0 \tag{25.15}$$

The characteristics of this equation are the lines $t = \text{const}$ in the x,t plane. It is easy to see that if one specified Cauchy boundary conditions along a characteristic, one would not have a well-posed problem.

Suppose that one specifies the value of u at a given time $t = 0$

$$u(x,0) = b(x) \tag{25.16}$$

The condition 25.16 determines $\partial^2 u/\partial x^2$ for $t = 0$, which in turn determines the initial value of $\partial u/\partial t$, for from the equation itself, one has

$$\left.\frac{\partial u}{\partial t}\right|_{t=0} = a^2 \frac{d^2 b(x)}{dx^2}$$

It is evident that we no longer have the freedom to specify $\partial u/\partial t\,|_{t=0}$ and therefore if Cauchy conditions were imposed along the line $t = 0$, they would overdetermine the problem.

We consider an example that shows to what extent a single initial condition completely determines the solution.

Suppose that

$$u|_{t=0} = \begin{cases} 1 & \text{for } x > 0 \\ 0 & \text{for } x < 0 \end{cases} \tag{25.17}$$

We note that both the differential equation and the initial condition are invariant under the scale transformation

$$\begin{aligned} x &\to sx \\ t &\to s^2 t \end{aligned} \tag{25.18}$$

for an arbitrary s. Therefore, the solution must also be invariant under this transformation

$$u(x,t) = u(sx,s^2 t)$$

For $t > 0$, we put

$$s = \frac{1}{\sqrt{t}}$$

and find that $u(x,t)$ is in fact a function of the single variable $y = x/\sqrt{t}$.

In terms of y, Eq. 25.15 becomes an ordinary differential equation

$$\frac{d^2 u}{dy^2} + \frac{y}{2a^2}\frac{du}{dy} = 0 \tag{25.19}$$

and the boundary conditions are

$$u(-\infty) = 0; \quad u(+\infty) = 1 \tag{25.20}$$

Since one solution of Eq. 25.19 is $u_1 = \text{const}$, the other solution is immediately found from Eq. 2.17

$$u(y) = \text{const} \int_{y_0}^{y} dy' e^{-y'^2/4a^2} \tag{25.21}$$

The constants are obtained from the boundary conditions 25.20

$$u(y) = \frac{2a}{\sqrt{\pi}} \int_{-\infty}^{y} dy' e^{-y'^2/4a^2}$$

or in terms of x and t

$$u(x,t) = \frac{2a}{\sqrt{\pi}} \int_{-\infty}^{x/\sqrt{t}} dy e^{-y^2/4a^2} \qquad t > 0 \tag{25.22}$$

which is the solution of the problem for $t > 0$.

It is important to note that there do not exist any solutions of the diffusion equation for $t < 0$ which would satisfy the boundary condition (25.17) at $t = 0$. Setting $s = 1/\sqrt{-t}$, one obtains the same equation as Eq. 25.19, but with a minus sign, and this leads to an expression similar to Eq. 25.21, but with a positive exponent. Such an expression cannot satisfy the boundary conditions.

This behavior can be given a simple physical interpretation. A diffusion equation describes the "disorganization" of physical systems, and it is clear that while we can describe the manner in which a given system evolves in time, we cannot have a situation wherein a system that has been disorganized an infinite amount of time becomes organized at $t = 0$.

25.3 The Two-dimensional Laplace Equation

Let our problem be to find a function $u = u(x,y)$, which is a solution of Laplace's equation

$$\frac{\partial^2 u}{\partial x^2} + \frac{\partial^2 u}{\partial y^2} = 0$$

within some closed curve C in the x,y plane and which satisfies Dirichlet boundary conditions on C.

For an elliptic equation, Eqs. 24.9 and 24.10 have only complex solutions and there are no characteristic curves in the x,y plane. We shall therefore use quite different arguments from those developed in the preceding two examples.

Assume first that our problem has two solutions u_1 and u_2. Then their difference, $u_1 - u_2$, will also satisfy Laplace's equation in the domain delimited by C, and it will vanish on C.

Notice now that if a function satisfies Laplace's equation throughout the region enclosed by C, it may be regarded in that region as the real part of a certain analytic function of $z = x + iy$ (see Sec. 4 of Chapter I; in particular, Eq. 4.6). But we proved in Sec. 14 of Chapter I that the real part of an analytic function cannot have either a local maximum or a local minimum; therefore, it must take its maximum and minimum values on the boundary of a region where it is analytic. It follows that if the real part of a function vanishes on the boundary of a region where the function is analytic, it must vanish everywhere throughout the region. Hence, $u_1 - u_2 \equiv 0$ and the solution of our problem must be unique. The existence of the solution is in general more difficult to prove; however, we did find it in the particular case when C is a circle of radius R. Then the solution is given by Poisson's formula (Chapter I, Eq. 19.9), which can be rewıitten as

$$u(r\cos\theta, r\sin\theta) = \frac{1}{2\pi}\int_0^{2\pi} u|_C \frac{R^2 - r^2}{R^2 - 2Rr\cos(\theta - \theta') + r^2}\, d\theta'$$

Since the Dirichlet conditions on C are sufficient to uniquely determine the solution, it is clear that had we prescribed Cauchy conditions on C, the problem would be in general overdetermined. Even if we were exceptionally lucky in prescribing $u|_C$ and $\partial u/\partial n|_C$, i.e., if the derivatives $\partial u/\partial n|_C$ in the Cauchy conditions were accidentally equal to $\partial u/\partial n|_C$ as determined from $u|_C$ alone, it is obvious that an infinitesimal change of $u|_C$ (or of $\partial u/\partial n|_C$) would lead to an overdetermined problem.

If the solution does not depend continuously on the boundary conditions, one says that it is **unstable**. A problem that leads to an unstable solution is, at least from the physicist's point of view, not properly stated.

We have given a few simple examples which have illustrated how, with equations of different types, one can associate well-posed boundary conditions. The following "rules of thumb" indicate general correlations between the types of equations and the types of boundary conditions that may lead to a stable solution of the differential equation:

(a) Elliptic equations: Dirichlet or Neumann conditions on a closed hypersurface.
(b) Parabolic equations: Dirichlet or Neumann conditions on an open hypersurface. A stable solution exists on one side of the hypersurface only.
(c) Hyperbolic equations: Cauchy conditions on an open hypersurface.

In the last case, if the hypersurface is finite, the Cauchy conditions must be supplemented by Dirichlet or Neumann conditions on another hypersurface.

26 · MULTIDIMENSIONAL FOURIER TRANSFORMS AND δ FUNCTION

The results of Chapter III, Secs. 12 and 13, can be extended to the case of functions of several variables. The following notation for multiple integrals is commonly used

$$\iint\cdots\int dx_1\, dx_2 \cdots dx_N \equiv \int d_N x$$

Let $f(x_1,x_2,\cdots,x_N)$ be a function of N variables. The reciprocal Fourier transforms are obtained by an N-fold application of the one-dimensional formulae

$$F(k_1,k_2,\cdots,k_N) = \frac{1}{(2\pi)^{N/2}}\int_{-\infty}^{+\infty} d_N x\, f(x_1,\cdots,x_2,x_N)e^{i(k_1x_1+\cdots+k_Nx_N)} \tag{26.1}$$

and

$$f(x_1,x_2,\cdots,x_N) = \frac{1}{(2\pi)^{N/2}}\int_{-\infty}^{+\infty} d_N k\, F(k_1,k_2,\cdots,k_N)e^{-i(k_1x_1+\cdots+k_Nx_N)} \tag{26.2}$$

We shall not enter into the details of the validity of Eqs. 26.1 and 26.2. The conditions are similar to those stated in connection with the one-dimensional case.

Similarly, one can generalize the notion of distributions to many variables by defining them as sequences of good functions of several variables. Clearly, the restriction to simple integrals in developing the theory of distributions was not an essential one, and one could repeat the same arguments, using multiple integrals.

Here we shall only generalize the notion of the δ function and introduce the δ function of N variables as a product of N δ functions of one variable

$$\delta_N(x - x') \equiv \delta(x_1 - x'_1)\delta(x_2 - x'_2)\cdots\delta(x_N - x'_N) \tag{26.3}$$

The definition 26.3 does not contradict the statement made in Chapter III, Sec. 13.3, that products of distributions are not meaningful, for the variables in each product in Eq. 26.3 are different. Since

$$\delta(x_j - x'_j) = \frac{1}{2\pi}\int_{-\infty}^{\infty} e^{ik_j(x_j - x'_j)}dk_j$$

we have

$$\begin{aligned}\delta_N(x - x') &= \frac{1}{(2\pi)^N}\prod_{j=1}^{N}\int_{-\infty}^{\infty} e^{ik_j(x_j - x'_j)}dk_j \\ &= \frac{1}{(2\pi)^N}\int_{-\infty}^{+\infty} d_Nk e^{i[k_1(x_1 - x'_1) + k_2(x_2 - x'_2) + \cdots + k_N(x_N - x_N')]}\end{aligned} \tag{26.4}$$

It is often useful to know how δ functions transform under a change of coordinates. To see this, we note that

$$f(x_1,x_2,\cdots,x_N) = \int d_Nx'\ \delta_N(x - x')f(x'_1,\cdots,x'_N) \tag{26.5}$$

Under the change of coordinates

$$x_i = X_i(y_1,y_2,\cdots,y_N) \qquad (i = 1, 2, \cdots, N)$$

with Jacobian

$$J = \frac{\partial(x'_1,x'_2,\cdots,x'_N)}{\partial(y'_1,y'_2,\cdots,y'_N)}$$

Eq. 26.5 becomes

$$f(X_1,\cdots,X_N) = \int d_Ny'\ \delta_N(X - X')f(X'_1,X'_2,\cdots,X'_N)J$$

Thus when $J \neq 0$

$$\delta_N(X - X') = J^{-1}\cdot\delta_N(y - y') \tag{26.6}$$

EXAMPLE

In spherical coordinates

$$x = r\sin\theta\cos\Phi; \quad y = r\sin\theta\sin\Phi, \quad z = r\cos\theta$$

one has

$$J = \begin{vmatrix} \sin\theta\cos\Phi & r\cos\theta\cos\Phi & -r\sin\theta\sin\Phi \\ \sin\theta\sin\Phi & r\cos\theta\sin\Phi & r\sin\theta\cos\Phi \\ \cos\theta & -r\sin\theta & 0 \end{vmatrix} = r^2\sin\theta$$

In terms of the variables r, θ, and Φ, the three-dimensional δ function is

$$\frac{1}{r^2 \sin\theta}\,\delta(r-r')\,\delta(\theta-\theta')\,\delta(\Phi-\Phi') \tag{26.7}$$

i.e., upon changing from Cartesian to spherical coordinates, one has

$$\delta(x-x')\,\delta(y-y')\,\delta(z-z')\,dx\,dy\,dz$$
$$\rightarrow \delta(r-r')\,\delta(\theta-\theta')\,\delta(\Phi-\Phi')\,dr\,d\theta\,d\Phi$$

where

$$x' = r'\sin\theta'\cos\Phi'$$
$$y' = r'\sin\theta'\sin\Phi'$$
$$z' = r'\cos\theta'$$

27 · GREEN'S FUNCTIONS FOR PARTIAL DIFFERENTIAL EQUATIONS

From here on we shall be concerned with linear partial differential equations only.

The method of Green's functions that was used to solve ordinary differential equations will be extended to partial differential equations. It is here that the method of Green's functions finds its most useful applications in physics.

Let L_{x_i} be a formal partial differential operator of second order. The formal adjoint $L^+_{x_i}$ of this operator will be defined by the following relation, which is a straightforward generalization of the Lagrange identity to the case of N variables*

$$[\bar{v}L_{x_i}u - u\overline{(L^+_{x_i}v)}] = \sum_{i=1}^{N} \frac{\partial}{\partial x_i}[Q_i(u,\bar{v})] \tag{27.1}$$

where $Q_i(u,\bar{v})(i = 1, 2, \cdots, N)$ depends bilinearly on $u(x_1, \cdots, x_N)$, $v(x_1, \cdots, x_N)$ and their first-order partial derivatives.

We use Gauss' theorem, which, in N dimensions reads

$$\int_V \sum_{i=1}^{N} \frac{\partial}{\partial x_i} F_i(x_1,x_2,\cdots,x_N)d_Nx = \int_S \sum_{i=1}^{N} F_i n_i\, dS \tag{27.2}$$

where S is a closed hypersurface, V the volume enclosed by S, and n_i are the projections on the coordinate axis of the unit vector in the direction of the outward normal to the surface element dS.

Integrating Eq. 27.1 over a volume V and using Eq. 27.2, we obtain the generalized Green's identity

$$\int_V d_Nx[\bar{v}L_{x_i}u - u\overline{(L^+_{x_i}v)}] = \int_S \sum_{i=1}^{N}[Q_i(u,\bar{v})n_i]ds \tag{27.3}$$

Given homogeneous boundary conditions imposed on u on the surface S, we define the adjoint boundary conditions as those conditions imposed on the function v which make the integrand on the RHS in Eq. 27.3 vanish identically on S

$$\sum_{i=1}^{N} Q_i(u,\bar{v})n_i = 0 \qquad \text{on S} \tag{27.4}$$

* The Lagrange identity in one dimension contains a weight function which, for simplicity, we shall henceforth take to be unity.

When Eq. 27.4 is satisfied, we obtain Green's identity

$$\int_V d_N x[\bar{v}L_{x_i}u - u\overline{(L^+_{x_i}v)}] = 0 \tag{27.5}$$

As in the one-dimensional case, a formal differential operator L_{x_i}, together with homogeneous boundary conditions, defines a differential operator L

$$\langle x_1,x_2,\cdots,x_N| L |u\rangle = L_{x_i}u(x_1,x_2,\cdots,x_N)$$

where the vectors $|x_1,x_2,\cdots,x_N\rangle$ labeled with N continuous "indices" $x_1,x_2,\cdots,x_N$, are a basis of a vector space whose vectors are represented by the functions of N variables $x_1,x_2,\cdots,x_N$. These vectors have the property

$$\langle x_1,x_2,\cdots,x_N|x'_1,x'_2,\cdots,x'_N\rangle = \delta_N(x-x') \tag{27.6}$$

The Green's functions associated with the partial differential equations

$$L_{x_i}u(x_1,x_2,\cdots,x_N) = f(x_1,x_2,\cdots,x_N) \tag{27.7}$$

$$L^+_{x_i}ux_1,(x_2,\cdots,x_N) = h(x_1,x_2,\cdots,x_N) \tag{27.8}$$

satisfy

$$L_{x_i}G(x_1,x_2,\cdots,x_N;x'_1,\cdots x'_N) = \delta_N(x-x') \tag{27.9}$$

$$L^+_{x_i}g(x_1,x_2,\cdots,x_N;x'_1,\cdots,x'_N) = \delta_N(x-x') \tag{27.10}$$

$G(x_1,\cdots,x_N;x'_1,\cdots,x'_N)$ and $g(x_1,\cdots,x_N;x'_1,\cdots,x'_N)$ obey the same boundary conditions as $u(x_1,\cdots,x_N)$ and $v(x_1,\cdots,x_N)$, respectively, but always in their homogeneous form. Using the methods of Sec. 8, it is easy to verify the following properties of the Green's functions.

(i)
$$G(x_1,\cdots,x_N;x'_1,\cdots,x'_N) = g(x'_1,\cdots,x'_N;x_1,\cdots,x_N) \tag{27.11}$$

(ii) If L is Hermitian

$$G(x_1,\cdots,x_N;x'_N,\cdots,x'_N) = \bar{G}(x'_1,\cdots,x'_N;x_1,\cdots,x_N)$$

(iii) If L is Hermitian and the coefficients in L_{x_i} are real

$$G(x_1,\cdots,x_N;x'_1,\cdots,x'_N) = G(x'_1,\cdots,x'_N;x_1,\cdots,x_N)$$

From Eqs. 27.9, 27.10, and 27.11(i) it follows that solutions of Eq. 27.7 and 27.8 are given respectively by

$$u(x_1,\cdots,x_N) = \int d_Ny\, G(x_1,\cdots,x_N;y_1,\cdots,y_N)f(y_1,\cdots,y_N) \quad +(\text{surface terms}) \tag{27.12}$$

$$v(x_1,\cdots,x_N) = \int d_Ny\, g(x_1,\cdots,x_N;y_1,\cdots,y_N)h(y_1,\cdots,y_N) \quad +(\text{surface terms}) \tag{27.13}$$

In Eqs. 27.12 and 27.13, the surface terms are present only when the boundary conditions associated with Eqs. 27.7 and 27.8 contain inhomogeneities.

EXAMPLE

Consider the Laplace equation

$$\vec{\nabla}^2 u(x,y,z) = f(x,y,z) \tag{27.14}$$

where

$$\vec{\nabla}^2 = \frac{\partial^2}{\partial x^2} + \frac{\partial^2}{\partial y^2} + \frac{\partial^2}{\partial z^2}$$

is the Laplace operator in three dimensions. The following relation is an immediate consequence of Gauss' theorem in three dimensions (see also Sec. 29.2)

$$\int_V d_3x(u\vec{\nabla}^2 v - v\vec{\nabla}^2 u) = \iint_S dS[u\vec{\nabla}v - v\vec{\nabla}u] \cdot \vec{n}$$

This relation is a special case of Eq. 27.3, the RHS being the surface term, and shows that $\vec{\nabla}^2$ is self-adjoint.

With the following boundary conditions, $\vec{\nabla}^2$ defines a Hermitian differential operator

$u = 0$ on S (homogeneous Dirichlet condition)

$\vec{\nabla}u \cdot \vec{n} = 0$ on S (homogeneous Neumann condition)

Before entering into a discussion of the construction of a Green's function for partial differential equations, let us review briefly some of the characteristic properties of the Green's functions for an ordinary differential equation (Sec. 9). In particular, we call the attention of the reader to Eq. 9.7. It can be seen from this equation that Green's function is the sum of two terms. The first term, G_s, satisfies by itself the differential equation for the Green's function

$$L_x G_s = L_x G = \frac{\delta(x-y)}{w(x)}$$

but it does not satisfy the boundary conditions. The second term is a solution of the homogeneous equation and its *raison d'être* is to make the total Green's function satisfy the boundary conditions. The part G_s of the Green's function is differentiable only in the sense of generalized functions and gives rise to the δ function in the defining equation for G.

In the case of partial differential equations, one can also split the Green's function into two parts

$$G = G_s + G_0$$

where G_s will satisfy the equation

$$L_{x_i} G_s = L_{x_i} G = \delta_N(x - x')$$

and G_0 is a solution of the homogeneous equation

$$L_{x_i} G_0 = 0$$

which is so chosen that G satisfies the homogeneous boundary conditions associated with the operator L.

The marked difference between partial and ordinary differential equations is that in the former case, G_s will be a truly singular function itself, while in the latter case, G_s is a continuous function (although its first derivative is discontinuous and its second derivative is proportional to a δ function).

In the next few sections we shall give methods for constructing the singular part G_s of the Green's function for some of the most important differential equations that occur in physics. The question of finding a Green's function satisfying the prescribed boundary conditions and which requires the knowledge of a solution of the homogeneous differential equation will be discussed subsequently.

28 · THE SINGULAR PART OF THE GREEN'S FUNCTION FOR PARTIAL DIFFERENTIAL EQUATIONS WITH CONSTANT COEFFICIENTS

28.1 The General Method

The method of constructing the singular part of the Green's function which we shall describe applies generally to partial differential equations whose coefficients are constants.

Consider the equation

$$L_{x_i}u(x_1,x_2,\cdots,x_N)=f(x_1,x_2,\cdots,x_N) \tag{28.1}$$

where L_{x_i} is an operator of the form

$$L_{x_i}=a_0+a_1\frac{\partial}{\partial x_1}+\cdots+a_N\frac{\partial}{\partial x_N}+a_{N+1}\frac{\partial^2}{\partial x_1^{\,2}}+\cdots+a_{N(N+3)/2}\frac{\partial^2}{\partial x_N^2} \tag{28.2}$$

the coefficients $a_0,\cdots,a_{N(N+3)/2}$ being constants. The equation for the Green's function associated with Eq. 28.1 is

$$L_{x_i}G(x_1,\cdots,x_N;x'_1,\cdots,x'_N)=\delta_N(x-x') \tag{28.3}$$

We use the representation (26.4) of the N-dimensional δ function

$$\delta_N(x-x')=\frac{1}{(2\pi)^N}\int_{-\infty}^{+\infty}d_Nk\exp\left(i\sum_{j=1}^{N}k_j(x_j-x'_j)\right) \tag{28.4}$$

The singular part of the Green's function, G_s, is given by

$$G_s(x_1,\cdots,x_N;x'_1,\cdots,x'_N)$$
$$=\frac{1}{(2\pi)^N}\int_{-\infty}^{+\infty}\frac{d_Nk\exp\left(i\sum_{j=1}^{N}k_j(x_j-x'_j)\right)}{a_0+ia_1k_1+\cdots+ia_Nk_N+a_{N+1}(ik_{N+1})^2+\cdots+a_{N(N+3)/2}(ik_N)^2} \tag{28.5}$$

as can be immediately verified by applying the operator 28.2 to this expression and interchanging the order of differentiation and integration*; this gives a numerator that just cancels the denominator, and because of Eq. 28.4, the result is the N-dimensional δ function.

28.2 An Elliptic Equation: Poisson's Equation

It is shown in electrostatics that the potential $u(\vec{r})$ due to a free-charge distribution density in space $\rho(\vec{r})$ satisfies Poisson's equation

$$\vec{\nabla}^2u(\vec{r})=-4\pi\rho(\vec{r}) \tag{28.6}$$

* This is justified because the differentiations are understood in the sense of generalized functions, and the result is a Fourier transform of a generalized function that always exists and is unique.

where $\vec{\nabla}^2$ is the Laplace operator. The Green's function for Eq. 28.6 satisfies

$$\vec{\nabla}^2 G(\vec{r};\vec{r}') = \delta_3(\vec{r} - \vec{r}') \tag{28.7}$$

and $G_s(\vec{r};\vec{r}')$ can be immediately written down from Eq. 28.5 as

$$G_s(\vec{r};\vec{r}') = -\frac{1}{(2\pi)^3}\int_{-\infty}^{+\infty} d_3k \, \frac{e^{i\vec{k}\cdot(\vec{r}-\vec{r})}}{\vec{k}^2}$$

Using spherical coordinates

$$k_1 = k \sin\theta \cos\Phi$$

$$k_2 = k \sin\theta \sin\Phi$$

$$k_3 = k \cos\theta$$

we have

$$\vec{k}^2 = k^2$$

and taking the k_3 axis in the direction of $(\vec{r} - \vec{r}')$ we also have

$$\vec{k}\cdot(\vec{r} - \vec{r}') = kR\cos\theta \qquad (R = |\vec{r} - \vec{r}'|)$$

Of course

$$d_3k = k^2 \sin\theta \, dk \, d\theta \, d\Phi$$

Hence

$$G_s(\vec{r};\vec{r}') = -\frac{1}{(2\pi)^3}\int_0^\infty dk \int_0^{2\pi} d\Phi \int_0^\pi d\theta \sin\theta e^{ikR\cos\theta}$$

But

$$\int_0^\pi d\theta \sin\theta e^{ikR\cos\theta} = -\left.\frac{e^{ikR\cos\theta}}{ikR}\right|_0^\pi = \frac{2\sin kR}{kR}$$

Hence

$$G_s(\vec{r};\vec{r}') = \frac{-4\pi}{(2\pi)^3}\int_0^\infty dk \frac{\sin kR}{kR} = \frac{-1}{4\pi R} = \frac{-1}{4\pi|\vec{r} - \vec{r}'|} \tag{28.8}$$

A keen reader will have recognized in Eq. 28.8 the Coulomb potential. This is not astonishing, since (as explained in Chap. III, Sec. 13.1) the δ function may be regarded as a mathematical device for describing the idealized point charge picture.

28.3 A Parabolic Equation: The Diffusion Equation

The phenomenon of heat flow, or of the flow of a compressible fluid, or of the diffusion of thermal neutrons through matter can all be described by a diffusion equation, which in one dimension reads

$$L_{x_i} u(x,t) = -\sigma(x,t) \tag{28.9}$$

where

$$L_{x_i} = \frac{\partial^2}{\partial x^2} - \frac{1}{a^2}\frac{\partial}{\partial t} \tag{28.10}$$

Here x denotes the space coordinate, t stands for the time coordinate, and a is a real constant called the **diffusion coefficient**.

The singular part of the Green's function can be written down by inspection from Eq. 28.5 as

$$G_s(x,t;x',t') = \frac{ia^2}{4\pi^2}\int_{-\infty}^{\infty} dk_1 \int_{-\infty}^{\infty} dk_2 \frac{e^{ik_1(x-x')+ik_2(t-t')}}{k_2 - ia^2k_1^2} \tag{28.11}$$

The integral over k_2 can be evaluated by a contour integration. If $(t - t') > 0$, the contour must be closed in the upper half of the complex k_2 plane, and it will enclose the singularity of the integrand at $k_2 = ia^2k_1^2$. For $(t - t') < 0$, the contour must be closed in the lower half-plane and it will not include any singularities

$$G_s(x,t;x',t') = \begin{cases} \dfrac{-a^2}{2\pi}\displaystyle\int_{-\infty}^{\infty} dk_1\, e^{ik_1(x-x_1)-a^2k_1|t-t'|} & \text{for } t > t' \\ 0 & \text{for } t < t' \end{cases}$$

The integral given above has already been evaluated (Chapter I, Eq. 22.20). Thus

$$G_s(x,t;x',t') = \begin{cases} \dfrac{-a}{2\sqrt{\pi|t-t'|}} \exp\left(\dfrac{-(x-x')^2}{4a^2|t-t'|}\right) & \text{for } t > t' \\ 0 & \text{for } t < t' \end{cases} \tag{28.12}$$

28.4 A Hyperbolic Equation: The Time-dependent Wave Equation

The wave equation that describes electromagnetic phenomena in three-dimensional space is

$$L_{x_i}u(\vec{r},t) = -4\pi\rho(\vec{r},t) \tag{28.13}$$

where

$$L_{x_i} = \vec{\nabla}^2 - \frac{1}{c^2}\frac{\partial^2}{\partial t^2} \tag{28.14}$$

in which $\vec{\nabla}^2$ is the Laplace operator, c is the velocity of propagation of the wave, t stands for the time coordinate, and $\rho(\vec{r},t)$ is a time-dependent source density function.

The Green's function associated with Eq. 28.13 satisfies

$$L_{x_i}G(\vec{r},t;\vec{r}',t') = \delta_3(\vec{r} - \vec{r}')\delta(t - t') \tag{28.15}$$

We write

$$\delta_3(\vec{r} - \vec{r}') = \frac{1}{(2\pi)^3}\int d_3k\, e^{ik\cdot\vec{R}}$$

$$\delta(t - t') = \frac{1}{2\pi}\int_{-\infty}^{\infty} d\omega\, e^{-i\omega T}$$

where

$$\vec{R} = \vec{r} - \vec{r}' \qquad \text{and} \qquad T = t - t'$$

The singular part of the Green's function is obtained as usual

$$G_s(\vec{r},t;\vec{r}',t') = \frac{-c^2}{(2\pi)^4}\int_{-\infty}^{+\infty} d_3k\, e^{i\vec{k}\cdot\vec{R}} \int_{-\infty}^{\infty} d\omega \frac{e^{-i\omega T}}{c^2k^2 - \omega^2} \tag{28.16}$$

Consider the second integral

$$\Delta \equiv \int_{-\infty}^{\infty} d\omega \, \frac{e^{-i\omega T}}{c^2k^2 - \omega^2} \tag{28.17}$$

Since the integrand has poles on the path of integration at $\omega = \pm ck$, it is a meaningless integral unless one specifies the proper contours that avoid these poles. This was done in detail for precisely this type of integral in the example of Chapter I on p. 62, where we saw that there are essentially four solutions but only two independent ones. We consider the two solutions corresponding to Δ_1 (Fig. 22(a), Chapter I) and Δ_3 (Fig. 22(c), Chapter I). These are usually called "retarded" and "advanced" Δ functions. Thus

$$\Delta^{\text{ret}} = \begin{cases} \dfrac{2\pi \sin ck(t-t')}{ck} & \text{for } t > t' \\ 0 & \text{for } t < t' \end{cases} \tag{28.18}$$

and

$$\Delta^{\text{adv}} = \begin{cases} 0 & \text{for } t > t' \\ \dfrac{-2\pi \sin ck(t-t')}{ck} & \text{for } t < t' \end{cases} \tag{28.19}$$

Δ^{adv} leads to the physical consequence that an electromagnetic signal is emitted at a later time t' than the time t at which it arrives. Δ^{ret} on the contrary, leads to a proper causal behavior: A signal is emitted at an earlier or retarded time t' compared to the time of arrival t. Thus, on purely physical grounds, we choose the solution 28.18. Inserting Eq. 28.18 into Eq. 28.16, we have for the **retarded Green's function**

$$G_s^{\text{ret}}(\vec{r},t;\vec{r}',t') = \begin{cases} \dfrac{-(2\pi)c}{(2\pi)^4} \displaystyle\int_{-\infty}^{+\infty} d_3k \, e^{i\vec{k}\cdot\vec{R}} \frac{\sin ckT}{k} & \text{for } T > 0 \\ 0 & \text{for } T < 0 \end{cases} \tag{28.20}$$

The angular integration can easily be carried out:

$$\int_{-\infty}^{+\infty} d_3k e^{i\vec{k}\cdot\vec{R}} \frac{\sin ckT}{k} = \frac{4\pi}{|\vec{R}|} \int_0^{\infty} dk \sin k|\vec{R}| \sin ckT$$

The sines on the RHS above can be converted into exponentials and combining terms, one has:

$$\int_{-\infty}^{+\infty} d_3k e^{i\vec{k}\cdot\vec{R}} \frac{\sin ckT}{k} = \frac{-\pi}{|\vec{R}|} \int_{-\infty}^{\infty} dk \{e^{ik[|\vec{R}|+cT]} - e^{ik[|\vec{R}|-cT]}\}$$

$$= \frac{-2\pi^2}{|\vec{R}|} [\delta(|\vec{R}| + cT) - \delta(|\vec{R}| - cT)]$$

Since the first δ-function does not contribute for $T > 0$ and since (cf. Chap. III, Sec. 13.17)

$$\delta(|\vec{R}| - cT) = \frac{1}{c} \delta\left(\frac{|\vec{R}|}{c} - T\right)$$

we obtain:

$$G^{\text{ret}}(\vec{r},t;\vec{r}',t') = \begin{cases} -\dfrac{1}{4\pi|\vec{R}|}\,\delta\!\left(\dfrac{|\vec{R}|}{c} - T\right) & T > 0 \\ 0 & T < 0 \end{cases} \tag{28.21}$$

29 · SOME UNIQUENESS THEOREMS

29.1 Introduction

We showed in the preceding section how one can find the singular part of the Green's function in the case where the differential equation has constant coefficients. From the point of view of applications to physics, this is not a very severe limitation, since a great many of the equations that one encounters either have constant coefficients or their coefficients are at least approximatively constant in certain domains of variation of the independent variables. In general, when the coefficients depend strongly on the independent variables, the analytical methods of solution fail, and one must have recourse to numerical methods.

In most cases of interest to us, therefore, the main problem in constructing a Green's function for a partial differential equation will consist in finding the nonsingular part of the Green's function, which is a solution of the homogeneous differential equation and which is such that the total Green's function, the sum of the singular and nonsingular parts, obeys the prescribed homogeneous boundary conditions of the problem.

Conversely, if we have succeeded in constructing the complete Green's function, we have at hand a solution of both the homogeneous and inhomogeneous equations, as can be seen from the generalized Green's identity.

In the subsequent sections we shall describe some of the more standard methods for finding solutions of homogeneous partial differential equations. Since some of these methods will be based on a certain amount of ingenious guessing, it is of the utmost importance to know if, and under what conditions, the solution obtained is unique.

The theorem of Cauchy-Kovalevska is a uniqueness theorem and, moreover, it guarantees the existence of a solution for a class of equations that are very general but for very particular boundary conditions; also, these solutions will usually be valid only "in the small," i.e., in an immediate neighborhood of a point in the space of their arguments. It turns out that the physicist may very well restrict his attention to special types of equations only, but he requires on the one hand more elasticity in the imposition of boundary conditions, and on the other hand, that the uniqueness theorems be valid "in the large."

We shall discuss some typical problems, leaving aside the difficult question of the existence of a solution.

29.2 The Dirichlet and Neumann Problems for the Three-dimensional Laplace Equation

The Dirichlet problem for the Laplace equation consists in finding a solution of

$$\vec{\nabla}^2 u = 0 \tag{29.1}$$

in the interior of a closed surface S and which takes on prescribed values on S. The

Neumann problem is analogous except that there the values of the directional derivatives of u along the outward normal to S are prescribed, instead of the values of the function itself.

We have defined the so-called **interior** Dirichlet and Neumann problems. One could also define the **exterior** problems by looking for solutions of Laplace's equation outside of a given surface S.

Before formulating the uniqueness theorems, we need to derive some elementary properties of harmonic functions.*

In three-dimensions, Gauss' formula (Eq. 27.2) reads as

$$\int_V d_3x\, \vec{\nabla}\cdot\vec{A} = \iint_S dS\, \vec{A}\cdot\vec{n} \tag{29.2}$$

Putting successively

$$\vec{A} = u\vec{\nabla}v \qquad \text{and} \qquad \vec{A} = u\vec{\nabla}v - v\vec{\nabla}u$$

in Eq. 29.2 and using the obvious relation

$$\vec{\nabla}\cdot(v\vec{\nabla}u) = v\vec{\nabla}^2u + (\vec{\nabla}v)\cdot(\vec{\nabla}u)$$

we obtain the so-called **first** and **second Green's identities**

$$\int_V d_3x\, v\vec{\nabla}^2u = \iint_S dS\, v\vec{\nabla}u\cdot\vec{n} - \int_V d_3x(\vec{\nabla}u)\cdot(\vec{\nabla}v) \tag{29.3}$$

and

$$\int_V d_3x(u\vec{\nabla}^2v - v\vec{\nabla}^2u) = \iint_S dS(u\vec{\nabla}v - v\vec{\nabla}u)\cdot\vec{n} \tag{29.4}$$

With the help of Eq. 29.3, we obtain the following properties of a function u which is harmonic within the volume V.

(i)

$$\iint_S dS\, \vec{\nabla}u\cdot\vec{n} = 0 \tag{29.5}$$

This relation is obtained by putting $v = 1$ into Eq. 29.3.

(ii) Let S_R be a sphere of radius R within V and centered at a point x,y,z. Then we have the mean-value theorem for harmonic functions

$$u(x,y,z) = \frac{1}{4\pi R^2}\iint_{S_R} dS\, u \tag{29.6}$$

To prove Eq. 29.6, we set

$$v = \frac{1}{r} = \frac{1}{\sqrt{x^2+y^2+z^2}}$$

* A harmonic function was defined in Chapter I, and is a function that satisfies Laplace's equation.

in Eq. 29.4, and remembering that

$$\vec{\nabla}^2\left(\frac{1}{r}\right) = -4\pi\,\delta(x)\,\delta(y)\,\delta(z) \tag{29.7}$$

which follows from Eqs. 28.8 and 28.7, we obtain

$$-4\pi u(x,y,z) = \iint_{S_R} dS\left[u\left(\frac{-\vec{r}}{r^3}\right)\cdot\vec{n} - \frac{1}{r}\vec{\nabla}u\cdot\vec{n}\right]$$

Since r is constant on S_R, we immediately get Eq. 29.6, using Eq. 29.5.

(iii) If u is continuous in V + S, then either it is a constant or it takes its maximum and minimum values on the boundary S.

To see this, let (x_0,y_0,z_0) be an arbitrary point within V (not on S) and let S_R be an immediate neighborhood of this point. From Eq. 29.6, we have

$$u(x_0,y_0,z_0) \leq \frac{1}{4\pi R^2}\iint_{S_R}\{\max u\}_{S_R} dS = \{\max u\}_{S_R} \tag{29.8}$$

and

$$u(x_0,y_0,z_0) \geq \frac{1}{4\pi R^2}\iint_{S_R}\{\min u\}_{S_R} dS = \{\min u\}_{S_R} \tag{29.9}$$

If $u(x,y,z)$ had a relative maximum at (x_0,y_0,z_0), Eq. 29.8 would be contradicted, and if it had a relative minimum there, Eq. 29.9 would be contradicted. The preceding statement follows at once.

We are now in a position to state the uniqueness theorem for the interior Dirichlet problem. Let u_1 and u_2 be two solutions of Laplace's equation which take on the same values on S. Then $\tilde{u} \equiv u_1 - u_2$ also satisfies Laplace's equation and vanishes on S. But since $\tilde{u}$ is zero on S, it must vanish everywhere in V because it can have neither a relative maximum nor minimum within V. Thus, $u_1 \equiv u_2$ and the solution is unique.

To prove the uniqueness theorem for the interior Neumann problem, we put $v = u$ into Eq. 29.3 and take u to be a harmonic function; then we have

$$\int_V d_3x\,(\vec{\nabla}u)\cdot(\vec{\nabla}u) = \iint_S dS\,u(\vec{\nabla}u)\cdot\vec{n} \tag{29.10}$$

Let u_1 and u_2 be two solutions of Laplace's equation within V which satisfy Neumann conditions. Then $\tilde{u} = u_1 - u_2$ is a harmonic function and satisfies a homogeneous Neumann condition. It follows from Eq. 29.10, since the surface term vanishes, that

$$\int_V d_3x\,(\vec{\nabla}\tilde{u})\cdot(\vec{\nabla}\tilde{u}) = 0 \tag{29.11}$$

which means, since the length of a vector is a positive number, that $\vec{\nabla}\tilde{u} = 0$ and therefore

$$\tilde{u} = \text{const}$$

Thus

$$u_1 = u_2 + \text{const} \tag{29.12}$$

and the solution is unique to within an arbitrary additive constant.

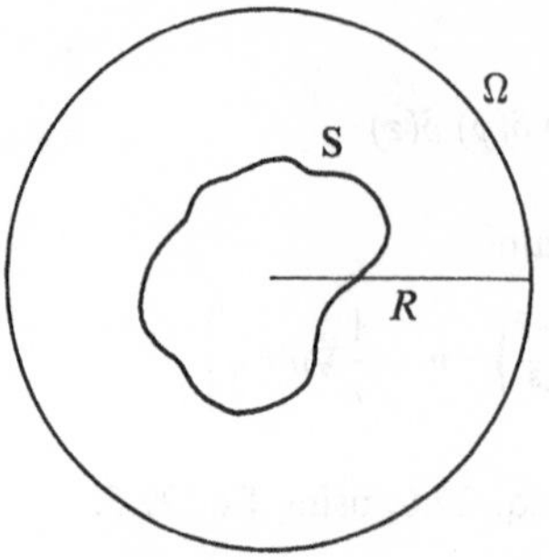

Fig. 51. The multiply-connected region for the exterior Dirichlet and Neumann problems.

The reader should notice that had we used Eq. 29.10 to prove the uniqueness theorem for Dirichlet conditions, we would have had to assume the existence of $\vec{\nabla}u \cdot \vec{n}$ on S, which is not ensured.

The results just derived are valid for the interior Dirichlet and Neumann problems. For the exterior problems, one has to add conditions about the behavior of the solution of Laplace's equation at infinity.

For the exterior Dirichlet problem, one requires in addition that the solutions of Laplace's equation tend uniformly to zero as $r = \sqrt{x^2 + y^2 + z^2}$ tends to infinity. In that case, $\tilde{u} = u_1 - u_2$ tends uniformly to zero on a large sphere Ω of radius R which surrounds S, as $R \to \infty$ (Fig. 51). Of course it also vanishes on S. Therefore, the maximum-minimum principle applied to the multiply-connected region between S and Ω implies that in the limit $R \to \infty$, $\tilde{u} \equiv 0$.

For the exterior Neumann problem, the additional requirement is that both ru and $r^2 |\vec{\nabla}u|$, where u is a solution of Laplace's equation, be bounded as $r \to \infty$. The formula 29.10 applied* to the multiply-connected region between S and Ω leads again to Eq. 29.12, but the constant now must be zero, since both u_1 and u_2 are assumed to vanish as $r \to \infty$.

29.3 The One-dimensional Diffusion Equation

We shall prove that a function $u(x,t)$ continuous in the closed rectangle (see Fig. 52)

$$x_1 \le x \le x_2, \qquad 0 \le t \le T \tag{29.13}$$

satisfying the diffusion equation

$$\frac{\partial^2 u}{\partial x^2} - \frac{1}{a^2}\frac{\partial u}{\partial t} = 0 \qquad \text{for } x_1 < x < x_2, \quad 0 < t \le T$$

and the boundary conditions

$$\begin{aligned} u|_{t=0} &= a(x) \qquad \text{(initial condition)} \\ u|_{x=x_1} &= b(t) \\ u|_{x=x_2} &= c(t) \end{aligned}$$

is unique.

---

* Because of the regularity conditions imposed on u, Eq. 29.10 is valid even when the radius R of Ω tends to infinity.

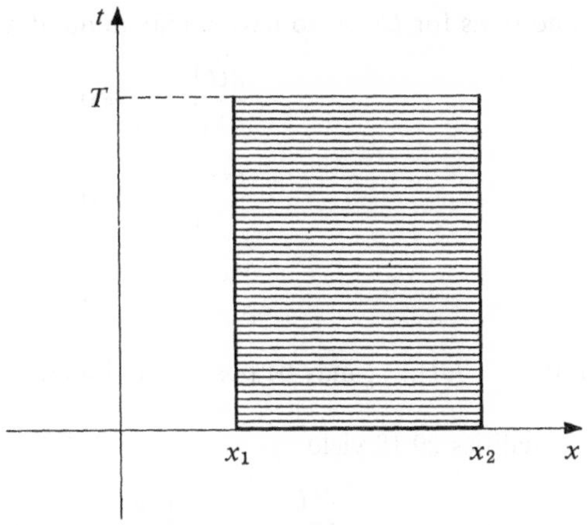

Fig. 52. The shaded area is the rectangle (Eq. 29.13). The boundary conditions are specified along the heavy lines, i.e., on the set B.

As in the case of the Dirichlet problem of the preceding section, the proof of uniqueness is based on a maximum-minimum property of the solutions of the diffusion equation.

We shall show that $u(x,t)$ can take its maximum or minimum values only on one of the three sides of the rectangle defined by inequalities 29.13

$$x = x_1, \qquad x = x_2, \qquad t = 0$$

These three edges of the rectangle will be denoted by B.

Suppose first that there exists a point x_0,t_0 not on B where $u(x,t)$ has a maximum. Then

$$u(x_0,t_0) = \Gamma + \gamma, \qquad \gamma > 0 \tag{29.14}$$

where Γ denotes the maximum of $u(x,t)$ on B.

We introduce an auxiliary function

$$U(x,t) \equiv u(x,t) + \eta \cdot (t_0 - t) \qquad \eta > 0 \tag{29.15}$$

and choose η such that

$$|\eta(t_0 - t)| \leq \frac{\gamma}{2} \tag{29.16}$$

Then

$$U(x_0,t_0) = \Gamma + \gamma$$

and

$$U(x,t) \leq \Gamma + \frac{\gamma}{2} \qquad x,t \in B \tag{29.17}$$

$U(x,t)$ is continuous, since by hypothesis $u(x,t)$ is continuous, and therefore it must have a maximum at some point x'_0,t'_0 in the rectangle (29.13). Then

$$U(x'_0,t'_0) \geq U(x_0,t_0) = \Gamma + \gamma$$

and therefore

$$t'_0 > 0; \qquad x_1 < x'_0 < x_2$$

because on B the inequality 29.17 is satisfied.

The conditions for $U(x,t)$ to have a maximum at x'_0,t'_0 are

$$\left.\frac{\partial U}{\partial x}\right|_{x=x'0} = 0$$
$$\left.\frac{\partial U}{\partial t}\right|_{t=t'0} \geq 0 \tag{29.18}$$
$$\left.\frac{\partial^2 U}{\partial x^2}\right|_{x=x'0} \leq 0$$

The reason that $\partial U/\partial t\,|_{t=t'_0}$ may be positive and not simply zero is that one may have $t'_0 = T$.

The inequalities 29.18 yield

$$\left.\frac{\partial^2 U}{\partial x^2}\right|_{x=x'_0} = \left.\frac{\partial^2 u}{\partial x^2}\right|_{x=x'_0} \leq 0 \tag{29.19}$$

and

$$\left.\frac{\partial U}{\partial t}\right|_{t=t'_0} = \left.\frac{\partial u}{\partial t}\right|_{t=t'_0} - \gamma \geq 0$$

or

$$\left.\frac{\partial u}{\partial t}\right|_{t=t'_0} > 0 \tag{29.20}$$

The inequalities 29.19 and 29.20 are incompatible with the assumption that $u(x,t)$ satisfies the diffusion equation. This proves that u must assume its maximum value on B. Analogously, one can prove that $u(x,t)$ must take its minimum value on B.

The remainder of the proof is straightforward. If there existed two solutions u_1,u_2 satisfying the conditions of the theorem, their difference $u_1 - u_2$ would also satisfy these conditions and, moreover, would vanish on B. Therefore, it would vanish everywhere within the rectangle and so $u_1 \equiv u_2$.

It can be shown that when one of or both points x_1,x_2 goes to infinity, the corresponding conditions on $u\,|_{x=x_1}$ or on $u\,|_{x=x_2}$, or on both become superfluous, and the uniqueness theorem holds, provided that $|u|$ is finite for any x and for $t \geq 0$. A particular example of the case where $x_1 = -\infty$ and $x_2 = +\infty$ was given in the example in Sec. 25.

29.4 The Initial Value Problem for the Wave Equation

We consider the equation

$$\vec{\nabla}^2 u - \frac{1}{c^2}\frac{\partial^2 u}{\partial t^2} = 0 \tag{29.21}$$

with the boundary (initial) conditions

$$u(x,y,z,t)\,|_{t=0} = a(x,y,z)$$
$$\left.\frac{\partial u(x,y,z,t)}{\partial t}\right|_{t=0} = b(x,y,z) \tag{29.22}$$

Integrating the second Green's formula (Eq. 29.4) with respect to the time, we have

$$\int_0^T dt' \int_V d_3x(v\,\vec{\nabla}^2 u - u\,\vec{\nabla}^2 v) = \int_0^T dt' \iint_S dS'(v\,\vec{\nabla}u - u\,\vec{\nabla}v)\cdot\vec{n}$$

In this formula we take u to satisfy the wave equation (Eq. 29.21), we replace v by the retarded Green's function (Eq. 28.21)

$$v \equiv G^{\text{ret}} = \frac{-c}{4\pi}\,\frac{\delta(|\vec{r}-\vec{r}'| - ct + ct')}{|\vec{r}-\vec{r}'|} \tag{29.23}$$

and we choose $T > t$. We find

$$-u(x,y,z,t) + \frac{1}{c^2}\int_0^T dt' \int_V d_3x'\left(G^{\text{ret}}\frac{\partial^2 u}{\partial t^2} - u\,\frac{\partial^2 G^{\text{ret}}}{\partial t^2}\right)$$

$$= \int_0^T dt' \iint_S dS'(G^{\text{ret}}\,\vec{\nabla}u - u\,\vec{\nabla}G^{\text{ret}})\cdot\vec{n}$$

We shall consider the infinite domain problem. Then the surface integral on the RHS vanishes because the argument of the δ function in G^{ret} is never zero when $|r'|$ is sufficiently large.

Hence, we have

$$\begin{aligned} u(x,y,z,t) &= \frac{1}{c^2}\int_0^T dt' \int d_3x'\left(G^{\text{ret}}\frac{\partial^2 u}{\partial t'^2} - u\,\frac{\partial^2 G^{\text{ret}}}{\partial t'^2}\right) \\ &= \frac{1}{c^2}\int d_3x' \int_0^T dt'\,\frac{\partial}{\partial t'}\left[G^{\text{ret}}\frac{\partial u}{\partial t'} - u\,\frac{\partial G^{\text{ret}}}{\partial t'}\right] \\ &= \frac{1}{c^2}\int d_3x'\left[G^{\text{ret}}\frac{\partial u}{\partial t'} - u\,\frac{\partial G^{\text{ret}}}{\partial t'}\right]\Bigg|_0^T \end{aligned}$$

Since $t < T$, the upper limit gives a vanishing contribution, again because of the δ function in G^{ret}. Therefore

$$u(x,y,z,t) = \frac{-1}{c^2}\int d_3x'\left[G^{\text{ret}}\frac{\partial u}{\partial t'} - u\,\frac{\partial G^{\text{ret}}}{\partial t'}\right]_{t'=0} \tag{29.24}$$

We integrate Eq. 29.24 in spherical coordinates and choose the origin of the prime coordinate system at the point x,y,z

$$x' - x = r'\sin\theta'\cos\Phi'$$

$$y' - y = r'\sin\theta'\sin\Phi'$$

$$z' - z = r'\cos\theta'$$

Using Eq. 29.23, we have

$$u(x,y,z,t) = \frac{1}{4\pi c}\int d\Omega' \int_0^\infty dr'\,r'\Big\{\delta(r'-ct)b(r',\Omega')$$

$$- \left[\frac{\partial}{\partial t'}\,\delta(ct' - ct + r')\right]_{t'=0} a(r',\Omega')\Big\}$$

Since

$$\left[\frac{\partial}{\partial t'}\,\delta(ct' - ct + r')\right]_{t'=0} = c\,\frac{\partial}{\partial r'}\,\delta(r' - ct)$$

after a partial integration we find

$$u(x,y,z,t) = \frac{1}{4\pi c}\int d\Omega'\left[r'b(r',\Omega') + c\,\frac{\partial}{\partial r'}\,r'a(r',\Omega')\right]_{r'=ct}$$

or

$$u(x,y,z,t) = \frac{1}{4\pi}\int_0^{2\pi} d\Phi'\int_{-1}^{1} d(\cos\theta')\left[tb(\xi,\eta,\zeta) + \frac{\partial}{\partial t}\,ta(\xi,\eta,\zeta)\right] \tag{29.25}$$

where

$$\xi = x + ct\sin\theta'\cos\Phi'$$

$$\eta = y + ct\sin\theta'\sin\Phi'$$

$$\zeta = z + ct\cos\theta'$$

Equation 29.25, known as **Poisson's formula**, is the solution to our problem. The uniqueness of the solution is evident, for the difference between two solutions of the wave equation would itself be a solution of that equation, but it would obey homogeneous boundary conditions; therefore, by virtue of Poisson's formula, it would be identically zero.

We shall not prove a uniqueness theorem for finite spatial boundaries, but only state the result. If the initial conditions are prescribed in some limited region of three-dimensional space, and if they are supplemented by Dirichlet or Neumann conditions on the boundary of the region, then the solution to the problem is unique under fairly general conditions on the shape of the boundary.

30 · THE METHOD OF IMAGES

The uniqueness theorems, some examples of which where presented in the preceding section, find an immediate application in the so-called **method of images**. The method is based on the fact that the singular Green's functions found in Sec. 28 are solutions of the corresponding homogeneous equations everywhere except at the isolated singular points. For a problem with a not too complicated geometry, it is often possible to construct the complete Green's function by using a singular Green's function with singularities located outside the domain of interest.

Consider the problem of finding a solution of Laplace's equation within a sphere V_R of radius R, which satisfies Dirichlet conditions on the surface S_R of the sphere.

The problem will be solved once we have obtained a Green's function for the Laplace equation which vanishes on the sphere. In Sec. 28.2 we found the singular part G_s of the Green's function for the Laplace equation

$$G_s(\vec{r},\vec{r}\,') = \frac{-1}{4\pi}\,\frac{1}{|\vec{r} - \vec{r}\,'|} \tag{30.1}$$

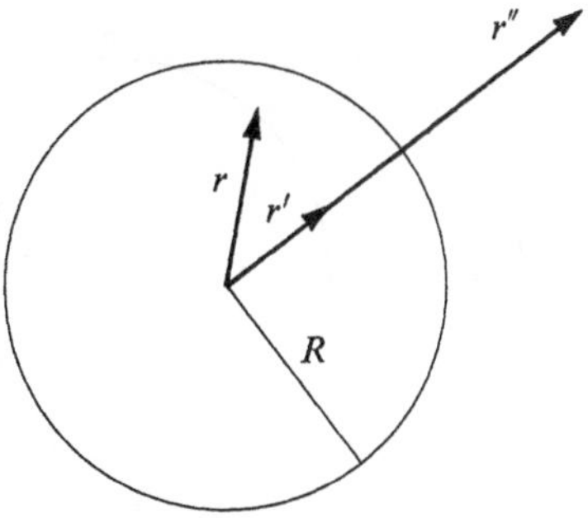

Fig. 53. $\vec{r}$ defines an observation point. The collinear vectors $\vec{r}'$ and $\vec{r}''$ that satisfy the relation 30.4 determine two points that are inverses of each other with respect to the sphere.

and there remains to find the nonsingular part G_0 of the Green's function which is a solution of the homogeneous Laplace equation and such that the total Green's function

$$G = G_s + G_0$$

vanishes on S_R.

From the theorem of Sec. 29.2 (interior Dirichlet problem), a function G_0, which is harmonic in V_R and which takes on prescribed values on the surface of the sphere, is unique.

Notice now that if $\vec{r}''$ locates any point outside V_R (Fig. 53) and $\vec{r}$ a point within V_R, the function

$$\frac{k}{|\vec{r} - \vec{r}''|} \tag{30.2}$$

where k is a constant, will be harmonic in V_R. The reason is that 30.2 has the same form as G_s but r is never equal to $\vec{r}''$. We take G_0 in the form 30.2 with $\vec{r}''$ along the same radius vector as $\vec{r}'$. The constants $|\vec{r}''|$ and k must be determined so that

$$G = \frac{-1}{4\pi}\left\{\frac{1}{|\vec{r} - \vec{r}'|} + \frac{k}{|\vec{r} - \vec{r}''|}\right\} \tag{30.3}$$

vanishes at $|\vec{r}| = R$. Since

$$\vec{r}'' = \frac{|\vec{r}''|}{|\vec{r}'|}\vec{r}'$$

Eq. 30.3 can be written as

$$G = \frac{-1}{4\pi}\left\{\frac{1}{|\vec{r} - \vec{r}'|} + \frac{k}{\left|\vec{r} - \dfrac{|\vec{r}''|}{|\vec{r}'|}\vec{r}'\right|}\right\}$$

The preceding expression vanishes at $|\vec{r}| = R$ if

$$|\vec{r}''| = \frac{R^2}{|\vec{r}'|} \tag{30.4}$$

and

$$k = -\frac{|\vec{r}''|}{R} \tag{30.5}$$

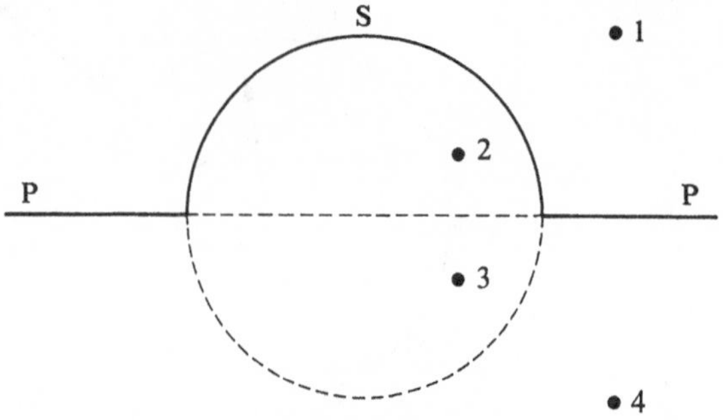

Fig. 54. The charges 1, 4 and 2, and 3 are symmetrical with respect to the plane P. Charges 1, 2 and 3, and 4 are inverse images of each other with respect to the sphere and satisfy relations of the type of Eq. 30.4. Thus, the hemisphere and plane are at zero potential, and the effect of the boundary S + P is replaced by the effect of three point charges, 2, 3, and 4.

Two points that lie on the same radius vector and which satisfy Eq. 30.4 are said to be **inverses** of each other with respect to the sphere. We have, therefore, the unique solution

$$G(\vec{r},\vec{r}'') = G_s + G_0 = \frac{-1}{4\pi}\left\{\frac{1}{|\vec{r}-\vec{r}'|} - \frac{R}{|\vec{r}'|}\frac{1}{\left|\vec{r} - \dfrac{R^2}{|\vec{r}'|^2}\vec{r}'\right|}\right\} \qquad |\vec{r}'| \le R \qquad (30.6)$$

It is now understandable why the method used in this example is called the method of images. We noted that $G_s(\vec{r},\vec{r}')$ has the physical significance of the potential due to a point charge located at $\vec{r} = \vec{r}'$. From the physical point of view, our problem is to find the potential due to a point charge enclosed in a conducting sphere. It is seen that the effect of the conductor is the same as that of an "image" point charge located as in Fig. 53.

We leave as an exercise to the reader the verification that the field due to a point charge above a conducting hemisphere that rests on a conducting plane is the same as the field due to the four point charges located as in Fig. 54.

31 · THE METHOD OF SEPARATION OF VARIABLES

31.1 Introduction

Our relative familiarity with ordinary differential equations leads us quite naturally to inquire into the possibility of reducing a partial differential equation to a system of ordinary differential equations. This is feasible only in a number of very limited cases, but it is very fortunate that included among these cases are some of the most important equations of mathematical physics. The method (called the method of **separation of variables**) is easily explained. Suppose that L_{x_i} is a formal partial differential operator that depends on N variables $x_1,x_2, \cdots ,x_N$. One seeks a solution of the equation

$$L_{x_i}u(x_1,x_2, \cdots ,x_N) = 0 \qquad (31.1)$$

in the product form

$$u(x_1,x_2, \cdots ,x_N) = R(x_1)S(x_2) \cdots V(x_N) \qquad (31.2)$$

where each factor depends on one variable only. If the method succeeds, Eq. 31.2 will transform Eq. 31.1 into a system of N ordinary differential equations, each depending on a single variable x_i; in that case, the operator is said to be **separable** with respect to these variables. An operator will usually be separable only in special coordinate systems, and then one takes advantage of the circumstance and solves the equation in one of these coordinate systems. Among the possible coordinate systems in which the operator is separable, one chooses the one that is best adapted to the geometry of the problem considered.

The solutions of the separate ordinary differential equations will not immediately yield a general solution of the original partial differential equation. However, one can construct from them the required solution satisfying the boundary conditions.

The method should become clear as we apply it to the example given below. This particular example will give us the means not only of illustrating the method of separation of variables, but also of introducing an important set of functions, the spherical harmonics.

31.2 The Three-dimensional Laplace Equation in Spherical Coordinates

We solve Laplace's equation in spherical coordinates.

$$\begin{aligned} x &= r \sin\theta \cos\Phi \\ y &= r \sin\theta \sin\Phi \\ z &= r \cos\theta \end{aligned} \tag{31.3}$$

In these coordinates, the Laplace equation is

$$\vec{\nabla}^2 u = \frac{1}{r^2}\frac{\partial}{\partial r}\left[r^2 \frac{\partial u}{\partial r}\right] + \frac{1}{r^2 \sin\theta}\frac{\partial}{\partial\theta}\left[\sin\theta \frac{\partial u}{\partial\theta}\right] + \frac{1}{r^2 \sin^2\theta}\frac{\partial^2 u}{\partial\Phi^2} = 0 \tag{31.4}$$

We shall look for solutions of Eq. 31.4 in the product form

$$u(r,\theta,\phi) = R(r)P(\theta)T(\Phi) \tag{31.5}$$

and require that the solution be finite and single-valued within any sphere of finite radius.

Substituting Eq. 31.5 in Eq. 31.4 and multiplying through by $r^2[R(r)P(\theta)T(\Phi)]^{-1}$, we get

$$\frac{1}{P(\theta)\sin\theta}\frac{d}{d\theta}\left[\sin\theta\frac{dP}{d\theta}\right] + \frac{1}{\sin^2\theta}\frac{1}{T(\Phi)}\frac{d^2T}{d\Phi^2} = \frac{-1}{R(r)}\frac{d}{dr}\left[r^2\frac{dR}{dr}\right] \tag{31.6}$$

The RHS of Eq. 31.6 is a function of r only, whereas the LHS is a function of θ and Φ only. In order for this to be possible, both sides must be equal to a constant, which we denote by λ. Hence, we have the two equations

$$\frac{1}{P\sin\theta}\frac{d}{d\theta}\left[\sin\theta\frac{P}{d\theta}\right] + \frac{1}{\sin^2\theta}\frac{1}{T}\frac{d^2T}{d\Phi^2} + \lambda = 0 \tag{31.7}$$

and

$$\frac{d}{dr}\left[r^2\frac{dR}{dr}\right] - \lambda R = 0 \tag{31.8}$$

Multiplying through the first equation by $\sin^2\theta$, we have

$$\frac{\sin\theta}{P}\frac{d}{d\theta}\left[\sin\theta\frac{dP}{d\theta}\right]+\lambda\sin^2\theta=\frac{-1}{T}\frac{d^2T}{d\Phi^2} \tag{31.9}$$

Since the RHS depends only on Φ, while the LHS is a function of θ only, both sides must again be equal to a constant, which for convenience we set equal to m^2. Thus, Eq. 31.9 splits into the two equations

$$\frac{d^2T}{d\Phi^2}+m^2T=0 \tag{31.10}$$

and

$$\frac{1}{\sin\theta}\frac{d}{d\theta}\left[\sin\theta\frac{dP}{d\theta}\right]+\left[\lambda-\frac{m^2}{\sin^2\theta}\right]P=0 \tag{31.11}$$

It is convenient to write Eq. 31.11 in terms of the variable

$$x=\cos\theta$$

An elementary calculation yields

$$(1-x^2)\frac{d^2P}{dx^2}-2x\frac{dP}{dx}+\left[\lambda-\frac{m^2}{1-x^2}\right]P=0 \tag{31.12}$$

Thus, the method of separation of variables has succeeded, since Eqs. 31.8, 31.10, and 31.12 are three ordinary differential equations which replace Laplace's equation in spherical coordinates. These equations contain the "separation constants" m^2 and λ, which, owing to the requirement that the solution of Laplace's equation be finite and single-valued, will not be altogether arbitrary.

Before pursuing these questions, however, we need to sidetrack from our immediate problem and study in detail a particular case of Eq. 31.12, namely, when $\lambda=l(l+1)$ with $l=0,1,2,\cdots$ and when m is an integer such that $|m|\le l$.

31.3 Associated Legendre Functions and Spherical Harmonics

We consider the equation

$$(1-x^2)\frac{d^2P}{dx^2}-2x\frac{dP}{dx}+\left[l(l+1)-\frac{m^2}{1-x^2}\right]P=0 \tag{31.13}$$

where m is an integer, l a non-negative integer, and $|m|\le l$. Since Eq. 31.13 does not depend on the sign of m, it will be sufficient to assume that $0\le m\le l$.

The substitution

$$P(x)=\text{const}(1-x^2)^{m/2}C(x) \tag{31.14}$$

reduces Eq. 31.13 to the form

$$(1-x^2)\frac{d^2C}{dx^2}-2(m+1)x\frac{dC}{dx}+(l-m)(l+m+1)C=0 \qquad 0\le m\le l \tag{31.15}$$

For $m=0$, this is exactly the Legendre equation (see Eq. 18.7), one of whose solutions is the Legendre polynomial $P_l(x)$

$$(1-x^2)\frac{d^2P_l}{dx^2}-2x\frac{dP_l}{dx}+l(l+1)P_l=0$$

Differentiating the Legendre equation m times, one obtains*

$$(1-x^2)\frac{d^2}{dx^2}\left[\frac{d^mP_l}{dx^m}\right]-2(m+1)x\frac{d}{dx}\left[\frac{d^mP_l}{dx^m}\right]+(l-m)(l+m+1)\frac{d^mP_l}{dx^m}=0$$

which has just the form of Eq. 31.15. Hence

$$C(x)=\text{const}\,\frac{d^mP_l(x)}{dx^m} \tag{31.16}$$

is a solution of Eq. 31.15. Combining Eqs. 31.14 and 31.16 and using the conventional normalization, we define the **associated Legendre functions** $P_l^m(x)$, which are solutions of Eq. 31.13

$$P_l^m(x)=(-1)^m(1-x^2)^{m/2}\frac{d^m}{dx^m}P_l(x) \qquad (0\le m\le l) \tag{31.17}$$

It is obvious that the associated Legendre functions reduce to the Legendre polynomials for $m=0$

$$P_l^0(x)=P_l(x)$$

Using the Rodriguez formula for $P_l(x)$ (Chapter III, Sec. 10.6), Eq. 31.17 can also be expressed as

$$P_l^m(x)=\frac{(-1)^{m+l}}{2^l l!}(1-x^2)^{m/2}\frac{d^{l+m}}{dx^{l+m}}(1-x^2)^l \tag{31.18}$$

Comparing Eq. 31.15 to the Gegenbauer equation (Chapter III, Sec. 10.6), it is seen that the polynomials d^mP_l/dx^m must be proportional to the Gegenbauer polynomials. One has, in fact, the relation

$$\frac{d^mP_l(x)}{dx^m}=\frac{\Gamma(2m+1)}{2^m\Gamma(m+1)}C_{l-m}^{m+1/2}(x) \qquad 0\le m\le l \tag{31.19}$$

Using the Rodriguez formulae for Gegenbauer polynomials (Chapter III, Sec. 10.6) and inserting Eq. 31.19 into Eq. 31.17, we find

$$P_l^m(x)=\frac{(-1)^l}{2^l l!}\frac{(l+m)!}{(l-m)!}(1-x^2)^{-m/2}\frac{d^{l-m}}{dx^{l-m}}(1-x^2)^l \tag{31.20}$$

This formula is meaningful also for negative m, provided $|m|<l$, and permits an extension of the definition of $P_l^m(x)$ to negative values of m. Comparing Eqs. 31.17 and 31.20, we see that

$$P_l^{-m}(x)=(-1)^m\frac{(l-m)!}{(l+m)!}P_l^m(x) \qquad -l\le m\le l \tag{31.21}$$

For a given value of m, the functions $P_l^m(x)$ form an orthogonal set with respect to the index l on $[-1,1]$

$$\int_{-1}^{1}dx\,P_l^m(x)P_{l'}^m(x)=0 \qquad \text{for } l\ne l' \tag{31.22}$$

---

* This can be most easily achieved by using the well-known Laplace formula

$$\frac{d^m}{dx^m}(f\cdot g)=\sum_{k=0}^{m}\binom{m}{k}\frac{d^k}{dx^k}f\cdot\frac{d^{m-k}}{dx^{m-k}}g$$

To see this, we use Eq. 31.17 and integrate by parts m times (because of Eq. 31.21 we need to consider only positive values of m)

$$\int_{-1}^{1} dx\, P_l^m(x) P_{l'}^m(x) = \int_{-1}^{1} dx (1 - x^2)^m \left[\frac{d^m P_l}{dx^m}\right]\left[\frac{d^m P_{l'}}{dx^m}\right]$$

$$= \int_{-1}^{1} dx \frac{d^m}{dx^m}\left\{(1 - x^2)\frac{d^m P_l}{dx^m}\right\} P_{l'}(x)$$

Without any loss of generality, we can assume that $l' > l$, and since the derivative in the integrand is a polynomial of order $\leq l$, the orthogonality properties of the Legendre polynomials lead at once to Eq. 31.22.

The functions $P_l^m(x)$ are not normalized to unity. One has

$$\int_{-1}^{1} dx[P_l^m(x)]^2 = \frac{2}{2l+1}\frac{(l+m)!}{(l-m)!} \tag{31.23}$$

We first prove Eq. 31.23 for $m > 0$. Using Eq. 31.18 we have

$$\int_{-1}^{1} dx\,[P_l^m(x)]^2 = \int_{-1}^{1} dx \frac{(1 - x^2)^m}{(2^l l!)^2}\left[\frac{d^{m+l}}{dx^{m+l}}(1 - x^2)^l\right]\left[\frac{d^{m+l}}{dx^{m+l}}(1 - x^2)^l\right]$$

Integrating by parts $m + l$ times, we find, since the integrated terms vanish

$$\int_{-1}^{1} dx\,[P_l^m(x)]^2 = \frac{(-1)^{m+l}}{(2^l l!)^2}\int_{-1}^{1} dx\,(1 - x^2)^l\left[\frac{d^{l+m}}{dx^{l+m}}\left\{(1 - x^2)^m \frac{d^{l+m}}{dx^{l+m}}(1 - x^2)^l\right\}\right]$$

But since $m \leq l$

$$\frac{d^{l+m}}{dx^{l+m}}\left\{(1 - x^2)^m \frac{d^{l+m}}{dx^{l+m}}(1 - x^2)^l\right\} = (-1)^{m+l}\frac{d^{l+m}}{dx^{l+m}}\left\{x^{2m}\frac{d^{l+m}}{dx^{l+m}}x^{2l}\right\}$$

$$= (-1)^{m+l}(2l)!\frac{(l+m)!}{(l-m)!}$$

Hence

$$\int_{-1}^{1} dx\,[P_l^m(x)]^2 = \frac{(l+m)!}{(l-m)!}\frac{(2l)!}{(2^l l!)^2}\int_{-1}^{1} dx\,(1 - x)^l(1 + x)^l$$

After l more partial integrations and a final simple integration, we find

$$\int_{-1}^{1} dx\,[P_l^m(x)]^2 = \frac{2}{2l+1}\frac{(l+m)!}{(l-m)!}$$

Because of Eq. 31.21, this formula also holds for $m < 0$.

We now introduce functions of the two angular variables θ,Φ, called **spherical harmonics** and defined for $-l \leq m \leq l$ as

$$Y_l^m(\theta,\Phi) = (-1)^m\left[\frac{2l+1}{4\pi}\frac{(l-m)!}{(l+m)!}\right]^{1/2} P_l^m(\cos\theta)e^{im\Phi} \qquad \begin{matrix} 0 \leq \theta \leq \pi \\ 0 \leq \Phi \leq 2\pi \end{matrix} \tag{31.24}$$

Using Eqs. 31.22 and 31.23, the reader can easily verify that the spherical harmonics form an orthonormal set of functions

$$\int_0^{2\pi} d\phi \int_{-1}^{1} d(\cos\theta)\,\overline{Y}_{l'}^{m'}(\theta,\Phi)Y_l^m(\theta,\Phi) = \delta_{ll'}\,\delta_{mm'} \tag{31.25}$$

The first few spherical harmonics are

$$Y_0^0(\theta,\Phi) = \frac{1}{2\sqrt{\pi}}$$

$$Y_1^0(\theta,\Phi) = \frac{1}{2}\sqrt{\frac{3}{\pi}}\cos\theta$$

$$Y_1^{\pm 1}(\theta,\Phi) = \mp\frac{1}{2}\sqrt{\frac{3}{2\pi}}\sin\theta e^{\pm i\Phi}$$

$$Y_2^0(\theta,\Phi) = \frac{1}{4}\sqrt{\frac{5}{\pi}}(2\cos^2\theta - \sin^2\theta)$$

$$Y_2^{\pm 1}(\theta,\Phi) = \mp\frac{1}{2}\sqrt{\frac{15}{2\pi}}\cos\theta\sin\theta e^{\pm i\Phi}$$

$$Y_2^{\pm 2}(\theta,\Phi) = \frac{1}{4}\sqrt{\frac{15}{2\pi}}\sin^2\theta e^{\pm 2i\Phi}$$

The importance of spherical harmonics stems from the fact that under very general conditions, a function $f(\theta,\Phi)$ of the two angular variables θ,Φ can be expanded in terms of these functions

$$f(\theta,\Phi) = \sum_{l=0}^{\infty}\sum_{m=-l}^{l} f_{lm}Y_l^m(\theta,\Phi) \tag{31.26}$$

where

$$f_{lm} = \int_0^{2\pi} d\Phi \int_{-1}^{1} d(\cos\theta)\overline{Y}_l^m(\theta,\Phi)f(\theta,\Phi) \tag{31.27}$$

More descriptively, one says that the spherical harmonics form a complete orthonormal set of functions on a sphere, since the two angular variables θ,Φ define a point on a sphere.

To demonstrate the completeness of the set of spherical harmonics, we shall use arguments similar to those employed to prove the completeness of the set of trigonometrical functions (Chapter III, Sec. 11.1).

Consider a function

$$f(x,y,z) = rf(\theta,\Phi)$$

and assume that $f(\theta,\Phi)$ is a continuous function of its arguments. Then $f(x,y,z)$ is a continuous function for

$$-1 \leq x,y,z \leq 1$$

and according to the generalized version of the Weierstrass theorem (Chapter III, Sec. 9), it can be uniformly approximated in this domain by functions of the type

$$f_n(x,y,z) = \sum_{0\leq m_i,m_j,m_k\leq n} a_{ijk}^{(n)}x^{m_i}y^{m_j}z^{m_k} \tag{31.28}$$

We now relate the monomials $x^{m_i}y^{m_j}z^{m_k}$ to the spherical harmonics.

Given a monomial $x^{m_i}y^{m_j}z^{m_k}$, one calls the number $l = m_i + m_j + m_k$ the order of the monomial. We impose the constraint

$$x^2 + y^2 + z^2 = 1 \tag{31.29}$$

and ask: How many monomials of order l exist, which are not only linearly independent among themselves, but also linearly independent of all monomials of order less than l? We shall call such monomials, **irreducible monomials** of order l.

The constraint 31.29 allows us to express each monomial of order l that contains powers of x higher than the first, as a linear combination of monomials of the type

$$y^{n_j}z^{n_k} \qquad n_j + n_k = l \tag{31.30}$$

or of the type

$$x^{n_j}yz^{n_k} \qquad 1 + n_j + n_k = l \tag{31.31}$$

and of monomials of order less than l.

Hence, we can reformulate our question by asking: How many monomials are there of the type 31.30 or 31.31? The answer is immediate: There are $l+1$ monomials of the type 31.30 and l monomials of the type 31.31. Thus, in total, there are at most $(2l+1)$ irreducible monomials of order l.

EXAMPLE

There are ten monomials of order 3: $x^3, y^3, z^3, x^2y, x^2z, y^2z, y^2x, z^2y, z^2x, xyz$. Using Eq. 31.29, we have, for example

$$x^3 = x - xy^2 - xz^2$$

$$x^2y = y - y^3 - yz^2$$

$$x^2z = z - zy^2 - z^3$$

Thus, among the ten monomials, three are not irreducible, and there remains $10 - 3 = (2 \times 3 + 1)$ monomials that may be irreducible (in fact, they are!).

We shall show that there are exactly $(2l+1)$ irreducible monomials of order l. We notice that with the constraint 31.29 the spherical harmonics are linear combinations of monomials of order l. To see this, we recall that from Eqs. 31.24 and 31.21, one has

$$Y_l^m(\theta,\Phi) = \text{const } P_l^{|m|}(\cos\theta)\, e^{im\Phi}$$

Using Eq. 31.18, we get

$$\begin{aligned} r^l Y_l^{\pm|m|}(\theta,\Phi) &= \text{const } r^l \sin^{|m|}\theta \cos^{l-|m|}\theta e^{\pm i|m|\Phi} \\ &= \text{const } r^l \sin^{|m|}\theta \cos^{l-|m|}\theta \sum_{k=0}^{|m|} \binom{|m|}{k} (\pm i)^k \sin^k\Phi \cos^{|m|-k}\Phi \\ &= \text{const} \sum_{k=0}^{|m|} \binom{|m|}{k} (\pm i)^k x^{|m|-k} y^k z^{l-|m|} \end{aligned}$$

which is just a linear combination of monomials of order l. But because of the orthogonality properties of spherical harmonics (see Eq. 31.25), there are, for a given l, $(2l+1)$ functions $r^l Y_l^m(\theta,\Phi)$ $(m = 0, \pm1, \cdots, \pm l)$, which are linearly independent among themselves and also linearly independent of all the functions $r^l Y_l^m(\theta,\Phi)$ $(m = 0, \pm1, \cdots, \pm k)$ with $k \neq l$.

Thus, in Eq. 31.28, one can express the monomials in terms of spherical harmonics, and after a rearrangement of terms, one obtains $f_n(x,y,z)$ in the form

$$f_n(x,y,z) = \sum_{l=0}^{n} \sum_{m=-l}^{+l} a_{lm} r^l Y_l^m(\theta,\Phi)$$

On the unit sphere $r = 1, f(x,y,z) = f(\theta,\Phi)$ can be uniformly approximated by linear combinations of spherical harmonics, and this implies that they form a complete set of functions in the sense that any function $f(\theta,\Phi)$ with the property

$$\int_0^{2\pi} d\Phi \int_{-1}^{1} d(\cos\theta)|f(\theta,\Phi)|^2 < \infty$$

can be expanded in terms of them. As usual, the convergence of the series will be a convergence in the mean unless additional conditions are satisfied by $f(\theta,\Phi)$.

31.4 The General Factorized Solution of the Laplace Equation in Spherical Coordinates

After this lengthy digression, we return to the problem of finding a factorized, single-valued, and finite solution of Eq. 31.4.

The solutions of Eq. 31.10 are

$$T = \text{const } e^{\pm im\Phi}$$

The requirement that T be single-valued implies that m must be restricted to integer values.

We now show that the condition for the solution $P(x)$ of Eq. 31.12 to be finite in the closed interval $[-1, 1]$ is that $\lambda = l(l+1)$, where l is a non-negative integer. Equation 31.12 has the form

$$L_x P + \lambda P = 0 \tag{31.32}$$

where

$$L_x = (1 - x^2)\frac{d^2}{dx^2} - 2x\frac{d}{dx} - \frac{m^2}{1 - x^2} \tag{31.33}$$

is a self-adjoint formal differential operator (see Sec. 6). Denoting by $P(x,\lambda,m)$ the solution of Eq. 31.32 corresponding to a given value of λ and m, and using Eq. 6.5, we have

$$\begin{aligned}\int_{-1}^{1} dx\,\{\bar{P}(x,\lambda,m)[L_x P(x,\lambda',m] - P(x,\lambda',m)\overline{[L_x P(x,\lambda,m)]}\}\\ = \left\{(1 - x^2)\left[\bar{P}(x,\lambda,m)\frac{d}{dx}P(x,\lambda',m) - P(x,\lambda',m)\frac{d}{dx}\bar{P}(x,\lambda,m)\right]\right\}\Bigg|_{x=1}^{x=1}\\ = 0\end{aligned} \tag{31.34}$$

With the aid of Eq. 31.32, Eq. 31.34 becomes

$$(\bar{\lambda} - \lambda')\int_{-1}^{1} dx\,\bar{P}(x,\lambda,m)P(x,\lambda',m) = 0 \tag{31.35}$$

When $\lambda = \lambda'$, the integrand is positive definite, and therefore λ must be real. On the other hand, when $\lambda \neq \lambda'$, the integral must vanish, which means that the solutions of Eq. 31.12 corresponding to different values of λ are orthogonal. If $P(\cos\theta,\lambda,m)$ were continuous in $[-1, 1]$ and if $\lambda \neq l(l+1)$, l being a non-negative integer, the function $P(\cos\theta,\lambda,m)e^{im\Phi}$ would be expandable in the set of spherical harmonics

$$P(\cos\theta,\lambda,m)e^{im\Phi} = \sum_{l,m'} a_{lm'} Y_l^{m'}(\theta,\Phi)$$

where

$$a_{lm'} = \int_0^{2\pi} d\Phi \int_{-1}^{1} d(\cos\theta)\bar{Y}_l^{m'}(\theta,\Phi)P(\cos\theta,\lambda,m)e^{im\Phi}$$

$$= \text{const}\int_0^{2\pi} d\Phi \int_{-1}^{1} d(\cos\theta)P_l^{m'}(\cos\theta)P(\cos\theta,\lambda,m)e^{i(m-m')\Phi}$$

This integral, however, vanishes when $m \neq m'$, on account of the Φ integration, and when $m = m'$, it vanishes because of the orthogonality (Eq. 31.35) of the solutions of Eq. 31.32. Thus, $P(\cos\theta,\lambda,m)$ would be identically zero.

On the other hand, if $P(\cos\theta,\lambda,m)$ were the second linearly independent solution of Eq. 31.32 corresponding to $\lambda = l(l+1)$, and if it were finite, it would be proportional to P_l^m, since it is orthogonal to all other functions P_k^m with $k \neq l$. This would contradict the fact that it is a linearly independent solution of Eq. 31.32. Hence, the second solution of Eq. 31.32 cannot be finite everywhere in $[-1, 1]$ (it is, in fact, infinite at $x = \pm 1$).

Summarizing: The only continuous and finite solutions of Eq. 31.32 in $[-1, 1]$ are the associated Legendre functions.

Putting $\lambda = l(l+1)$ in Eq. 30.8, we obtain the equation

$$\frac{d}{dr}\left[r^2\frac{dR}{dr}\right] - l(l+1)R = 0$$

whose independent solutions are

$$R = r^l \qquad \text{and} \qquad R = r^{-l-1} \tag{31.36}$$

Only the first solution is finite at the origin, and therefore the general factorized solution of the Laplace equation, which is finite and single-valued within any sphere of finite radius, is

$$u(r,\theta,\Phi) = r^l[P_l^{|m|}(\theta)][be^{im\Phi} + ce^{-im\Phi}] \tag{31.37}$$

which is just a linear combination of the solutions

$$r^l Y_l^{\pm m}(\theta,\Phi) \tag{31.38}$$

If we had looked for a solution of Laplace's equation **outside** a sphere, the boundedness condition for $r \to \infty$ would have led to the other choice for $R(r)$ in Eq. 31.36 and thus to a factorized solution of the form

$$r^{-l-1}Y_l^{\pm m}(\theta,\Phi)$$

31.5 General Solution of Laplace's Equation with Dirichlet Boundary Conditions on a Sphere

Consider now the problem of finding the solution of the Laplace equation, which is finite and single-valued throughout the interior of a sphere of radius ρ with prescribed Dirichlet conditions on the sphere

$$u(\rho,\theta,\Phi) = h(\theta,\Phi) \tag{31.39}$$

where $h(\theta,\Phi)$ is a given function of the angular variables.

Obviously, the particular solution $r^l Y_l^m(\theta,\Phi)$ found in the previous section already has a well-defined angular dependence, and therefore cannot satisfy arbitrarily prescribed boundary conditions. However, and this is a crucial point, the general solution can be found with the aid of the factorized solution, for owing to the linearity of the equation, any linear combination of functions satisfying the equation also satisfies the equation.

We look for solutions in the form

$$u(r,\theta,\Phi) = \sum_{l=0}^{\infty} \sum_{m=-l}^{l} a_{lm} \left(\frac{r}{\rho}\right)^l Y_l^m(\theta,\Phi) \tag{31.40}$$

The boundary condition 31.39 requires that

$$h(\theta,\Phi) = \sum_{l=0}^{\infty} \sum_{m=-l}^{l} a_{lm} Y_l^m(\theta,\Phi) \tag{31.41}$$

Therefore, we see that the coefficients a_{lm} in the solution 31.40 are just the coefficients of the expansion of $h(\theta,\Phi)$ in spherical harmonics. If the boundary conditions are smooth enough, such an expansion will exist and the coefficients will be given by

$$a_{lm} = \int_0^{2\pi} d\Phi \int_{-1}^{1} d(\cos\theta) \bar{Y}_l^m(\theta,\Phi) h(\theta,\Phi) \tag{31.42}$$

It is evident that the general solution outside a sphere of radius ρ, which vanishes as $r \to \infty$, is

$$u(r,\theta,\Phi) = \sum_{l=0}^{\infty} \sum_{m=-l}^{l} a_{lm} \left(\frac{\rho}{r}\right)^{l+1} Y_l^m(\theta,\Phi)$$

with the coefficients a_{lm} again given by Eq. 31.42.

BIBLIOGRAPHY

General References

1. R. Courant and D. Hilbert, *Methods of Mathematical Physics*, Vols. I–II, Interscience Publishers, New York, 1953.
2. A. Erdélyi *et al.*, *Higher Transcendental Functions*, Vols. I–III, McGraw-Hill Book Co., New York, 1953.
3. E. Goursat, *A Course of Mathematical Analysis*, Vols. I–III, (trans. by E. R. Hedrick), Dover Publications, Inc., New York, 1959.
4. P. M. Morse and H. Feshbach, *Methods of Theoretical Physics*, Vols. I–II, McGraw-Hill Book Co., New York, 1953.
5. V. I. Smirnov, *A Course of Higher Mathematics*, Vols. I–V, Pergamon Press, New York, 1964.
6. E. T. Whittaker and G. N. Watson, *A Course of Modern Analysis, 4th ed.*, Cambridge University Press, New York, 1958.

References for Chapter I

1. S. Bochner and W. T. Martin, *Several Complex Variables*, Princeton University Press, Princeton, N.J., 1948.
2. A. Erdélyi, *Asymptotic Expansions*, Dover Publications, New York, 1965.
3. E. Kamke, *Theory of Sets*, Dover Publications, New York, 1950.
4. K. Knopp, *Theory of Functions*, Vols. I–II, Dover Publications, New York, 1947.
5. F. Leja, *Analytic and Harmonic Functions* (in Polish), Monografie Matematyczne, Warsaw, 1952.
6. G. Sansone and J. Gerretsen, *Lectures on the Theory of Functions of a Complex Variable*, Vol. I, P. Noordhoff, Groningen, 1960.
7. E. C. Titchmarsh, *The Theory of Functions*, Oxford University Press, New York, 1964.

References for Chapter II

1. G. Birkhoff and S. MacLane, *A Survey of Modern Algebra*, Macmillan, New York, 1953.
2. P. R. Halmos, *Finite-Dimensional Vector Spaces*, D. Van Nostrand Co., Princeton, N.J., 1958.
3. G. Julia, *Introduction Mathématique aux Théories Quantiques* (in French), Vol. I, Gauthier-Villars, Paris, 1955.
4. O. Schreier and E. Sperner, *Modern Algebra and Matrix Theory*, Chelsea Publishing Co., New York, 1959.

References for Chapter III

1. P. R. Halmos, *Introduction to Hilbert Space and the Theory of Spectral Multiplicity*, Chelsea Publishing Co., New York, 1957.
2. M. J. Lighthill, *Introduction to Fourier Analysis and Generalized Functions*, Cambridge University Press, New York, 1958.

3. G. Julia, *Introduction Mathématique aux Théories Quantiques* (in French), Vol. II, Gauthier-Villars, Paris, 1955.
4. F. Riesz and B. Sz.-Nagy, *Functional Analysis*, Frederick Ungar Publishing Co., New York, 1955.
5. G. Sansone, *Orthogonal Functions*, Interscience Publishers, New York, 1959.
6. G. E. Shilov, *Introduction to the Theory of Linear Spaces* (in Russian), G.I.T.–T.L., Moscow, 1956.
7. G. Szegö, *Orthogonal Polynomials*, American Mathematical Society, New York, 1959.
8. E. C. Titchmarsh, *The Theory of Functions*, Oxford University Press, New York, 1964.
9. E. C. Titchmarsh, *Introduction to the Theory of Fourier Integrals*, Oxford University Press, New York, 1937.
10. F. G. Tricomi, *Vorlesungen über Orthogonalreihen*, Springer, Berlin, 1955.

References for Chapter IV

1. B. Friedman, *Principles and Techniques of Applied Mathematics*, John Wiley & Sons, New York, 1957.
2. S. Goldberg, *Introduction to Difference Equations*, John Wiley & Sons, New York, 1958.
3. E. L. Ince, *Ordinary Differential Equations*, Dover Publications, Inc., New York, 1926.
4. C. Lanczos, *Linear Differential Operators*, D. Van Nostrand Co., Princeton, N.J., 1961.
5. T. M. MacRobert, *Spherical Harmonics*, Dover Publications, New York, 1948.
6. I. G. Petrovsky, *Lectures on Partial Differential Equations*, Interscience Publishers, Inc., New York, 1957.
7. A. Sommerfeld, *Partial Differential Equations in Physics*, Academic Press, Inc., New York, 1949.
8. A. N. Tikhonov and A. A. Samarski, *Partial Differential Equations of Mathematical Physics*, Pergamon Press, New York, 1963.
9. G. N. Watson, *A Treatise on the Theory of Bessel Functions*, Cambridge University Press, New York, 1952.

INDEX

A CATALOG OF SELECTED

DOVER BOOKS

IN SCIENCE AND MATHEMATICS

Astronomy

BURNHAM'S CELESTIAL HANDBOOK, Robert Burnham, Jr. Thorough guide to the stars beyond our solar system. Exhaustive treatment. Alphabetical by constellation: Andromeda to Cetus in Vol. 1; Chamaeleon to Orion in Vol. 2; and Pavo to Vulpecula in Vol. 3. Hundreds of illustrations. Index in Vol. 3. 2,000pp. 6⅛ x 9¼.
Vol. I: 0-486-23567-X
Vol. II: 0-486-23568-8
Vol. III: 0-486-23673-0

EXPLORING THE MOON THROUGH BINOCULARS AND SMALL TELESCOPES, Ernest H. Cherrington, Jr. Informative, profusely illustrated guide to locating and identifying craters, rills, seas, mountains, other lunar features. Newly revised and updated with special section of new photos. Over 100 photos and diagrams. 240pp. 8¼ x 11. 0-486-24491-1

THE EXTRATERRESTRIAL LIFE DEBATE, 1750–1900, Michael J. Crowe. First detailed, scholarly study in English of the many ideas that developed from 1750 to 1900 regarding the existence of intelligent extraterrestrial life. Examines ideas of Kant, Herschel, Voltaire, Percival Lowell, many other scientists and thinkers. 16 illustrations. 704pp. 5⅜ x 8½. 0-486-40675-X

THEORIES OF THE WORLD FROM ANTIQUITY TO THE COPERNICAN REVOLUTION, Michael J. Crowe. Newly revised edition of an accessible, enlightening book recreates the change from an earth-centered to a sun-centered conception of the solar system. 242pp. 5⅜ x 8½. 0-486-41444-2

A HISTORY OF ASTRONOMY, A. Pannekoek. Well-balanced, carefully reasoned study covers such topics as Ptolemaic theory, work of Copernicus, Kepler, Newton, Eddington's work on stars, much more. Illustrated. References. 521pp. 5⅜ x 8½.
0-486-65994-1

A COMPLETE MANUAL OF AMATEUR ASTRONOMY: TOOLS AND TECHNIQUES FOR ASTRONOMICAL OBSERVATIONS, P. Clay Sherrod with Thomas L. Koed. Concise, highly readable book discusses: selecting, setting up and maintaining a telescope; amateur studies of the sun; lunar topography and occultations; observations of Mars, Jupiter, Saturn, the minor planets and the stars; an introduction to photoelectric photometry; more. 1981 ed. 124 figures. 25 halftones. 37 tables. 335pp. 6½ x 9¼. 0-486-40675-X

AMATEUR ASTRONOMER'S HANDBOOK, J. B. Sidgwick. Timeless, comprehensive coverage of telescopes, mirrors, lenses, mountings, telescope drives, micrometers, spectroscopes, more. 189 illustrations. 576pp. 5⅝ x 8¼. (Available in U.S. only.)
0-486-24034-7

STARS AND RELATIVITY, Ya. B. Zel'dovich and I. D. Novikov. Vol. 1 of *Relativistic Astrophysics* by famed Russian scientists. General relativity, properties of matter under astrophysical conditions, stars, and stellar systems. Deep physical insights, clear presentation. 1971 edition. References. 544pp. 5⅝ x 8¼. 0-486-69424-0

Chemistry

THE SCEPTICAL CHYMIST: THE CLASSIC 1661 TEXT, Robert Boyle. Boyle defines the term "element," asserting that all natural phenomena can be explained by the motion and organization of primary particles. 1911 ed. viii+232pp. 5⅜ x 8½. 0-486-42825-7

RADIOACTIVE SUBSTANCES, Marie Curie. Here is the celebrated scientist's doctoral thesis, the prelude to her receipt of the 1903 Nobel Prize. Curie discusses establishing atomic character of radioactivity found in compounds of uranium and thorium; extraction from pitchblende of polonium and radium; isolation of pure radium chloride; determination of atomic weight of radium; plus electric, photographic, luminous, heat, color effects of radioactivity. ii+94pp. 5⅜ x 8½. 0-486-42550-9

CHEMICAL MAGIC, Leonard A. Ford. Second Edition, Revised by E. Winston Grundmeier. Over 100 unusual stunts demonstrating cold fire, dust explosions, much more. Text explains scientific principles and stresses safety precautions. 128pp. 5⅜ x 8½. 0-486-67628-5

THE DEVELOPMENT OF MODERN CHEMISTRY, Aaron J. Ihde. Authoritative history of chemistry from ancient Greek theory to 20th-century innovation. Covers major chemists and their discoveries. 209 illustrations. 14 tables. Bibliographies. Indices. Appendices. 851pp. 5⅜ x 8½. 0-486-64235-6

CATALYSIS IN CHEMISTRY AND ENZYMOLOGY, William P. Jencks. Exceptionally clear coverage of mechanisms for catalysis, forces in aqueous solution, carbonyl- and acyl-group reactions, practical kinetics, more. 864pp. 5⅜ x 8½. 0-486-65460-5

ELEMENTS OF CHEMISTRY, Antoine Lavoisier. Monumental classic by founder of modern chemistry in remarkable reprint of rare 1790 Kerr translation. A must for every student of chemistry or the history of science. 539pp. 5⅜ x 8½. 0-486-64624-6

THE HISTORICAL BACKGROUND OF CHEMISTRY, Henry M. Leicester. Evolution of ideas, not individual biography. Concentrates on formulation of a coherent set of chemical laws. 260pp. 5⅜ x 8½. 0-486-61053-5

A SHORT HISTORY OF CHEMISTRY, J. R. Partington. Classic exposition explores origins of chemistry, alchemy, early medical chemistry, nature of atmosphere, theory of valency, laws and structure of atomic theory, much more. 428pp. 5⅜ x 8½. (Available in U.S. only.) 0-486-65977-1

GENERAL CHEMISTRY, Linus Pauling. Revised 3rd edition of classic first-year text by Nobel laureate. Atomic and molecular structure, quantum mechanics, statistical mechanics, thermodynamics correlated with descriptive chemistry. Problems. 992pp. 5⅜ x 8½. 0-486-65622-5

FROM ALCHEMY TO CHEMISTRY, John Read. Broad, humanistic treatment focuses on great figures of chemistry and ideas that revolutionized the science. 50 illustrations. 240pp. 5⅜ x 8½. 0-486-28690-8

Engineering

DE RE METALLICA, Georgius Agricola. The famous Hoover translation of greatest treatise on technological chemistry, engineering, geology, mining of early modern times (1556). All 289 original woodcuts. 638pp. 6¾ x 11. 0-486-60006-8

FUNDAMENTALS OF ASTRODYNAMICS, Roger Bate et al. Modern approach developed by U.S. Air Force Academy. Designed as a first course. Problems, exercises. Numerous illustrations. 455pp. 5⅜ x 8½. 0-486-60061-0

DYNAMICS OF FLUIDS IN POROUS MEDIA, Jacob Bear. For advanced students of ground water hydrology, soil mechanics and physics, drainage and irrigation engineering and more. 335 illustrations. Exercises, with answers. 784pp. 6⅛ x 9¼. 0-486-65675-6

THEORY OF VISCOELASTICITY (Second Edition), Richard M. Christensen. Complete consistent description of the linear theory of the viscoelastic behavior of materials. Problem-solving techniques discussed. 1982 edition. 29 figures. xiv+364pp. 6⅛ x 9¼. 0-486-42880-X

MECHANICS, J. P. Den Hartog. A classic introductory text or refresher. Hundreds of applications and design problems illuminate fundamentals of trusses, loaded beams and cables, etc. 334 answered problems. 462pp. 5⅜ x 8½. 0-486-60754-2

MECHANICAL VIBRATIONS, J. P. Den Hartog. Classic textbook offers lucid explanations and illustrative models, applying theories of vibrations to a variety of practical industrial engineering problems. Numerous figures. 233 problems, solutions. Appendix. Index. Preface. 436pp. 5⅜ x 8½. 0-486-64785-4

STRENGTH OF MATERIALS, J. P. Den Hartog. Full, clear treatment of basic material (tension, torsion, bending, etc.) plus advanced material on engineering methods, applications. 350 answered problems. 323pp. 5⅜ x 8½. 0-486-60755-0

A HISTORY OF MECHANICS, René Dugas. Monumental study of mechanical principles from antiquity to quantum mechanics. Contributions of ancient Greeks, Galileo, Leonardo, Kepler, Lagrange, many others. 671pp. 5⅜ x 8½. 0-486-65632-2

STABILITY THEORY AND ITS APPLICATIONS TO STRUCTURAL MECHANICS, Clive L. Dym. Self-contained text focuses on Koiter postbuckling analyses, with mathematical notions of stability of motion. Basing minimum energy principles for static stability upon dynamic concepts of stability of motion, it develops asymptotic buckling and postbuckling analyses from potential energy considerations, with applications to columns, plates, and arches. 1974 ed. 208pp. 5⅜ x 8½. 0-486-42541-X

METAL FATIGUE, N. E. Frost, K. J. Marsh, and L. P. Pook. Definitive, clearly written, and well-illustrated volume addresses all aspects of the subject, from the historical development of understanding metal fatigue to vital concepts of the cyclic stress that causes a crack to grow. Includes 7 appendixes. 544pp. 5⅜ x 8½. 0-486-40927-9

ROCKETS, Robert Goddard. Two of the most significant publications in the history of rocketry and jet propulsion: "A Method of Reaching Extreme Altitudes" (1919) and "Liquid Propellant Rocket Development" (1936). 128pp. 5⅜ x 8½. 0-486-42537-1

STATISTICAL MECHANICS: PRINCIPLES AND APPLICATIONS, Terrell L. Hill. Standard text covers fundamentals of statistical mechanics, applications to fluctuation theory, imperfect gases, distribution functions, more. 448pp. 5⅜ x 8½. 0-486-65390-0

ENGINEERING AND TECHNOLOGY 1650–1750: ILLUSTRATIONS AND TEXTS FROM ORIGINAL SOURCES, Martin Jensen. Highly readable text with more than 200 contemporary drawings and detailed engravings of engineering projects dealing with surveying, leveling, materials, hand tools, lifting equipment, transport and erection, piling, bailing, water supply, hydraulic engineering, and more. Among the specific projects outlined-transporting a 50-ton stone to the Louvre, erecting an obelisk, building timber locks, and dredging canals. 207pp. 8⅜ x 11¼. 0-486-42232-1

THE VARIATIONAL PRINCIPLES OF MECHANICS, Cornelius Lanczos. Graduate level coverage of calculus of variations, equations of motion, relativistic mechanics, more. First inexpensive paperbound edition of classic treatise. Index. Bibliography. 418pp. 5⅜ x 8½. 0-486-65067-7

PROTECTION OF ELECTRONIC CIRCUITS FROM OVERVOLTAGES, Ronald B. Standler. Five-part treatment presents practical rules and strategies for circuits designed to protect electronic systems from damage by transient overvoltages. 1989 ed. xxiv+434pp. 6⅛ x 9¼. 0-486-42552-5

ROTARY WING AERODYNAMICS, W. Z. Stepniewski. Clear, concise text covers aerodynamic phenomena of the rotor and offers guidelines for helicopter performance evaluation. Originally prepared for NASA. 537 figures. 640pp. 6⅛ x 9¼. 0-486-64647-5

INTRODUCTION TO SPACE DYNAMICS, William Tyrrell Thomson. Comprehensive, classic introduction to space-flight engineering for advanced undergraduate and graduate students. Includes vector algebra, kinematics, transformation of coordinates. Bibliography. Index. 352pp. 5⅜ x 8½. 0-486-65113-4

HISTORY OF STRENGTH OF MATERIALS, Stephen P. Timoshenko. Excellent historical survey of the strength of materials with many references to the theories of elasticity and structure. 245 figures. 452pp. 5⅜ x 8½. 0-486-61187-6

ANALYTICAL FRACTURE MECHANICS, David J. Unger. Self-contained text supplements standard fracture mechanics texts by focusing on analytical methods for determining crack-tip stress and strain fields. 336pp. 6⅛ x 9¼. 0-486-41737-9

STATISTICAL MECHANICS OF ELASTICITY, J. H. Weiner. Advanced, self-contained treatment illustrates general principles and elastic behavior of solids. Part 1, based on classical mechanics, studies thermoelastic behavior of crystalline and polymeric solids. Part 2, based on quantum mechanics, focuses on interatomic force laws, behavior of solids, and thermally activated processes. For students of physics and chemistry and for polymer physicists. 1983 ed. 96 figures. 496pp. 5⅜ x 8½. 0-486-42260-7

Mathematics

FUNCTIONAL ANALYSIS (Second Corrected Edition), George Bachman and Lawrence Narici. Excellent treatment of subject geared toward students with background in linear algebra, advanced calculus, physics and engineering. Text covers introduction to inner-product spaces, normed, metric spaces, and topological spaces; complete orthonormal sets, the Hahn-Banach Theorem and its consequences, and many other related subjects. 1966 ed. 544pp. 6⅛ x 9¼. 0-486-40251-7

ASYMPTOTIC EXPANSIONS OF INTEGRALS, Norman Bleistein & Richard A. Handelsman. Best introduction to important field with applications in a variety of scientific disciplines. New preface. Problems. Diagrams. Tables. Bibliography. Index. 448pp. 5⅜ x 8½. 0-486-65082-0

VECTOR AND TENSOR ANALYSIS WITH APPLICATIONS, A. I. Borisenko and I. E. Tarapov. Concise introduction. Worked-out problems, solutions, exercises. 257pp. 5⅝ x 8¼. 0-486-63833-2

AN INTRODUCTION TO ORDINARY DIFFERENTIAL EQUATIONS, Earl A. Coddington. A thorough and systematic first course in elementary differential equations for undergraduates in mathematics and science, with many exercises and problems (with answers). Index. 304pp. 5⅜ x 8½. 0-486-65942-9

FOURIER SERIES AND ORTHOGONAL FUNCTIONS, Harry F. Davis. An incisive text combining theory and practical example to introduce Fourier series, orthogonal functions and applications of the Fourier method to boundary-value problems. 570 exercises. Answers and notes. 416pp. 5⅜ x 8½. 0-486-65973-9

COMPUTABILITY AND UNSOLVABILITY, Martin Davis. Classic graduate-level introduction to theory of computability, usually referred to as theory of recurrent functions. New preface and appendix. 288pp. 5⅜ x 8½. 0-486-61471-9

ASYMPTOTIC METHODS IN ANALYSIS, N. G. de Bruijn. An inexpensive, comprehensive guide to asymptotic methods–the pioneering work that teaches by explaining worked examples in detail. Index. 224pp. 5⅜ x 8½ 0-486-64221-6

APPLIED COMPLEX VARIABLES, John W. Dettman. Step-by-step coverage of fundamentals of analytic function theory–plus lucid exposition of five important applications: Potential Theory; Ordinary Differential Equations; Fourier Transforms; Laplace Transforms; Asymptotic Expansions. 66 figures. Exercises at chapter ends. 512pp. 5⅜ x 8½. 0-486-64670-X

INTRODUCTION TO LINEAR ALGEBRA AND DIFFERENTIAL EQUATIONS, John W. Dettman. Excellent text covers complex numbers, determinants, orthonormal bases, Laplace transforms, much more. Exercises with solutions. Undergraduate level. 416pp. 5⅜ x 8½. 0-486-65191-6

RIEMANN'S ZETA FUNCTION, H. M. Edwards. Superb, high-level study of landmark 1859 publication entitled "On the Number of Primes Less Than a Given Magnitude" traces developments in mathematical theory that it inspired. xiv+315pp. 5⅜ x 8½. 0-486-41740-9

TENSOR CALCULUS, J.L. Synge and A. Schild. Widely used introductory text covers spaces and tensors, basic operations in Riemannian space, non-Riemannian spaces, etc. 324pp. 5⅜ x 8¼. 0-486-63612-7

ORDINARY DIFFERENTIAL EQUATIONS, Morris Tenenbaum and Harry Pollard. Exhaustive survey of ordinary differential equations for undergraduates in mathematics, engineering, science. Thorough analysis of theorems. Diagrams. Bibliography. Index. 818pp. 5⅜ x 8½. 0-486-64940-7

INTEGRAL EQUATIONS, F. G. Tricomi. Authoritative, well-written treatment of extremely useful mathematical tool with wide applications. Volterra Equations, Fredholm Equations, much more. Advanced undergraduate to graduate level. Exercises. Bibliography. 238pp. 5⅜ x 8½. 0-486-64828-1

FOURIER SERIES, Georgi P. Tolstov. Translated by Richard A. Silverman. A valuable addition to the literature on the subject, moving clearly from subject to subject and theorem to theorem. 107 problems, answers. 336pp. 5⅜ x 8½. 0-486-63317-9

INTRODUCTION TO MATHEMATICAL THINKING, Friedrich Waismann. Examinations of arithmetic, geometry, and theory of integers; rational and natural numbers; complete induction; limit and point of accumulation; remarkable curves; complex and hypercomplex numbers, more. 1959 ed. 27 figures. xii+260pp. 5⅜ x 8½. 0-486-63317-9

POPULAR LECTURES ON MATHEMATICAL LOGIC, Hao Wang. Noted logician's lucid treatment of historical developments, set theory, model theory, recursion theory and constructivism, proof theory, more. 3 appendixes. Bibliography. 1981 edition. ix + 283pp. 5⅜ x 8½. 0-486-67632-3

CALCULUS OF VARIATIONS, Robert Weinstock. Basic introduction covering isoperimetric problems, theory of elasticity, quantum mechanics, electrostatics, etc. Exercises throughout. 326pp. 5⅜ x 8½. 0-486-63069-2

THE CONTINUUM: A CRITICAL EXAMINATION OF THE FOUNDATION OF ANALYSIS, Hermann Weyl. Classic of 20th-century foundational research deals with the conceptual problem posed by the continuum. 156pp. 5⅜ x 8½. 0-486-67982-9

CHALLENGING MATHEMATICAL PROBLEMS WITH ELEMENTARY SOLUTIONS, A. M. Yaglom and I. M. Yaglom. Over 170 challenging problems on probability theory, combinatorial analysis, points and lines, topology, convex polygons, many other topics. Solutions. Total of 445pp. 5⅜ x 8½. Two-vol. set.
Vol. I: 0-486-65536-9 Vol. II: 0-486-65537-7

INTRODUCTION TO PARTIAL DIFFERENTIAL EQUATIONS WITH APPLICATIONS, E. C. Zachmanoglou and Dale W. Thoe. Essentials of partial differential equations applied to common problems in engineering and the physical sciences. Problems and answers. 416pp. 5⅜ x 8½. 0-486-65251-3

THE THEORY OF GROUPS, Hans J. Zassenhaus. Well-written graduate-level text acquaints reader with group-theoretic methods and demonstrates their usefulness in mathematics. Axioms, the calculus of complexes, homomorphic mapping, *p*-group theory, more. 276pp. 5⅜ x 8½. 0-486-40922-8

History of Math

THE WORKS OF ARCHIMEDES, Archimedes (T. L. Heath, ed.). Topics include the famous problems of the ratio of the areas of a cylinder and an inscribed sphere; the measurement of a circle; the properties of conoids, spheroids, and spirals; and the quadrature of the parabola. Informative introduction. clxxxvi+326pp. 5⅜ x 8½. 0-486-42084-1

A SHORT ACCOUNT OF THE HISTORY OF MATHEMATICS, W. W. Rouse Ball. One of clearest, most authoritative surveys from the Egyptians and Phoenicians through 19th-century figures such as Grassman, Galois, Riemann. Fourth edition. 522pp. 5⅜ x 8½. 0-486-20630-0

THE HISTORY OF THE CALCULUS AND ITS CONCEPTUAL DEVELOPMENT, Carl B. Boyer. Origins in antiquity, medieval contributions, work of Newton, Leibniz, rigorous formulation. Treatment is verbal. 346pp. 5⅜ x 8½. 0-486-60509-4

THE HISTORICAL ROOTS OF ELEMENTARY MATHEMATICS, Lucas N. H. Bunt, Phillip S. Jones, and Jack D. Bedient. Fundamental underpinnings of modern arithmetic, algebra, geometry and number systems derived from ancient civilizations. 320pp. 5⅜ x 8½. 0-486-25563-8

A HISTORY OF MATHEMATICAL NOTATIONS, Florian Cajori. This classic study notes the first appearance of a mathematical symbol and its origin, the competition it encountered, its spread among writers in different countries, its rise to popularity, its eventual decline or ultimate survival. Original 1929 two-volume edition presented here in one volume. xxviii+820pp. 5⅜ x 8½. 0-486-67766-4

GAMES, GODS & GAMBLING: A HISTORY OF PROBABILITY AND STATISTICAL IDEAS, F. N. David. Episodes from the lives of Galileo, Fermat, Pascal, and others illustrate this fascinating account of the roots of mathematics. Features thought-provoking references to classics, archaeology, biography, poetry. 1962 edition. 304pp. 5⅜ x 8½. (Available in U.S. only.) 0-486-40023-9

OF MEN AND NUMBERS: THE STORY OF THE GREAT MATHEMATICIANS, Jane Muir. Fascinating accounts of the lives and accomplishments of history's greatest mathematical minds–Pythagoras, Descartes, Euler, Pascal, Cantor, many more. Anecdotal, illuminating. 30 diagrams. Bibliography. 256pp. 5⅜ x 8½. 0-486-28973-7

HISTORY OF MATHEMATICS, David E. Smith. Nontechnical survey from ancient Greece and Orient to late 19th century; evolution of arithmetic, geometry, trigonometry, calculating devices, algebra, the calculus. 362 illustrations. 1,355pp. 5⅜ x 8½. Two-vol. set. Vol. I: 0-486-20429-4 Vol. II: 0-486-20430-8

A CONCISE HISTORY OF MATHEMATICS, Dirk J. Struik. The best brief history of mathematics. Stresses origins and covers every major figure from ancient Near East to 19th century. 41 illustrations. 195pp. 5⅜ x 8½. 0-486-60255-9

Physics

OPTICAL RESONANCE AND TWO-LEVEL ATOMS, L. Allen and J. H. Eberly. Clear, comprehensive introduction to basic principles behind all quantum optical resonance phenomena. 53 illustrations. Preface. Index. 256pp. 5⅜ x 8½. 0-486-65533-4

QUANTUM THEORY, David Bohm. This advanced undergraduate-level text presents the quantum theory in terms of qualitative and imaginative concepts, followed by specific applications worked out in mathematical detail. Preface. Index. 655pp. 5⅜ x 8½. 0-486-65969-0

ATOMIC PHYSICS (8th EDITION), Max Born. Nobel laureate's lucid treatment of kinetic theory of gases, elementary particles, nuclear atom, wave-corpuscles, atomic structure and spectral lines, much more. Over 40 appendices, bibliography. 495pp. 5⅜ x 8½. 0-486-65984-4

A SOPHISTICATE'S PRIMER OF RELATIVITY, P. W. Bridgman. Geared toward readers already acquainted with special relativity, this book transcends the view of theory as a working tool to answer natural questions: What is a frame of reference? What is a "law of nature"? What is the role of the "observer"? Extensive treatment, written in terms accessible to those without a scientific background. 1983 ed. xlviii+172pp. 5⅜ x 8½. 0-486-42549-5

AN INTRODUCTION TO HAMILTONIAN OPTICS, H. A. Buchdahl. Detailed account of the Hamiltonian treatment of aberration theory in geometrical optics. Many classes of optical systems defined in terms of the symmetries they possess. Problems with detailed solutions. 1970 edition. xv + 360pp. 5⅜ x 8½. 0-486-67597-1

PRIMER OF QUANTUM MECHANICS, Marvin Chester. Introductory text examines the classical quantum bead on a track: its state and representations; operator eigenvalues; harmonic oscillator and bound bead in a symmetric force field; and bead in a spherical shell. Other topics include spin, matrices, and the structure of quantum mechanics; the simplest atom; indistinguishable particles; and stationary-state perturbation theory. 1992 ed. xiv+314pp. 6⅛ x 9¼. 0-486-42878-8

LECTURES ON QUANTUM MECHANICS, Paul A. M. Dirac. Four concise, brilliant lectures on mathematical methods in quantum mechanics from Nobel Prize-winning quantum pioneer build on idea of visualizing quantum theory through the use of classical mechanics. 96pp. 5⅜ x 8½. 0-486-41713-1

THIRTY YEARS THAT SHOOK PHYSICS: THE STORY OF QUANTUM THEORY, George Gamow. Lucid, accessible introduction to influential theory of energy and matter. Careful explanations of Dirac's anti-particles, Bohr's model of the atom, much more. 12 plates. Numerous drawings. 240pp. 5⅜ x 8½. 0-486-24895-X

ELECTRONIC STRUCTURE AND THE PROPERTIES OF SOLIDS: THE PHYSICS OF THE CHEMICAL BOND, Walter A. Harrison. Innovative text offers basic understanding of the electronic structure of covalent and ionic solids, simple metals, transition metals and their compounds. Problems. 1980 edition. 582pp. 6⅛ x 9¼. 0-486-66021-4